THE GLOBAL OIL & GAS INDUSTRY

MANAGEMENT, STRATEGY & FINANCE

THE GLOBAL OIL & GAS INDUSTRY

MANAGEMENT, STRATEGY & FINANCE

Andrew Inkpen
Thunderbird School of Global Management

Michael H. Moffett
Thunderbird School of Global Management

Disclaimer: The recommendations, advice, descriptions, and the methods in this book are presented solely for educational purposes. The author and publisher assume no liability whatsoever for any loss or damage that results from the use of any of the material in this book. Use of the material in this book is solely at the risk of the user.

PennWell Corporation
1421 South Sheridan Road
Tulsa, Oklahoma 74112-6600 USA
800.752.9764
+1.918.831.9421
sales@pennwell.com
www.pennwellbooks.com
www.pennwell.com

Marketing: Jane Green
National Account Executive: Barbara McGee
Director: Mary McGee
Managing Editor: Marla Patterson
Production Manager: Sheila Brock
Production Editor: Tony Quinn
Book Designer: Susan E. Ormston
Cover Designer: Karla Pfeiffer

Permissions: RightsHouse, Inc.

Library of Congress Cataloging-in-Publication Data

Inkpen, Andrew C.
The global oil & gas industry : management, strategy, and finance / Andrew Inkpen, Michael H. Moffett.
p. cm.
ISBN 978-1-59370-239-7
1. International business enterprises--Management. 2. Petroleum industry and trade--Management. 3. Gas industry--Management. 4. International finance. 5. International trade. I. Moffett, Michael H. II. Title. III. Title: Global oil and gas industry.
HD62.4.I515 2011
665.5068--dc22
2011006878

Printed in the United States of America
1 2 3 4 5 15 14 13 12 11

Contents

Preface

The oil and gas industry is one of the world's largest and most important global industries. In this book, we describe and analyze the global oil and gas industry with a focus on the strategic, financial, and business aspects of the industry. Our goal is to provide a single source for anyone interested in how the world's largest industry actually works: business executives, students, government officials and regulators, people working in the industry, and the general public.

A basic premise underlying the book is that despite the size and importance of the oil and gas business, there is a basic lack of knowledge about the industry. This lack of knowledge is surprising given how important the industry is in the global economy and how the industry touches our daily lives in so many ways. Although there are thousands of books written about the industry, most are technical guides with very narrow audiences, or populist diatribes on the coming end of society as we know it. Among the books that deal with the business side of the industry, many are written by technical experts for nontechnical readers, such as a nontechnical guide to refining or a nontechnical guide to petroleum geology. Our approach in this book is the opposite: we have written a nontechnical business book that should help readers with technical backgrounds better understand the business of oil and gas. Some readers may recall that this was a major premise of the MBA degree many years ago—a curriculum focused on understanding business management for corporate employees who had reached management positions following careers in technical fields.

We address a wide range of topics, such as how resource nationalism and national security drive the competitive behavior of national oil companies (NOCs), how complex projects are planned and executed, how fiscal regimes are created, how crude oil is bought and sold, why cost management is so critical in the production of oil and gas, why some oil and gas firms are integrated across a diverse set of activities and others have a narrow focus on a single sector, why refining is not as profitable as the upstream, how fuels marketers compete with each other, and how the chemical industry supports a range of different business models. And this is just a small sample of our topic areas—this is in fact a big book!

Organization

The book is organized around the oil and gas industry value chain. The industry value chain starts with exploration and ends with products sold to consumers, such as gasoline, heating oil, natural gas for heating and power generation, and thousands of petrochemical products. We explain the different business segments in the value chain, such as exploration, development, production, crude oil marketing, refining, refined product marketing, the natural gas and liquefied natural gas businesses, and the petrochemicals sector. For each segment, we discuss the business and competitive aspects, with the goal of understanding competitive dynamics, key business and financial drivers, bases for competitive advantage, main competitors, and industry competitive challenges. We also discuss a number of NOCs and their evolving role in the industry and provide some predictions about the future of oil and gas.

We have tried to make the book as relevant, timely, and accessible as possible. Each chapter includes many real industry examples and case studies. The examples are drawn from all over the world because this industry is perhaps the most global of all industries. Each chapter also includes several "Industry Insights" to provide interesting and unique examples of different industry business practices, competitive actions, or managerial decisions.

Audience

The book should be of interest to several different audiences. The first is the people working in the oil and gas industry and especially engineers and scientists who seek greater understanding of the oil and gas business. The book should prove very useful for those making the transition from technical roles to managerial positions, where knowledge of the business is essential for their career development. The oil and gas industry does an excellent job in turning engineers into businessmen, and this book was written with that objective in mind. The book should also be useful for new employees with strong technical training as petroleum or chemical engineers but limited knowledge of the business of oil and gas. A second audience is business students studying the oil and gas industry and looking for a comprehensive reference text. Some of these students will be taking specialized courses in undergraduate and graduate degree programs focused on the global oil and gas industry. The third audience is general readers interested in learning more about the business dynamics of arguably the world's largest and most important industry.

The Global Oil & Gas Industry

Chapter 1

THE GLOBAL OIL AND GAS INDUSTRY

Oil is like a wild animal. Whoever captures it, has it.

—John Paul Getty, oil billionaire and founder of Getty Oil

Drill for oil? You mean drill into the ground to try and find oil? You're crazy.

—Drillers whom Edwin Drake tried to enlist for his project in 1859

The oil and gas industry is one of the largest, most complex, and important global industries. The industry touches everyone's lives with products such as transportation, heating, and electricity fuels; asphalt; lubricants; propane; and thousands of petrochemical products from carpets to eyeglasses to clothing. The industry impacts national security, elections, geopolitics, and international conflicts. The prices of crude oil and natural gas are probably the two most closely watched commodity prices in the global economy. In recent years, the industry has seen many tumultuous events, including the continuing efforts from oil-producing countries like Kazakhstan, Russia, and Venezuela to exert greater control over their resources; major technological advances in deepwater drilling and shale gas; Chinese firms acquiring exploration rights at record high prices; ongoing strife in Sudan, Nigeria, Chad, and other oil-exporting nations; continued heated discussion about global warming and nonhydrocarbon sources of energy; and huge movements up and down in crude prices. All of this comes amid predictions that the global demand for energy will increase by 30% to 40% by 2030.

In this chapter, we provide an overview of the industry. We begin with some historical background and key industry concepts. We then discuss the supplies of oil and gas, the major producing nations, and the major industry competitors. We also identify the major segments of the industry and introduce the oil and gas industry value chain. The chapters in this book are organized around the major value chain activities. Each chapter explores a major value chain activity and its competitive dynamics.

Oil and Gas Industry Background

When Colonel Edwin Drake struck oil in northwestern Pennsylvania in 1859, the first phase of the oil industry began. John D. Rockefeller emerged in those early days as a pioneer in industrial organization. When Rockefeller combined Standard Oil and 39 affiliated companies to create Standard Oil Trust in 1882, his goal was not to form a monopoly, because these companies already controlled 90% of the kerosene market. His real goal was the economy of scale, which was achieved by combining all the refining operations under a single management structure. In doing so, Rockefeller set the stage for what historian Alfred Chandler called the "dynamic logic of growth and competition that drives modern capitalism."[1]

With the discovery of oil at Spindletop in East Texas in 1901, a new phase of the industry began. Before Spindletop, oil was used mainly for lamps and lubrication. After Spindletop, petroleum would be used as a major fuel for new inventions, such as the airplane and automobile. Ships and trains that had previously run on coal began to switch to oil. For the next century oil, and then natural gas, would be the world's most important sources of energy.

Since the beginning of the oil industry, petroleum producers and consumers have feared that eventually the oil would run out. In 1950, the US Geological Survey estimated that the world's conventional recoverable resource base was about 1 trillion barrels. Fifty years later, that estimate had tripled to 3 trillion barrels. In recent years, the concept of *peak oil* has been much debated. The peak oil theory is based on the fact that the amount of oil is finite.

After peak oil, according to the Hubbert Peak Theory, the rate of oil production on earth will enter a terminal decline. In the United States, oil production peaked in 1971 and some analysts have argued that on a global basis, the peak has also occurred. Others argue that peak oil is a myth. An article in the journal *Science* argued:

> *Although hydrocarbon resources are irrefutably finite, no one knows just how finite. Oil is trapped in porous subsurface rocks, which makes it difficult to estimate how much oil there is and how much can be effectively extracted. Some areas are still relatively unexplored or have been poorly analyzed. Moreover, knowledge of in-ground oil resources increases dramatically as an oil reservoir is exploited. . . . To "cry wolf" over the availability of oil has the sole effect of perpetuating a misguided obsession with oil security and control that is already rooted in Western public opinion—an obsession that historically has invariably led to bad political decisions.*[2]

Regardless of whether the peak has or has not been reached, oil and natural gas are an indispensable source of the world's energy and petrochemical feedstocks and will be for many years to come. The difficulty in determining oil and gas reserves is that "true reserves" are a complex combination of technology, price, and politics. While technical change continues to reveal new sources of oil and gas, prices have demonstrated more volatility than ever, and governments have sought more control over resource information and access than ever. As prices rise, reserves once considered noneconomic to develop may become feasible.

As illustrated by figure 1–1, crude oil prices ranged between $2.50 and $3.00 per barrel from 1948 through the end of the 1960s. The Arab oil embargo of 1974 resulted in a large price increase. Events in Iran and Iraq led to another round of crude oil price increases in 1979 and 1980. The 1990s saw another spike in prices that ended with the 1997 Asian financial crisis. Prices then started back up, only to fall after September 11, 2001. After 9/11, prices rose until the recession at the end of the decade.

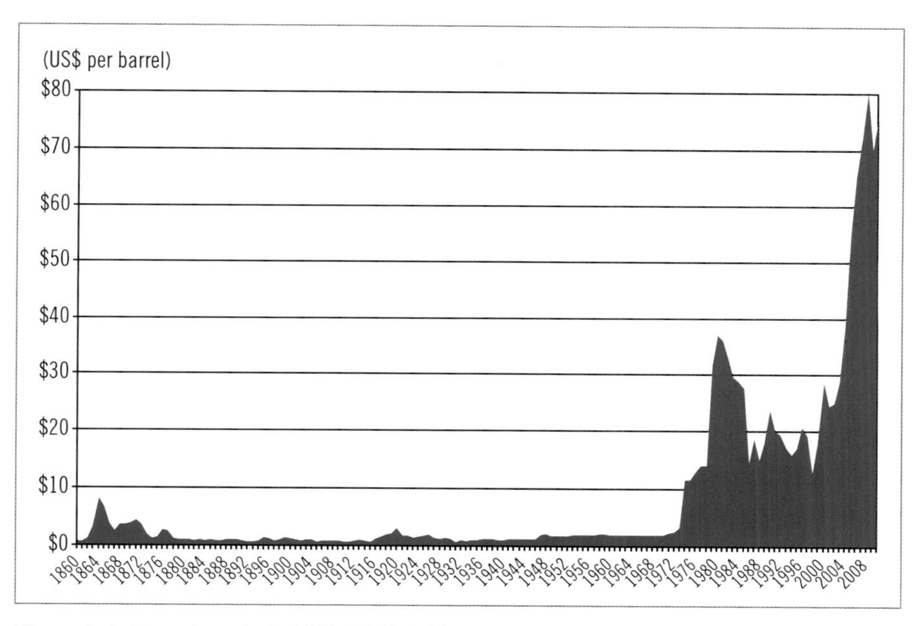

Figure 1–1. The price of oil, 1860–2010 (US$ per barrel)

Source: Annual average prices in US$ per barrel. Based on "BP Statistical Review of World Energy," June 2009. 2010 price estimated by authors, April 2010.

Oil and Gas Reserves

Discovering new oil and gas reserves is the lifeblood of the industry. Without new reserves to replace oil and gas production, the industry would die. However, measuring and valuing reserves is a scientific and business challenge because reserves can only be measured if they have value in the marketplace.

The oil sands of Alberta, Canada are a good illustration of how difficult it is to accurately measure oil and gas reserves. Oil sands are deposits of *bitumen*, a molasses-like viscous oil that will not flow unless heated or diluted with lighter hydrocarbons. Although the Alberta oil sands are now considered second only to the Saudi Arabia reserves in the potential amount of recoverable oil, for many years these were not viewed as real reserves because they were not economical to develop. By the mid-2000s, the main town in the oil sands region, Fort McMurray, was in the midst of a boom not unlike the gold rush booms of the 1800s. Housing and labor were scarce and the infrastructure was struggling to keep pace with the influx of people, companies, and capital. The development of the oil sands occurred because of a combination of rising oil prices and technological innovation. There were estimates that oil sands production could reach 3 million barrels per day (b/d) by 2020 and possibly even 5 million b/d by 2030.

Oil and Gas in the Global Economy

Oil and gas play a vital role in the global economy. The International Energy Agency (IEA) predicts that energy demand will rise by an average of 1.5% each year through 2030. Demand in 2030 will be about 60% higher than in 2000. Demand in the non-OECD (Organization for Economic Cooperation and Development) nations will account for approximately 80% of the global increase. Most of the world's growing energy needs through 2030 will continue to be met by oil, gas, and coal. With increased energy efficiency, energy as a percentage of the total gross domestic product (GDP) has fallen and is expected to continue to fall.

Oil and gas supply

One of the fascinating aspects of the industry is the fact that all countries are consumers of products derived from the oil and gas industry, but only a small set of nations are major producers of oil and gas. Over the past decades, the large developed economies of the world have become net importers of oil and gas, giving rise to challenging geopolitical issues involving a diverse set of oil consumers and producers.

Table 1–1 shows the major oil- and gas-producing nations and their change in output over a decade. Countries like Angola, Brazil, and Kazakhstan have made their way into the top tier of oil producers, whereas the United States, Mexico, and Venezuela, for different reasons, are on their way down. In natural gas, newcomers like Qatar and Turkmenistan are now major players. Unlike oil, the United States continues to increase its production of gas. Of the 28 different countries that make up the oil and gas lists, all but seven (Argentina, Brazil, Canada, China, Egypt, the Netherlands, and the United Kingdom) have national budgets that are highly dependent on exports of oil and gas.

Industry financial performance

The oil and gas industry has been widely criticized by politicians and the media for its high profits of recent years. In the US, talk of an excess profits tax prompted Lee Raymond, former ExxonMobil CEO, to comment in 2005: "I can't remember any of these people seven years ago, when the price was $10 a barrel, coming forward and saying, are you guys going to have enough money to be able to continue to invest in this business? I don't recall my phone ringing and anybody asking me that question."[3]

The oil and gas industry is highly cyclical, and the cycles can last many years. In the 1990s, crude oil prices fell steadily and in the new millennium, the first few years saw steadily rising prices. The Great Recession put a damper on some experts' prediction of $200 per barrel prices. Although the oil industry is highly profitable in some years, its long-term profitability is not much higher than the average profitability across many industries. In the US, the oil and gas industry has earned return on sales (net income divided by revenue) of about 8% compared to an average of about 6% for all US manufacturing, mining, and wholesale trade corporations. As evidence of the cyclical nature of the industry, some years ago *Fortune* magazine reported that the oil industry ranked 30th out of 36 industries in return to investors over the 1985–1995 period, 34th out of 36 US industries in return on equity in 1995, and 32nd in return on sales.[4]

The role of OPEC

The oil and gas industry has seen a remarkable bevy of government regulations and interventions over the past century, from heavy taxation of petrol in Europe to US price controls on domestic production in the 1970s. The creation of the Organization of the Petroleum Exporting Countries (OPEC) represents government intervention on a global scale. OPEC was founded in 1960 with the objective of shifting bargaining power to the producing countries and away from the large oil companies. In 2006, Angola became the 12th member of OPEC, and there was speculation that Sudan might be next.

Table 1–1a. Major oil producing nations

Country	Percent of World Production, 2009	Output Change Since 1999
Russia	12.9%	62.4%
Saudi Arabia	12.0%	9.7%
United States	8.5%	–6.9%
Iran	5.3%	17.0%
China	4.9%	18.0%
Canada	4.1%	23.4%
Mexico	3.9%	–10.9%
Venezuela	3.3%	–22.0%
United Arab Emirates	3.2%	3.5%
Kuwait	3.2%	19.0%
Iraq	3.2%	–4.9%
Norway	2.8%	–25.4%
Nigeria	2.6%	–0.3%
Brazil	2.6%	79.1%
Angola	2.3%	139.4%
Algeria	2.0%	19.5%
Libya	2.0%	15.9%
Kazakhstan	2.0%	166.3%
United Kingdom	1.8%	–50.2%
Qatar	1.5%	85.9%
Total	84.1%	

OPEC's mission is "to coordinate and unify the petroleum policies of Member Countries and ensure the stabilization of oil prices in order to secure an efficient, economic and regular supply of petroleum to consumers, a steady income to producers and a fair return on capital to those investing in the petroleum industry."[5] Despite being a cartel, OPEC's ability to control prices is questionable. Surging oil prices in the 1980s resulted in energy conservation and increased exploration outside OPEC. Maintaining discipline among OPEC members has been a major problem (as is typical in all cartels). Massive cheating was blamed for the oil price crash of 1986, and in the 1990s Venezuela was considered one of the bigger OPEC cheats by regularly producing more than its quota.

Figure 1–2 shows OPEC production and crude oil prices. Although it is difficult to identify any clear continuing relationship between OPEC's production over time and the movement of crude oil prices, the organization has clearly been instrumental in periodic "shocks to the system" as characterized by one analyst.

Table 1–1b. Major gas producing nations

Country	Percent of World Production, 2009	Output Change Since 1999
United States	20.1%	11.3%
Russia	17.6%	–1.5%
Canada	5.4%	–8.7%
Iran	4.4%	132.8%
Norway	3.5%	113.4%
Qatar	3.0%	305.0%
China	2.8%	238.0%
Algeria	2.7%	–5.3%
Saudi Arabia	2.6%	67.6%
Indonesia	2.4%	2.7%
Uzbekistan	2.2%	28.1%
Malaysia	2.1%	53.4%
Netherlands	2.1%	4.1%
Egypt	2.1%	273.2%
United Kingdom	2.0%	–39.8%
Mexico	1.9%	56.8%
Argentina	1.4%	19.6%
Trinidad & Tobago	1.4%	246.1%
United Arab Emirates	1.6%	26.9%
Turkmenistan	1.2%	76.3%
Total	82.5%	

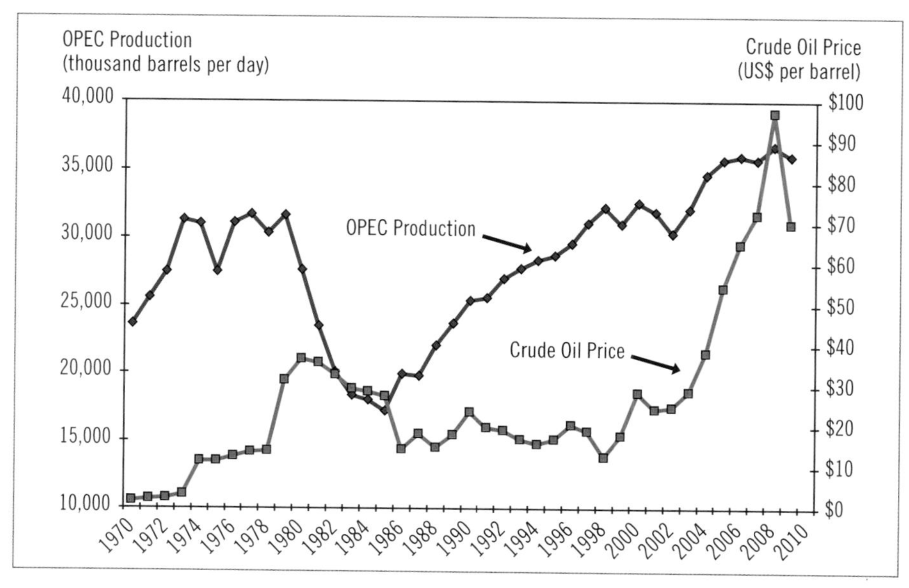

Figure 1–2. OPEC production and crude oil prices

Source: Data drawn from "BP Statistical Review of World Energy," June 2009. All data is annual average.

The resource curse

The *resource curse* is a paradox of the oil and gas industry. Despite high resource prices the living standards in many oil-producing countries are low. This condition has led to the inability of countries rich in natural resources to use that wealth to strengthen their economies and, counterintuitively, to have lower economic growth than countries without an abundance of natural resources.[6] When times are good and oil prices are high, oil-rich countries may prosper. When oil prices fall, as they inevitably do, an overreliance on the oil sector can leave a country in a perilous situation. Moreover, the oil industries of the petroleum-nationalistic countries often suffer from a lack of investment and heavily subsidized domestic petroleum products.

Iran, although second only to Saudi Arabia in the size of its reserves, is one such country. Its oil industry today is, quite honestly, in shambles. Iran's 2009 production was only about two-thirds of the level reached under the government of the former shah of Iran in 1979. Iran imports about 40% of its gasoline and is unable to produce sufficient crude to meet its OPEC quota. In June 2007, Iran introduced gasoline rationing, which reduced imports and resulted in widespread black marketeering. Some experts predicted that without huge foreign direct investment in the industry, Iran's oil production would decline precipitously over the next few decades. According to one analyst:

> *Iran burns its candle at both ends, producing less and less [oil] while consuming more and more. Absent some change in Iranian policy, a rapid decline in exports seems likely. Policy gridlock and a Soviet-style command economy make practical problem-solving almost impossible.*[7]

Mexico also has declining production and significant imports of refined products. The Mexican constitution does not allow foreign direct investment in the oil and gas industry. After many years of underinvestment and of Mexican governments using the oil industry as their primary source of revenue, the industry is in dire straits. Without major investment and new technology, Mexico's oil production is poised to fall. For example, production at the Cantarell oil field, one of the largest fields in the world, fell from more than 2 million b/d in 2004 to substantially less than 1 million b/d in 2009.

The Players

The global oil and gas industry is made up of thousands of firms of all shapes, sizes, and capabilities. The industry may suffer from an overabundance of terminology when describing these players, so here is some clarification of names and identities:

- **Independent.** A nonintegrated company generating nearly all its revenue from either oil and gas production or downstream activities. The term *independent* is sometimes used more narrowly to refer only to oil and gas producers and not downstream firms.

- **Integrated oil company (IOC).** A company that competes in the upstream, midstream, downstream, and perhaps petrochemicals. *IOC* is a term usually used in reference to large oil and gas companies—BP, Chevron, ConocoPhillips, ExxonMobil, Shell, and Total—and could also include smaller firms such as Eni and Marathon.

- **International oil company (IOC).** An oil and gas company that competes across borders. More generally, the term is used to describe the largest oil and gas companies that compete globally and often operate in partnership with NOCs in the NOC's home country. Because most IOCs are involved in oil and gas, a more appropriate term would be *international energy company*. Confusingly, international oil companies and integrated oil companies both use the acronym IOC. For our purposes, when we use the acronym IOC, we are referring to the largest international oil companies: BP, Chevron, ConocoPhillips, ExxonMobil, Shell, and Total.

- **Junior.** These are small oil and gas firms producing between 500 and 10,000 oil equivalent b/d. In spite of the connotation, they are the critical lifeblood of the global industry in terms of operations and execution.

- **National oil company (NOC).** A company controlled by a national government, usually formed to manage the country's hydrocarbon resources. Many NOCs, such as Gazprom, Petrobras, and Sinopec, are majority owned by the state and partially owned by private investors. NOCs are usually an arm of a government ministry, such as the ministry of petroleum or ministry of oil and gas. Some NOCs operate only in their home country (e.g., Pemex), and others compete globally across multiple sectors much like an IOC (e.g., Gazprom, Petrobras, and Statoil). As the NOCs get larger and more global, and list their shares, the boundaries are blurring between IOCs and NOCs.

- **Oil major.** These are the large non-state-owned oil and gas companies. Although they are typically publicly traded companies, they may also be privately owned. The terms *oil majors* and *IOCs* are often used interchangeably.

- **Supermajor.** A term used to describe the largest IOCs/oil majors, usually BP, Chevron, ConocoPhillips, ExxonMobil, Shell, and Total.

The organizations that have dominated the global oil and gas industry for more than a century have changed dramatically over time—in who they are, what they do, and of critical significance for the future of the industry, what they want. Figure 1–3 lists the largest oil and gas companies by *market capitalization* (share price times number of shares outstanding). The list includes both IOCs (international oil companies) and NOCs (national oil companies) and is evidence of two factors: mergers and acquisitions, and the global nature of the industry in production and ownership. Based on market capitalization, the top 15 publicly traded (and in some cases, government-controlled) companies include a diverse and global set of firms such as Petrochina (China), Gazprom (Russia), Sinopec (China), Petrobras (Brazil), Total (France), and Eni (Italy).

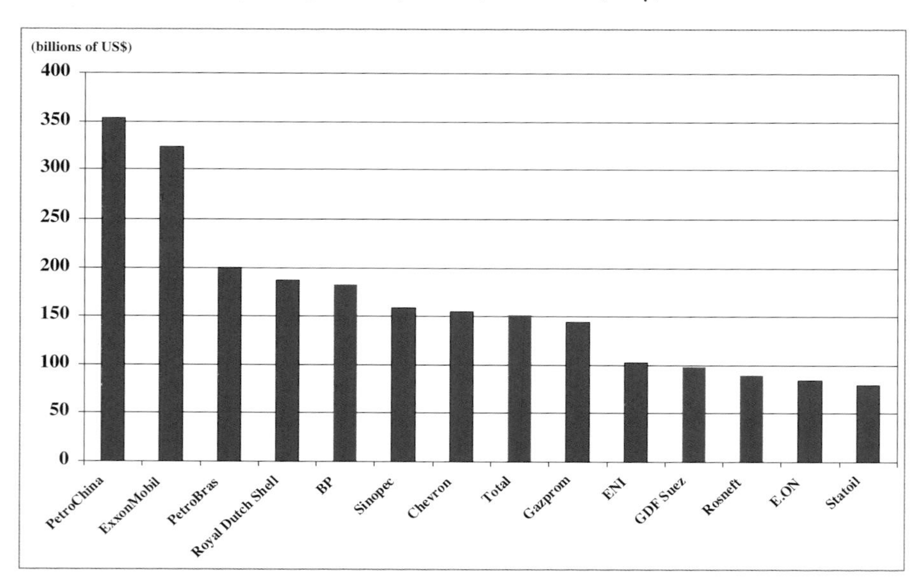

Figure 1–3. World's largest energy firms by market capitalization (billions US$)

Source: PFC Energy. Eni, Gazprom, Petrochina, Petrobras, Sinopec, Rosneft, and StatoilHydro have both publicly traded shares and government owned shares. The government ownership ranges from 90% for Petrochina to 32% for Petrobras. Gazprom is an integrated natural gas company. The other companies on the list are involved in oil and/or natural gas. Market cap as of end-of-year 2009.

IOCs

The global oil and gas industry has long been dominated by vertically integrated multinational oil companies known as IOCs. The IOCs include the largest oil and gas companies such as BP, ExxonMobil, and Shell. Their control lies in the hands of private investors, not governments, and their objectives have always been to generate the greatest sustainable profitability over time. The term IOC is a bit confusing in practice, sometimes meaning international oil companies, sometimes integrated oil companies. Regardless of the words behind the acronym, IOCs are profit-oriented organizations that are global in reach and vertical in structure.

In the early days of the industry, a few oil companies were truly vertically integrated—producing, refining, and marketing nearly 100% of their own product. In today's industry IOCs operate in many industry segments and also buy oil and gas for their refineries, sell crude oil and gas to other firms, and buy and sell finished products (later in the chapter, we discuss the industry segments in detail using the value chain concept). Thus, the integrated nature of today's large oil and gas firms looks more like industry sector diversification than classic vertical integration. Regardless, the term integrated oil company still applies.

Given the long product life cycles and the huge capital investment required in the oil industry, the large IOCs are often described as stodgy and conservative. Before bankruptcy, Enron executives regularly derided the oil majors as dinosaurs that were too slow moving and that would eventually become extinct. The reality, of course, is very different. Oil majors like BP, Shell, ExxonMobil, and their predecessor companies have been around for more than a century. Through experience that is occasionally painful, the IOCs have learned how to deal with the enormous financial and political risks of the oil and gas industry. The IOCs take a long-term view and recognize that cycles and uncertainty are an inherent part of the industry. In the words of Lee Raymond, former ExxonMobil CEO:

> *We're in a commodity [business]. We go through peaks and valleys, but our business is to level out the peaks and valleys, so that, over the cycle, our shareholders see an adequate return on their investment.*[8]

On the surface, the IOCs look similar in terms of the activities they perform. All appear to be integrated from exploration to retail distribution. However, there are fundamental organizational and financial differences among the firms. The IOCs use various organizational designs to deal with vertical integration. ExxonMobil, for example, is organized around global businesses and global functions, with common global operating processes, global enterprise back-office systems, such as SAP, and integrated operating structures at major sites. BP announced in 2007 that it would adopt a global structure organized around different businesses, and Shell is moving in the same direction. The other IOCs tend to use more regional processes and regional management structures.

NOCs

One of the most important trends of the new century has been the growing importance of the NOCs. Although ExxonMobil, BP, and Shell are among the largest publicly traded companies in the world, they do not rank in the top 10 of the world's largest oil and gas firms measured by reserves. The largest oil and gas firms based on reserves are, by a large margin, NOCs partially or wholly state-owned. NOCs control about 90% of the world's oil and gas, and most new oil is expected to be found in their territories.

Viewed from a business perspective, the NOCs have a mixed reputation. The national oil company of Indonesia, Pertamina, was described a few years ago as a bloated and inefficient bureaucracy:

> *[Pertamina] operated almost as a sovereignty unto itself, ignoring transparent business practices, often acting independently of any ministry, and increasingly taking on the role of a cash cow for then-President Suharto and his cronies. During the 32-year tenure of President Suharto, Pertamina awarded 159 contracts to companies linked to his family and cronies. These contracts were awarded without formal bidding or negotiation processes. . . . Indonesian petroleum law dictated that every aspect of operation in the country was subject to approval by Pertamina's foreign contractor management body, BPPKA. Dealing with the incomprehensible BPPKA bureaucracy on simple matters, such as acquiring work permits for expatriate personnel, can take hours of filling in applications and months of waiting.*[9]

Venezuela nationalized its oil industry in the 1970s and created Petróleos de Venezuela (PDVSA). PDVSA developed a reputation for professionalism and competence and was relatively free from the corruption and cronyism that pervaded, and continues to pervade, so many of the NOCs.[10] By 1998, 36 foreign oil firms were operating in Venezuela and PDVSA had ambitious expansion plans. In 1999 Hugo Chávez was elected president and almost immediately began to question the management and autonomy of PDVSA. After a bitter strike in 2002, PDVSA lost about two-thirds of its managerial and technical staff. From a peak of 2.9 million b/d in 1998, output was estimated by OPEC to be 2.3 million b/d at the end of the decade, as PDVSA imported a significant amount of gasoline.

As a company today, PDVSA is indistinguishable from the government. Its CEO, Rafael Ramírez, is also minister of energy. The company is required to spend a tenth of its investment budget on social programs, which includes sending low-cost heating oil to poor Americans. Company hiring policy is based on social and political goals; e.g., candidates from larger families are given priority. In 2006, the Venezuelan Congress approved new guidelines to turn 32 privately run oil fields over to state-controlled joint ventures. ExxonMobil and ConocoPhillips rejected the new joint venture agreements. The Venezuelan government subsequently expropriated the Cerro Negro heavy oil project, leading to an arbitration claim from ExxonMobil.

According to the *Economist*, nationalization has failed to live up to expectations almost everywhere. All NOCs suffered to some extent from government intervention. Many NOCs operated as the de facto treasury for the country. In Nigeria, for example, oil revenues represented more than 90% of hard currency earnings and about 60% of GDP. Nigeria's economic and financial crimes commission estimated that more than $380 billion of government revenues had been stolen or misused since 1960.[11] Some of the Middle Eastern NOCs are required to hire large numbers of locals, leaving them heavily overstaffed. Others, for example in India and Russia, must sell their products at subsidized prices. Underinvestment is a chronic problem for many NOCs, resulting in countries like Indonesia and Iran, with huge reserves, having to import petroleum. Monopoly positions held by many NOCs contribute to underinvestment. In Russia, Gazprom controls the pipeline network, making it difficult for other Russian gas producers, such as TNK-BP, to expand their production. Russia increasingly is using its NOCs as agents of foreign policy. A dispute between Belarus and Russia in early 2007 resulted in disruption of oil shipments to Western Europe. This prompted speculation in Germany that the government might rethink its decision to phase out nuclear power because of uncertainty about oil supplied from Russia.

Some NOCs are well-run and profitable enterprises. Statoil of Norway is considered to be among the best of the NOCs. In 2007, Statoil acquired Norsk Hydro in a $30 billion deal. According to analysts, the motivation for the deal was that a larger company would make it easier for expansion outside Norway. The NOCs of Brazil (Petrobras) and Malaysia (Petronas) are also viewed as well run companies. Petrobras has developed leading technology in deepwater drilling and has a market capitalization rivaling that of the IOCs. All three

NOCs are growing and diversifying. For example, Petronas has acquired some lubricants firms and is actively sponsoring Formula 1 racing. The desire to get larger and more integrated can be seen in comments from the ONGC chairman. ONGC, an Indian state-controlled firm and primarily an upstream company, had made public its commitment to participate in the entire petroleum value chain. According to the former chairman of ONGC:

> *We have to be an integrated oil company. Every major global oil company is an integrated player. I'm not being arrogant, but oil and gas is big business where the big boys play. You can survive in this business only if you are integrated, otherwise you will be out.*[12]

The role that NOCs will play in the future is not clear. Some analysts see the NOCs as inefficient and corrupt arms of government that will never compete in a true economic sense. Others raise different issues, suggesting that the NOCs are in a period of transition and will become competitive forces to be reckoned with. Regardless of what happens, the NOCs and their sovereign owners control most of the world's oil and gas reserves. As Paolo Scaroni, the chairman of Eni, the Italian IOC, commented:

> *Big Western oil firms are like addicts in denial. . . .The oil giants are trying to do business as usual as if nothing was wrong. Yet they are, in fact, having trouble laying their hands on their own basic product. State-owned national or state-controlled oil companies are sitting on as much as 90% of the world's oil and gas and are restricting outsiders' access to it. Worse, the best NOCs are beginning to expand beyond their own frontiers and to compete with the oil majors for control over the remaining 10% of resources. The first step in overcoming this predicament is admitting that it is a problem.*[13]

The strategic goals of IOCS and NOCs

One way to view the differences between publicly traded IOCs and state-controlled NOCs is to consider their strategic goals. Figure 1–4 positions large oil and gas firms based on the degree to which they are motivated by shareholder maximization or public policy goals. As publicly traded firms, the IOCs must be responsive to the expectations and demands of their private shareholders. These expectations are largely concerned with wealth creation, which means the IOCs must be very focused on cost control and financial performance. Maximizing shareholder value, both in general financial profitability and the (hopeful) increase in share value, is clearly the primary objective of large IOCs such as BP and Shell. They are private industry concerns, owned and operated on behalf of private individuals, and not a government. As a result, they have very limited public policy goals, although as part of a global trend toward emphasizing the so-called *triple bottom line* (financial returns, social responsibility, and environmental sustainability), they do include nonfinancial objectives in their business and strategic decision making.

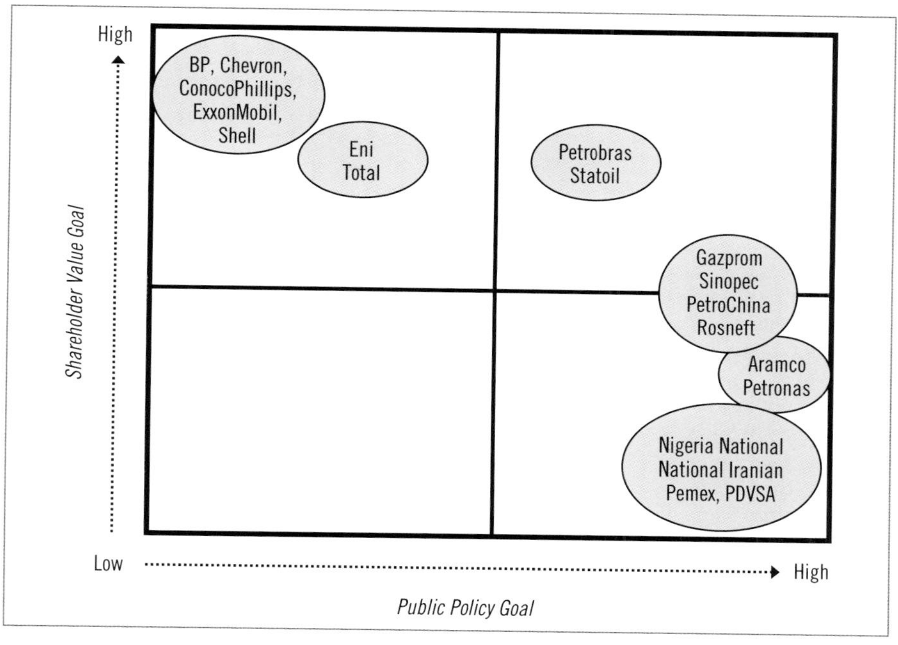

Figure 1–4. The strategic goals of NOCs and IOCs

Two IOCs are shown separately, Total (France) and Eni (Italy), because both firms are publicly traded and also tightly connected to their respective national governments. A small group of firms is termed *hybrid*: well-run publicly traded firms with government control. Petrobras and Statoil fit in the hybrid category. Some state-owned firms have a small amount of their ownership traded on stock exchanges, including Gazprom, Rosneft, and PetroChina. Another set of firms is 100% government owned and controlled but employs strong financial discipline and stewardship. Petronas, Aramco, and a few other state-owned firms fall into this category. Finally, there is a set of government-owned and government-controlled NOCs that seems to exist primarily as public policy arms of their government owners. This set includes the NOCs of Iran, Mexico, Nigeria, and Venezuela. These firms have limited shareholder value goals [i.e., goals tied to financial metrics such as return on investment (ROI) and return on capital employed (ROCE)]. In chapter 2, we provide a much deeper examination of NOCs and their goals.

Independents

Independents are the non-government-owned companies that focus on either the upstream or the downstream. Many of these companies are sizable players and rank in the top 50 of all non-government-owned oil and gas companies. In the following chapters, we will note the growing role of these firms in some of the more high-risk and innovative oil and gas areas, in terms of geography, products, and technology.

As shown in figure 1–5, the largest independent exploration and production company is Occidental, followed by Canadian Natural (Canada), Apache (US), Devon (US), OGX (Brazil), BHP Billiton (Australia), and Woodside (Australia). In the downstream refining and marketing area, the largest independents are scattered around the world's largest energy consuming countries, as illustrated in figure 1–6. The downstream independents, outside of Reliance, generally have lower market capitalizations than the upstream independents.

Other firms

In addition to the IOCs, NOCs, and independents, the oil and gas industry includes a huge number of others firms that perform important functions. Upstream oil and gas producers that are too small to be labeled *independents* are termed *juniors*. The largest oilfield services firms are listed in figure 1–7, the largest being Schlumberger (87,000 employees), Halliburton (51,000 employees), Weatherford (50,000 employees), and Baker Hughes (35,000 employees). These firms play a critical role throughout the exploration, development, and production phases by providing both products and services that, according to Baker Hughes, help oil and gas producers "find, develop, produce, and manage oil and gas reservoirs." Because the oil field service firms do not seek ownership rights to oil and gas reserves, many analysts predict that their role will become increasingly important in the future as partners to the NOCs.

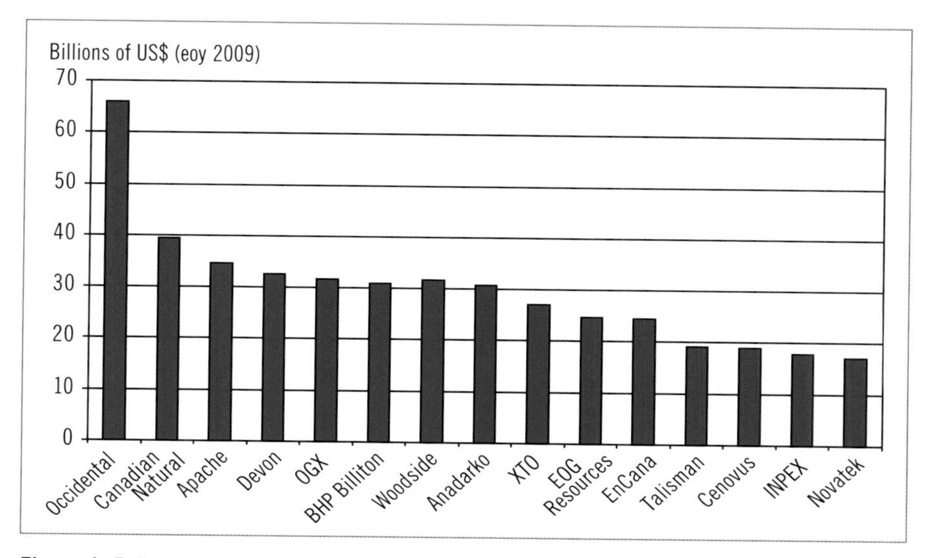

Figure 1–5. Largest independent upstream oil and gas companies based on market capitalization

Source: PFC Energy, 2010. BHP Billiton is a diversified company primarily focused on minerals. The value of its oil and gas E&P business was estimated by PFC to be $25–$28 billion.

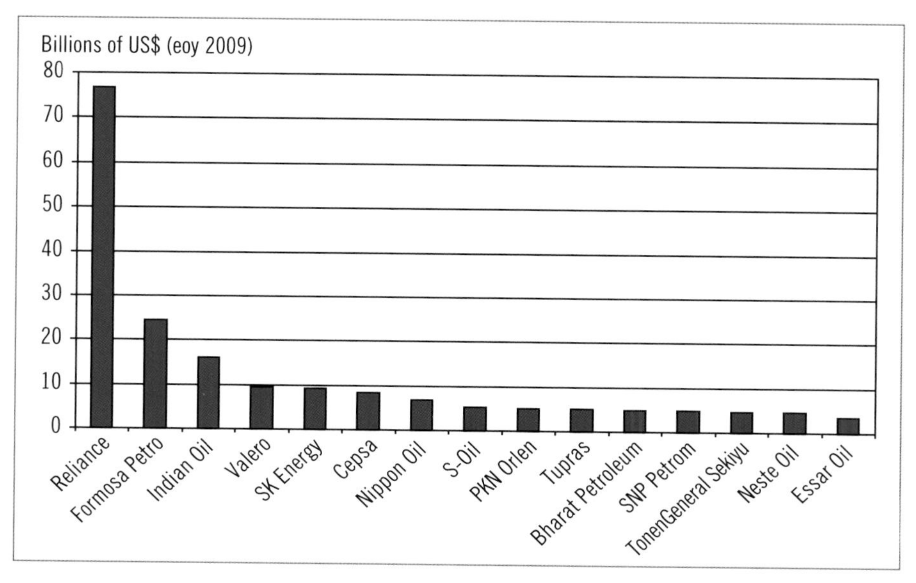

Figure 1–6. Largest independent downstream oil and gas companies based on market capitalization

Source: PFC Energy, 2010. Besides refining, Reliance is also involved in exploration and production, chemicals, and textiles.

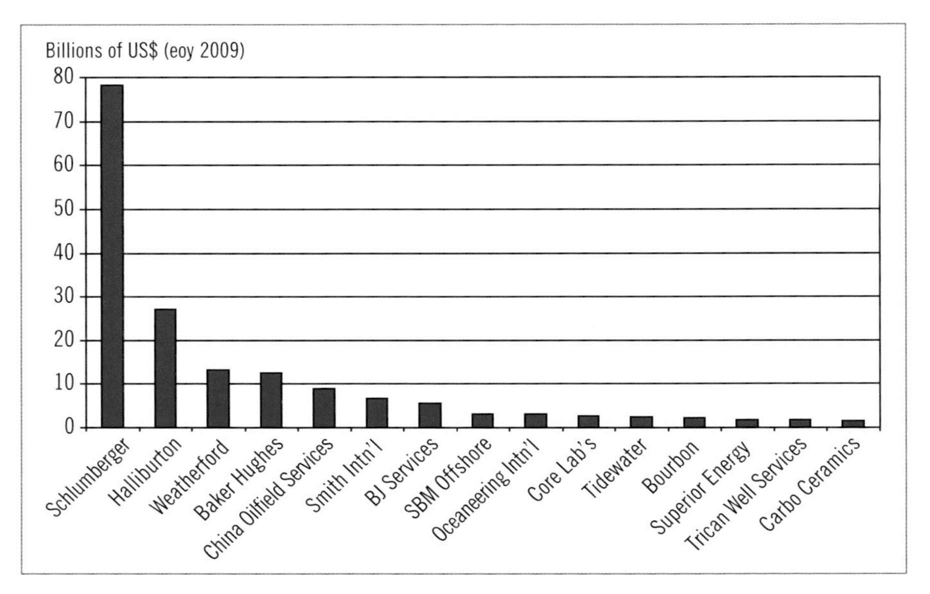

Figure 1–7. Largest independent oilfield services firms based on market capitalization
Source: PFC Energy, 2010. Besides refining, Reliance is also involved in exploration and production, chemicals, and textiles.

Thousands of other firms provide a vast array of services and products for the industry. For example, gas utilities such as Gaz de France and Tokyo Gas are major customers for gas producers. Pipeline companies distribute gas, crude oil, and petroleum products. The firms involved in drilling and seismic services provide drilling rigs and expertise for onshore and offshore wells.

The Oil and Gas Industry Value Chain

In every industry, there are various activities that must take place to transform inputs of raw materials, knowledge, labor, and capital into end products purchased by customers. A *value chain* is a device that helps identify the independent, economically viable segments of an industry.[14] *Value* refers to what customers are willing to pay for, and so the value chain helps to identify the specific activities that create value throughout the chain. Companies can use value chains to determine where they are strong and where they have limited competitive strength. All industries have upstream (close to raw materials and basic inputs) and downstream (close to the customer) segments. In the oil and gas industry, the terms upstream, downstream, and midstream are important descriptors of the industry activities. In fact, these terms have existed far longer than the value chain concept, which emerged in the 1980s.

The oil and gas industry value chain is shown in figure 1–8. There are three main segments: *upstream*, *midstream*, and *downstream*. At the far upstream end, the industry starts with exploration rights. At the downstream end products are sold to end users. Each of the different segments could be performed by a stand-alone firm. The IOCs such as BP and Shell perform activities throughout the value chain. They also rely heavily on other firms for many different activities. For example, consider a deepwater upstream development project. An IOC may do the exploration and then manage the development and production of an oil field. The development will involves many other firms to perform activities such as drilling, ship or rig-building, subsea pipe design, production support and equipment installation, and the supply of many different types of equipment and services. Overall, the oil and gas industry value chain incorporates thousands of firms. Some are specialists or niche players, and others perform many different activities from exploration to retail fuels marketing.

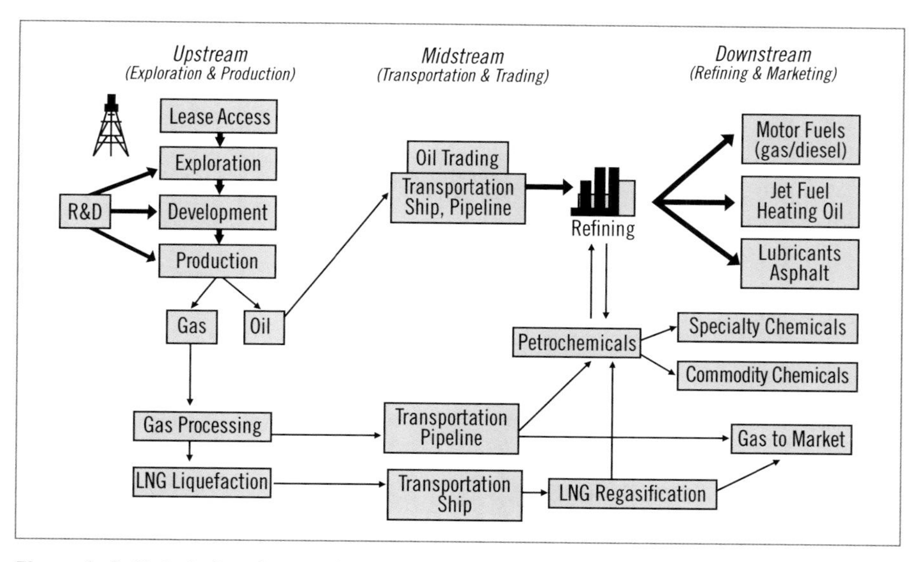

Figure 1–8. Global oil and gas value chain

Upstream: Exploration, development, and production

Upstream activities include exploration, development, and production. In simple terms, after a lease is obtained, oil and gas are discovered during exploration; the discovery requires development; and production is the long-term process of drilling and extracting oil and gas. Since exploration and development must take place where resources are located and most oil ownership regimes are based on state sovereignty, companies have to deal with very complex government policies and regulations. Most countries grant oil and gas development rights to private companies through a process of either negotiation or bidding. The main aim of the private company is profit maximization whereas the host country government is interested in maximizing revenue. Not surprisingly, these two aims often conflict. Most agreements between oil companies and governments come under the term *production-sharing agreements.*

The method used to bid for, grant, and then renew or extend oil and gas rights varies from country to country. Once the rights to explore are acquired, a well is drilled. A financial analysis is a determining factor in the classification of a well as an *oil well, natural gas well,* or *dry hole.* If the well can produce enough oil or gas to cover the cost of completion and production, it will be put into production. Otherwise, it is classified as a dry hole even if oil or gas is found.

The percentage of wells completed is a widely used measure of success. Immediately after World War II, 65% of the wells drilled were completed as oil or gas wells. This percentage declined to about 57% by the end of the 1960s. It then rose steadily during the 1970s to reach 70% at the end of that decade, primarily because of the rise in oil prices. A plateau or modest decline followed through most of the 1980s. Beginning in 1990, completion rates increased dramatically to 77%. The increases of the 1990s had more to do with new technology than higher prices.[15]

Most upstream projects are done in some type of partnership structure. For example, a production sharing agreement (PSA) for the Azeri, Chirag, and Gunashli development in Azerbaijan was signed in September 1994. BP is the operator with a 34.1% stake; the partners were Chevron with 10.3%; SOCAR, 10%; Inpex, 10%; Statoil, 8.56%; ExxonMobil, 8%; TPAO, 6.8%; Devon, 5.6%; Itochu, 3.9%; and Hess, 2.7%.

Reservoir management

For companies involved in the upstream, reservoir management is an essential skill. Reservoir management involves ensuring that reserves are replaced and that existing oil and gas fields are efficiently managed. Asset acquisition, divestiture, and partnering are key aspects of reservoir management. Upstream companies try to replace more than 100% of the oil and gas produced. Determining the level of *proved reserves* (the amount of oil and gas the firm is reasonably certain to recover under existing economic and operating conditions) is a complex process. Consider the following comment on the auditing of reserves:

> *Though the word "audit" is customarily used for these evaluations, oil and gas reserves cannot be "audited" in the conventional sense of a warehouse inventory or a company's cash balances. Rather, "proved reserves" are an approximation about formations thousands and even tens of thousands of feet below ground. Their size, shape, content and production potential are estimated in a complex combination of direct evidence and expert interpretation from a variety of scientific disciplines and methodologies. Added to the science is economics; if it costs more to produce oil from a reservoir than one can sell it for profitably, then one cannot "book it" as a reserve. Reserves are "proved" if there is a 90% chance that ultimate recovery will exceed that level. . . . As perverse as it may sound, under the "production sharing agreements" that are common in many oil-producing countries, when the price goes up, proved reserves go down.*[16]

Matthew Simmons, founder of the energy-focused investment bank Simmons and Company, commented that "95% of world 'proven reserves' are in-house guesses," "most reserve appreciation is exaggerated," and "95% of the world's 'proven reserves' are unaudited."[17] The pressure to replace reserves has on occasion resulted in some unintended behaviors. In 2004, Shell's CEO left earlier than anticipated after revelations that the company had overstated its reserves by nearly 25%.

Upstream profitability

Profitability is largely a function of costs and commodity prices. According to Simmons and Company, Saudi Arabia's oil producers could make a profit if the price of crude oil fell to $10/barrel; the Canadian oil sands company Suncor could be profitable at $25/barrel with existing facilities, North Sea oil producers could be profitable at $25/barrel with existing facilities, Venezuelan heavy oil required a price of $25–30/barrel for profitability, new facilities in the Canadian oil sands would need a price of at least $50/barrel to make a profit, and for US ethanol production to be competitive, the price of crude had to be at least $50/barrel.[18]

Midstream: Trading and Transportation

The midstream in the value chain comprises the activities of storing, trading, and transporting crude oil and natural gas. As shown in figure 1–7, once oil and gas are in production, there is a divergence in the value chain. Crude oil that is produced must be sold and transported from the wellhead to a refinery. Natural gas must also be moved to markets via pipeline or ship; we provide an overview of the gas business in a later section. And as described in the *Industry Insight: The Oil Industry's Fixers*, trading itself can be a unique business.

Industry Insight: The Oil Industry's Fixers

"In many African countries, a Western entrepreneur might hand over money to a fixer or middleman, who would then pass it on to a political leader in exchange for support for a business venture."

—Robin Urevich, "Chasing the Ghosts of a Corrupt Regime," *Frontline*, January 8, 2010

The global oil industry has long been the source of much power and wealth. In an industry of such importance, the role of a select few middlemen, *fixers*, who have relationships, access, and occasionally influence, has been one of the global industry's key lubricants. They were for many years the major midstream institution in the global oil industry.

It is difficult to actually categorize what these fixers do. They act in some cases as liaisons, middlemen, brokers, or influence peddlers between firms and governments, all in the pursuit of developing some of the largest oil and gas plays in the world. They do have one common characteristic: they are all in it for the profit. A partial list of fixers would include the following:

- **Gilbert Chagoury.** Born in Nigeria to Lebanese parents, Chagoury acted as a close associate and financial and oil adviser to Nigerian president Sani Abacha for many years. A close friend and major financial contributor to President Bill Clinton, Chagoury today is a diplomatic representative of the Caribbean island of St. Lucia.

- **John Deuss.** Johannes Christiaan Martinus Augustinus Maria 'John' Duess, a Dutch oil trader and sometimes banker, was a global player in the oil industry for nearly two decades. Deuss owned a fleet of oil tankers, was accused of smuggling arms to South Africa's apartheid regime, and was integrally involved in Oman's royal family investment in Kazakhstan.

- **Samuel Dossou-Aworet.** Born in Benin and educated in France, Dossou-Aworet has acted as financial and oil adviser to Gabon's president Omar Bongo for many years. He has served as Gabon's representative on the OPEC Governing Board, where he has also acted as chairman. He is owner and founder of Petrolin, a private exploration and production company operating in Africa and the Middle East.

- **James Giffen.** Founder of Mercator Corporation, Giffen is an expert on American-Soviet trade, organizing the American Trade Consortium, which expedited the entry by major US-based multinationals into the Soviet Union in the 1980s and 90s. He also served as oil adviser to the president of Kazakhstan, Nursultan Nazarbayev. Although accused of funneling more than $80 million from American oil interests to the Kazakh president and associates, he was found innocent of all but minor charges in August 2010.

- **Mark Rich.** An international commodities trader and founder of the oil trading firm Glencore, he was convicted of illegally trading oil with Iran during the 1970s and 80s. Although convicted *in absentia*, having never returned to the US, he received a presidential pardon from President Bill Clinton upon Clinton's departure from office in 2001.

- **Hany Salaam.** Lebanese by birth, Salaam is a powerful and influential global traveler who arranged numerous deals for Armand Hammer and his oil firm Occidental. He was a purported insider at various times to presidents and kings in the Middle East, including King Hussein of Jordan. His son, Mohamed, has been accused of attempting to lead a coup to take over the tiny oil-rich country of Equatorial Guinea.

- **Oscar Wyatt.** One of a group of wealthy and powerful Houston oil men and founder of Coastal Corporation, Wyatt has been characterized as a corporate raider and deal maker. He was convicted in 2007 of illegal trading in the oil-for-food scandal with Iraq, serving one year in prison.

Source: Based on a number of sources including "Invisible hands: The secret world of the oil fixer," Ken Silverstein, *Harper's Magazine*, March 2009; "Chasing the Ghosts of a Corrupt Regime," by Robin Urevich, *Frontline*, January 8, 2010; *The Oil and the Glory*, Steven Levine, New York: Random House, 2007.

Crude oil has little or no value until it is refined into products such as gasoline and diesel. Thus, producers of crude oil must sell and transport their product to refineries. The market for crude oil involves many players, including refiners, speculators, commodities exchanges, shipping companies, IOCs, NOCs, independents, and OPEC. Market-making activities in the oil business have become front page news, and the daily price of crude oil is as frequently reported in the news as the weather.

The ease by which liquids can be transported is a key reason why crude oil has become such an important source of energy. Although pipelines, ships, and barges are the most common transportation platforms for crude oil, railroads and tank trucks are also used in some parts of the world. The shipping industry is very fragmented and, because oil tankers travel for the most part in international waters, largely unregulated. New technologies in ship building in recent decades have allowed ships to become larger and safer.

Pipelines in Alaska, Chad and Cameroon, Russia, and other countries have allowed oil to be transported from very remote locations to markets. The construction and management of pipelines is fraught with geopolitical intrigue, which means the pipeline development process takes many years or even decades. Pipelines that cross national borders are enormously complex to negotiate and build. Countries with pipelines that cross their territory have been known to use them as bargaining chips. Terrorists often sabotage pipelines and in some countries, such as Nigeria and Iraq, oil theft from pipelines, and the associated environmental and safety issues are daily occurrences.

Downstream: Oil Refining and Marketing

The refining of crude oil produces a variety of products, including gasoline, diesel fuel, jet fuel, home heating oil, and chemical feedstocks. In the United States, about 60% of refinery product volume is gasoline. Products are sold directly to end users through retail locations, directly to large users, such as utilities and commercial customers, and through wholesale networks. A *merchant refinery* is a stand-alone refinery not part of an integrated distribution system. Increasingly, NOCs such as Saudi Aramco are jumping into the merchant refining business as a means of capturing additional value added from their crude production. Although it is more economical to transport crude oil versus refinery products such as gasoline, the United States imports about 10% of its gasoline supply. The volume of imported refinery products is a function of regional arbitrage opportunities due to short-term swings in local supply and demand balances.

The financial performance of the refining industry has always been volatile. The primary measure of industry profitability is the *refining margin*, which is the difference between the price of crude oil and that of the refined products. Crude prices can fluctuate for many reasons. Weather in the Gulf Coast states, political instability in oil-producing countries, or OPEC actions, for example, all influence the price of crude oil. These fluctuations were not always accompanied by matching changes in the price of finished products, leading to large expansions or contractions of the margin.

Figure 1–9 shows that profits on refining are usually lower than profits in other lines of business for petroleum companies. To put the downstream business in perspective, Lee Raymond, former ExxonMobil CEO, said in 1997, "I've been pessimistic on refining for 30 years, and I've run the damn places."[19] In 1999, BP CEO John Browne announced an aggressive plan to improve returns at BP by sharply reducing global refining capacity in the expectation of persistently weak profit margins.

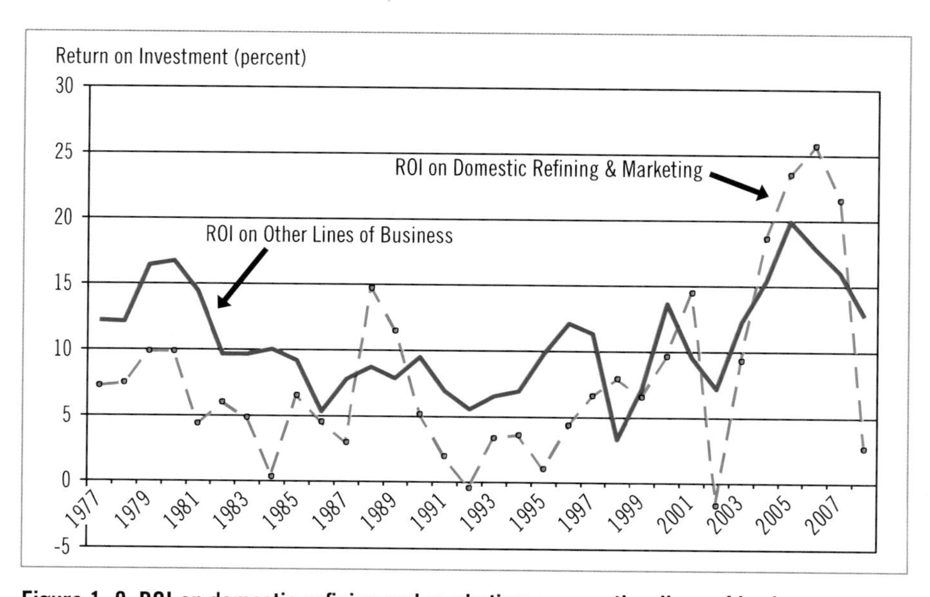

Figure 1–9. ROI on domestic refining and marketing versus other lines of business
Source: "Return on Investment in U.S. and Foreign Refining and Marketing and All Other Lines of Business for U.S. Major Oil and Gas Companies 1982–2008," United States Energy Information Agency (EIA), December 2009.

Shell's head of downstream operations described the business as, "Grubbing [i.e., begging] for pennies in a street. . . . If this industry, and especially the downstream, were to let its cost base slip, then we're going to have difficulty getting through those down low cycles."[20]

There are a number of reasons why the price of finished products does not track that of the crude inputs. According to the New York Mercantile Exchange:

> *A petroleum refiner, like most manufacturers, is caught between two markets: the raw materials he needs to purchase and the finished products he offers for sale. The prices of crude oil and its principal refined products, heating oil and unleaded gasoline, are often independently subject to variables of supply, demand, production economics, environmental regulations, and other factors. As such, refiners and nonintegrated marketers can be at enormous risk when the prices of crude oil rise while the prices of the finished products remain static, or even decline. Such a situation can severely narrow the crack spread—the margin a refiner realizes when he procures crude oil while simultaneously selling the products into an increasingly competitive market. Because refiners are on both sides of the market at once, their exposure to market risk can be greater than that incurred by companies who simply sell crude oil at the wellhead, or sell products to the wholesale and retail markets.*[21]

What this means is that profitability of refining is set by two factors:

1. The supply and demand for refinery products (i.e., if refining capacity is tight the refining margins are high and refineries make a lot of money)
2. Refinery product prices, which are set by a combination of the supply and demand of refinery products and crude oil prices

Gasoline prices can be high because of high crude prices, but refining margins and refining profitability can be weak if the demand for refinery products is also weak. In 2005 and 2006, US refining experienced an unusual situation with both high crude prices and high refining margins.

The number of operating US refineries dropped from 195 in 1987 to 141 in 2009, but during that period, US production capacity increased from less than 15 million b/d to more than 17 million b/d.[22] The increased refining capacity came from debottlenecking and expanding existing refineries, which is much cheaper than building new ones. Refinery capacity utilization and profitability is cyclical and highly dependent on overall economic activity. In the early 1980s, US refinery utilization was about 70%. In 2007, capacity utilization was 90% and profit margins were high. By 2009, utilization was about 85% and margins were falling.

In contrast to the situation in the United States and Europe, new refineries are being built in other countries. In 2009, Reliance Industries completed the world's largest refinery complex at Jamnagar in India. The Jamnagar complex has a capacity of 1.24 million b/d, and the number of construction workers at the site reached about 150,000. In the near term, Jamnagar is expected to focus on export markets. The largest market for Jamnagar is in the Middle East, followed by Africa, Europe, and the United States. Shipping costs are only pennies per gallon for finished products.

Gasoline retailing

In the gasoline retail sector, competition is intense and margins have eroded over the past 10 to 15 years. For the IOCs, returns on capital employed are much lower in retail than in other business areas. The entry of hypermarkets/supermarkets into retail gasoline sales in Western Europe had displaced small dealer networks, and national players found they could make good money from convenience store sales. That said, Shell's head of downstream dismissed the notion that convenience store sales should be the focus for the fuels marketing business:

> *The industry thought it could save itself with Coke . . . we found out that maybe the fuels game is more our game than the convenience store game. . . . It's not a saviour for our industry. The important thing in retail is that you need to keep on changing things: that you keep different customer value propositions and you keep changing them all the time.*[23]

In the US, supermarket and "petropreneur" entry into gasoline sales is also occurring, although not with the same speed as in Europe. In most countries gasoline is seen as a commodity product, which means spending money on brand development has questionable results. The weakness of brands favored the entry of supermarkets because they compete on price and proximity and sell fuel as a loss leader. With traditional retail barriers to competition gone, the largest IOC retailers are selling company-owned stores. In the US, new entrants, such as Tosco (subsequently part of ConocoPhillips) and Valero, were able to buy refinery and retail assets and knit together profitable retail networks integrated with their refinery acquisitions.

Natural gas

Natural gas, an important global energy source, is a naturally occurring fossil fuel found by itself or near crude oil deposits. Like oil, the largest gas reserves are found in countries such as Russia, Venezuela, Iran, and throughout the Middle

East. In the United States, gas accounts for approximately a quarter of the energy consumed, and the OECD average is 22%. Natural gas represented less than 4% of China's energy consumption in 2009, but demand is rising by more than 20% per year.

For many years, natural gas was a niche product because, unlike crude oil, natural gas is not easily transported. Without a pipeline infrastructure, natural gas in its gas form cannot be transported far from its source. In some parts of the world, such as Canada, the United States, and Western Europe, a network of pipelines allows gas to be distributed efficiently. In the US there are 160 gas pipeline companies operating more than 285,000 miles of pipe. In other parts of the world, such as offshore Africa or Aceh Province in Indonesia, pipelines to customers are not feasible. To transport the "stranded" gas, it must be converted to liquid natural gas (LNG). To liquefy natural gas, impurities such as water, carbon dioxide, sulfur, and some of the heavier hydrocarbons are removed. The gas is then cooled to about −259°F (−162°C) at atmospheric pressure to condense the gas to liquid form. LNG is transported by specially designed cryogenic sea vessels and road tankers.

Historically, the costs of LNG treatment and transportation were so huge that development of gas reserves was slow. In recent years, LNG has moved from being a niche product to a vital part of the global energy business. As more players take part in investment, both in upstream and downstream, and as new technologies are adopted, the prices for construction of LNG plants, receiving terminals, and ships have fallen, making LNG a more competitive energy source. LNG ships are also getting much larger. The larger ships, plus larger LNG trains (i.e., plants to convert the gas to LNG), are expected to result in a 25% reduction in delivery cost relative to the cost in 2000. In addition, natural gas to liquid technology provides an alternative to LNG and converts gas to liquid products, such as fuels and lubricants, that can be easily transported.

Major structural changes are occurring in the gas business. A short-term LNG market was virtually nonexistent a decade ago. Long-term contracts were sought to ensure security of supply for the buyer and security of revenue for the producer. Recent changes in the LNG market and in LNG shipping have increased flexibility for producers and consumers, and contracts are being negotiated for shorter periods of time. The agreement to develop the huge Qatargas 2 project, jointly owned by ExxonMobil and Qatar Petroleum, was finalized in 2002 without contracts for gas sales in place. An LNG ship can deliver its gas anywhere there is an LNG terminal, making LNG almost as flexible in delivery as crude oil (although a reluctance of many communities to allow terminals to be developed has been a growth constraint).

There is also speculation that the rapid growth in Middle East LNG supply could lead to a global convergence in gas pricing and markets, with LNG someday becoming a traded commodity. As well, buyers and sellers have been taking on new roles. Buyers have been investing in the upstream, including liquefaction plants (e.g., Tokyo Gas and the Tokyo Electric Power Company have invested in the Darwin liquefaction plant in Australia). Producers, such as BP and Shell, have leased capacity at terminals and are extending their role into trading. New buyers have been emerging, including independent power producers. Finally, gas produced from shale is becoming increasingly important as an energy source.

Petrochemicals

Although all of the major IOCs are involved in chemicals to some degree, they have different strategic approaches. ExxonMobil Chemical, one of the world's largest chemical businesses, includes cyclical commodity type products, such as olefins and polyethylene, as well as a range of less cyclical specialty businesses. Many of ExxonMobil's refineries and chemical plants are colocated, providing opportunities for shared knowledge and support services and the creation of product-based synergies. In the past, BP and Shell had chemical businesses that were among the largest in the world.[24] In 2005, BP decided that its chemical business was noncore and divested the majority of the business. BP's remaining chemicals businesses became part of the refining and marketing division and were no longer considered a separate corporate division. Shell also downsized its chemicals business. The rising players in chemicals are in the Middle East and Asia and included NOCs, such as SABIC (Saudi Arabia) and Sinopec (China), and non-state-owned companies, such as Reliance (India). There is some concern in the industry that excess capacity is being created in Asia and especially in commodity products in China.

Fundamentals of Business: What Is Strategy?

A primary goal of this book is to help readers understand the critical business decisions necessary to compete and survive in the oil and gas industry. To understand business decisions, readers must be familiar with the concept of strategy. Strategy can be viewed from two perspectives: corporate strategy and business level strategy. Corporate strategy is concerned with the scope and breadth of the diversified firm. The key issues at the corporate strategy level are "what business should we be in?" and "how should we allocate resources to the various businesses?" For an IOC, decisions as to whether or not the firm should compete in fuels marketing or petrochemicals are corporate strategy decisions. In the oil and gas industry, the degree of corporate vertical integration (i.e., upstream to downstream activities) will be tied to corporate strategy decisions. The issue of integration and corporate strategy decisions is discussed in various chapters throughout the book. For example, chapter 12 considers the advantages and disadvantages for a refiner that does not have an upstream business.

Business level strategy involves the choices and tradeoffs about how to compete in a specific industry or business. The oil and gas industry is a collection of many different businesses, as explained earlier. Thus, an IOC such as BP or Shell is competing in many businesses, each of which would have its own competitive strategy. Collectively, the businesses of BP and Shell drive their respective corporate strategies. In reality, there will always be overlap between business and corporate strategies in a diversified firm. For example, the decision to build a petrochemical plant may be linked with a refinery expansion because the refinery provides feedstock for the chem plant.

The business level strategy must address three main questions:[25]

1. The first question is, what is the strategic objective for the firm or business? Without a clear objective, it is impossible to evaluate the success of a strategy. There are various possible objectives, such as maximizing net income or return on capital employed or increasing market share.
2. The second choice involves scope: Where will the business compete? What products and services will be offered? What geographic locations, customers, and market segments will be served?
3. A third choice deals with what is necessary to ensure that the business is distinctive and different from competitors. To address this choice businesses must consider:
    - How will the business create and capture value? (Note: value is *created* when a customer is willing to pay for a good or service produced by the firm; value is *captured* when the firm retains some portion of the sales revenue after all operating expenses are paid for.)
    - What will ensure that suppliers or customers do not appropriate all the value created?
    - What is the customer value proposition? What is the customer willing to pay for (i.e., what drives value creation)?
    - What are the unique activities that allow the business to deliver the value proposition?

Collectively, the set of choices constitutes a business strategy. The last choice involving unique activities refers to the execution of the strategy. It is not enough to choose a corporate scope or customer value proposition. An organization has to be created that can execute the strategy. Ultimately, strategy is a disciplining device that helps sort out the opportunities that should be pursued from those that should be ignored. The outcome of successful strategic choices is a unique position in an industry and a competitive advantage relative to competitors.

In the following chapters, the oil and gas industry value chain is examined. The strategic choices and drivers for creating competitive advantage are discussed for the different industry sectors. For example, chapter 3 considers the role of technology in creating a unique competitive position. Chapter 5 discusses the importance of achieving a low-cost position given that crude oil is sold into a commodity market and product differentiation is impossible. Chapter 13 examines how firms with traditional advantages in fuels marketing have seen those advantages erode over time as gasoline and diesel shifted from a consumer brand to a commodity-like product.

Other important strategic issues are examined, including:

- Why are some E&P firms much more productive than their competitors?
- What are the major barriers to entry for newcomers to E&P?
- Does an integrated refiner have an advantage over a stand-alone refiner?
- Can refined products command a premium price by being differentiated?
- How much control over transportation is necessary for an E&P firm?
- To execute their strategies, do the NOCs really need the IOCs?
- What are the strategic synergies between refining and chemicals?
- Will the IOCs be able to regain a strong competitive position in fuels marketing?

Evolution of the Industry

There are a number of major forces that have driven evolutionary change in the oil and gas industry over the past century. We begin with innovation.

Innovation and technology

Innovation plays a key role in all parts of the oil and gas industry. Innovations in areas such as deepwater drilling and LNG shipping were discussed earlier. In the upstream, several key technological improvements have been developed in the past few decades, including increased use of 3-D seismic data to reduce drilling risk, and directional and horizontal drilling to improve production in reservoirs.[26] Innovations in financial instruments have been used to limit exposure to resource price movements. In oilfield management, wireless technologies allow for faster and cheaper communication than the traditional wired underground infrastructure. In refining, nanotechnology has enabled refiners to tailor refining catalysts to accelerate reactions, increase product volumes, and remove impurities, which has led to increased refining capacity. In retailing, innovations such as unmanned stations have reduced retail costs.

Mergers and acquisitions

Mergers and acquisitions (M&A) have been an important element in the oil and gas industry since its inception. Although the megamergers, such as BP-Amoco, Total-PetroFina, Chevron-Texaco, and ExxonMobil, receive much of the press, there are also many smaller deals. Note also that many of the major acquisitions in recent years have been by firms from emerging markets.

In looking at the mega-M&A deals done over the past few decades, one might conclude that eventually there will only be a handful of oil companies in the world. The reality is different. Research shows that the oil industry is much less concentrated today than it was 50 years ago.[27] Opportunities exist for new entrants despite the huge size of the largest IOCs and NOCs. In the downstream in the 1990s, new entrants, such as Tosco, Premcor, and Petroplus, had a significant impact on industry structure. In chemicals, Ineos, the privately held British company, grew through a series of related acquisitions to become the world's third-largest chemical company, with sales of about $33 billion. In the upstream, the huge financial scale of projects such as Sakhalin I and II or Qatargas 2 makes it unlikely that a new entrant could challenge the IOCs. However, if NOCs in China and India continue to acquire and grow, they may develop the technological and financial skills to compete for the largest and most complex upstream projects.

China and India

In 1998, China became a net importer of oil for the first time. In 2006, China overtook Japan to become the world's second largest importer. By 2030, China will likely be importing about 80% of its oil. Clearly, China and Chinese companies are going to be major players in the oil and gas industry. Thousands of gas stations are being built, and Chinese companies are aggressively investing in upstream projects around the world.

Unlike the United States and Europe, China has no qualms about allowing its oil industry (the big three Chinese NOCs to start with) to invest in countries like Sudan and Iran. On the retail side in China, prices are regulated, resulting in unintended consequences. If the government increases prices, especially for diesel, there might be social unrest. Because refiners lose money on diesel, they cut back on diesel production, which can lead to diesel shortages and increases in diesel imports. State-owned refiners have little capital available for upgrades and modernization and often purchase low-quality crudes high in sulfur content. China has much less stringent environmental regulations than the developed world. More stringent regulations would mean higher fuel costs. As a comparison, the United States allowed maximum sulfur concentrations of 15 parts per million for most diesel fuels while China allowed up to 2,000 parts per million.[28] China's cities are among the most polluted in the world.

India is also a force to be reckoned with in the global oil and gas industry. India, the fifth largest oil consumer, needs energy to feed its rapidly growing and industrializing economy. Companies such as Reliance are moving aggressively into the upstream, and stodgy state-owned companies such as ONGC, Oil India Limited, and Gas Authority of India are slowly becoming more productive. Like China, India is far from self-sufficient in energy and must find new energy sources.

Industry substitutes and alternative fuels

The role and future of non-hydrocarbon-based fuels and energy sources has become a critical issue for policy makers and energy companies. Various factors are contributing to a large investment flow into alternative fuel projects, including the rapid rise in oil and gas prices in recent years, concerns about global climate change, perceived competitive opportunities by energy companies (new entrants and entrenched players), and government subsidies.

Forecasts by the International Energy Agency suggest biofuels output could rise to the equivalent of more than 5 million barrels of crude oil a day by 2011, close to triple the output of 2005. Deutsche Bank issued a provocative industry report in 2009 called "The Peak Oil Market" that says, "We forecast a game change. US and then global oil demand will fall dramatically once the high efficiency fleet hits critical mass; competing structurally cheaper natural gas will exacerbate the pace of demand decline. In our view global oil demand peaks in 2016, with oil prices, before a long, tandem, decline."[29]

What's next for the global oil and gas industry?

> *There are two times in a man's life when he should not speculate: when he can't afford it and when he can.*
>
> —Mark Twain, "Following the Equator,"
> *Pudd'nhead Wilson's New Calendar*

A few predictions for the industry are very safe: the global demand for oil and gas will continue to rise over the next few decades; the industry will remain one of the most vital for the global economy; and despite the high prices of recent years, the industry will continue to go through up and down cycles. Oil and gas firms will continue to do what they have done for more than a century: take a long-term view, invest for the future, push the boundaries of technology, and seek new resources and markets in every corner of the world. In doing so, firms will face a variety of technological, regulatory, environmental, and geopolitical challenges.

Notes

1. Alfred D. Chandler, "The Enduring Logic of Industrial Success," *Harvard Business Review*, 1990, March-April, Vol. 68, Issue 2, pp. 130–140.
2. Leonardo Maugeri, "Oil: Never Cry Wolf—Why the Petroleum Age Is Far from Over," *Science*, 2004, Vol. 304, pp. 1114–1115.
3. Fox News, Transcript: ExxonMobil's Lee Raymond, October 17, 2005, http://www.foxnews.com.
4. The Fortune 500 Medians, *Fortune*, April 29, 1996, pp. 23–25.
5. www.opec.org /opec_web/en/press_room/178.htm.
6. Richard Auty, *Sustaining Development in Mineral Economies: The Resource Curse*, Thesis, 1993, London: Routledge.

7. Roger Stern, "Iran Actually Is Short of Oil: Muddled Mullahs," *International Herald Tribune*, January 8, 2007, www.iht.com.
8. http://www.foxnews.com/story/0,2933,172527,00.html.
9. "Indonesia Considers Legislation That Would End Pertamina's 30-year Petroleum Monopoly," *Oil & Gas Journal*, July 26, 1999, pp. 27–32.
10. "Special Report, National Oil Companies," *The Economist*, August 12, 2006, pp. 55–57.
11. Dino Mahtani, "Nigeria Struggles to Eliminate Corruption from Its Oil Industry," *Financial Times*, January 11, 2007 p. 8.
12. "We Have to Be An Integrated Oil Company," *Hindu Business Line*, August 10, 2003, www.thehindubusinessline.com.
13. "Face Value: Thinking Small," *Economist*, July 22, 2006, p. 64.
14. The value chain concept was developed by Harvard Professor Michael Porter and is the main theme of the book *Competitive Advantage: Creating and Sustaining Superior Performance* (Free Press, 1985). Porter used the concept to explain how firms created competitive advantage. Porter's generic value chain included primary and support activities. Primary activities included: inbound logistics, operations (production), outbound logistics, marketing and sales (demand), and services (maintenance). Support activities included: administrative infrastructure management, human resource management, technology (R&D), and procurement. The extension of the firm value chain to the industry is logically consistent, especially in the oil and gas industry where the IOCs compete across most of the major industry segments.
15. Oil Price History and Analysis from WTRG Economics, http://www.wtrg.com/prices.htm.
16. Daniel Yergin, "How Much Oil Is Really Down There?" *Wall Street Journal*, April 27, 2006, p. A.18.
17. http://www.simmonsco-intl.com/files/HBS%20Energy%20Forum.pdf.
18. http://www.simmonsco-intl.com.
19. Richard Teitelbaum, "Exxon: Pumping Up Profits," *Fortune*, April 28, 1997.
20. Ed Crooks, "Interview: Rob Routs: You Have to Keep Changing," *Financial Times*, October 20, 2006, Special Report Energy, p. 10.
21. New York Mercantile Exchange, *Crack Spread Handbook*, 2000, p. 4.
22. http://tonto.eia.doe.gov.
23. Ed Crooks, *Financial Times*.
24. Peter Partheymuller, "Chemicals," *Hoover's*, http://premium.hoovers.com.
25. David J. Collis and Michael G. Rukstad, "Can You Say What Your Strategy Is?" *Harvard Business Review*, April, 2008, pp. 82–90.
26. WTRG Economics, http://www.wtrg.com/prices.htm.

27. Pankaj Ghemawat and Fariborz Ghadar, "The Dubious Logic of Global Megamergers," *Harvard Business Review*, July–August, 2000, Vol. 78, Issue 4, pp. 65–72.
28. Keith Bradsher, "Trucks Power China's Economy, at a Suffocating Cost," *New York Times*, December 8, 2007, www.nytimes.com.
29. Deutsche Bank Securities Inc., "The Peak Oil Market: Price Dynamics at the End of the Oil Age," October 4, 2009, p. 2.

Chapter 2

NATIONALISM, NATIONAL OIL COMPANIES, AND THE CURSE OF OIL

The relationship between governments and energy companies will be increasingly important as the world confronts two great challenges—providing secure and economic energy supplies to meet expanding needs, and responding to climate change. Much more energy will have to be supplied across borders and much more invested to develop and deliver it. Governments must recognise that this is an increasingly complex, demanding and risky business.

—Jeroen van der Veer, chairman and president, Royal Dutch Shell, August 2004[1]

The first point is that Schlumberger probably for the last 40 or 50 years has not had a nationality. It's not a French company. It's not an American company. When the Schlumberger brothers formed a holding company, it was to avoid family death duties, or taxation on succession. And they formed a company in the Netherland Antilles. So we've never domiciled in one of the large countries.

—Andrew Gould, CEO Schlumbeger, April 2010[2]

The interests and power of all stakeholders in the development and exploitation of oil and gas resources are forever intertwined. A firm's license to operate is granted by the state. State governments hold title to most of the world's oil and gas. Governments are responsible for looking after civil society, the true constituents of any country. Holding title to such a valuable resource as a nation's oil and gas reserves involves complex issues of national security, national wealth, nationalism, and geopolitics.

We begin this chapter by exploring how the oil and gas industry is "different" in terms of its development and management. We then explore the interests of both government and oil and gas firms, followed by an in-depth analysis of the evolving relationships between IOCs and NOCs. The formation and role of a collective of IOCs, independent oil companies, NOCs, and OPEC is also examined. The impact of oil and gas development on economic and social development is considered along with a discussion of the "curse of oil." Finally, we consider the role that the oil and gas industry plays in domestic politics.

The Role and Value of Oil and Gas

Oil and gas are more than a valuable product of subterranean mining. They power the world and are, in many ways, the lifeblood of the modern global economy. Without oil and gas energy and the thousands of modern petrochemical products, most of the world's population would have a very different lifestyle. As a result, oil and gas must be viewed as integral to the national security and national wealth of almost all countries.

National security

Beginning a century ago, ready access to and interest in oil reserves were becoming critical to a nation's security. As the world shifted from horses to cars, oil became the fuel for the global economy. Oil also became the energy for a nation's war machine—the ships, tanks, trucks, and aircraft that provided military transportation and fighting capability. Access to oil played a significant role in the wars of the 20th century, and in today's world, keeping the Persian Gulf free from conflict is vital to the global economy. The economic infrastructure of all countries is heavily dependent on oil and, increasingly, on natural gas. While that dependence will likely evolve over the next century, oil and gas will remain the dominant source of energy for decades to come.

A country cut off from petroleum products will quickly grind to a halt. In 2008, when oil prices reached record levels, many of the world's poorest countries struggled to manage their economies because, unlike the cost of local labor or locally manufactured products, oil is a global commodity. Regardless of national income levels, countries must pay world prices for oil. In recognition of oil's impact on economic wealth, many countries have made the politically expedient and fiscally disastrous decision to subsidize petroleum products. Even some of the largest oil producers, such as Iran and Venezuela, have struggled to maintain their high subsidies. Unfortunately, cutting subsidies is politically dangerous for ruling parties, again emphasizing the critical role that oil plays in the world of politics and economics.

Along with oil's role in the wars over the past century (a role that has been much debated and one that provokes strong reactions, especially with respect to the two Gulf wars), there is no question that petroleum plays a critical role in national security. Whether one believes in peak oil arguments or not, oil and gas are finite resources, and at the current time, there are no viable energy substitutes. No nation can afford to dismiss the significance of oil and gas in their national economy. China is perhaps the extreme case. As a nation with insufficient domestic oil and gas reserves to support its rapidly growing economy, the country's government has established ownership in international oil and gas reserves as strategic priority for the government. In contrast, the energy needs of the United States, Japan, and the large industrial nations of Western Europe have to date been met by market-oriented approaches.

National wealth

> *Oil kindles extraordinary emotions and hopes, since oil is above all a great temptation. It is the temptation of ease, wealth, strength, fortune, power. It is a filthy, foul smelling liquid that squirts obligingly into the air and falls back to the earth as a rustling shower of money. . . . Oil creates an illusion of a completely changed life, life without work, life for free. Oil is a resource that anesthetizes thought, blurs vision, corrupts.*
>
> —Ryszard, Kapuściński, *Shah of Shahs*, 1992,
> Vintage International, New York

For nations with oil and gas reserves, petroleum represents an asset of immense potential national value. Much of the value associated with oil and gas arises from two dimensions—extraction and scarcity.[3] (Unfortunately, as we discuss later in the section, "The Curse of Oil," not all resource owners are able to capture this value.)

Oil and gas are not produced in the sense of being manufactured; they need to be extracted from the ground. This is not meant to suggest that oil and gas exploration and production are easy or not technologically challenging. What it means is that a natural resource like oil can be extracted from the earth independently of other economic development that has occurred within a country. Often, in fact, oil and gas are developed and produced from a country that has limited economic infrastructure or capabilities, such as Chad or Papua New Guinea. The rapid economic development of Asian countries like Japan, South Korea, and Taiwan resulted from the combined productivity of technology and critical masses of highly skilled labor in sectors such as microelectronics and telecommunications. Moreover, these three countries possess limited oil and gas reserves. Oil extraction does not require a high level of economic development because oil and gas producers can bring all the necessary skills to the country. In contrast, if a nation wants to produce computers or cars, various economic elements must exist, such as skilled labor, support industries, financial and legal services, and educational institutions.

Because oil and gas are extractive, they have been described as being *enclaved*, meaning oil and gas resources may be effectively separable from the physical and economic conditions and capabilities of the country owning the reserves. Oil and gas extraction can occur independently of the political processes of a country. Oil and gas development does not have to have the explicit cooperation of the citizenry and often requires little support from state institutions. In fact, in some very poor countries, oil and gas companies are almost a substitute for the state because they provide support for education, roads, health, and security.

The second dimension is scarcity. Oil and gas reserves are assets, not incomes, and they are scarce nonrenewable natural resources. When the value of scarce assets rises in the marketplace, the margin between price and the cost of their extraction also rises, generating what is called *economic rent*. This rent is sustainable because the barriers to entry are extremely high for new competitors. In many competitive markets, high degrees of profitability attract new entrants. If barriers to entry can be overcome and competitive products are introduced, the margins for incumbents will decline or vanish. Not so in the oil and gas industry. To compete against the owner of scarce oil and gas reserves requires new entrants to find their own reserves, something that is becoming exceedingly difficult. Moreover, the pursuit and capture of rents from oil and gas can create destructive behaviors on the part of many individuals, organizations, and governments that comprise civil society.

Government and Corporate Interests

Given oil's unique characteristics, what explicit interests do government and companies have regarding its development? We begin by explaining the role of the corporation in society.

The corporation

> *A* CORPORATION *then, or a body politic, or body incorporate, is a collection of many individuals united into one body, under a special denomination, having perpetual succession under an artificial form, and vested, by policy of the law, with the capacity of acting, in several respects, as an individual, particularly of taking and granting property, of contracting obligations, and of suing and being sued, of enjoying privileges and immunities in common, and of exercising a variety of political rights, more or less extensive, according to the design of its institution, or the powers conferred upon it, either at the time of its creation, or at any subsequent period of its existence.*
>
> —Stewart Kyd, *A Treatise on the Law of Corporations*, 1793, p. 13

The creation of the *corporation*, as described by Stewart Kyd in 1793, solved a number of fundamental dilemmas faced by small businesses created by individuals in the 18th century:

- **A corporation is legally separate from its owner(s).** If it loses money, acquires debts, or incurs legal liabilities, the individual owners can lose no more than what they have invested in the company. The owners are therefore not legally liable for debts or obligations to creditors beyond their ownership investment. This is what is meant by the oft-used phrase *limited liability*.

- **A corporation is perpetual.** As opposed to individuals, a corporation cannot die in the normal human sense of the word. A corporation may go bankrupt, it may be acquired, it may be intentionally closed, but it can continue to operate in perpetuity beyond the lives of its creators or immediate owners.

- **A corporation is a legal creation deemed to have rights and responsibilities like a person.** Corporations are sometimes described as having a "legal personality." Corporations are actually given a charter by the government to undertake a business activity. The original corporate charters often established limits for the life span and nature of the corporation's activities. A corporation can buy or sell, sue or be sued, borrow and lend, and conduct business, all as a member of society. With these legal rights it has the responsibility to follow all laws of the state and can be held accountable for failing to act in a legal manner.

The creation of the corporation, a legally separate but legally responsible entity, made it possible to raise capital independent of the financial health of any of the individual owners, to contract for purchases and sales, and to survive the working life spans of its human creators. It may also have freed the business entity to act without the social, cultural, and moral concerns of human individuals. The ethics and interests of the organization's stakeholders must then fill that void.

Corporate stakeholders

So who are the stakeholders in the corporation, and what is the nature of their interests?[4] A stake, in theory, is an interest or a share in some organization or undertaking. Stakes themselves may be differentiated as to whether they are interests, rights, or ownership stakes. Figure 2–1 provides an overview of the potential stakeholders in the firm.

- An *interest* is when a person or group is potentially affected by the actions of an organization or firm. The community in which a firm operates and from which it employs people may oppose the closing of the firm, citing the community's interest in jobs or taxes.

- A *right* is when a person or group has a legal claim, a contract, or some other existing arrangement to be treated in a certain way or have a particular right protected. A right may be expressed in a variety of ways, including a right to be treated fairly, or an expectation that a firm will stand behind the quality of its products.

- *Ownership*, the strongest legal form of stake, is when a person or group holds legal title to an asset such as a firm. Of course this raises the question of whether the owner of a firm has dictatorial power or must also consider the interests of nonowners.

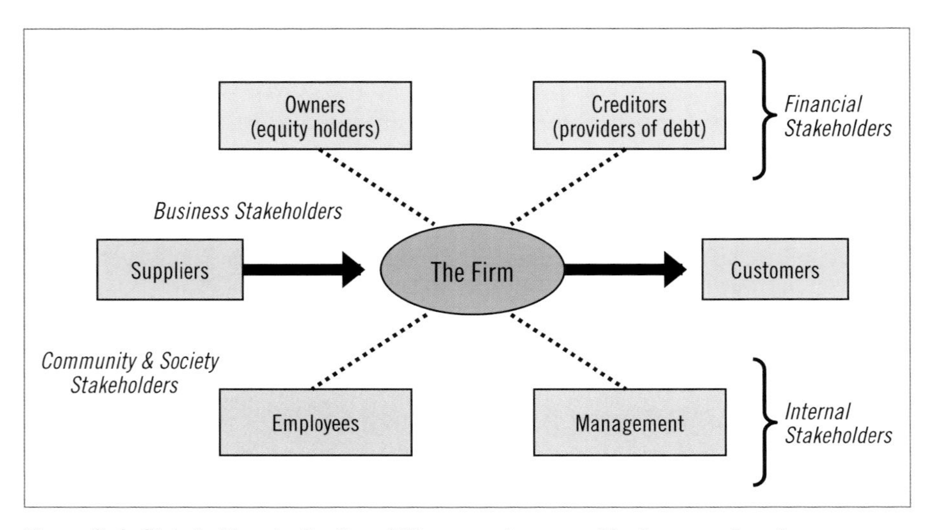

Figure 2–1. Stakeholders in the firm. Differences in ownership, law, country of incorporation and activity, and culture lead to differences in interest and power among stakeholder groups (business, financial, internal, and society).

Any analysis of corporate stakeholders must include all three levels of stakeholder interest. Depending on the country and culture of the firm, stakeholders will have different voices in the vision, strategy, and operation of the firm. As described in the following sections, the power and influence of those having an interest versus a right versus legal title (ownership) often serve to differentiate the business and political environments of countries.

Financial stakeholders. The two primary *financial stakeholders* of the firm, the providers of capital for both investment and operation, are shareholders and creditors. *Shareholders* are the actual owners of the firm, holding title to the firm's assets and operations as a result of their capital investment. *Creditors*, typically bank loans (Europe and Asia) or debt markets (Anglo-American markets), lend capital to the firm in exchange for promised repayment of both the principal amount and a return (interest). In the event that the firm cannot repay its debt obligations in a timely manner, the creditors may hold a legal right over specific assets of the firm. Note that there is no assumption that the owners are the operators of the firm; the two groups are often completely separate. Financial stakeholders hold ownership stakes.

Business stakeholders. The *business stakeholders* in the firm, its suppliers and customers, are the core of what business strategists term the value chain (see chapter 1 for a discussion of the oil and gas industry value chain). The company purchases products and services from suppliers, transforms the inputs into its own products and services, and sells them as finished products and services to its customers. Most of these transactions are conducted under contract. How the company conducts its business, including its financial health, the ethics of its business practices, and its business development prospects, all impact the operations and livelihood of its business stakeholders. Business stakeholders hold rights.

Among business stakeholders, customers have always enjoyed an area of particular focus. Peter Drucker, an influential academic and writer, advocated in one of his earliest writings that the customer and not the stockholder should be the focus of management action:

> *If we want to know what business is, we have to start with its purpose. And the purpose must lie outside the business itself. In fact, it must lie in society, since a business enterprise is an organ of society. There is only one valid definition of business purpose: to create a customer. The customer is the foundation of a business and keeps it in existence. He alone gives employment. And it is to supply the customer that society entrusts wealth-producing resources to the business enterprise.*[5]

This belief in the customer as the focal point of management action has a long list of believers. Without customers willing to buy what a firm produces, there will be no firm. One of the more recent advocates of customers first is the former CEO of Porsche, who stated:

> *Yes, of course we have heard of shareholder value. But that does not change the fact that we put customers first, then workers, then business partners, suppliers and dealers, and then shareholders.*[6]

Advocates for a customer focus are, in principle, arguing that the best way to run and grow the business is to focus solely on the needs and satisfaction of the customer and not the corporate owner. They are not, however, arguing that the benefits of the ongoing business are for distribution to the customer.

Proponents for the primacy of shareholder value as the objective of the firm take a different tack and argue that shareholder value as the objective function will lead to decisions that enhance outcomes for multiple stakeholders.[7]

Internal stakeholders. The *internal stakeholders* of the firm are the employees of the firm, including management and leadership. Workers and managers are hired to conduct the product and service procurement—purchasing, processing, production, and distribution of the company's products and services. Leadership and management may provide most of the strategic vision and direction of the operations of the firm. Leadership may hold no ownership, partial ownership, or even complete ownership in the firm. Internal stakeholders hold both rights and interest stakes.

Social stakeholders. Although the corporation is a legal construction, it will influence and impact local communities and society. Communities and society are the *social stakeholders*. Social stakeholders are more distant from the actual operations and transactions of the firm than other stakeholders but can still be impacted by the success or failure of the corporation, its investments and divestments, its employment and involvement, and its longevity and environmental stewardship. Social stakeholders hold interest stakes.

Beyond this set of participants are others who may be stakeholders, including different government organizations and regulators, equity exchanges, social and environmental interest groups, and nongovernmental organizations (NGOs). As discussed in the following section, depending on the general and business cultures in which the corporation operates, the voices of different stakeholder groups are more or less pronounced.

The state and civil society

Comparisons with private enterprises showed that SOEs [state owned enterprises] tended to be less efficient, less profitable, more highly leveraged, and prone to influence/capture, in particular by politicians who used SOEs to reward their supporters and maintain their patronage networks.

—The World Bank Group[8]

Since the mid-1950s the power of oil, its development, and its proceeds have been shifting from the IOCs to the countries owning the oil, their governments, and their national oil companies. In most of the world, oil and gas that lies beneath the surface of the earth is owned by the state (the United States is one of the last countries in which a private citizen may own natural resources). As a resource owned by the state, the oil and its development is now subject to a varied list of priorities different from the traditional stockholder wealth or corporate stakeholder wealth maxims of the capitalist markets.

The future of any upstream oil or gas company is its reserves in the ground. The control of reserves by the state has had enormous implications for the goals and strategies of the oil majors. For many large oil- and gas-exporting countries, oil and gas produces one of the largest, if not the largest, source of export revenues and hard-currency earnings. In countries such as Angola, Libya, and Venezuela, oil makes up more than 50% of the government's budget revenues. When oil prices fall precipitously as they did in 2008, government budgets and governments themselves are usually not prepared for the reduced budgetary cash flows. This is a problem that members of the Organization of Petroleum Exporting Countries (OPEC) have been dealing with since the 1970s.

In contrast to the corporation (i.e., firms in the private sector), the state's objectives are varied and may target a nearly universal set of stakeholders. Figure 2–2 identifies what might be called a "noble" set of potential objectives and their associated rationales.

It is possible that the government has little interest in objectives such as maintaining employment or sustaining economic development. Data shows that the track record of government-run businesses—as noted in the above quotation—is not particularly good, and state-owned oil and gas firms are no exception. Perhaps the objectives of the government running the state-owned firms are more concerned with self-interest. A recent book entitled *The End of the Free Market* makes this point, arguing that governments in countries like Russia and China use markets "to create wealth that can be directed as political officials see fit."[9] The goal of the governments is not about economic development or maximizing growth. The goal is political and self-interested: government-run businesses help maximize the state's power and help finance and sustain the leadership's chances of survival. State-owned oil and gas companies can be enormous cash cows, and as the Venezuela case shows, the oil industry provides money for redistribution and political survival.

Overcome Market Failure	**Market failure can occur in economic activities that include:** • Natural monopolies (electricity, water) • Public goods (law and order, national security) • Merit goods (education and health) • Externalities (positive or negative) • Information asymmetry
Overcome Regulatory Failure	**State ownership is desirable if and when:** • The state does not have the capacity to regulate effectively • The economic activity renders the drafting of contracts incomplete • The state cannot credibly promise not to confiscate or tax excessively
Industrial Economics	**Industrial economics goals:** • Sustain industrial sectors of particular interest for the national economy • Safeguard employment • Launch emerging industries with significant start-up costs when future property rights are uncertain • Control the decline of senile industries • Help the private sector carry risk
Development Economics	**Development economics goals:** • Boost the economy of the less-developed region(s) of the country • Pursue equality and social goals
Fiscal Policy and Redistributive Objectives	**Fiscal policy and redistributive objectives:** • Invest in a sector, control entry, impose monopoly prices, then use the revenues as fiscal income • Sell at reduced prices to targeted populations and distribute subsidies • Maintain employment and supplement welfare systems

Figure 2–2. Rationale for state ownership

Oil and gas and government

One way to approach the role of government in the oil and gas industry and the relationship between business, government, and society is to focus on how government affects the global oil and gas industry. According to Jeroen van der Veer, chairman and president of Shell Oil Company, there are three issues:[10]

1. Decisions as to who can access energy resources and markets
2. The regulatory framework within which companies must meet energy needs
3. Agreements on how to respond to shared challenges such as climate change

Deciding access. Governments can regulate the *who*, *what*, and *when* of all economic activities within their sovereign boundaries. In the oil and gas industry, control over access has critical significance since most governments own the resources under their land and water and hold all subsurface rights to access. The entire future of the industry lies in the hands of governments that will allow or deny organizations to explore, develop, and produce hydrocarbon resources.

Regulating. Regulatory frameworks for the industry follow the full array of civil society concerns, including health, safety, and environment. Regulation in most markets is then expanded to include pricing, taxation, and possibly redistribution of economic resources associated with every stage of the oil and gas value chain. The regulations are dictated by governments as they pursue their political, economic, and social agendas.

Sharing challenges. With the growth, maturity, and complexity of the global oil and gas industry has come an acknowledgment that government alone cannot achieve many of its security and wealth objectives without partnering with the private sector. Although still not universally accepted or well developed in the oil and gas industry, there is a growing consensus that oil companies and government need to work together.

The shared challenge debate has often obscured a more controversial and core debate: the degree to which the interests of civil society are addressed by the government. Separating the interests of civil society (a country's people, groups, communities, and organizations), its government, and oil companies is extremely complex. The interests of one group may conflict with those of other groups. For example, one group may want to protect an area of land for recreation or farming, whereas another group may seek economic development. As illustrated in figure 2–3, the people and communities impacted both positively and negatively by oil and gas and the IOCs must coexist within a web of complex and competing interests.

Wealth and security continue to drive government and its actions on behalf of civil society within the global oil industry. Although this book focuses on the business of oil and gas, understanding how the industry operates is not possible without understanding the interests of the oil-producing nations, their governments, and the executors of their estates, the national oil companies. That is our next topic.

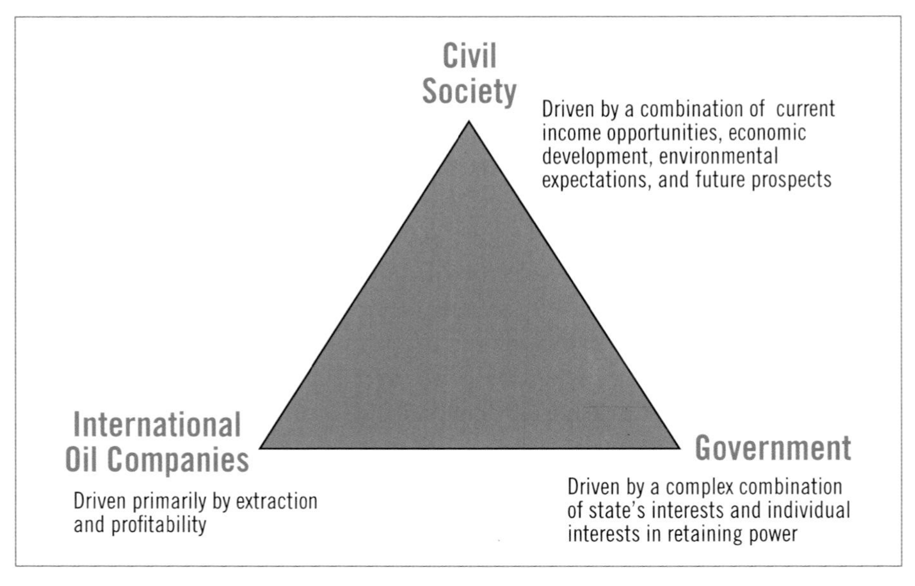

Figure 2–3. Oil development interests

Evolution of National Oil Companies

When activists, journalists and others speak of "Big Oil," you know exactly what they mean: companies such as ExxonMobil, Chevron, BP, and Royal Dutch Shell. These titans have been making lots of money for their shareholders; their bosses enjoy vast pay packets; and their actions affect us all.

Yet Big Oil is pretty small next to the industry's true giants: the national oil companies (NOCs) owned or controlled by the governments of oil-rich countries, which manage over 90% of the world's oil, depending on how you count. Of the 20 biggest oil firms, in terms of reserves of oil and gas, 16 are NOCs. Saudi Aramco, the biggest, has more than ten times the reserves that Exxon does. Those with misgivings about oil—that its price is too high, that reserves are running out, that it damages the environment, that it is more a curse than an asset for countries that produce it—must look to NOCs for reassurance.

—"National Oil Companies: Really Big Oil," *The Economist*, August 10, 2006

In the early years when production was centered in Pennsylvania and Baku, the oil industry was about individual entrepreneurs gaining rights to drill and develop oil from property owners. The property owners were a mix of private and public landowners. Once rights to access were obtained, the entrepreneurs developed and produced the oil, garnering much of the proceeds themselves. Their rapid success, growth, and power led to the intrusion of governments. The near-monopoly power attained by John D. Rockefeller and Standard Oil over refining and transportation eventually led to the forced breakup of the Standard Oil Trust (creating what became Exxon, Mobil, Chevron, SOHIO, and Amoco, as shown in figures 2–4a and 2–4b). In Baku, Azerbaijan the government sanctioned a small set of oil entrepreneurs like the Nobels and Rothschilds. The huge profitability of the entrepreneurs quickly led to the government's aggressive regulation of oil's societal and economic impact, with a parallel objective of capturing some of the wealth.

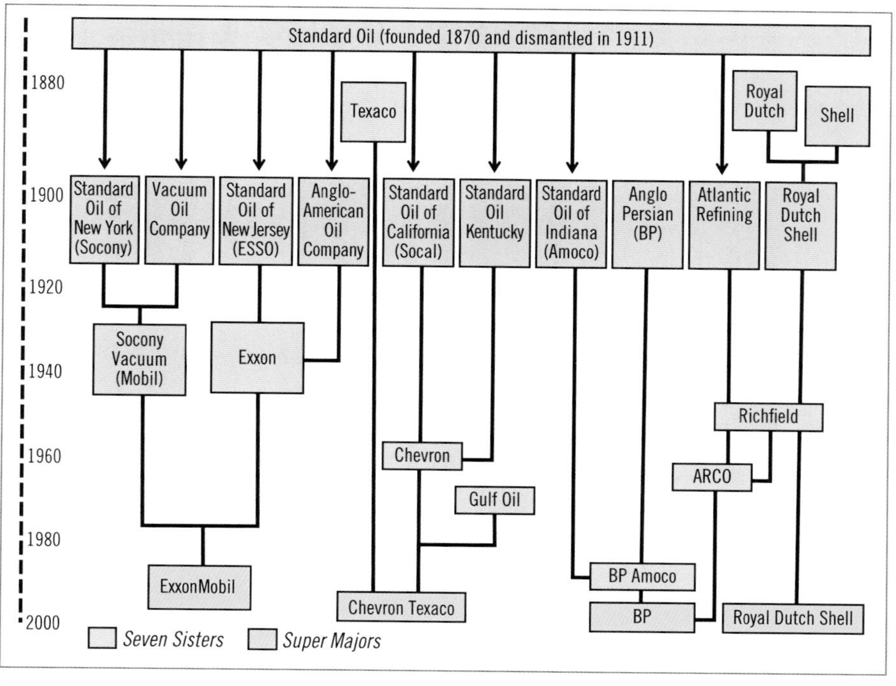

Figure 2–4a. Evolution of today's IOCs

Source: Deutschebank, 2008.

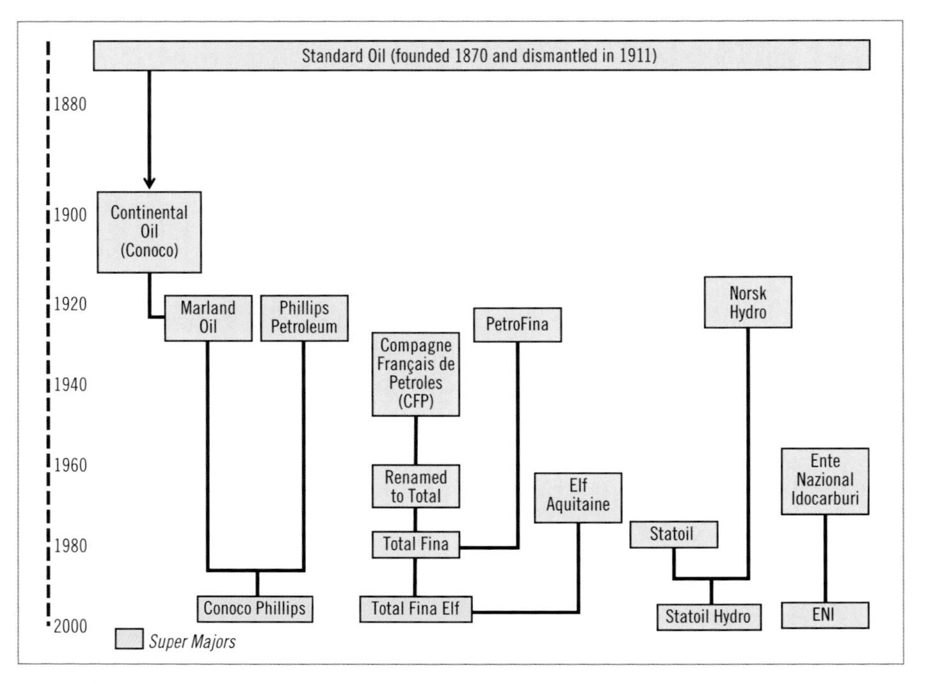

Figure 2–4b. Evolution of today's IOCs (cont. from figure 4–1a)
Source: Deutschebank, 2008.

Birth of the national oil company

The first national oil company (NOC) was born in 1908.[11] Emperor Franz Joseph approved the development in 1908 of a topping plant owned and operated by the government of Austria-Hungary in order to fill a widening gap between rapid production and inadequate product development for oil products. Other countries and governments followed in creating partial or wholly government-owned enterprises that would assure them of petroleum-based fuels for their military and civilian fleets. An NOC was created in the United Kingdom in 1914 and in France in 1924. The national security of the major industrialized consuming countries, not the desire to control petroleum resources, was what drove the initial wave of NOC development.

The nature of NOCs changed in the mid-1920s. The rapidly expanding oil industry within Latin America set the stage for NOCs to become the device for wealth capture and recapture by the state. Several companies were captured or recaptured (nationalized) from private corporate development. Argentina's Yacimientos Petrolíferos Fiscales (YPF) was the first in 1922, followed by the creation of NOCs in Chile, Uruguay, Mexico, Peru, Venezuela, and Bolivia. The first nationalization was the Bolivian government's takeover of Standard Oil of New Jersey's operations in Bolivia in 1937. This initiated a string of government actions across Latin America, including the nationalization of foreign oil company operations in Mexico in 1938, that spread to the Middle East in the 1940s.

With massive oil discoveries in the Middle East in the 1940s and 1950s, the growth and development of NOCs was established globally. Most of the early Middle East properties were first developed by a consortium of companies, often combining US, British, and other country company interests to share risks associated with development and political tensions. These consortia were gradually absorbed by the state in creeping nationalizations over the following decades.

Saudi Aramco is a prime example of the evolution of the NOC. Standard Oil of California (Socal), one of the former pieces of Standard Oil, obtained a concession to explore and develop oil in Saudi Arabia in 1933. The concession was assigned to a wholly owned subsidiary of Socal called the California-Arabian Standard Oil Company (Casoc). After several years of unsuccessful exploration, Socal sold 50% interest in its concession to the Texas Oil Company (Texaco) in 1936. Oil was finally discovered in 1938, and in 1944 the company's name was changed to the Arabian-American Oil Company (Aramco).

From there it was a slippery slope to nationalization. Beginning in 1950, the company agreed to share 50% of all profits with the government of Saudi Arabia, one of the so-called *Fifty-Fifty* arrangements in the 1940s and 1950s globally. (This will be discussed further in chapter 3.) The Saudi government then took a 24% ownership share in 1973. The government stake was expanded to 60% in 1974, during the height of Middle East tensions. Aramco was finally nationalized in total in 1980. Aramco today is the largest oil company in the world based on reserves and production. In 2008, Aramco produced more than three times more oil per day than ExxonMobil. Aramco's financial results are not public but given its low cost of production and huge oil output, it is surely the most profitable oil and gas firm in the world.

Evolution of resource-rich NOCs

NOCs take many forms and perform various roles depending on their country of origin. As discussed, most NOCs exist to control and manage a country's oil and gas resources. There are also NOCs in resource-poor countries like India and China that exist to manage the energy security needs of their countries.

Figure 2–5 provides a framework for viewing the different levels of investment and activity for resource-rich NOCs. Although there are NOCs at all three levels depicted in figure 2–5, few are at the highest level of both investment and activity.

Most Level I NOCs are small enterprises with little capital or assets of their own other than ownership interests in oil and gas assets. They also have limited skills in upstream activities of exploration, development, and production. Their primary purpose is to collect taxes and royalties from the various private upstream oil and gas firms that manage the petroleum sector. NOCs in countries such as Trinidad and Tobago, Equatorial Guinea, Brunei, and Gabon would fit into this category. It is unlikely that these NOCs will ever grow much beyond their home countries. Figures 2–6a and 2–6b show the historical evolution of the largest NOCs.

As the state invests more in NOCs in terms of roles and activities, NOCs take a more active interest in the conduct of oil and gas development within the country. NOCs at this level are often partners in many specific field developments and other midstream and downstream derivative products and services. At this level, NOCs are roughly equivalent to a small- to medium-sized independent oil company in terms of assets and activities. Sonangol (Angola) and most of the Middle Eastern oil and gas firms such as QatarGas would be examples of level II NOCs.

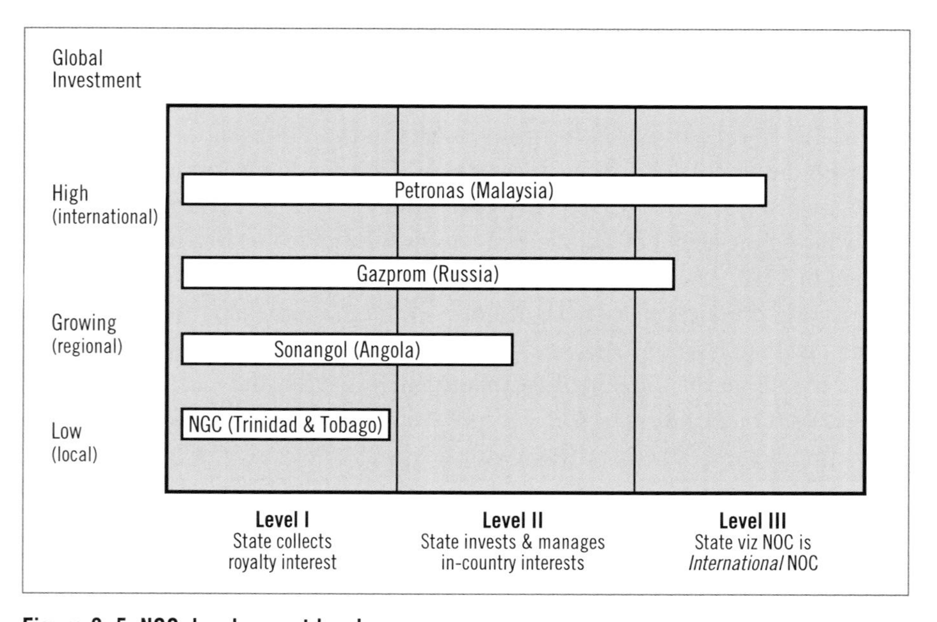

Figure 2–5. NOC development levels

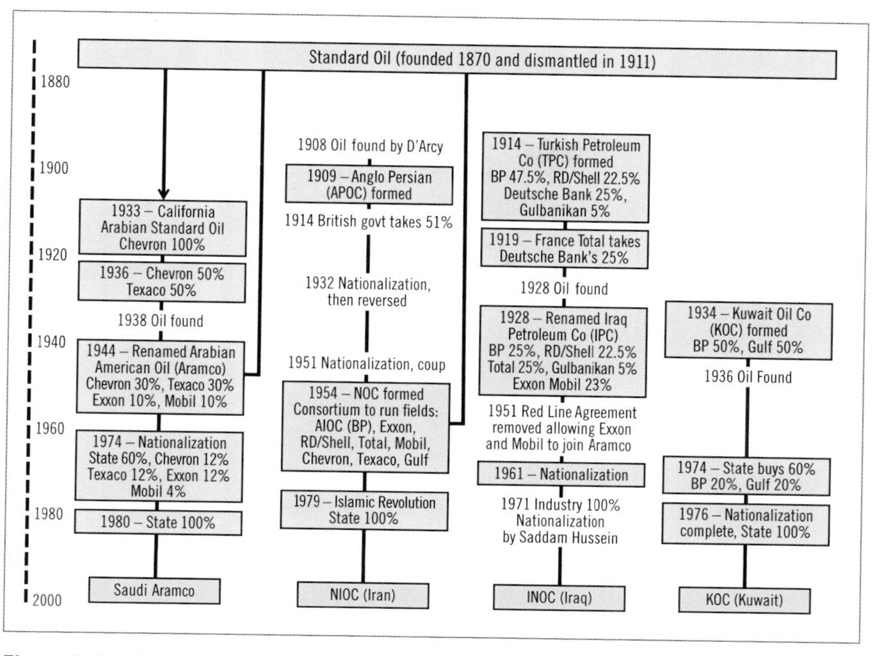

Figure 2–6a. Evolution of today's NOCs

Source: Deutschebank, 2008.

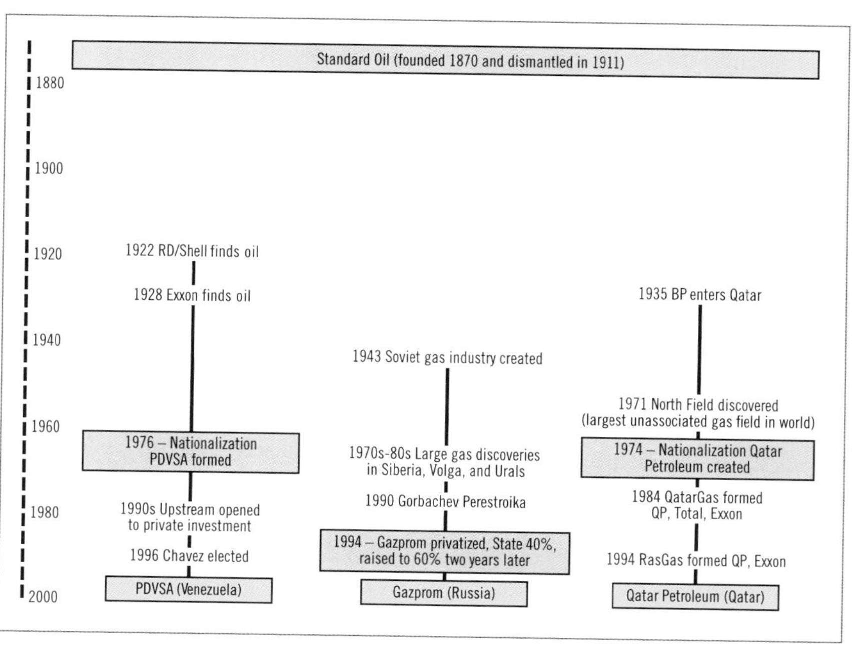

Figure 2–6b. Evolution of today's NOCs (cont. from figure 2–6a)

Source: Deutschebank, 2008.

Once the NOC leaves its home country and enters level III development, it is well capitalized and must deal with the competitive marketplace as it leaves the comfort of home. Companies like Petrobras (Brazil), Petronas (Malaysia) and Statoil (Norway) are representative of this highly sophisticated and capital-needy level of IOC. Aramco is on its way to becoming a major international oil company, although unlike Statoil and Petronas, it has made international investments only in the downstream. Level III NOCs have expanded outside their home countries and appear to have strategic goals of becoming fully integrated IOC-type firms. Petronas, for example, has acquired some lubricants companies and in 2010 became a sponsor of the Mercedes Formula 1 racing team.

As with all attempts at classification, some firms do not neatly fit the categories. An example of a difficult-to-fit firm is Petroleos Mexicanos (Pemex). Pemex is the biggest industrial firm in Mexico and has about 140,000 employees. Mexico is one of the world's largest oil producers with almost 4 million b/d of production. In 1938, Mexico nationalized its oil sector, creating Pemex as the country's sole oil operator. The Mexican constitution provides that the Mexican nation owns all hydrocarbon resources in the country, and private investment in the oil industry is banned. Unfortunately, after years of underinvestment in the industry, Mexico must import refined products, and some analysts predict that without major investment in the oil sector, Mexico will become a net importer of oil in about a decade. In recent years, Pemex has had to borrow heavily because it has paid out an unsustainable level of royalties and taxes. The irony is that Mexico has huge untapped reserves, needs investment, and has many potential investors. Unfortunately, the political environment in Mexico is such that allowing ExxonMobil, Chevron, and other oil companies access to Mexico oil is simply not possible.

Gazprom, a firm with a huge presence on the world stage, also does not neatly fit the resource-rich NOC categories. Based on number of employees, Gazprom is the largest NOC. Gazprom is also diversified far beyond the oil and gas sector, with interests in banking, insurance, media, construction, and agriculture. Gazprom's stated strategic objective is "to become a leader among global energy companies by developing new markets, diversifying business activities and securing the reliability of supplies."[12] Paradoxically, even though Gazprom controls the world's largest gas resources, the company faces a looming crisis as production in its major fields continues to decline and it fails to invest adequately in new fields. Gazprom pipelines also need new capacity and upgrading because about 15% are beyond their useful life.

Russia also uses Gazprom as a proxy for its foreign policy, especially with the former Soviet states. For example, for several years Ukraine and Russia have been locked in a bitter battle over gas prices, as the following incident shows:

> *Even a disparager of Ukraine's independence could not have plotted the farce that played out on April 27th, when the country's parliament ratified a deal to keep Russia's Black Sea fleet in Sebastopol in exchange for cheaper gas. Eggs flew at the speaker, who sheltered under umbrellas. . . . Fisticuffs broke out beside a giant national flag stretched over the seats. . . . Under the agreement, Russia's fleet will stay in Sebastopol until 2042. Ukraine will get a 30% discount on its gas.*[13]

Resource-poor NOCs

The NOCs in China and India are an entirely different type of firm relative to the resource-rich NOCs. China has three large NOCs: China Petroleum & Chemical Corporation (Sinopec), China National Petroleum Corporation (PetroChina), and Chinese National Offshore Oil Corporation (CNOOC). The three Chinese NOCs control almost all domestic Chinese oil and gas production and almost all downstream operations. Sinopec and PetroChina are huge companies in terms of operations, employees, and market capitalization.

In 1988, a reorganization of China's petroleum industry created Sinopec and PetroChina. Sinopec was originally focused on downstream and chemicals and PetroChina on upstream. With a more recent reorganization that resulted in the three NOCs listing their shares, Sinopec and PetroChina seem intent on becoming global integrated oil companies. (The ownership structure is much more complex than can be explained here. For example, Sinopec is controlled by a state-owned enterprise called Sinopec Group.) The two companies have been trying to strip away noncore businesses and reduce their huge payrolls. Sinopec has more than 600,000 employees and Petrochina has close to 500,000. CNOOC is much smaller than the other two firms and is focused on the upstream offshore sector. CNOOC was thrust into the international public relations arena during its unsuccessful attempt to acquire the US firm Unocal, subsequently acquired by Chevron. The Chinese government has allowed some competition in the offshore. Several international companies have PSAs with CNOOC as nonoperating partners.

In theory, the public listings of CNOOC, PetroChina, and Sinopec allow the firms to operate autonomously from the government. In practice, the top executives of all three companies are appointed by the state and are closely tied to the Communist Party.[14] The government is the majority shareholder in the companies, and it is clear that the companies play a key role in China's energy security. The state provides access to low-cost capital and controls wholesale and retail petroleum prices. It appears that the Chinese government is opposed to competitive upstream or downstream market, within its domestic sectors.

Because of the imbalance between China's oil needs and domestic production, Chinese firms have been expanding aggressively outside China in recent years through acquisitions and partnerships. Chinese firms operate in upstream activities in many countries and regions, including Sudan, Iran, Indonesia, Canada, Brazil, Gulf of Mexico, Middle East, and Central Asia. China's NOC access to international projects often starts with the diplomacy by China's government and could be tied to other infrastructure projects and foreign aid. In that respect, Chinese NOCs operate very differently from their IOC competitors. Unlike huge energy importers like the United States, Japan, and Germany, the Chinese government appears unwilling to trust markets for its energy supplies.

The Chinese government has announced a number of major international deals designed to obtain access to supply. For example, in 2009, China announced a deal to lend $25 billion to two Russian oil companies in exchange for two decades of oil supply at discounted prices. Whether the deal actually happens remains to be seen. The important point is that it shows that the Chinese government and the NOCs are working together to implement a Chinese energy policy. In the competitive market for access to oil and gas reserves, Chinese companies have shown a willingness to bid much higher than their IOC and independent upstream competitors. According to one observer, "Sinopec makes its investment decisions based on $100 dollar oil and a 10% hurdle rate." If that is true, it will be very difficult for the oil majors to compete against the Chinese firms for access if the decision is based on price.

Unlike the resource-rich NOCs, Sinopec and PetroChina are similar to the super majors in that they must find overseas reserves to replace their declining domestic assets. They have access to capital and are showing a willingness to operate in tough environments like Sudan and Iran (countries that are closed to US firms). That said, all three Chinese NOCs have production and operating inefficiencies similar to those found in many of the resource-rich NOCs. Were China to allow real competition and full privatization of oil and gas firms, the country would likely be better served in terms of both allocation of capital and development of oil and gas projects. However, a competitive marketplace might conflict with the state's energy objectives.

In India, the industry is much less government controlled. For many years, two large state-owned firms, Oil and Natural Gas Corporation (ONGC) and Indian Oil Corporation (IndianOil) were the dominant players in India's domestic oil industry. Both companies have publicly listed their shares but remain government controlled. Like their Chinese counterparts, ONGC and IndianOil are trying to shake off decades of inefficiencies as state-owned enterprises and transform themselves into world-class integrated oil companies. IndianOil's vision is to become "an integrated, diversified, transnational energy major" and according to ONGC's 2009 Annual Report, the company is "aggressively pursuing for overseas E&P projects for equity oil and gas." While less tightly linked to state policy than the Chinese NOCs, ONGC and IndianOil have clear societal objectives. For example, Indian Oil has its "country first" objectives and its goal of being "the energy of India."

Access to capital

NOC access to capital, a necessary and critical element for expansion beyond domestic markets and for efficient operations in domestic markets, has been a challenge. During good times, when crude oil prices are high and it appears the world will run out of oil tomorrow (which it seemingly has done a half dozen times since 1973), NOCs have had little problem raising capital in the international marketplace.

But during down cycles, when crude prices have dropped to $10 per barrel or less and many oil companies, whether IOCs or NOCs, can barely break even, the ability of NOCs to raise capital has disappeared along with their operating margins. In a number of the down cycles, not only has the state behind the NOC altered fiscal terms to attract international capital and expertise held by IOCs, but it has moved towards monetizing some of its NOC equity stake. *Privatization*, the process of selling state assets to private interests to both raise capital and change the mission and strategy of the state-owned organization, has occurred frequently, as shown in table 2–1.

The global oil and gas industry is a highly capital intensive industry. Access to affordable capital in sizeable quantities for long periods of time is critical for the development of any firm hoping to succeed in this industry. Some of the NOCs have better access to capital than others, as illustrated by figure 2–7. As noted previously, the Chinese NOCs have excellent access to capital from the Chinese government and seemingly few capital constraints on their ability to grow. Petrobras and Statoil have outstanding operational records and are able to tap into international capital markets. Other NOCs have struggled to raise capital and build capacity, especially in their domestic downstream sector. Pemex, NIOC, and NNPC are unable to meet the refining needs of their respective countries and, as a result, Mexico, Iran, and Nigeria must import motor fuels.

Table 2–1. Selective NOC privatizations

Company	Country	Privatization	% Private	% State
Sinopec	China	2000, 2001	24%	76%
PetroChina	China	2000, 2007	14%	86%
CNOOC	China	2001	29%	71%
ONGC	India	1993–1999	16%	84%
IndianOIl	India	1995	11%	89%
Total	France	1992, 1998	30%, 100%	70%, 0%
Eni	Italy	1995, 2001	15%, 70%	85%, 30%
Statoil	Norway	2001	38%	62%
Gazprom	Russia	1994	62%	38%
Repsol YPF	Spain	1989–1997	100%	0%
BP	United Kingdom	1979–1995	100%	0%
Petrobras	Brazil	1995	49%	51%
Lukoil	Russia	1994–2004	100%	0%

Source: Compiled by authors.

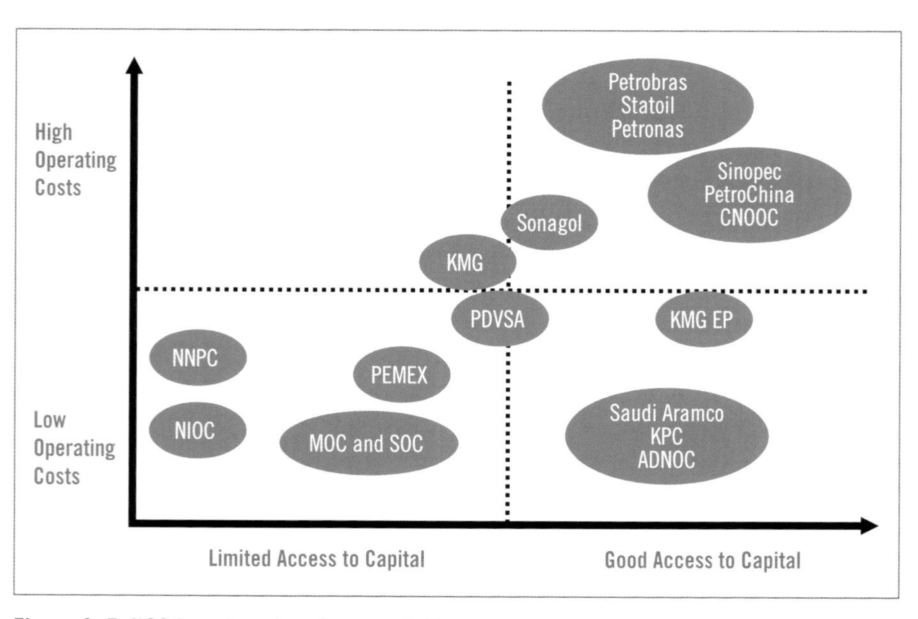

Figure 2–7. NOC investment and segmentation

The NOC/IOC Relationship

As NOCs gain experience and knowledge, some of them are asking a question the IOCs would prefer not get asked: Can we buy or rent the IOC's technical expertise rather than ceding control over our oil and gas resources to an IOC? For the IOC interested in risk taking, booking reserves, and participating in upside returns, selling expertise for a negotiated fee is contrary to the idea of putting capital at risk in a project. If an IOC is just selling expertise at finding and producing oil and gas, the firm becomes a contractor similar to an oil field service firm or one of the many other contractors that work in the industry. This type of transactional relationship is very different from the long-term partnerships and equity oil typically sought by the IOCs.

Historically, the IOCs were able to offer a range of skills and know-how that ensured their participation in major developments and their ability to acquire an equity stake (i.e., book reserves) in the projects. When acquiring "equity oil," the IOCs become full partners in the upside potential of risky capital investment. The IOCs brought unique skills in project management and integration, financial analysis, upstream technologies, contractor relationships, and logistical support. The IOCs also brought access to capital.

In the 1970s, international oil companies held approximately 85% of the world's known hydrocarbon reserves.[15] Three decades later that number has fallen to less than 10%. The NOCs and their national governments control virtually all known reserves. In addition, NOCs such as Petrobras and Statoil have access to huge pools of domestic capital and are expanding to markets outside their home countries. Ironically, many of the skills that the NOCs have acquired over the past few decades have come directly from learning via their IOC partnerships.

NOCs are no longer willing to be silent commercial partners to IOCs who call all the shots. In order for IOCs to continue to gain access to equity oil they must offer a value proposition that the NOCs are willing to pay for and are unable to duplicate. That value proposition will likely involve a combination of several attributes: the ability to manage complex large-scale projects; unique technologies; and project integration beyond discrete oil and gas project development and production. The IOCs must also develop greater in-country capabilities and enhance the quality of their local workforce.

Ultimately, some NOCs will improve their skills and technological know-how and may even approach the IOCs in terms of capabilities. However, given that NOCs are state owned and state controlled, the absence of a shareholder value mentality in most NOCs limits their productivity relative to the

IOCs. Thus, the more poorly managed NOCs will need the IOCs as much as they ever did. The IOCs' managerial and technical expertise and access to international capital markets will continue to make their role valuable. As one oil company executive noted:

> *The continuing need of oil governments for the big international companies is the difference between economics and business. In economics, all firms are the same. In business—in the real world—all firms are not created equal.*

The strategic interests of IOCs and NOCs

Using the NOC categories developed earlier, figure 2–8 presents a view of the strategic interests of IOCs and NOCs. Simply put, IOCs are in search of upstream opportunities and access, while most NOCs need technology, expertise, and access to technology and intellectual property. The stakeholders, interests, and objectives of IOCs are clearly narrower and more focused than those of NOCs.

In the highly complex and dynamic oil and gas industry, IOCs and NOCs have always coexisted. Increasingly, common ground is becoming difficult to find. The tension between NOCs and IOCs has historically been measured by crude oil prices and relative bargaining power. As illustrated in figure 2–9, during periods of low crude oil prices, when government margins from crude oil and gas production decline, NOCs often are in a relatively weaker negotiating position when it comes to attracting international capital and competency from IOCs. It is during these periods that IOCs have often been able to negotiate the access to reserves they need to grow and compete globally.

When crude oil prices rise, government revenues also rise. Increases in wealth move many NOCs, sometimes on their own initiative and sometimes on the urging of government and civil society, to pursue resource nationalism. This nationalism takes several forms, mostly in terms of the proportion of profits accruing to the state versus the IOC in existing developments, or in the more extreme, the elimination or denial of IOC access to state-controlled reserves. The changing roles and relative bargaining power of IOCs and NOCs are inevitably cyclical. In a nutshell, during up cycles resource nationalism increases and during down cycles it declines.

Ironically, this continuing relationship has put many of the world's IOCs in the forefront of the battle over "the curse of oil." Whereas the governments of the producing states are not "reachable" in a political sense, the publicly traded IOCs are. And IOCs continue to bear the brunt of growing concern over the curse of oil and are increasingly expected to act in some fashion to solve the problem. That is a difficult task given that a sovereign state is just that—*sovereign*, and if it has a standing elected government, that government speaks for the people of that country.

IOCs Seek

- Returns on invested capital that meet shareholder expectations
- Long-term growth in income
- Access to reserves and reserve replacements
- Partnerships with NOCs and other IOCs
- Safe and environmentally responsible operations
- Efficient and disciplined management
- Strong corporate reputation
- Proprietary technology
- Talented employees
- Minimal political involvement

Level III NOCs Seek (e.g., Petronas, PetroBras, Statoil)

- Control over their domestic reserves
- Domestic employment
- International expansion
- Integration across the value chain
- Return on invested capital that meets various stakeholder expectations
- Safe and environmentally responsible operations
- Efficient and disciplined management
- Partnerships with IOCs
- Proprietary technology

Level II NOCs Seek (e.g.,Sonagol, NNPC)

- Control over domestic reserves
- Domestic employment
- Sufficient cash flow to provide a significant share of government revenues
- Social contributions for the nation
- Access to technology from partners and contractors
- Partnerships with IOCs (but limited to contracting if possible)
- Improved efficiency and management know-how

Level I NOCs Seek (e.g.,Trinidad and Tobago, Brunei)

- Control over their domestic reserves
- Domestic employment
- Sufficient cash flow to provide a significant share of government revenues
- Social contributions for the nation
- IOCs as operators
- Improved efficiency and management know-how

Resource-Poor NOCs Seek (China, India)

- Control over their domestic reserves
- Long-term domestic energy security
- Access to reserves globally
- International expansion
- Breakeven returns on invested capital
- Proprietary technology
- Partnerships with IOCs in international projects
- Improved efficiency and management know-how

Figure 2–8. Strategic interests of IOCs and NOCs

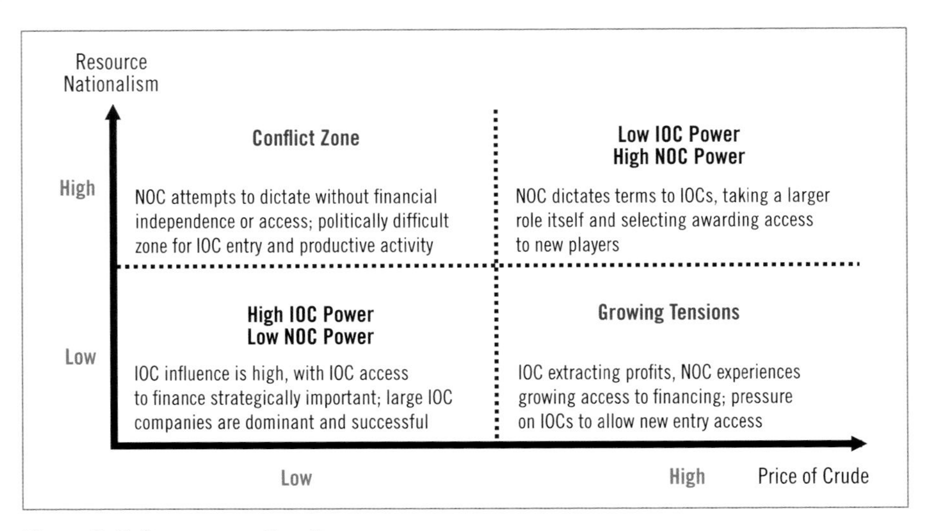

Figure 2–9. Resource nationalism

Looking forward, we see a multitude of questions over the future of the IOC/NOC relationship:

- Has the sun set for IOCs and risen for NOCs?
- Are there any unique competencies offered by IOCs that NOCs cannot obtain from contractors?
- Since NOCs control access to reserves, will they crowd IOCs out?
- Does it make business sense for NOCs to expand to competitive international markets?
- Does it make sense for upstream NOCs to expand into downstream and chemical markets?
- Without the discipline of markets and shareholders, can NOCs ever be as efficient, competitive, and responsive to changing markets as the IOCs?

Organization of Petroleum Exporting Countries (OPEC)

The Organization of Petroleum Exporting Countries (OPEC), founded in Baghdad in September 1960, has proven to be the most influential noncorporate organization in oil markets. OPEC's organization and objectives, summarized in the following text, are focused on the management of oil prices and the revenue interests of its members—"the coordination and unification of the petroleum policies of Member Countries."

Articles 1 and 2 of the OPEC Statute

Article 1

The Organization of the Petroleum Exporting Countries (OPEC), hereinafter referred to as the "Organization", created as a permanent intergovernmental organization in conformity with the Resolutions of the Conference of the Representatives of the Governments of Iran, Iraq, Kuwait, Saudi Arabia and Venezuela, held in Baghdad from September 10 to 14, 1960, shall carry out its functions in accordance with the provision set forth hereunder.

Article 2

A. The principal aim of the Organization shall be the coordination and unification of the petroleum policies of Member Countries and the determination of the best means for safeguarding their interests, individually and collectively.

B. The Organization shall devise ways and means of ensuring the stabilization of prices in international oil markets with a view to eliminating harmful and unnecessary fluctuations.

C. Due regard shall be given at all times to the interests of the producing nations and to the necessity of securing a steady income to the producing countries; and efficient, economic and regular supply of petroleum to consuming nations; and a fair return on their capital to those investing in the petroleum industry.

Outside of the founding five members (Iran, Iraq, Kuwait, Saudi Arabia, and Venezuela), OPEC's membership has varied over the years with various entries and exits. As of 2010, the organization totaled 12 members, with Algeria, Angola, Ecuador, Libya, Nigeria, Qatar, and the United Arab Emirates added to the original five. Ecuador suspended its membership from 1992 until late 2007; Gabon withdrew from the organization in 1995. Indonesia, as a result of becoming a net oil importer, ended its membership in January 2009.[16] The addition of Angola in 2002 was the first major addition since the early 1970s, but will most likely prove to be a very important one in extending and establishing OPEC's power in the sub-Saharan West Africa producing region.

Founding principles

OPEC was the brainchild of Juan Pérez Alfonzo, former Minister of Development and widely considered *the* expert on energy in the Venezuelan government.[17] Alfonzo's ideas grew from his extensive studies of various oil development strategies and authorities around the world, including the Railroad Commission of Texas. The concept of regulating the supply of crude to the market and managing the market price through quantity restrictions was central to his thinking. The lessons originating from the Railroad Commission of Texas are described in the *Industry Insight: The Railroad Commission of Texas.*

Industry Insight: The Railroad Commission of Texas

Founded in 1871 to regulate railroad traffic in the state of Texas, the powers of the Railroad Commission of Texas were expanded in the following years to include pipelines (1917), all oil and gas development (1919), and other utilities (1920) within the state. When oil was discovered in East Texas in the 1930s, the rapid production of the field resulted in a number of major problems. First and foremost was the reservoir itself. With no unified development plan, the reservoir's natural pressure was being depleted at such a rapid rate that ultimate oil recovery would be a fraction of the oil in place. The commission's regulations required a unified production plan for the field. A second problem was the massive glut of oil coming onto the market. The market price of oil in East Texas plummeted, resulting in widespread economic failure. Thus, a second role for the commission was price regulation.

OPEC has often been characterized as a *cartel*, but theoretically speaking, it is not. A true cartel controls all quantities of a commodity, effectively creating a monopoly.[18] OPEC, however, has never controlled more than roughly 50% of all crude oil volumes. From its very beginnings it has operated by agreeing upon production volumes per state, therefore setting and restricting the flow of the commodity to the market. That said, its ability to control a major portion of the market's total volume in different periods of time has proven dramatic and highly influential on price.

In the 1960s, the major IOCs held most of the pricing power. Although it is not clear if there was ever actual overt collusive behavior, the *Seven Sisters* wielded significant pricing power over the producing states. As a result, the price per barrel of crude to the major producing states was held at very low levels although prices in retail markets were consistently rising. By the early 1970s, global crude oil supply and demand were increasingly tight, increasing OPEC's potential leverage. OPEC continued to try to collectively negotiate with the IOCs, but the IOCs, fearing loss of pricing power, refused to negotiate with the collective. In October 1973, the OPEC nations unilaterally announced they were increasing the posted price of the major benchmark crude, Arab Light crude, from $3.65 per barrel to $5.119, on all production from member states. Two months later, in December 1973, OPEC raised the price once again, this time to $11.651 per barrel. A new era began.

Collusion among competitors, an activity usually illegal in competitive markets, is a fundamental principle underlying OPEC's effectiveness. OPEC has always walked a fine line between individual quotas and the independent actions of constituents and the action of the group as a whole, in attempting to restrict the overall flow of crude oil to the market. Internal tensions among members are often much more serious than perceived in the marketplace. Widespread cheating on quotas has been a chronic problem.

Market share

At the time of its birth, OPEC comprised only about 28% of world production. This share rose dramatically over the 1960s and reached 50% in 1970. As illustrated in figure 2–10, OPEC's share of global oil production has varied significantly over time. The periods around the first oil price shock of 1973 and the second shock of 1979 were the high points of OPEC's pricing influence. That said, given the organization's massive production capabilities on the low end of the lifting costs scale, OPEC members have benefitted mightily from the higher price of oil since the late 1990s.

Immediately following the 1979 price shock, several factors contributed to a fall in global oil prices and a decline in OPEC's influence. A global recession drove down oil demand. Major discoveries in Alaska, the North Sea, Africa, and other areas resulted in new volumes of oil reaching the global market and undermining OPEC influence. At the same time various factors drove OPEC's share of global production down: the internal dynamics of the organization itself, the Iran-Iraq War, quota cheating by members, and countries such as Indonesia having to withdraw as they became net oil importers rather than exporters. OPEC's share reached its low point in 1986 at 30% of global oil production.

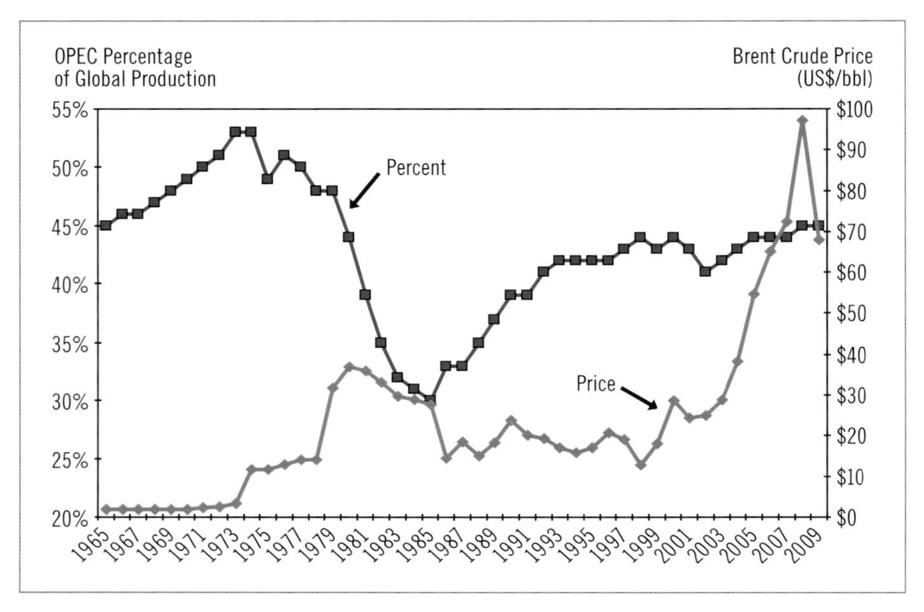

Figure 2–10. OPEC's share of global oil production

Source: *BP Statistical Review of World Energy*, 2010.

Since the mid-1980s, OPEC has worked to reestablish its market share and corresponding market influence. As of 2009, OPEC directly controlled 45% of global oil production. For our purposes, it is important to recognize that OPEC is comprised of both countries and companies. The relationship between OPEC-based NOCs and other oil companies is never simple. Many oil companies, including the IOCs, produce within the OPEC countries in partnership with NOCs and in support of the production interests of cartel members.

Emerging strategy

> *The cartel's true strength lies in holding back a flood of cheap oil. OPEC has extremely low production costs and holds most of the world's oil reserves. The large Persian Gulf producers have costs of less then $5 per barrel, and OPEC member costs outside the Persian Gulf average less than $9 per barrel.*
>
> —"OPEC Strategy and Oil Price Volatility," US Congress, February 2007[19]

Although we are not privy to OPEC's inner thoughts, there are a few basic principles at work that can be used as divining rods when considering the organization's strategic future:[20]

- OPEC's pricing power has varied and will continue to vary over time. Changing markets and changing global political tides periodically push OPEC's power back and forth, and the ebb and flow is likely to continue.
- Depending on market conditions and competitive reactions, OPEC's power and influence may increase in both profitable and unprofitable oil markets. Interestingly, during many down markets, the fall in profits often induces underinvestment and contractions in non-OPEC competitive producers, increasing OPEC's pricing influence.
- Global oil and gas pricing and markets have become more complex with the role of futures markets, speculator behavior, and government intervention often at work. (This will be discussed in more depth in later chapters on oil and gas markets and pricing.)

In the beginning and in the end, with OPEC, everything is still about the price of oil.

The Curse of Oil

An unusual meeting took place in October at St Matthew's church in Baltimore. After the sermon, some parishioners stayed behind to hear two emissaries from Africa explain the harm that America's gasoline guzzling does to the poor in faraway lands. An elderly parishioner raised his hand: "I know Africa is very rich in diamonds, gold, and oil, but the people are very poor. Why are your governments so bad at managing that wealth?" Austin Onuoha, a human-rights activist from Nigeria, smiled and conceded, "You hit the nail right on the head."

—"The Curse of Oil: The Paradox of Plenty," *The Economist*, December 20, 2005

One of the most controversial topics related to the development of petroleum resources is what is commonly referred to as *the curse of oil* or *the paradox of plenty*. The curse of oil is a concept developed to explain how the exploitation of oil and gas resources slows the rate of economic development in a country and increases poverty, income inequality, corruption, and fraud. Proponents of the curse point to countries that low have levels of economic development despite earning billions of dollars from their oil.

Countries afflicted by the curse

There is no definitive agreement as to which countries have or have not been afflicted by the curse of oil. Figure 2–11 shows six countries that have earned sizable amounts of oil revenue but have not seen significant increases in the economic and social health of their citizens.

Consider Equatorial Guinea, a country with one of world's highest oil revenues per capita and a GDP per capita of about $37,000 in 2009, which ranks 30th highest in the world and the highest in sub-Saharan Africa.[21] Despite this high level of national income, the *African Economic Outlook's* analysis of Equatorial Guinea found that the country had "major governance challenges, notably a high perceived level of corruption." The country ranked among the bottom 13 countries in the world on the Corruption Perception Index of Transparency International. . . . Widespread poverty and the persistence of poor health and low levels of other human development indicators raise questions about the extent to which the country's oil wealth benefits the majority of the population. The most recent statistics indicate that about 77% of the population fell below the poverty line in 2006. Maternity and infant mortality rates are among the highest in the world."[22] A *Foreign Policy* article described Equatorial Guinea as follows:[23]

> *Imagine a tiny country flush with oil money, where the wealth per person is on par with that of Spain or Italy. Now picture a place quite the opposite, where nearly two-thirds of the population lives in extreme poverty and infant and child mortality rates are on par with those of the war-ravaged Democratic Republic of the Congo. . . . Equatorial Guinea is a textbook case of the resource curse: The country's leaders have squandered its oil wealth while its people have languished. . . . Rather than benefitting the people, vast sums of the country's oil revenues have gone to bankroll personal purchases for President Obiang, including two mansions in suburban Washington.*

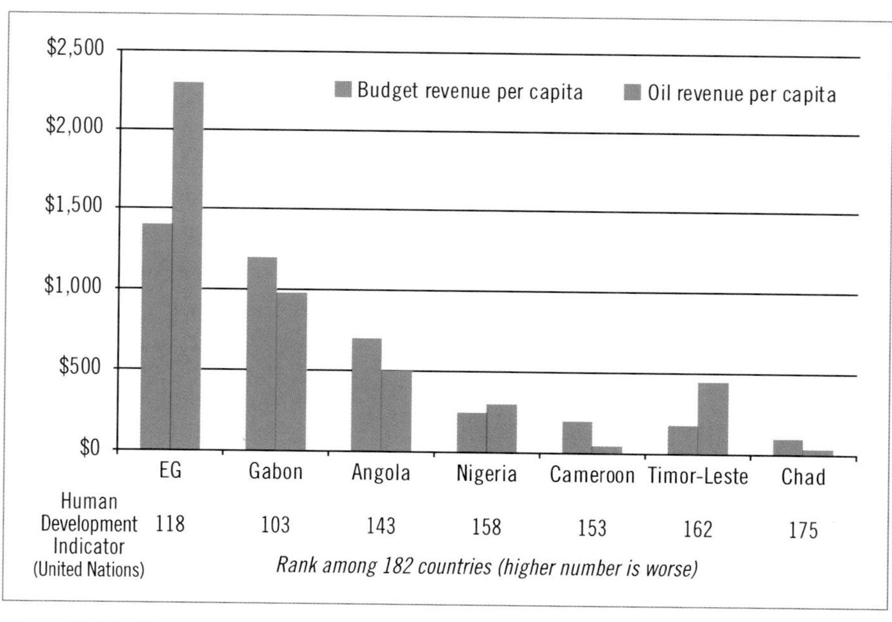

Figure 2–11. Oil revenue and development

Source: Developed by authors based on the analysis of Vidar Ovensen, "Petroleum Fiscal Regimes and Revenue Management Systems," Norwegian Ministry of Finance, African Development Bank Seminar, Kampala, July 9, 2008. Budget and oil revenue values for 2005. HDI indicators updated to 2009.

For the average Equatorial Guinean citizen, oil and gas production has not had much impact on quality of life. As figure 2–11 shows, several other major oil producers in sub-Saharan Africa also have low scores on the United Nations Human Development Index. That said, economic development rarely happens quickly. In Equatorial Guinea and Timor-Leste, significant oil production did not start until the late 1990s, and in Chad, production started in 2003. Although the three countries have yet to spread the oil wealth widely in their countries, that could change in the future.

The economics of the curse of oil

Three arguments underpin the curse. The first is that petroleum developments in emerging markets are usually done under contract or license (*fiscal regimes*) with foreign companies. These companies possess the managerial and technical knowledge necessary for development. They do not typically pass this knowledge to a local or in-country partner for both business and skill reasons. The companies generate income that, like the oil and gas, is extracted from the country. Significant employment opportunities for national residents are usually limited to the construction phase (three years or less in most cases). Therefore, there is limited development in long-term human resource capacity.

The second argument behind the curse is a derivative argument of what *Economist* magazine labeled *Dutch Disease* many years ago, which is the impact of natural resource development on the country's exchange rate and consequent decrease in the nonresource sector. Dutch Disease, named after the development of North Sea gas properties by the Netherlands in the 1970s, describes how the rapid growth of a natural resource for export purposes may drive the value of the country's currency up as the demand for the natural resource rises. The appreciation of the domestic currency then makes other export products and commodities from the country increasingly expensive, reducing the health, earnings, and investment in, and future of, those other industries.

In recent years the Dutch Disease concept has been altered, moving away from the impact on exchange rates and focusing more specifically on the economic opportunities—or lack thereof—in competing sectors. For example, Nigeria was at one time a major exporter of a number of highly profitable agricultural commodities. The movement of labor and capital to the oil industry is one of the factors that has contributed to a decline in the agriculture industry. Unfortunately, when the oil and gas have been fully developed and largely depleted, the country is left with little in the way of internationally competitive industries.

The third argument for the curse is the impact major increases in income and wealth have on the government and politically powerful entities in the country. The income earned by the government of any country as a result of the development of a new oil or gas field may be enormous. Money flows from big oil to the *big man*, as government is frequently labeled in Africa. The sudden surge in income may result in enrichment of the politicians currently in power in the country. The inflow of earnings from concessions or production sharing agreements (PSAs) used for oil development often fill the government's coffers and are sufficient to run the government. As a result, the health of the nonoil domestic economy becomes unimportant, as does the need for sales, excise, value-added, or personal and corporate income taxes. Since people are not paying taxes, they have limited interest in how the government is spending money. The result is a government increasingly distant from its own people. To minimize citizen discontent, the government often provides subsidies for fuel and heating oil, and in the case of Venezuela, uses the NOC (PDVSA) as a social benefits organization for the poor (who happen to make up the majority of voters).

The exact determinants of the curse of oil are not easy to decipher because every country's experience with oil and gas is different. Coming up with a common set of causal factors for countries as different as Equatorial Guinea, Venezuela, and Nigeria is not easy. Moreover, interesting questions arise: Why has Algeria's economy grown so differently from Libya's? Is oil and gas the reason for Iran's stagnant economic development? Why have Middle Eastern states like Qatar and Abu Dhabi been so much more successful in their economic development than Iran? Will Angola go the way of Nigeria, or will it be able to build a vibrant and diversified economy that lifts the personal wealth of all Angolans?

Regardless of why the curse occurs, there are many different factors in the mix: IOCS, sovereign states, governments, NOCS, the people, and their communities. As power shifts away from the IOCs, the solutions to the curse of oil will have to come from within the countries themselves.

Can the curse of oil be avoided?

> *This report argues that transparency in revenues, expenditure and wealth management from extractive industries, is crucial to defeating the resource curse. Achieving transparency requires a higher profile in US diplomacy and foreign policy. When oil revenue in a producing country can be easily tracked, that country's elite are more likely to use revenues for the vital needs of their citizens and less likely to squander newfound wealth for self-aggrandizing projects.*[24]
>
> —"The Petroleum Poverty Paradox: Assessing US and International Community Efforts to Fight the Resource Curse," US Senate Foreign Relations Committee, September 9, 2008

The primary proposal to reduce the negative impacts of oil and gas is to increase the transparency of all payments to the parties involved in petroleum development. Exposing these payments to the light of day (and the global press) will force governments and oil companies to be accountable to the rule of law. Two of the more visible transparency initiatives are the *Extractive Industries Transparency Initiative*, a voluntary program combining governments, IOCs, and NOCs and the *Publish What You Pay* program, supported by global philanthropist George Soros. Both programs have achieved some degree of success.

Beyond the two transparency programs, real solutions require temptation to be removed from the hands of the government. Several oil and gas projects in the past 15 years have been designed with requirements for the state to either distribute oil proceeds in a specific mix that includes the needs of the people, or set aside large portions of the wealth for future projects and generations. One development showing some success is the Chad-Cameroon Petroleum Development in Africa. This project stipulates that certain oil revenues are managed transparently and committed to specific government programs. The creation of the transparency initiative involved various players: governments, a consortium of IOCs, and the World Bank. Strangely, and probably unfortunately, the World Bank withdrew in August 2008 after five full years of petroleum production and requested that the Chadian government pay off its debt obligation, which it promptly did. In essence, the World Bank exited the Chad oil business. In the eyes of many, the World Bank's withdrawal was a confusing and disappointing outcome. The project itself continues to reap benefits for the remaining parties.

In conclusion, at the heart of any debate about solving the curse of oil is the question of what power external parties like the British or American governments or agencies like the United Nations have over sovereign states. In a number of oil developments in the past decade, distribution/saving arrangements like that in Chad were requirements for gaining access to the capital, debt, and equity for petroleum development. Forcing revenue distribution in sovereign states has proven to be a blunt instrument for global policy, and results have been mixed at best.

We must also emphasize that not all oil-rich countries are victims of the curse of oil. Norway and Malaysia clearly have not been cursed, and oil and gas in the Middle East have resulted in positive economic development in several nations. Over the next few years, Ghana, Uganda, and several other countries will become oil exporters and interesting test cases. Will Ghana, one of the most democratic and well run countries in sub-Saharan Africa, fall victim to the curse, or will oil revenues lead to economic development that is widely shared by the people of Ghana?

Oil and Gas and Domestic Politics

The choice for Western policy makers is simple: They can clamp down on their resource industries for domestic political reasons and hand pricing and supply power to nonmarket and nondemocratic governments. Or they can allow their own companies to run a truly global resource market capable of meeting the world's need for things.

—Joseph Sternberg, "The West's Wrong Turn on Natural Resources," *Wall Street Journal*, June 1, 2010, A17

This chapter has focused on the global environment of the oil and gas industry. Oil and gas issues, upstream and downstream, also have an important position in many domestic political debates, which is consistent with the important societal role occupied by oil and gas products. These political debates differ depending on the country or even the region within a country. For example, a constant political issue in the United States is the price of gasoline. It is said that every US incumbent politician wants falling gasoline prices when he or she comes up for reelection. In contrast, motor fuels in Western Europe are so highly taxed that the market price of crude oil has limited impact on retail prices. Because European motor fuels prices tend to be relatively stable, price issues do not occupy the same political space in Europe as in the United States.

Within countries, political viewpoints about oil and gas can differ depending on the region, city, or community. The issue of climate change is a particularly complex political issue. Producing petroleum from the oil sands in Alberta, Canada is energy intensive and produces large amounts of greenhouse gases. Outside Alberta there have been calls for new regulations on emissions in the oil sands area, something not particularly popular within Alberta. The state of Alaska also has complex political issues tied to oil and gas. The oil and gas industry is by far the largest and most important industry in Alaska. Like an oil company, there must be new reserves in Alaska if the state is to maintain its economic position. One of the potential areas for development is the Alaska National Wildlife Refuge (ANWR). Whether there are commercial quantities of oil and gas in ANWR may never be known because opposition to exploration in ANWR is extremely strong outside Alaska. Global warming issues are also contentious in Alaska. Coastal communities in Alaska are seeing less winter ice and more shoreline erosion, which has been attributed to global warming. These same communities may have local economies that are heavily reliant on the oil and gas industry.

A significant challenge for the oil and gas industry in national and regional politics is its unfavorable reputation. The oil and gas industry in most of the major developed nations ranks at or near the bottom in surveys of the most favorable industries. Oil and gas firms are blamed for high motor fuel prices, as discussed in chapter 13, and are vilified for a perceived indifference to environmental concerns. At regular intervals, events such as the *Exxon Valdez* or the BP Gulf oil spill become hugely visible and are touted as more evidence as to the "evils" of the oil industry. When pressed for solutions about too much reliance on fossil fuels or too much reliance on the Middle East for oil supplies, politicians retreat to standard rhetoric. The reality that they inevitably must confront is that the world will be heavily dependent on petroleum products for decades to come. As a result, energy and oil and gas issues are sure to remain highly visible political issues in Canada, the United States, and many other countries.

The Future

By the middle of this decade, the picture had changed dramatically. International oil companies only have full access to seven percent of the world's oil reserves today, mostly in the United States and the North Sea. The rest is either controlled by Russian companies or by national oil companies that offer limited access to foreign investments. Saudi Arabia, which holds a quarter of the world's known oil reserves, does not allow any foreign investments; its oil industry is controlled by Saudi Aramco.

— Jad Mouawad, "Oil: OPEC and Oil Nationalism," *The New York Times*, April 30, 2010

There are two certainties with regard to the intersection between the oil and gas industry, governments, and politics: the issues will continue to evolve, and they will continue to be challenging for all stakeholders. The competitive landscape is changing rapidly. New oil producing countries will be coming on-stream over the next few years. IOCs will continue to push into new regions of the world, such as the Arctic and deeper water (the *Deepwater Horizon* disaster notwithstanding). Energy demands in China and India will have implications for the entire oil and gas industry. The role of NOCs as a group will grow in depth and breadth. A recent analysis of the growing NOC role by the consulting firm Accenture argues that in the coming years IOCs will be interacting with NOCs along two major fronts:[25]

1. In open international markets where NOCs play the dual role of potential partner and commercial competitor

2. In closed resource-holding nation markets, where the NOC fulfills its traditional role as custodian and assurance vehicle for the state

The first of these, open international markets, has seen activities such as Saudi Aramco's downstream activities in China, Statoil interests in the Gulf of Mexico, and PetroChina in Angola. With IOCs competing more frequently with NOCs in international markets, there is sure to be some impact on the historical partnerships that have existed within NOCs' home markets. For the second front, the key question is whether NOCs, in managing resources for the state, will continue to need the services of the IOCs and large independent oil companies.

Summary Points

- Oil and gas are different than other commodities. They are the lifeblood of the modern global economy and tightly connected to the national security and national wealth of most countries.

- Any analysis of an organization must include an assessment of its stakeholders and include three levels of stakeholder interest: interests, rights, and ownership.

- Since the mid-1950s the power of oil, its development, and its value have been shifting from IOCs to the countries owning the oil, their governments, and their NOCs. In most of the world, the oil and gas that lies beneath the surface of the earth is owned by the state (the United States is one of the last countries in which a private citizen may own natural resources).

- As a resource owned by the state, oil and gas and its development is subject to stakeholder priorities different from the traditional stockholder or corporate stakeholder wealth maxims of capitalist markets.

- As opposed to the corporation, the state has a much wider set of objectives and a nearly universal set of stakeholders. Among its powers is its ability to determine access, to regulate, and to determine if and to what degree it wishes to share its resources with outside interests.

- NOCs vary dramatically in character and capability. An NOC may act as government's tax and fee collecting agent, regulator, or production partner for its domestic oil and gas industry. Some NOCs have acquired competencies that rival those of IOCs and have accessed capital and markets far beyond their countries of origin. These global NOCs are becoming legitimate competitor IOCs.

- The extent of NOC power, influence, and financial resources has cycled up and down with the price of oil. NOC power has increased steadily and dramatically, however, over the past decade with the prolonged increase in the price of crude and the advancement of NOC organizational competencies. As a result, IOCs will need a new mindset that ensures continued relevance in a changing environment.
- OPEC, once feared and then marginalized, appears to be growing in influence as oil and gas resources outside OPEC's sphere of influence decline and OPEC resources continue to grow.
- That some countries have suffered from the development of oil has given rise to the curse of oil concept. Whether the curse of oil exists, the reasons for the curse remain complex questions.
- Higher sustained crude oil prices have led to greater levels of resource nationalism and could fundamentally change the future of the oil and gas industry.
- In addition to the geopolitics of the oil and gas industry, domestic political issues have a major impact on the strategies and management of oil and gas companies.

Notes

1. Jeroen van der Veer, "The Geopolitics of Oil and Energy," Offshore Northern Seas Conference, Stavanger, Norway, August 24, 2004.
2. "McKinsey Conversations with Global Leaders: Andrew Gould of Schlumberger," *McKinsey Quarterly*, April 2010.
3. *Escaping the Resource Curse*, Macartan Huymphreys, Jeffrey D. Sachs, Joseph E. Stiglitz, editors, New York: Columbia University Press, 2007, p. 4.
4. The stakeholder analysis of the corporation is often credited to Edward Freeman, *Strategic Management: A Stakeholder Perspective*, Upper Saddle River: Prentice-Hall, 1984.
5. Peter Drucker, *The Practice of Management*, New York: Harper & Row Publishers, 1954.
6. Dr. Wendelin Wiedeking, CEO, Porsche, Die Zeit, April 17, 2005.

7. Anant Sundaram and Andrew Inkpen, "The Corporate Objective Revisited," *Organization Science*, 2004, Vol. 15: 350–363.
8. "Overview of the Political and Economic Arguments in Favor of and Against the Establishment of a NOC," The World Bank Group, June 2009.
9. Ian Bremmer, *The End of the Free Market: Who Wins the War Between States and Corporations?* New York: Portfolio, 2010.
10. Jeroen van der Veer, August 24, 2004.
11. The World Bank Group, June 2009, p. 8.
12. http://www.gazprom.com/strategy/exploration/.
13. "Ukraine and Russia: A Normal Day's Debate in Kiev," *The Economist*, May 1, 2010, p. 51.
14. Sizhi Guo, "The Business Development of China's National Oil Companies: The Government to Business Relationship in China," The James A. Baker III Institute for Public Policy, Rice University, March 2007.
15. Rob Jessen, "IOC Challenge: Providing Value Beyond Production," *Oil & Gas Journal*, February 2, 2009, Vol. 107, pp. 24–26.
16. Organization of the Petroleum Exporting Countries, http://www.opec.org/opec_web/en/17.htm.
17. Daniel Yergin, *The Prize: The Epic Quest for Oil, Money, & Power*, New York: Free Press, 1991, p. 492.
18. "OPEC's Action to Manage Prices and Relations with Non-OPEC Nations," Pijush Paul, unpublished, p. 4.
19. "OPEC Strategy and Oil Price Volatility," Joint Economic Committee, US Congress, February 2007.
20. This section draws from Bassam Fattouh, "OPEC Pricing Power: The Need for a New Perspective," Oxford Institute for Energy Studies, WPM 31, March 2007.
21. *CIA World Factbook*, https://www.cia.gov/library/publications/the-world-factbook.
22. http://www.africaneconomicoutlook.org/en/countries/central-africa/equatorial-guinea/.
23. T. Alicante and L. Misol, "Resource Cursed," *Foreign Policy*, August 26, 2009.
24. "The Petroleum Poverty Paradox: Assessing US and International Community Efforts to Fight the Resource Curse," Report of the Minority Staff of the US Senate, Foreign Relations Committee, September 9, 2008, draft, p. 3.
25. "The National Oil Company—Transforming the Competitive Landscape for Global Energy," *Accenture*, April 2006, p. 10.

Chapter 3

ACCESS, LEASING, AND EXPLORATION

Exploration for oil always costs money; production of oil always makes money.

—Anonymous

If you're only going after elephants, you'll never hunt.

—Ali Moshiri, president of Chevron Africa and Latin America Exploration and Production Company

Finding, developing, and producing crude oil and natural gas are the core activities in the upstream end of the oil and gas value chain. Access, leasing, and exploration activities are the starting point in the value chain because if a firm does not acquire new oil and gas reserves, there will be no new development and production opportunities.

Finding new oil and gas reserves is not just about the technology and cost of seismic analysis and deepwater drilling, but also about the laws, regulations, leases, auctions, and permits required by authorities and mandated by governments globally. It is also about forming and managing partnerships, developing innovative new technologies to find oil and gas, negotiating complex deals, and working within the complex geopolitics of the industry. This chapter examines the processes involved in gaining access to new reserves.

The Oil Project Life Cycle

Opportunity, time, expertise, money, and a little luck are often identified as the five basic elements of oil and gas success. Luck notwithstanding, all successful petroleum developments involve many steps:

- Identification of oil and gas field targets
- Gaining legal access to exploration and development targets
- Generation and analysis of appraisal data
- Analysis of exploration discoveries
- Commercial arrangements such as project finance and negotiating contracts for production output
- Final investment decision (FID), when a company commits to recovering the oil and/or gas
- Field development
- Field production

Although the steps look like a logical sequential process, the reality is much different. The days of intrepid geologists and wildcatters discovering major oil fields, negotiating leases, drilling exploratory wells, and then developing and producing oil are over. As the industry has matured, unexplored areas on land have become scarce, and the NOCs have become increasingly reluctant to share ownership of their resources with exploration and production (E&P) firms. Today, E&P firms are as likely to acquire new reserves through negotiation, acquisition, or partnership as they are through exploration and geology. New fields will be found in offshore locations, but the cost and technological requirements to access the oil and gas mean that these discoveries will be made primarily by the large IOCs. E&P firms will also need to work closely with NOCs to gain access to the already discovered oil that has yet to be developed. For example, Iraq has been opening up its known oil fields to the IOCs. To date, the signed contracts in Iraq are primarily technology sharing deals, but the IOCs clearly expect that there is the potential to acquire what they call *equity oil*—oil that can be booked as reserves.

This chapter is primarily concerned with how targets are identified, access is obtained, and legal rights to explore and develop are bid for and negotiated. The following chapters extend the value chain discussion to the financial and managerial decisions associated with development and production of oil and gas.

Oil and Gas Formation

Crude oil and natural gas occur in nature and are complex combinations of carbon and hydrogen atoms. Both are formed from organic matter—"dinosaurs and diatoms" as the old expression goes, although geologists now believe that most oil was formed by tiny organisms at sea. The oil and gas was formed under intense pressure and heat through time in a source rock, typically sedimentary rock. The most common form of source rock is called *black shale*, an organic-rich sedimentary rock that was created in ancient sea bottoms (figure 3–1).

Heat is by far the most critical element in oil and gas creation. The heat within the earth's crust increases with depth. The minimum temperature for the creation of oil is 150°F (65°C) and occurs at a depth of approximately 7,000 feet (2,130 meters) below the surface of the earth. Oil is typically formed in the depths between 7,000 and 18,000 feet; below that depth, the higher heat levels—now exceeding 300°F (150°C)—can result in the formation of natural gas.

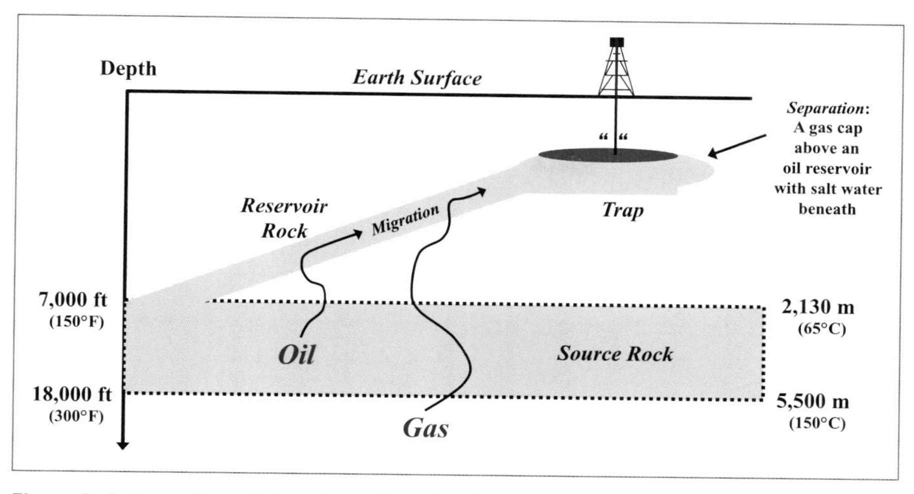

Figure 3–1. Formation of oil and gas deposits. Crude oil and natural gas are formed in a source rock, which is typically 7,000 to 18,000 feet below the earth's surface. The oil and gas then rise as a result of buoyancy to a reservoir rock formation. The oil and gas molecules then migrate upward until reaching a caprock formation, which traps the oil and gas, forming a reservoir, in which the gas forms a cap above the oil.

Subterranean oil and gas deposits are formed as the oil and gas rise through cracks and fractures in the layers of rock as a result of their buoyancy and lighter density. Following the path of least resistance, the oil and gas rise through some type of reservoir rock, often a sedimentary rock formation, and then follow the reservoir rock upwards in a process termed *migration*. The oil and gas finally collect in the reservoir rock when they become trapped under a geologic formation, a *caprock*, which prohibits the oil and gas from rising further. Within the trap, the oil and gas flow to the top, above any water in the trap, eventually separating according to density. Gas is the lightest in density and rises above the oil to form a gas cap. The oil fills the middle level, rising above any water, often salt water, because of its density. Although there are a number of different trap formations, the most common are called *domes* or *anticlines* and result from a natural arch in the reservoir rock (figure 3–2).

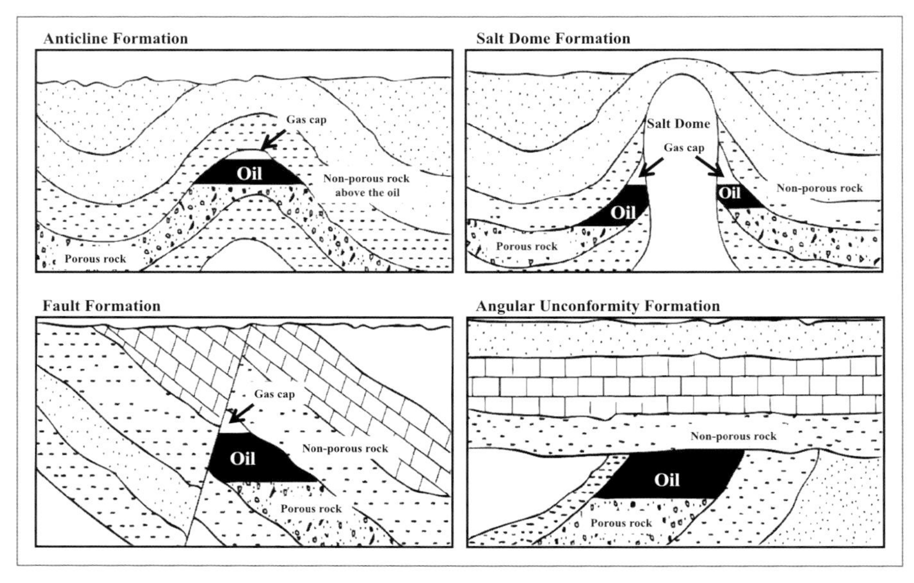

Figure 3–2. Oil reservoirs and cap structures
Courtesy *Oil & Gas Journal*, PennWell.

Finding Oil

The business of finding oil—oil exploration—is in and of itself a difficult and nearly always expensive business venture. Lying hundreds or thousands of feet below the surface, the earliest oil and gas deposits were located by geologic seepage to the surface. Some of the earliest production sites for oil were the results of seepages, including Pennsylvania in the United States, Baku in what is Azerbaijan today, and Trinidad in the Caribbean. Submarine seepage, oil seepage occurring naturally from the ocean floor, was also responsible for discovery of some of the earliest oil production areas in the ocean, including those in California off Santa Barbara and the vast deposits of offshore Venezuela. Marine seepage continues today in different areas, including the Gulf of Mexico, where the Department of Energy has estimated that there may be as many as 5,000 active seeps.

Most of the earliest oil drilling was in and around the seepages. For a number of years in the late 19th century, some drillers had luck drilling in sites that benefitted, luckily, from fractures and seepages associated with ground seepage. But towards the start of the 20th century, the sophistication of both geologic knowledge and drilling technology allowed a more scientific approach to be taken in determining where and how deep to drill. Two major developments were:

- **Surface topography.** The major development in these early years was the increased mapping of sedimentary rock formations on the surface, and then projecting much of this surface mapping to subsurface sedimentary rock formation maps.

- **Seismology.** The logical next steps in subsurface mapping and geologic analysis was in the development of seismic technology, in which sound waves are initiated from a source (such as dynamite or an air cannon), bounced off subsurface geologic structures, and subsequently recorded by a detector of some sort on the surface. This led to an increasingly sophisticated and detailed mapping of subsurface rock layers and formations.

Oil exploration today is much more science than art. The use of technology in determining the locations of likely oil traps continues to develop rapidly. Geologists, the true "rock hunters" who wander the earth in search of oil, use a variety of increasingly sophisticated techniques ranging from surface topography, soil types, satellite imaging, magnetometers, and seismology.

Access and Development Rights

E&P firms will strategically identify target areas where they believe hydrocarbon resources exist, such as the Gulf of Mexico or offshore West Africa. However, before any real exploration activities can begin, the firm must gain legal access to the target area. The right to explore and develop, involving mineral rights or subsurface rights, is typically held by the state.

The right to explore and develop

Gaining access to property in order to explore for oil and gas is first about ownership and who owns the rights. Are the rights owned by the person who owns the land that sits atop the oil? The local community? The government of the community? The federal government capital many miles away? With few exceptions, most notably Canada and the United States, most of the world's subsurface mineral and resource rights are held by a state government. Most countries have a legal structure covering both the responsibilities of the developer (lessee) and the mineral rights owner (lessor). These agreements, generically called *fiscal regimes for international petroleum agreements* (IPAs), have become synonymous with the financial "split" between the two parties over the life of the reservoir.

Much of the early oil exploration began with the right to explore. If oil was discovered, a separate and more detailed agreement would be negotiated between the developer and the rights holder (typically the state). A full development agreement, including distinctions of whether the developer takes ownership to the oil and gas or simply acts a service provider, as well as any applicable royalties, license fees, profit or cost oil splits, etc., are now included in the fiscal regime of the country. Although covered in detail in chapter 6, figure 3–3 provides an overview of the major fiscal regimes.

Fiscal Regimes, also termed *International Petroleum Agreements*, are the fundamental set of rules between the lessor and the lessee for the exploration, development, and production of an oil reservoir over its life.

Concession (Royalty/Tax System)

- The oil company developing the property takes title to the oil at the wellhead
- Lessee usually provides all exploration, development, and production capital
- Lessee pays the lessor a royalty (% of price) for the life of the project
- Lessee pays the host government corporate income taxes and possibly special oil taxes over the life of the project

Production Sharing Agreement (PSA)

- The oil company developing the property shares title to the oil at the wellhead
- Lessee splits exploration, development, and production capital with host country
- Lessee typically pays little or no royalty
- Lessee is limited in amount of capital and operating cost deemed deductible per year
- Lessee pays the host government corporate income taxes and possibly special oil taxes over the life of the project

Risk Service Contract

- The oil company developing the property is only a contractor, and takes no title to oil or gas produced
- Lessee provides little or no capital towards exploration, development, and production
- Lessee pays no royalties
- Lessee pays the host government corporate income taxes and possibly special oil taxes

Figure 3–3. Fiscal regimes, also known as international petroleum agreements

Concessions. The original agreements, whether Pennsylvania in 1857 or Kuwait in 1948, were termed *concessions*. The most important element of a concession, more commonly called a *royalty/tax system* today, is that the lessee (production company) takes title to the oil at the wellhead. The lessee produces and sells the oil, generally in any way it wishes, in exchange for a royalty payment to the lessor. In most concessions granted by governments, the lessee is liable for corporate income taxes and possibly special oil and gas taxes over time.

Production sharing agreements. First introduced by Indonesia in the 1960s, the production sharing agreement (PSA) or production sharing contract (PSC) was a major departure from the concession. The defining element of the PSA is that the host country retains partial ownership in the oil produced. The PSA allocates part of the oil produced to cover the costs of development and production (called *cost oil*) and allocates the residual oil (called *profit oil*) to be split between the host country and the producing company. PSAs are very complex in structure, often covering all aspects of which costs are deemed deductible, at what rate, and a variety of sliding scale terms that limit the total compensation to the producing company versus that of the country, the so-called *takes*.

Risk service contracts. Risk service contracts are essentially transactional service contracts. The producing company provides development, production, transportation, or other services to the mineral rights owner (host country) in return for a specific dollar per barrel rate. The mineral rights owner retains all ownership of the oil produced, thereby capturing all upside potential, but also carrying the weight of production capital and other production-related expenses.

Acquiring and managing the rights to exploration has become an industry of itself in the past 50 years, as resource owners have sought to capture more of the value of their oil and gas reserves. Much of this value lies in the information and knowledge about the individual properties. Because knowledge provides both power and value, many resource owners have monetized the value through exploration lease auctions.

In countries like the United States and Canada, where private individuals can own mineral rights, the role of the landman has a long history. The *landman* is an employee of a production company and performs two major functions under one hat: geologist and property lawyer. The landman may directly approach the landowner and seek a lease agreement. The landman often establishes the rate of production for other existing wells in the area to protect possible discovery. The landman may also propose a variety of partnership or joint operating agreement (JOA) structures. Although the landman is essentially a North American concept, the early "desert entrepreneurs" were the landmen of the Middle East in the 1920s and 1930s.

Historical Precedent: The Neutral Zone Concessions

The oil concessions of the first-half of the 20th century were typically rights to explore, produce, and market oil and gas with nearly all the risks and returns accruing to the foreign oil companies (FOCs). The government or the national oil company (NOC) risked little and gained little. A series of events in the late 1940s and early 1950s changed this situation dramatically, beginning with the development of a small piece of Middle Eastern sand called the *Neutral Zone*, shown in figure 3–4.

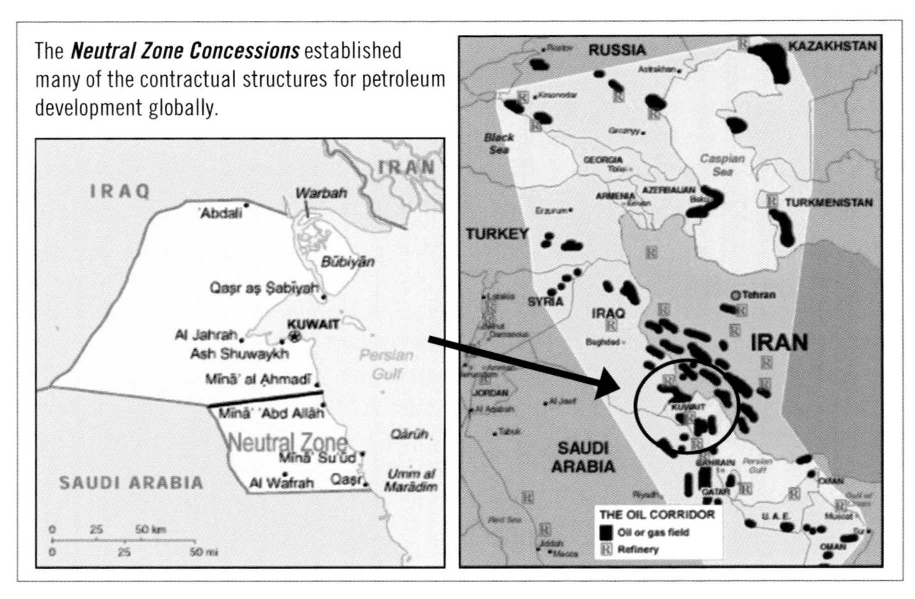

Figure 3–4. The Neutral Zone concessions

Source: Energy Information Administration. http://www.dia.doe.gov/cabs/Saudi_Arabia/images/neutzone.gif.

In 1922, the British, working in conjunction with Kuwait and Saudi Arabia, designated a 2,000 square-mile piece of land neighboring the Arabian Gulf (Persian Gulf today) as a neutral territory for the Bedouins, the nomadic tribe that wandered the area. The so-called Neutral Zone was administered by Kuwait and Saudi Arabia. Not surprisingly, the Zone was thought to hold substantial amounts of oil.

In 1947, Ralph Davies, a former director of the oil and gas division of the US Department of the Interior, as well as a former vice president of Standard Oil of California, came out of retirement to organize a consortium to bid on the Kuwaiti rights to the Neutral Zone. Nine companies, including Phillips, Ashland,

and Sinclair Oil, formed the American Independent Oil Company (Aminoil). The collective capital was just over $30 million.

The US State Department supported the formation of Aminoil as a way for US oil companies to break the major's Middle East presence and more importantly, to break up the growing British dominance of Arab oil. Aminoil surprised the global majors by quickly concluding a concession with Kuwait in June 1948 to develop the Neutral Zone. The terms of the agreement included:

- A $7.5 million signature bonus in cash.
- Exclusive development rights for 60 years.
- A royalty rate of $0.35/barrel, if oil was discovered (the highest rate in history at the time).
- A minimum annual royalty of $625,000, even if no oil was discovered. Though technically an advance upon royalties, the payment was due regardless of whether oil was ever discovered.
- 15% of all profits generated by Aminoil, *if* profits were ever produced.
- A yacht, approximately $1 million in value, for the amir of Kuwait.

The size of the bonus was unheard of, the royalty rate the highest known, and the guarantee of income regardless of oil discovery or production, astounding. Only a few years previous, Kuwait signed major concessions for $700,000, while Standard Oil of New Jersey paid only $200,000 for its concession on more than 440,000 square miles of Saudi Arabia. Many believed that Aminoil was making a huge mistake.

At the same time, J. Paul Getty was interested in the other half of the Neutral Zone. Getty dispatched a geologist to Saudi Arabia to explore for oil. On an aerial tour of the Zone, the geologist spotted a desert dome resembling a similar feature in Kuwait that had yielded oil. Getty was informed that there was a 50% probability of oil. Getty immediately initiated secret negotiations with the Saudi Arabian government.

Getty successfully won the concession. (Getty, known for a variety of peculiarities, was an extremely intense businessman. He prepared for the final concession discussions with Saudi Arabian King Ibn Saud by listening to "Teach Yourself Arabic" records.) The concession for the Saudi Arabian Neutral Zone, signed December 31, 1948, included:

- A signature bonus of $9.5 million.

- Exclusive rights for 60 years.
- A minimum annual royalty of $1 million, even if no oil was found.
- A royalty rate of $0.55/barrel (setting another new record).
- 16% of profits if oil was discovered.
- Getty's firm would underwrite the development of housing, schools, medical care, a mosque for oil workers' families, a telephone system, postal service, and a water utility (but no yacht).

The oil majors thought Getty had substantially overpaid, and one critic called the bid "completely insane." Many feared that the Aminoil and Getty deals would set dangerous precedents for purely speculative exploration and development agreements. The royalty rates particularly were considerably higher than the Middle East average and the up-front signature bonus was considered far too high.

Because of the shared ownership of the Neutral Zone, Aminoil and Getty (via his company Pacific Western) had to combine exploration and development operations. Although Aminoil was designated the operator, all costs and investments were shared equally. The partnership was difficult from the beginning, as Getty argued for development of *garbage oil*, high-sulfur, low-quality crude oil readily accessible at 800 to 1,000 feet below the surface. Although garbage oil was very cheap to produce, it was extremely expensive to refine. Aminoil refused, preferring to continue to search for the deeper, higher quality, higher value, low-sulfur crude.

Five years and more than $30 million later, the two companies had yet to make a major discovery. Pacific Western, continuing to struggle, suspended its dividend to shareholders in 1950 as a result of the debt burden arising from the payments on the Neutral Zone concession. The concession was being called a classic example of the "winner's curse." Getty, close to pulling out, demanded one last drill hole higher on the mound site identified five years earlier on the original aerial tours. The sixth hole at Wafra in the Burgan sand found oil at 3,500 feet in March 1953. The field, estimated at 13 billion barrels, was one of the largest petroleum fields yet identified in the entire Middle East.

Aramco's Middle East managing director sent a one-word cable to James MacPherson, Aminoil's CEO: "Hallelujah."[1] Within four years Getty alone had 60 wells producing 100,000 barrels of oil daily. The field eventually produced 16 million barrels per year, which contributed to making Getty the richest man in the world. The Neutral Zone concessions were no longer seen as extravagant but as bargains. Pacific Western's share price doubled in one month following the discovery, rising from $23.50 to $47.50.

The Neutral Zone deals left a lasting legacy on the industry:

- Neutral Zone oil fields became some of the largest and most profitable in the world.
- Aramco was propelled on its way to becoming the largest and most powerful NOC.
- Aramco's bloodlines were complicated. Standard Oil of California (Socal, later to become Chevron) had formed the California Arabian Standard Oil Company (Casoc) in 1933. In 1936, the Texas Oil Company (Texaco) acquired a 50% interest in Casoc. In 1944, Casoc's name was changed to the Arab American Oil Company (Aramco). In 1948, Standard Oil of New Jersey (later to become Esso then Exxon) and Socony-Vacuum (later Mobil) obtained interests in Aramco. In 1974, the Saudi Arabian government increased its interest in Aramco to 60% and in 1980 took 100% control. In 1988, Aramco was renamed the Saudi Arabian Oil Company (Saudi Aramco).
- What were considered outrageous and unrealistic concession terms became precedents for future deals.

Oil Leases

Oil and gas exploration and development formally begins with the acquisition of a lease. An *oil or gas lease* is a legal agreement between a *lessor*, the owner of access to minerals on a property and a *lessee*. The lease is the agreement for the Lessee to gain access for exploration on the property.

Deal terms

Most lease agreements have several common deal terms:[2]

- **Primary terms** define a specific period, from beginning to end, over which the lessee will have right of access. It may include the possibility of lease term extension.

- **Granted use** terms cover specific use and purposes rights of the lessee. These terms will be specific in defining the property and exact rights the lessee has for the term of the lease, including: minimum drilling requirements, nonperformance penalties, types of exploration techniques allowed, and drilling methods and distinctions.

- **Bonus and royalties** define up-front payments or bonus payments associated with results. The royalty clause will define the percentage or share of production proceeds that the Lessee will pay to the Lessor. Although there is no standard royalty rate, most lease agreements historically varied between 12.5% (1/8) and 25% (1/4). Royalties rates may differ between oil and gas. The lessor may have the right to receive royalties in monetary terms or the oil and gas itself (*in-kind*). There may also be a minimum annual royalty, regardless of discovery or development, and the possibility of a higher royalty rate post payout.

- **Delay rental** terms include conditions by which the lessee may delay drilling or other uses of the property in lieu of a delay rental payment, after completing required minimum exploration or drilling requirements.

- **Shut-in royalty** is a royalty owners receive in lieu of actual production when a well is shut in due to lack of a market, a lack of production facilities, or other situations defined in the lease.

- **Dry hole/continuous drilling** terms apply when a dry hole is drilled and involve the dry hole, cessation, and continuous drilling clauses.

- **Pooling** involves combining or pooling different leases held by the lessee for purposes of a single larger exploration, drilling, or production property.

- **Surrender and damage** detail the lessee's liabilities for specified damages, as well as what continuing responsibilities the lessee may have in the event of surrender.

Like most property agreements, the lease will detail how the parties to the lease, both lessee and lessor, may or may not transfer their rights associated with the lease. It will also include *force majeure* clauses clearly distinguishing that the lease is subject to state, provincial, and national laws and clear title to the property the lessor is leasing.

An added note is needed in distinguishing gas leases from oil leases. As opposed to oil, gas is not easily transported to market. Typically the gas must either be piped or compressed before transport to a market for sale and valuation, making explicit royalty rates and physical payment difficult for gas.

A note on royalty calculations

There are three methods used to establish royalty payments: market price, revenue received, and in-kind.

Market price. The most frequently used method of royalty calculation, the royalty is a set percentage of the market price of oil or gas in that area/locale. The market price is often determined at the wellhead (specific more to oil than gas). Lessors or landowners often prefer wellhead pricing as it allows upside royalty results from rising market prices. Many oil leases state royalty payments based on a market price of oil posted for similar quality crude (grade and gravity) on the same day of production within a specific geographic radius of, say, 100 miles.

Revenue received. This method links the royalty to the actual revenues received on sale of the oil and gas. The sale price of the specific oil or gas may not be the same as the more general prevailing "market price." This royalty calculation is used more frequently with natural gas. Since many gas sales are for longer term contracts, the sales price stated in the contract becomes the basis for the royalty calculation.

In-kind. This royalty calculation is based on the resource owner being paid through a share of the oil or gas produced, taking possession of a "share" at the wellhead. In-kind payments occur before the oil or gas has been marketed by the production company. Although risky for the resource owner, they provide the owner with the ability to effect sale and realize monetary returns. The resource owner may be able to obtain a clause in the lease to take royalty either "in-kind" or "in proceeds."

Reserves

One of the most unique valuation concepts in the oil and gas industry is the booking of reserves. No other industry puts greater emphasis on an asset that has yet to be developed, may never be developed, and cannot actually be seen or touched. Moreover, for an asset defined by a physical measure, barrels, it is strangely elastic in response to market price changes.

Physical features

The quantity of oil and gas that resides in any specific field or reservoir is uncertain. Conceptually, the total amount of oil in the reservoir is termed *oil in place*. But for business purposes, what is more meaningful is the proportion of oil in place that can be recovered (produced). The *recovery factor* for a reservoir is the percentage of oil in place that can be produced, for example, 25%.

Recovery is impacted by many factors, some in the hands of producers and some "in the hands of God." The recovery from an individual reservoir can be increased with greater in-fill of production wells, or the application of secondary recovery techniques like water, gas, or chemical reinjection. Technology itself, whether added technical investment in any individual field or technological changes such as horizontal drilling, can increase the ultimate recovery from the reservoir. Market economics also play a significant role in increasing recovery. An increase in crude price can make capital investments in oil recovery in an existing reservoir "economic." So in practice, the amount of reserves is a function of price.

The total amount of oil in a reservoir is always uncertain. The uncertainty declines over time as production from the reservoir reveals geologic and geophysical attributes of the reservoir. Production data is collected, sifted, and analyzed. Yet, the oil and gas industry has always preferred, at least officially, to take a very conservative approach in estimating oil likely to be recovered. This conservatism is inherent in what is generally regarded as prudent business management, but it is also the basis for what regulatory authorities prefer to think of as protecting investors. Simply put, a company that falsely claims to have extremely large recoverable reserves might raise capital illicitly from unsuspecting investors. As a result, standards have been constructed and accepted globally by most of the global oil and gas industry.

Defining Reserves

The total amount of oil that is estimated to exist within a reservoir is termed *oil-in-place* (OIP). Only a portion of this OIP is actually *recoverable*—capable of being produced. Of that which is physically recoverable, given the state of technology, the cost of employing extraction means, and current market prices, only a portion of that which is recoverable is economically feasible. It is only this last quantity, that which can be economically produced, which is classified as reserves.

A number of organizations, including the American Association of Petroleum Geologists (AAPG), the Society of Petroleum Engineers (SPE), and the World Petroleum Council (WPC), have combined to produce what they term a *petroleum resources management system.*

> *A petroleum resources management system provides a consistent approach to estimating petroleum quantities, evaluating development projects, and presenting results within a comprehensive classification framework.*[3]

A centerpiece of this system is the development of a comprehensive framework for classification, presented in figure 3–5. This framework maps oil quantities according to the range of uncertainty over their physical quantities against the possibility of those quantities becoming commercially produced. Reading from left to right, it begins with the total petroleum initially-in-place (before any production), which it then subdivides into discovered and undiscovered.

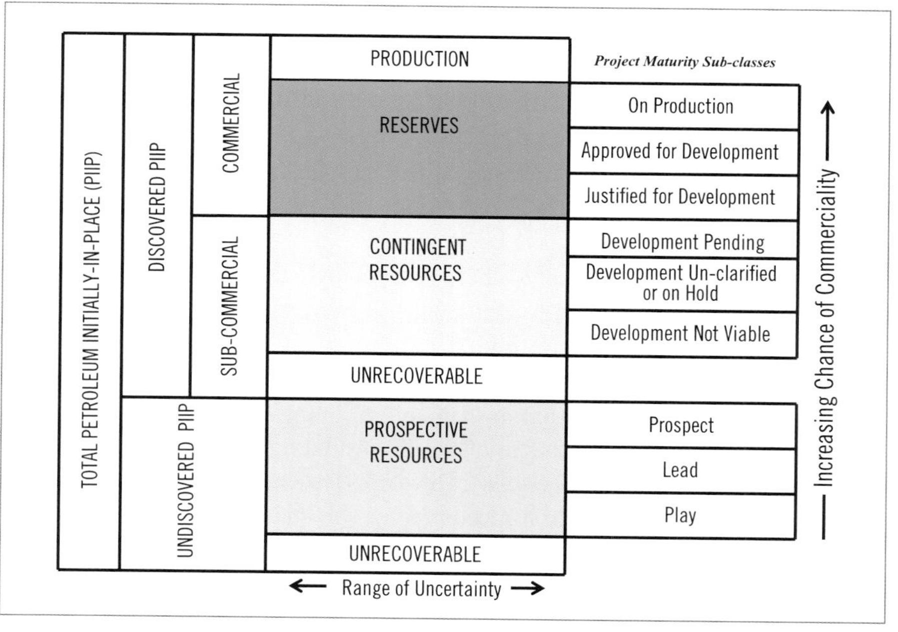

Figure 3–5. Resources classification framework

Undiscovered oil is essentially what all exploration activities are searching for. These prospective resources are then seen as prospects, leads, and plays. They are, in the words of Shakespeare, "such stuff as dreams are made of."[4]

Here we focus on discovered or known quantities. Discovered oil is divided into four categories: production, reserves, contingent resources, or unrecoverable. Production is measured as commercial activity proceeds over time. Unrecoverable oil is that deemed not producible even in the future given current technology and knowledge (though that can obviously change). Contingent resources are those petroleum quantities that are thought to be potentially recoverable, but the projects are not yet mature enough to be considered for commercial development. This "lack of maturity" arises from a variety of contingencies, including no viable commercial markets or needing technology not yet available.

This brings us back to reserves. There are two basic categories of reserves, proved and unproved (which includes probable and possible). *Proved reserves*, further subdivided into proved developed and proved undeveloped, should have a 90% or better certainty of being recovered on the basis of both their geologic features and economic factors. The distinction between developed and undeveloped depends on whether the actual recovery can occur now, with the field's existing infrastructure, or whether it will require substantial additional investment in wells and other recovery processes and equipment. Proved reserves, as described by a field production manager, will happen.

- **Proved.** Reserves that have been proved to a high degree of certainty by actual production or analysis based on geologic and engineering data. The reserves are considered to be actually commercial on the basis of actual production or detailed testing.

- **Developed.** Reserves that are capable of being produced (recoverable) from existing wells with state-of-the-art production methods and expected expenses are deemed *developed*. Developed reserves include wells that are currently producing and nonproducing. The quantity of reserves is based on current operating methods and expenses. *Developed nonproducing reserves* are those in reservoirs associated and accompanying currently producing wells, and are proved by the production of neighboring wells (or in some cases, detailed tests including core analyses). A final subcategory, *nonproducing reserves*, assumes that with only marginal additional investment and expense, the well can be productive.

- **Undeveloped.** Reserves that are in reservoirs that have not yet been drilled are classified as undeveloped. *Undeveloped reserves* are those reserves that are considered proved on the basis of geological data, but are not yet developed by any existing wells. The development of undeveloped reserves assumes that significant additional investment and expense will be required for drilling prior to their recovery. Reserves that may be produced using enhanced recovery methods (discussed in detail in the following chapter), where additional techniques or supplemental drives such as water injection are used, may be classified as proved developed or proved undeveloped depending on whether enhanced recovery processes are yet underway.

The recovery of oil from the unproved category is uncertain. Unproved reserves are further subdivided into probable and possible. Unproved reserves are highly speculative from both a production and business valuation basis. For this reason, the SEC allows firms subject to SEC oversight to report only proven reserves.[5] (A company is subject to SEC oversight if it has any publicly traded securities traded in the United States.) Although oil companies can internally measure and report unproved reserves, they cannot report this information. The objective of the SEC and similar regulatory bodies is to provide consistent guidelines for reporting reserves and a common metric for investors to use in assessing corporate value and prospects.

- **Probable.** Reserves susceptible of being proved that are based on reasonable evidence of producible hydrocarbons within the limits of a structure or reservoir above known or inferred fluid contacts but are defined to a lesser degree of certainty because of more limited well control and/or the lack of definitive production tests. Probable reserves may include extensions of proved reservoirs or other reservoirs that have not been tested at commercial rates of flow or reserves recoverable by enhanced recovery methods that have not yet been tested in the same reservoir or where there is reasonable uncertainty that the program will be implemented.

- **Possible.** Reserves that may exist but are less well defined by well control than probable reserves. Possible reserves include those based largely on log interpretation and other evidence of hydrocarbon saturation in zones behind the pipe in existing wells, possible extensions to proved and probable reserves areas where indicated by geophysical or geological studies, and those to be recovered by enhanced recovery methods where the data are insufficient to classify the reserves as proved or probable.

Differing definitions

Two organizations, the Society of Petroleum Engineers (SPE) and the United States Securities and Exchange Commission (SEC), have had the largest influence in the establishment of common reserve definitions and accounting practices. SEC guidelines are more restrictive than those of the SPE regarding the characteristics reserves must demonstrate to fulfill proved undeveloped status.

SEC guidelines are quite specific about two characteristics related to the economic evaluation of reserves. The first is the requirement that all firms use the price of crude on December 31 of the year as the basis for the *final investment decision (FID)* and the project's discounted cash flow (the SPE allows price forecasts to be used). Secondly, the SEC allows reserves reporting only for oil that can be recovered during the time interval of the existing lease agreement, again further restricting the total recoverable oil reserves booked. The FID is a practice that has helped in the process of reporting reserves. Most oil companies today will not book reserves until the FID has been made, indicating that the reserves are indeed considered recoverable and commercial in the eyes of the company.

In industry parlance, *proved reserves* are referred to as P90 (90% probability of recovery), *proved plus probable* as *2P*, and *proved plus probable plus possible* as *3P*. Again, although these may be helpful for a variety of internal assessments, they are not allowed to be reported by companies subject to SEC regulation. Figure 3–6 provides a summary to this bewildering array of reserves categories.

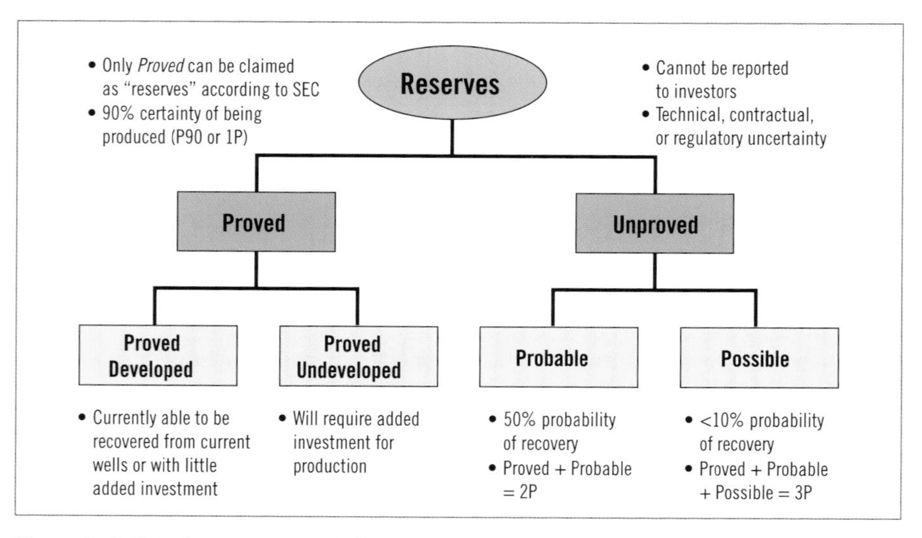

Figure 3–6. Petroleum reserves definitions

Privately held companies not subject to SEC or similar regulatory oversight, or NOCs with no publicly traded securities, may or may not report reserves. If they do report, they may or may not provide evidence or documentation (for example by a respected third-party reserves estimating firm) of their reserves by the same standards as those required by the SEC or SPE.

The corporate objective

Defining reserves returns us to our earlier discussion regarding the different objectives of publicly traded IOCs and NOCs. Companies with no public shareholders have no financial need or interest to publicly claim or report reserves. As discussed, NOC objectives are a balance between national security and public and social policy goals, including employment and economic support and stability in their respective economies.

Companies with publicly traded shares have very strong financial interests in growing their reportable reserves to demonstrate to existing and potential investors that the company's future profitability is promising and increasingly secure. One example of the role of reserves in publicly traded oil company valuation is the case of Rosneft of Russia's initial public offering (IPO) in 2006. The *Industry Insight: Rosneft's Reserves by SPE and SEC Standards*, presents the different measurements of reserves by category as determined by an independent consulting firm, DeGolyer & MacNaughton (a Houston-based firm), in the months preceding the company's first issuance of publicly traded shares.

Industry Insight: Rosneft's Reserves by SPE and SEC Standards

Table 3–1. Rosneft's Reserves by SPE and SEC

	SPE			SEC
	Oil (mbbl)	Gas (bcm)	Total (mboe)	Oil & Gas (mboe)
Proved developed	7,567	162	8,520	8,507
Proved undeveloped	7,310	528	10,422	3,306
Total proved	14,877	690	18,942	11,813
Probable	8,305	444	10,917	
Possible	7,219	435	9,778	

Source: DeGolyer and MacNaughton. Note 1 billion cubic meters (bcm) of gas equals 5.89 million barrels of oil equivalent (mboe).

The publicly traded firm clearly has the need to accumulate and report reserves. The type of fiscal regime that a company negotiates with a specific country like Indonesia or Angola is therefore extremely relevant. If a firm obtains a traditional concession agreement, the company will effectively own the oil and gas production and reserves. It can then book the reserves, reporting them to the investment public. If, however, the company signs a service agreement to provide drilling or production services, it will not own the oil and is unable to report or book the reserves accompanying the production field.

A second level of reporting interest by the publicly traded company is the value of *reserve replacement*: the reporting of reserves added through exploration each year as in-place replacements for the current year's production. Logically, a firm must succeed in finding at least as many barrels of oil each year as it produces if it wishes to maintain the same reserve life. *Reserve life* is the number of years the company can sustain its current production levels with its booked reserves. The longer the reserve life, the greater the value prospects and the greater the sustainability of the company. This poses a demanding expectation on E&P companies wishing to show real value growth over time. These firms must find and book ever higher reserves, either through organic growth (their own successful exploration) or through inorganic growth (acquisition). One recent study compiled its own estimates of reserve life by oil and gas company by dividing booked reserves by current year production. The results are presented in figure 3–7.

Lease Auctions

The majority of the world's potential oil and gas properties are held by governments. Even in the United States, where individuals may own mineral rights, all offshore rights lie with the US government. This section will detail both the theory and practice of allocating oil and gas exploration and development rights. Although the focus is on auctions, there are other methods of allocating exploration rights. The primary alternative method is *administrative* (often referred to as "beauty contests"), in which the government allocates rights on the basis of individual negotiations, relationships, political affiliations, or personal preferences. Although low in cost to conduct, this method fails to meet many of the longer term objectives of exploration leasing. Auctions have traditionally been favored for oil leasing because they both allocate and price scarce resources simultaneously, and do so in settings of uncertainty. We begin our discussion with fundamentals of business auctions and then analyze how the United States, Indonesia, and Libya have conducted their lease auctions in recent years.

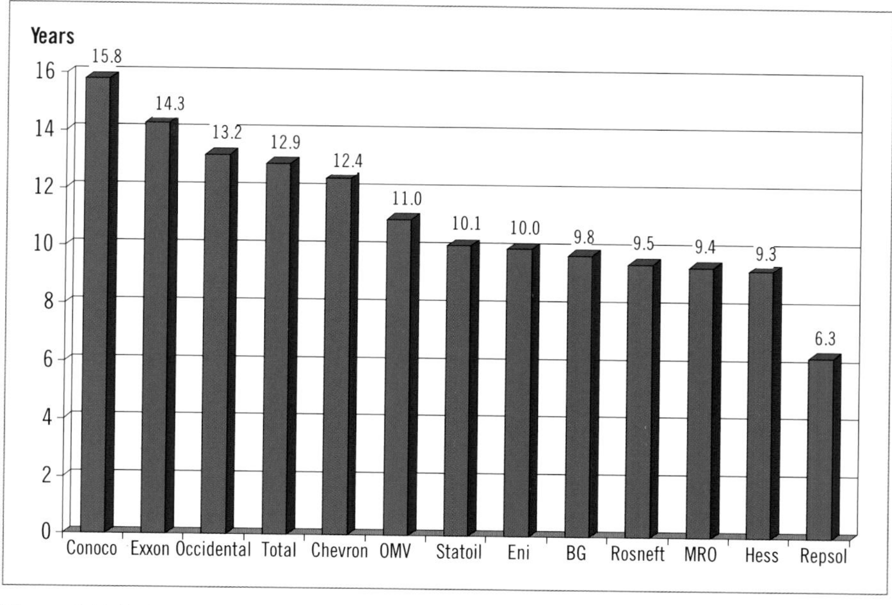

Figure 3–7. Estimated reserve life by company

Source: "Oil & Gas for Beginners," Deutsche Bank, January 7, 2008, fig. 87, p. 84. Copyright © 2008 Deutsche Bank AG.

Industry Insight: Art or Science? The Ormen Lange Gas Field

Although both the SPE and SEC guidelines on categorizing reserves are quite specific (the previous definitions and descriptions are only brief overviews), firms still have significant discretion in their determination and reporting of reserves. The Ormen Lange gas field of offshore Norway serves as a prime example of how reserve reporting may vary across firms.

The Ormen Lange gas field is on the Norwegian Continental shelf and is one of the largest gas fields currently under production. Exploration of the field in 2003 was undertaken by a consortia of five companies—Statoil (Norway), operator, Shell (UK), Norske Hydro (Norway), ExxonMobil (US), and BP (UK). Each of the companies booked their share of the Ormen Lange reserves following the signing of the final investment decision according to common industry practice. But a simple numerical analysis of what each firm booked, in an analysis conducted by Deutsche Bank and presented in figure 3–8, shows very different estimates of how many oil equivalent reserves the Ormen Lange field actually held. Production of the field started in 2007.

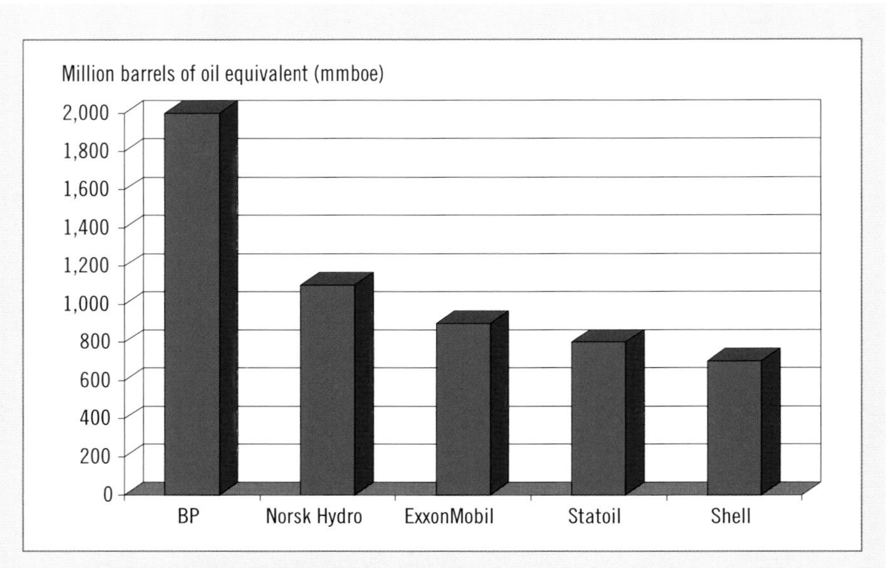

Figure 3–8. How big is the Ormen Lange field?

Source: "Oil & Gas for Beginners," Deutsche Bank, January 7, 2008, p. 79. Copyright © 2008 Deutsche Bank AG. Deutsche Bank back-calculated total field size by each party on the basis of the proportional share of reserves each company booked following the FID.

Auction theory

Oil lease auctions have several key objectives. First, the lessor, most frequently a host government, wishes to achieve an efficient assignment of exploration and development rights, and do so transparently. This will encourage confidence in the process and promote competitiveness among the participants. Second, like all sellers, the government desires an auction process that will generate maximum revenue. In general, good auction processes keep these two priorities in this order; poor processes often reverse the order of priority.

In order to compare and contrast auction methods, it is necessary to consider four basic questions:

1. *Do bidders value blocks individually or collectively?*

Block interdependency means that the value of one block is associated with another block. The primary issue here is whether the block will have the same value to all parties, regardless of the winning bidder. This depends on whether there are synergies or other efficiencies associated with E&P and whether the market price of oil and gas produced is the same for all parties.

2. *What is the basis for bidders' valuations?*

The basis for valuation will depend on bidders' private values, common values, or interdependent values. Do bidders have "common values," in which the block has an independent value, after auction and development, that is ultimately the same to all bidders? Do bidders have private information? Do bidders have interdependent values, in which the individual block's value is different before and after the auction and development process because of complementary or synergistic impacts of that block among other blocks?

3. *Should lease rights be sold simultaneously (in groups or bundles) or sequentially (by individual blocks)?*

If each block is valued independently and its value is not a function of rights obtained for associated blocks, then selling blocks simultaneously is not important. If the value of an individual block is dependent on gaining access and rights to associated blocks, the sequence of block auctions becomes critical.

4. *Should auctions be conducted statically (single bid) or dynamically (allowing rebidding)?*

This question concerns the value of price discovery. In an open dynamic bidding auction, the bidder has the right to continually revise their bid on the basis of competitive offers. To win, the bidder must pay only one dollar (or euro, riyal, etc.) more than the next highest bidder. Bids are known and there is open opportunity to bid higher.

In a static bid process, the bidder does not have knowledge of competitive bids (if they exist at all) and must make a bid that is not only believed to be superior to competitors, but is actually inflated to provide protection as a margin of error. The bidder may end up either winning or losing, by a substantial margin. The static bid is more likely to generate the "winner's curse," in which the winning bidder may have paid a price far above the next highest bid. The winner may, in effect, lose, in that the assigned value was much higher than any other bidder.

Auction methods

There are four standard methods used in auctioning a single item (block) such as an oil and gas lease: 1) first-price sealed-bid; 2) second-price sealed-bid; 3) English ascending; and 4) Dutch descending.[6] Each method has its strengths and weaknesses, depending on both the object of auction and the rules and practices adhered to in the auction process.

First-price sealed-bid. Sealed-bid auctions are static auctions (only a single bid is permitted) in which the bidders simultaneously submit bids in sealed envelopes to the auctioneer. The auctioneer then orders the bids from highest to lowest, awarding the winning bid to the highest. The winning bidder must pay the auctioneer the amount of its highest winning bid.

Second-price sealed-bid. The second-price sealed-bid auction is conducted identically to that of the first-price described, with the exception that the winning bidder pays the auctioneer the amount of the second-highest bid, not the winning bid. This is a method for partially correcting for the winner's curse, the old adage that the winner of an auction is typically the party willing to pay more than anyone else, therefore likely overpaying.

English ascending. This is a dynamic auction in which the auction is conducted with a series of bids or rounds, sometimes with a continuous price or "clock." A bidder has the right to continually increase and improve their bid until no bidder is willing to offer more. The winning bidder must pay the auctioneer the amount of its highest winning bid.

Dutch descending. This is a dynamic auction that is conducted in reverse. The auctioneer initiates the auction with a suggested price that is so high that no bidder is willing to accept/bid. The auctioneer then continuously reduces the offer price until a bid is received. This winning bidder then pays the auctioneer the amount of the winning bid.

Although there is a wealth of debate and theory over auction structures, the primary components of comparison are based on whether the mechanism promotes the highest price possible for each block, whether the mechanism results in a winner's curse, and whether the auction process is subject to or encourages collusion or collaboration among bidders.

Lease auctions in the United States

The US Mineral Management Service (MMS) of the US Department of the Interior controls and manages all oil and gas leasing on federal lands or the offshore Outer Continental Shelf (OCS).[7] The federal offshore leasing program was first established in 1954, and with a few adjustments, it has remained largely the same.[8]

Tract identification. In the early years, the MMS accepted nominations from industry regarding which federal tracts to make available for auction. Selection was largely based on demonstrated interest in the nomination process. This was changed later, with nearly all tracts available for auction on an ongoing basis.

Once the MMS announced that an area would be made available for exploration and drilling, the area was divided into blocks or tracks. The typical block is 5,000 acres or 5,760 acres, roughly 9 square miles in size (23 square km). All blocks are then carefully described and detailed by both latitude and longitude, with specific coordinates for all boundary points. For all offshore blocks, water depth is detailed (royalty rates and rental fees differ by depth under current MMS rules). The MMS will also note the minimum bid per block and any policy for rejecting high bids.

There are three kinds of oil and gas lease sales: wildcat sales, developmental sales, and drainage sales. A *wildcat sale* refers to the tracts whose geological and seismic characteristics are not well known. A *developmental sale* covers tracts previously sold and being reoffered, either because the winning bid was rejected or because the lease was relinquished. A *drainage sale* is the sale of tracts in which oil or gas deposits have been found.[9]

Bid preparation. Once the tracts available for bid are identified by the MMS, companies wishing to potentially bid on the blocks are free to research and conduct studies of all kinds in order to assess the blocks. Seismic analysis, either by in-house personnel or by other outside geophysical company contractors, is welcome, but drilling is not allowed. The older 2-D seismic analysis commonly used in the pre-1990 period was relatively simple and cheaper, averaging $200,000 to $300,000 per block. Currently, 3-D seismic is widely employed and is much more complex and expensive.

If drilling and production has occurred in the nearby area in previous years, that information is collected and compiled by the individual companies. A lot of information is available, even if the drilling or production was conducted by other companies. The reporting of production activity in the United States is quite detailed and leads to a high degree of transparency—not always consistent with company preferences. Wildcat sales of blocks or tracks that have not been previously explored have only seismic data to go on. The MMS has historically grouped wildcat sales into group blocks adjacent to current production areas to aid in company assessment and development.

Prior to the actual bidding, another important negotiation that takes place in many cases is the negotiation between potential bidding partners (partnership issues are discussed later in the chapter). Bidding consortia are widely used because bids have skyrocketed in price and individual companies are often not willing or able to invest massive quantities of capital in bids on their own. Partnering shares the investment, the risk, and subsequently the returns. Companies entering into bidding consortia often do so on a very block-specific basis, bidding together on one block while possibly bidding individually on an adjoining one. Bidding consortia are organized by legal contract, and membership is divulged in bidding. It is also interesting to note that most bidding consortia are organized only after the individual parties have conducted their own block assessments. An example of drilling rights bidding is shown in the following Industry Insight on the Chukchi Sea.

One restriction was added in 1975 to the MMS bid process. In an effort to prevent the major oil and gas companies from dominating oil lease bidding, joint bids involving two or more of the largest oil companies (first identified as eight in 1975) were banned. Although the majors could bid individually or jointly with other nonmajor partners, joint bidding by majors was prohibited.

Bid and auction. The actual auction process used by the MMS is a first-price sealed-bid format. *First price* refers to the simple award—the block or tract is typically awarded to the highest bidder (but not always) in exchange for the amount bid. Actual bids are termed *bonus payments*. Occasionally winning bids are deemed insufficient by the MMS and not accepted. As illustrated in the *Industry Insight: Record Bids for Leases in the Chukchi Sea, Alaska*, this may result in a quick exchange of much money.

Postauction lease. The winning bidder now has the exclusive rights, often referred to as *monopoly rights*, over the exploration, drilling, and production of the block or tract for the term of the lease, typically 5 or 10 years. The lessee is not obligated by law to conduct drilling. Lessees are required to make designated rental payments on the lease during the lease exploration period. If by the end of the lease period no development has occurred, the lease reverts back to the government. If development has and is occurring, the lease extends automatically and continues to confer monopoly rights over block production as long as the lessee stays current with all royalty payments.

Industry Insight: Record Bids for Leases in the Chukchi Sea, Alaska

US Minerals Management Service; Alaska Division of Oil & Gas

Oil companies bid nearly $2.7 billion on February 6, 2008 in an auction for drilling rights in the forbidding Chukchi Sea. The sum of winning bids is the most ever generated in an Alaska oil and gas lease sale, whether on land or offshore (table 3–2). The amount exceeds the $2.1 billion raised in a 1982 sale in the neighboring Beaufort Sea, and the $900 million in a 1969 land sale at Prudhoe Bay that was the beginning of Alaska's transition to a wealthy oil state. In the Chukchi Sea auction, all proceeds go to the US government, and none to the state, as the offshore acreage is under federal jurisdiction.

Table 3–2. Biggest Alaska oil lease sales

Year	Sponsor	Location	High Bids
2008	Federal	Chukchi Sea	$2.66 billion
1982	Federal	Beaufort Sea	$2.06 billion
1969	State	North Slope	$900 million
1984	Federal	Beaufort Sea	$867 million
1979	State	Beaufort Sea	$567 million
1976	Federal	Gulf of Alaska	$560 million
1984	Federal	Navarin Basin	$516 million
1979	Federal	Beaufort Sea	$489 million
1988	Federal	Chukchi Sea	$478 million
1983	Federal	St. George	$426 million

Source: U.S. Minerals Management Service; Alaska Division of Oil & Gas

According to the *Anchorage Daily News,*

> *Oil men, journalists and others packed a Loussac library auditorium in Anchorage and listened with anxiety and awe as officials announced hundreds of often jaw-dropping bids on tracts totaling 2.8 million acres in the Chukchi, a shallow and icy polar sea between Alaska and Russia.*

The competition was pitched as two global powers—Shell and Conoco Phillips—offered fortunes on some of the same tracts. Onlookers sometimes sounded like a football crowd, going "ohh!" or "aww!" when one company barely beat out the other. The day's highest bid for a single tract came from the Dutch company Shell at $105,304,581.

Source: Wesley Loy, "Record Bids for Oil, Gas Leases in Chukchi Sea," *Anchorage Daily News*, February 7, 2008, http://www.adn.com/oil/story/307588.html.

Observed bidding behavior

The more than 50 years of federal leasing in the US have resulted in an enormous data set of auctions, bids, and observed behavior. The average bid price over time has been roughly $2,500 per acre but has actually trended downward. Of all tracts leased, roughly 25% have qualified as productive at the end of their exploration period (hence 75% have reverted back to the federal government).

A multitude of academic studies, and some analyses by the MMS itself, have in recent years sought to glean insights into bidding behavior. Recent analysis has resulted in some very interesting observations, particularly about bid prices. Table 3–3 provides a synopsis of US auction history.[10]

Table 3–3. Analysis of offshore oil and gas lease sales (Gulf of Mexico), 1954–2006

Period	Tracts Offered	Tracts Bid	Bids Per Tract	Percent Bid	Tracts Sold	Percent Sold	Winning Bids	Mean Winning Bid
1954–1982	7,715	3,974	3.24	52%	3,525	46%	53,104	$15.07
1983–1987	71,243	3,763	1.47	5%	3,473	5%	9,424	$2.71
1988–1992	60,228	3,811	1.16	6%	3,701	6%	1,956	$0.53
1993–1997	52,563	5,183	1.52	10%	5,017	10%	2,501	$0.50
1998–2006	57,946	6,175	1.37	11%	5,951	10%	3,667	$0.62
1954–2006	249,695	22,906	1.71	9%	21,667	9%	70,652	$3.26

Source: Based on Table 1 of Philip Haile, Kenneth Hendricks, and Robert Porter, "Recent U.S. Offshore Oil and Gas Lease Bidding: A Progress Report," October 12, 2009, p. 16. Dollar figures are in millions of 1982 dollars.

US auction performance: 1954–1982. This was a period of stability—stable oil prices and stable technology for surveying and drilling. There were roughly two lease auctions per year on average, although these started rising toward the end of the period. The majority of blocks were wildcat sales, and the minimum bids set by the MMS were $15 per acre on wildcat tracts and $25 per acre on *drainage tracts* (tracks adjoining producing areas). All lease terms were five years, and the royalty rate was 1/6 (16.67%). The majority of bid winners were the major oil companies of the time—Shell, ARCO, Chevron, Gulf, Amoco, Exxon, Texaco, and Mobil. The 12 largest bidders accounted for about 75% of all bids and wins and 46% of all tracts offered for sale were sold.

The introduction of area wide leasing (AWL) expanded offshore tract leasing dramatically, opening up nearly all available tracts including deepwater. Table 3–3 shows how tracts offered increased by a factor of 10 in the periods following AWL's introduction. AWL extended the lease period from five years to either 8 or 10 years to encourage bidding. Royalty rates on deepwater—water depths greater than 400 meters—were lowered to 1/8 (12.5%), also to encourage bidding and drilling and development.

US auction performance: 1983–2007. Auction market behavior changed radically. With the enormous increase in tracts offered, few—only 5% or 6% in the 1980s—were bid on. The average number of bids per tract also plummeted, falling from 3.2 to 1.47 and 1.16 the following sub-periods. Participation in the bidding by the major oil companies also fell dramatically. Many more bidders participated in auctions held after 1983, but the mean high bid fell to $0.97 million in the 1983 to 2006 period. In the words of the authors, "It is not an exaggeration to say that bidding collapsed after 1983."[11]

This change in bidding behavior after the introduction of AWL is logical. Prior to AWL, tracts were only put up for auction if nominated by companies—meaning they were of commercial interest. When all properties were brought to the market, regardless of any preauction interest by industry, the number of tracts bid on would logically fall, as it did.

A number of other business environment issues contributed to changes in demonstrated behavior, including periods of low crude prices and the shift in recent years towards deepwater exploration and development. The opening of deepwater tracts to auction and exploration also saw a shift in the interest of the majors, as they shifted the majority of their bids to deepwater tracts following 1983. Table 3–4 provides a comparison of shallow versus deepwater bidding in the 1980 to 2002 period.

Table 3–4. Gulf of Mexico tract bidding

1980–2002	Shallow Water (<200 meters)	Deep Water (>800 meters)
Number of tracts sold	6,717	6,660
Number of tracts drilled	2,754	871
Percent drilled	41%	13%
Revenue per tract drilled	16.39	68.26
Cost per tract drilled	12.72	23.51
Mean bid per tract sold	1.39	0.66
Bid + royalty per tract sold	2.07	0.98
Profit per tract sold	(1.64)	3.76
Value per tract sold	0.43	4.74

Source: Based on Table 1 of Philip Haile, Kenneth Hendricks, and Robert Porter, "Recent U.S. Offshore Oil and Gas Lease Bidding: A Progress Report," October 12, 2009, p. 16. Dollar figures are in millions of 1982 dollars.

Indonesia's Oil and Gas Auctions

The auction process in the United States is one of the older and more developed in the world. The US process also has cultural and institutional features unique to the United States. The petroleum bidding practices of Indonesia, another country steeped in oil and gas industry history, provides a view of different auction practices and the nuances that make government/NOC-type deals and business transactions a challenge. Indonesia has two different petroleum bidding processes based on who identified the blocks for auction, regular tenders and direct tender offers.[12]

Regular tenders

Blocks originally identified by the Indonesian government for offer, *regular tenders*, follow a four-step process.

Step 1: Call for bids. The directorate determines blocks to be offered. A bid document is prepared for each block that includes a geological synopsis, tender procedures, and the draft PSC (production sharing contract). The tender process is then publicly announced. All companies wishing to submit a bid for a specific block, whether alone or as a member of a consortium, are required to purchase the bid document and data package for that block. In the 2008 Indonesian block auction, bid documents were priced at US$5,000 per block.

Step 2: Submission of bids. All bids, called "participating documents" must include the following:

1. The signature bonus, equipment or service bonus, and production bonuses.

2. The firm work commitment for the first three years of the PSC, which should include the type and quantity of seismic data to be acquired and the number and proposed location of exploratory wells to be drilled.

3. The work commitment for the second three years of the PSC, including the type and quantity of seismic data to be acquired and the number and proposed location of exploratory wells to be drilled.

4. Evidence of the firm's financial capability to undertake the work commitments described.

5. A letter confirming that the company is ready and able to directly pay all bonuses as a nonchargeable operational cost.

6. A letter confirming that the company accepts the terms and is ready to sign the draft PSC.

7. Other documents included in the bid document. One new addition here is a letter from the bidding company's parent company confirming responsibility for all corporate obligations under the PSC for the life of the agreement.

Step 3: Evaluation of bids. Bids are evaluated by committee on the basis of three criteria:

1. *Technical merit*—the proposed seismic and drilling commitment

2. *Financial merit*—the amount of the signature bonus offered, the financial ability of the bidder to meet the firm work commitment, and other financial liabilities associated with the contract

3. *Experience*—the bidder's industry experience and past compliance with Indonesian laws

The Indonesian ministry also notes that unless all criteria and documentation are met, a bid on a block will not be accepted, even if there is only one bid for the block.

Step 4: Award of bids. Bids are ranked by a cross-governmental evaluation committee on the basis of the criteria described in step 3. Final rankings are submitted to the ministry, which then recommends to the Minister the block be awarded to the company with the highest bid. Once awarded, the PSC is executed by the government in cooperation with the winning bidder.

Direct offers

Under Indonesian law, *direct offers*, in which an individual or company proposes new areas for exploration not slated for auction, a number of studies must be completed before the bid awarded process is instigated.

Step 1: Submission of direct offer proposal. The company must submit a proposal for consideration that includes the boundaries of the proposed block (including geographic coordinates), a short report on the area's geology and potential for commercial oil and gas deposits, all geological and geophysical data available to substantiate interest, and a company profile supporting its capabilities to fulfill development of the block.

Step 2: Evaluation of proposal. An evaluation committee reviews the submission. There is no specified time line for how long the evaluation process may take.

Step 3: Conduct of joint study. If the evaluation committee finds the proposal promising, the company is authorized to then conduct a joint study of the area with an Indonesian university or other specified entity. The company must provide a US$500,000 performance bond within 14 days of notification and the company must bear all costs associated with the conduct of the joint study.

Step 4: Evaluation of the joint study results. Once the joint study is completed, the evaluation committee reviews the joint study results. If favorable, it is up to the director general to determine whether the direct offer process is to continue. No time frame is specified for duration or completion.

Step 5: Offer and award of direct offer block. Procedures for bidding and award for direct offers are generally the same as those for regular tenders. If, however, no company has expressed an interest in a direct offer block within 45 days of the announced offer, the evaluation committee shall evaluate the offer submitted by the company that conducted the joint study of the block.

Industry Insight: Anadarko Wins Rights to Explore Offshore Indonesia

LONDON—Anadarko Petroleum Corporation (NYSE:APC) announced today that its subsidiary, Anadarko Indonesia Company, has been awarded exploration and production rights to North East Madura III Block in Indonesia's fourth licensing round. The approximate 1 million acre offshore block is 50 miles north of Madura Island and 150 miles northeast of Java in water depths of approximately 150 to 250 feet. Anadarko will operate the block with a 100% working interest.

Under terms of the standard production sharing contract to be entered into with BPMIGAS by year end, Anadarko will undertake a six-year exploration phase and 20-year production phase. During the initial three-year work program Anadarko plans to acquire a minimum of 2,560 square kilometers of 3-D seismic and drill six exploration wells.

"Exploration for oil and gas in Indonesia is a growth opportunity for Anadarko," said Anadarko Senior Vice President, Exploration and Production, Bob Daniels. "Following our analysis of an extensive 3–D seismic survey covering most of the block, as well as our in-depth regional evaluation, we are encouraged by its potential and we look forward to pursuing additional exploration and production opportunities in Indonesia to complement this acreage."

Anadarko Petroleum Corporation's mission is to deliver a competitive and sustainable rate of return to shareholders by developing, acquiring and exploring for oil and gas resources vital to the world's health and welfare. As of year-end 2003, the company had 2.5 billion barrels of oil equivalent of proved reserves, making it one of the world's largest independent exploration and production companies. Anadarko's operational focus extends from the deepwater Gulf of Mexico, up through Texas, Louisiana, the Mid-Continent, western US and Canadian Rockies, and onto the North Slope of Alaska. Anadarko also has significant production in Algeria, Venezuela, and Qatar, and smaller production or exploration positions in several other countries.

Source: "Anadarko Wins Rights to Explore Offshore Indonesia," *Business Wire*, Tuesday November 2, 2004.

The company that proposed the area or block has two significant advantages in the award process. First, the results of the joint study are not made available to other potential bidders. Second, the company has the right to match any higher bids received on the block. To date it appears that the direct offer process in Indonesia has proven both effective and popular. Between 2008 and end-of-year 2009, more than 170 proposals for joint studies were submitted.

Libya's Auction of 2005

Many oil industry analysts have long argued that Libya has the largest reserve base of any country in Africa, North or sub-Saharan. But, following the Libyan-backed terrorist bombing of a Pan American jet over Lockerbie, Scotland, in 1986, the country was isolated by much of the world's oil and gas industry. After Libya agreed to pay more than $2.7 billion to families of the victims of the Lockerbie disaster in 2003, most of the major sanctions were removed (the United Nations—2003, United Kingdom, European Union, and the United States—2004).

Libyan oil 2005

In the intervening years after 1986, the Libyan oil industry fell into severe decline, with total production falling from 3.1 mmb/d in 1970 to less than 1.3 mmb/d in 2004. Libya's oil industry needed—in the words of its newly appointed ambassador to the United States—"American technology and investment." Libya's goals were ambitious: to attract more than $30 billion in exploration and development capital, and increase annual production to over 3 mmb/d by 2010. It had already started lobbying OPEC to increase its allotted quota towards that goal.

The reopening of the Libyan market began with bids in January 2005 on 15 exploration areas (shown in figure 3–9), many composed of multiple blocks onshore and offshore. Libya's fiscal regimes began with a concession regime when oil was first discovered in 1959, but in 1974 changed to an exploration and production sharing agreement (EPSA).[13] All 163 prospective bidders had to prequalify, with 63 qualifying.

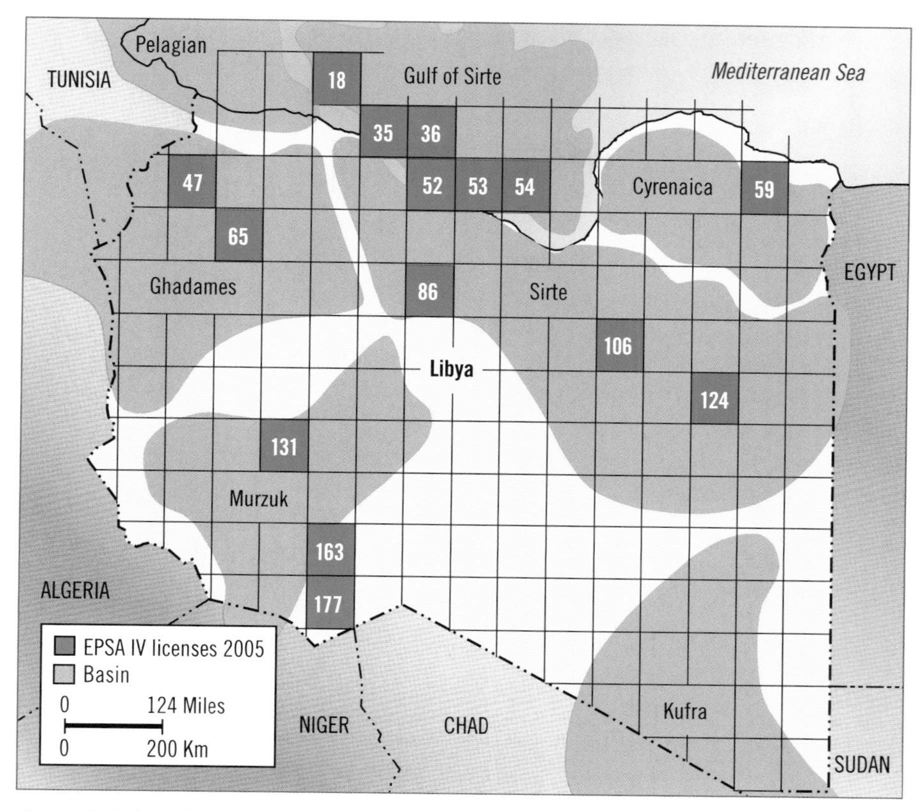

Figure 3–9. Libya's January 2005 auction blocks. After prequalifying, all companies submit bids on two criteria: 1) M–factor, the government share; and 2) signature bonuses. Offshore blocks 52, 53, and 54 garnered much of the attention.

Source: Daniel Johnston, "Impressive Libya Licensing Round," *Oil & Gas Journal*, April 18, 2005, Figure 1.

Bidding under EPSA-IV was based on two criteria. First, each bidder per area bid the so-called M-factor, the percentage of total production that would accrue to Libya.[14] The higher the percentage share assigned to Libya, the better the bid. Second, in the case of a tie on the production share bid, the tie-breaking criterion would be the signature bonuses bid. The bidding was a closed-bid first-price structure used frequently in the global oil industry in the post-2000 rising price environment. It was a simultaneous auction with all blocks up at once, rather than a sequential process of block after block. Table 3–5 shows the bids that were opened on what was considered the highlight area of the auction, Block 54. The bidding was aggressive, with the Libyan government garnering enormous share-bids.[15]

Table 3–5. Ranked bids on Libya's Block 54 (January 2005)

Bid Rank	Bidding Company/Consortium	M-Factor: Government Share (%)	Contractor Share (%)	Signature Bonus (million $)	Government Take (%)	Government Expected Value (million $)	Company Expected Value (million $)
1	Amerada Hess	87.6%	12.4%	6.18	92.8%	5,522	211
2	Santos/Nippon/Japex	85.5%	14.5%	1.50	92.3%	5,410	323
3	Marathon	80.5%	19.5%	1.12	88.4%	5,195	536
4	Woodside/Oxy/Liwa	80.3%	19.7%	10.51	88.2%	5,194	539
5	Wintershall/NorskHydro/Gazprom	80.0%	20.0%	0.50	88.0%	5,169	564
6	Total/Shell/Petronas	79.1%	20.9%	2.13	87.6%	5,137	595
7	Chevron/Texaco	76.9%	23.1%	1.60	86.2%	5,044	688
8	BG	75.8%	24.2%	1.51	85.4%	4,988	744
9	ExxonMobil	73.1%	26.9%	21.00	83.9%	4,900	832
10	Petro-Canada/Gas de France	72.6%	27.4%	1.12	83.4%	4,855	876
11	OMV	71.1%	28.9%	3.15	82.3%	4,784	948
12	BP	70.8%	29.2%	10.00	82.4%	4,791	941
13	Eni/Anadarko	70.1%	29.9%	1.11	82.1%	4,767	965
14	Petrobras	68.2%	31.8%	1.00	81.6%	4,689	999
15	BHP/Unocal	76.2%	23.8%	3.00	80.8%	5,011	1,043
	Average	76.5%	23.5%	4.36	85.7%	5,030	720

Source: Daniel Johnston, "Impressive Libya licensing round contained tough terms, no surprises," *Oil & Gas Journal*, April 18, 2005, Table 5. Financial analysis assumes $100 million exploration risk capital plus bonus, 50% chance of success.

Libya: The continuing story

Following the first auction, there was a second a month later and a third the following year. Both local political forces and global market forces negatively affected Libyan exploration activities. Falling crude prices in 2008 and 2009 meant that capital for Libya became tight for many E&P companies.[16]

The Indonesian and Libyan examples have a couple of implications for the exploration business:

- Governments of oil-rich nations are increasingly adding stringent and perhaps unrealistic conditions to auctions. For example, EPSA-IV required a high level of local sourcing, which proved a major impediment for many of the IOCs in executing their exploration and development programs.[17] There is a major shortage of skilled technical labor, particularly in oil field services. In response, the Libyan government has grown increasingly frustrated with the slow rate of exploration and development activity and has continued to pursue independent negotiations with E&P firms.

- The companies that bid so aggressively in the 2005 rounds found themselves looking at much lower crude oil prices a few years later. The aggressive and winning bid may have become the winner's curse.

Exploration Strategy and Technology

Finding oil requires, quite literally, turning over rocks. This task falls to petroleum geologists. The previous section on the fundamentals of oil and gas reservoir formation highlighted the critical capture point—the oil trap. The geologist searches the world over for that unique combination of source rock and entrapment that may hold oil and gas.

For an E&P firm, one of the most critical strategy decisions is scope: Where will the firm search for oil and gas? Onshore or offshore? In the US, Africa, Middle East, elsewhere? Will the firm seek new discoveries or acquire existing leases? The decision by firms to explore for oil in deepwater areas such as the Gulf of Mexico and offshore Brazil and West Africa has resulted in major scope expansion for many firms. Consider Chevron's experience in the Gulf of Mexico. Chevron began buying Gulf offshore leases in the 1990s before technology existed to drill wells in such deep water. After years of analyzing its holdings, Chevron spent $100 million to drill its first Gulf deepwater well in 2002. The well was considered a success, and further exploratory wells were drilled over the next three years. In 2005, a decision was made to develop the project, now called Tahiti. Tahiti produced its first oil in 2009. The costs of the project are staggering: the drill ship from Transocean costs $500,000 a day and the platform cost $650 million to build.[18] The following case study, "The Mukluk Story," details one of the most expensive exploration dry holes in history.

The Mukluk Story

We drilled in the right place. We were simply 30 million years too late.

—Richard Bray, president, Sohio Production Company, 1984

Exploration for oil and gas is a boom-and-bust story. After the second OPEC-induced price shock of the 1970s, a boom in the early 1980s ushered in massive investment in exploration. One such endeavor was Mukluk.[19]

Following the development of Prudhoe Bay on the North Slope of Alaska, a bidding frenzy for new exploration leases erupted in 1982 and 1983. In 1983, roughly $1.5 billion in lease sales were made on the North Slope. Mukluk, one of the highest profile projects,[20] involved a consortium that included BP, Standard Oil Company of Ohio (Sohio), and Diamond Shamrock.[21] The consortium drilled an exploration well that was particularly difficult.[22] Located 60 miles west of Prudhoe Bay and 20 miles north of the Colville River delta, Mukluk Well No. 1 was in 48 feet of water. Lying within sandstone formations similar to that of Prudhoe Bay, estimates of recoverable reserves ranged between 1 and 5 billion barrels of oil, only about 10% of what Prudhoe Bay promised but an enormous prospect by industry standards. BP's geologists classified the exploration as one of the lowest risk developments the company had ever been part of. But as the old oil field saying goes, "Only Dr. Drill knows for sure."

The lease sale agreement of October 1982 required drilling to take place within one year. Development of the exploration well started in the coldest days of the winter of 1983 and extended into the spring and summer. The drilling rig was mounted on a man-made island 350 feet in diameter, with a work surface 12 feet above sea level. The island required 1.29 million cubic yards of gravel to construct and roughly $100 million, with the drilling rig costing another $50 million. Although the summer of 1983 was a record year for difficult ice floes, the island was completed, and drilling began in early October 1983. By November drilling had reached the Ivishak Formation at 7,000 feet in depth and yielded few positive results. At 8,000 feet the drill hit salt water. Drilling pushed on towards the terminal target depth of 9,860 feet in the Lisburne Formation. There the consortium found only traces—*oil stain*—indicating that although oil had indeed once occupied and flowed through the formation, the reservoir had been breeched millions of years ago and drained of any commercial prospects.[23] On Friday, January 20, 1984, Sohio announced it was plugging its No. 1 Mukluk well in the Beaufort Sea.

Mukluk's message for years to come was interpreted in a variety of ways.[24] Mukluk became instantly known as "the most expensive dry hole in history."[25] That may or may not be so, given the numbers. But more importantly for the corporate strategies of the major oil and gas companies of the time was the reassessment of how to pursue the future. As Daniel Yergin argued in *The Prize*, Mukluk's message was clear:

Betting so heavily on exploration was too risky and too expensive. Managements would, in the future, be made to pay a penalty if they continued risking, and losing, money on such a scale. In the minds of many senior oil company executives, Mukluk sent a bracing message; they ought to shift from exploring for oil to acquiring proven reserves, in the form of either individual properties or entire companies. After Mukluk, they were much more in a buying mood.[26]

Technological innovation

Ultra deepwater exploration and development would not be possible without technological innovation. New technology has allowed petroleum exploration to radically change over the past century. Whereas early geologists focused primarily on surface features—the rise and fall of the terrain, the nature of soils and plant life, as well as proximity to naturally occurring features such as faults—the past 40 years have focused on subsurface analysis. A listing of the more sophisticated methods of exploration analysis includes: analysis of surface features and terrain, conducted by air or increasingly via satellite photography; gravity meters, which measure changes in the Earth's gravitational field caused by the movement of oil, water, and gas; magnetometers, which measure changes in the Earth's magnetic field caused by subterranean oil, water, or gas movements; "sniffers," which are electronic measurements of scent of hydrocarbon fuels; and last but not least, *seismology*, the mapping of subterranean layers by compiling data of reflected sound waves induced on the surface.

Even with these modern techniques, there is only a 20% or 25% chance on average of finding a commercial deposit of oil or gas. The process of exploration, analysis and appraisal of exploration data, and finally development involves significant risk. Figure 3–10 shows how much risk the business entails. It is clearly not work for the weak of heart. That said, the probability of dry holes has dropped significantly over the past few decades, which allows firms to commit $100 million or more to a single well.

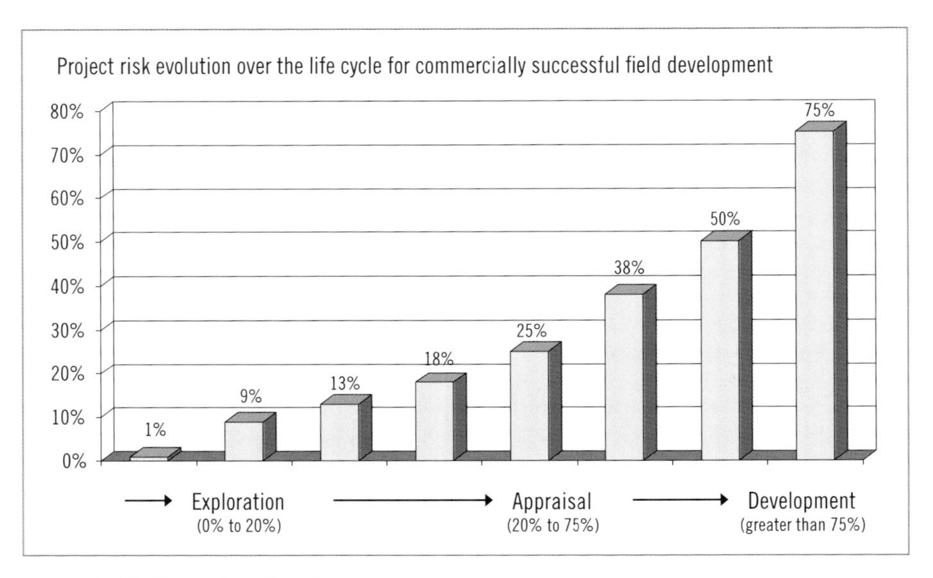

Figure 3–10. The probability of success

Source: Otis & Schneidermann, *AAPG Bulletin* 81, Deutsche Bank, p. 53.

Partnership and Farm-ins

As indicated previously, exploration is a very risky endeavor. Partnering to spread risk and costs in the upstream has been a key element of the oil and gas industry for a century or more. The creation of partnerships can occur at any stage in the value chain. For example, many refineries are run as joint ventures, such as Lukoil and ERG Company's jointly operated refinery in Sicily and ExxonMobil and PDVSA's refinery in Chalmette, Louisiana. In the upstream, large projects executed by a single company are a thing of the past. As projects get larger and more complex, even the IOCs are unwilling to shoulder the associated risks.

Upstream partnerships can be created in various ways. A very common occurrence is the sale of an interest in exploration permits to a firm that agrees to invest in the exploration project. This process is called a *farm-in* and is often used by small firms to raise capital and for large firms to gain access to new projects. The farm-in could be via competitive bidding or it may be noncompetitive between firms that have experience working together.

Brunei Darussalam

A project in Brunei Darussalam provides one example of how an exploration joint venture scenario could occur. Between 1955 and 1962, more than 600,000 barrels of oil were produced in Block L, a 2,200 square kilometer exploration and development area covering onshore and offshore areas in the northeastern portion of Brunei Darussalam.

Block L opened for bidding in October 2005. In 2006, following a joint bid by Loon Brunei and QAF-Brunei, Petroleum Brunei signed a production sharing agreement (PSA) with Loon Brunei (90%) and QAF-Brunei (10%). Loon Brunei is a Canada-based subsidiary of Kulczyk Oil Ventures, a company owned by an international holding company of Polish origin. QAF-Brunei is a Brunei-based firm. Loon Brunei's 90% interest in the PSA gave it the right to explore for and produce oil and gas from Block L. Under the terms of the PSA, there was an obligation to carry out a specific amount of seismic work during the first three years of the PSA.

In 2007, Nations Petroleum SE Asia Ltd., a subsidiary of a Canadian company, signed a 12-month option agreement with Loon Brunei. Under the agreement, Nations will fund 100% of costs up to $20.5 million related to the acquisition, processing, and interpretation of a 350-square-kilometer, 3–D seismic survey required in Brunei Block L. In return, Nations will earn 50% working interest in the Block L PSA. The following year, Nations exercised its option and became operator of Block L. Seismic survey by Nations Petroleum was completed in June 2009. In November of that same year, AED, an Australian company, signed a letter of intent to purchase Nations Petroleum's 50% interest in Block L.

The final result of these various transactions is a three-way joint venture for Block L between AED (50%), Loon Brunei (40%), and QAF Brunei (10%). In the example above, if the block yields a significant discovery, it is likely that the partnership will continue to evolve, with new partners buying into the consortium and existing partners selling out.

JV alignment and agreement

When a joint venture is created there are many managerial issues. First, a workable agreement must be created to deal with issues such as:

- Participating interests, operatorship, supervision of operations
- Duration of agreement and confidentiality

- Areas of mutual interest
- Assignment to third parties
- Joint operating agreement as to scope, operating committees, programs, rights of partners, transfer of equity, withdrawal, and default
- Unitization agreements (enlargement/reduction of the unit area, tract participation, redetermination, costs and production adjustments, cost sharing), unitization across borders

Second, the actual management of the operation has to be determined, which means agreements on: the organization, operating committee, subcommittees, work groups, technology, financial systems, exploration (surveys, licenses, drilling, improved technology, environment), project managers, schedules, and contracting. In the event that exploration leads to production, the joint venture will have to consider the full range of production tasks such as pipelines, terminals, field management, safety, transportation, abandonment, and so on. The finance and economics of the deal means a further set of issues to consider. In short, entering into a joint venture results in a new set of issues that must be managed in addition to the actual management of the exploration activities. In chapters 4 and 5, additional partnership management issues are discussed.

Summary Points

- Access to oil and gas is the lifeblood of an E&P firm. Without access, there are no oil and gas fields to discover and no projects to develop.
- Acquiring new oil and gas properties is as much about science and geology as it is about geopolitics, risk taking, partnership management, negotiation, and auction bidding.
- New oil and gas properties can be acquired in several ways: discovered through exploration; purchased from other firms, or conveyed from a property owner using a farm-in agreement.
- Gaining access to property in order to explore for oil and gas is first about ownership and who owns the rights. The rights owners will continue to apply stringent conditions to their exploration and lease agreements.

- Auctions have traditionally been favored for oil leasing because they allocate and price scarce resources simultaneously and do so in settings of uncertainty. There are various methods used in auctioning oil and gas leases, and auction practices vary across the different oil-producing nations.

- Oil exploration will become increasingly costly and technologically sophisticated as E&P firms move into deeper water, colder conditions, less-developed countries, and geopolitically sensitive areas.

- Because of the increased costs of exploration, the chance of success for new prospects will have to be high before firms are willing to invest.

Notes

1. "Allah Be Praised," *Time*, April 9, 1953.
2. This section draws from a variety of sources including "Description of an Oil Lease," oil-gas-leases.com; and John B. McFarland, "Checklist for Negotiating Oil and Gas Leases," 2006; R. H. Porter, "The Role of Information in US Offshore Oil and Gas Lease Auctions," *Econometrica,* Vol. 63, 1995; K. Hendricks and R. H. Porter, "The Timing and Incidence of Exploratory Drilling on Offshore Wildcat Tracts," *American Economic Review,* Vol. 86, 1996.
3. "Petroleum Resources Management System," sponsored by the Society of Petroleum Engineers (SPE), American Association of Petroleum Geologists (AAPG), World Petroleum Council (WPC), and Society of Petroleum Evaluation Engineers (SPEE), Preamble, p. 1.
4. William Shakespeare, *The Tempest*, Act 4, Scene 1.
5. http://www.sec.gov/interps/account/sabcodet12.htm.
6. Peter Cramton, "How to Best Auction Oil Rights," *Escaping the Resource Curse*, edited by Macartan Humphreys, Jeffery D. Sachs, and Joseph E. Stiglitz, New York: Columbia University Press, 2007.
7. Marc S. Robinson, "Oil Lease Auctions: Reconciling Economic Theory With practice," *UCLA Working Paper #271*, October 1982.
8. Phillip Haile, Kenneth Hendricks, and Robert Porter, "Recent US Offshore Oil and Gas Lease Bidding: A Progress Report," October 12, 2009 (unpublished).
9. Outer Continental Shelf (OCS) Auction Data, Pennsylvania State University, http://capcp.psu.edu/OCS/index.html.
10. Haile, et al., p. 5.
11. Haile, et al., p. 7.

12. "Procedures for Oil and Gas Blocks Allocation and Offering," Ministerial Regulation, Ministry of Energy and Mineral Resources, No. 40, 2006.
13. "Higher Shares for the Libyan Government as a Result of Transparency and Good Governance," Dr. Issa Tuweigiar, Secretariat of Energy Libya, undated.
14. Daniel Johnston, "Impressive Libya Licensing Round Contained Tough Terms, No Surprises," *Oil & Gas Journal*, April 18, 2005.
15. "Libya Awards 15 Exploration Blocks," *Oil & Gas Journal*, February 1, 2005.
16. Maher Chmaytelli, "Exxon, Chevron, BP among Companies Seeking Libya Oil Permits Share Business Exchange," Bloomberg, October 13, 2006.
17. *Oil and Gas Quarterly Bulletin*, Chadbourne & Parke LLP, Second Quarter 2005, p. 9.
18. Ben Casselman & Guy Chazan, "Cramped on Land, Big Oil Bets at Sea," *Wall Street Journal*, January 5, 2010, pp. A1, A17.
19. "The Offshore: Good Luck, Bad Luck and Mukluk," by Randy Udall and Steve Andrew, *Energy Bulletin*, September 11, 2008, www.energybulletin.net/node/46554.
20. *Barron's* reported that "Sohio (Cleveland, Ohio) won drilling rights for 47% of Mukluk with a $272 million offer." Lawrence J. Tell, "The Mukluk Prospect: Has Sohio Struck It Rich Again in Alaska?", *Barron's*, August 8, 1983, Vol. 63, Issue 32, p. 16.
21. Ibid.
22. "Mukluk Island Oil Well R111 V," *Toledo Blade*, December 7, 1983.
23. Ralph E. Winter, "Sohio Confirms Mukluk Well Is a Dry Hole—Firm Takes Pre-Tax Charge of $310 Million and BP Plans Major Write-Off," *Wall Street Journal* (Eastern Edition), January 23, 1984.
24. "A Lesson for Critics of the Oil Industry," Editorial, *Nation's Business*, March 1984.
25. Ken Wells, "Beautfort Sea Lease Sale Should Show Whether Mukluk Island Jitters Remain," *Wall Street Journal* (Eastern Edition), August 21, 1984.
26. Daniel Yergin, *The Prize: The Quest for Oil, Money, & Power*, New York: The Free Press, 1991, pp. 433–434.

Chapter 4

DEVELOPING OIL AND GAS PROJECTS

Increased project complexity in demanding environments and frontier areas, aggressive performance expectations, technological innovations, larger financial bets, competitive national and private indigenous companies, more sophisticated host countries—all coming at a time when the supply of oil and gas is constrained and demand is increasing—mean that management must make quantum shifts to be able to meet the future energy needs of the global economy.[1]

—Booz Allen Hamilton, 2006

The best-laid plans of mice and men go oft awry.

—Robert Burns, "To a Mouse"

Once an oil or gas field has been discovered, appraised as viable, and sanctioned for development, the project must be developed in order to create a producing asset. This chapter discusses the project development process with a focus on key decision areas such as the evaluation of project risks and returns. Although the focus of the chapter is on developing upstream projects, it is important to recognize that the decision-making and project management processes used in upstream development are applicable across the industry.

The oil and gas industry is very capital intensive, with investment decisions made in upstream exploration and development and postdevelopment during the production process and also in midstream and downstream areas such as shipping, pipelines, refining, and chemicals. Whether the project involves a $10 billion oil field development or a $10 million storage tank at a fuel terminal, the decision making and project development processes that occur are conceptually similar to the processes discussed in this chapter: project sponsors need to evaluate qualitative and quantitative risks, assess project economics, identify the source of capital, create contractor supply chains, manage the project design, construction, and execution, hand the project over to the operators, and integrate the project into a larger organization.

To convey some of the complexity and challenges in project development, we begin the chapter with a case example of the Frade Chevron project. We then discuss the various steps necessary to take an upstream project from concept to completion. In doing so, we caution the reader from assuming that development is always a straightforward linear process from start to on-time and on-budget finish. In reality, all major development projects involve many challenges, from time delays to partner disputes to government approval holdups to shortages of people, materials, and technology. For project development engineers and managers, working on a major project can be stressful, exhilarating, frustrating, and highly rewarding, often at the same time.

Project Development Case Example: Frade Chevron

The Frade field is located in the Northern Campos Basin, approximately 370 kilometers offshore Rio de Janeiro, Brazil, in 1,100 meters water depth. The 154 KM2 Frade concession area is adjacent to Petrobras' Albacore Leste and Roncador developments. The field was discovered in 1986 by Petrobras. Texaco negotiated a partnership with Petrobras and received Brazilian government oil and gas regulatory assignment in 2000. Conceptual engineering studies and the acquisition of 3-D seismic data were completed in 2000 by Texaco. Chevron inherited the project after its merger with Texaco in 2001. Chevron is the project operator with a 51.74% interest. Petrobras has a 30% interest and Frade Japao Petroleo Limitada (FJPL), a Japanese consortium has 18.26%. FJPL is made up of Japanese companies INPEX with 37.5%, government-owned Japan Oil, Gas & Metals National Corp. with 50%, and Sojitz with 12.5%.

The project is expected to produce between 200 and 300 million barrels over a period of 18 years. From the beginning, Chevron acknowledged that Frade was a marginal project compared with "elephants of the past." The project is geologically complex and deepwater and heavy oil elements add to the economic risks.[2]

The development plan involved 12 horizontal production wells and 7 vertical injection wells. The production wells utilize an open-hole gravel pack completion with gas lift and are tied back to a floating, production, storage, and offloading (FPSO) vessel. The FPSO provides the capacity to process 100,000 barrels, compress 106 million standard cubic feet of natural gas, store 1.5 million barrels of oil, and inject 150,000 barrels of water per day. Water injection is necessary from the beginning of production to maintain reservoir pressure and maximize life of field oil recovery. The subsea architecture is a series of production flowlines, gas lift and water injection pipelines, and umbilicals. The gas lift pipelines and production umbilicals supply up to four production wells. Surplus gas will be sent to shore via existing Petrobras pipelines in the area, while the processed oil will be sold on the world market, transported with conventional trading tankers.

The primary contractors for the project are:

1. SBM: Lease and operate the FPSO; manage the conversion of the VLCC to an FPSO
2. Transocean: Drilling rig
3. Acergy: 41 kilometer flexible gas import/export pipeline between the Frade FPSO and the Roncador pipeline; flexible risers; and well-control umbilicals (pipes that send chemicals and signals to the wells)
4. FMC Technologies: Subsea systems (enhanced horizontal subsea tree systems)
5. Wellstream International: 130 km of flowlines, risers, and jumpers
6. V&M: Casing and tubing
7. Marine: Umbilical and electric cables

8. Halliburton: Completion services

9. Shlumberger: Completion services

10. BJ Services: Well sand control

In 2004, the Brazilian government criticized Chevron for moving too slowly. A *Wall Street Journal* article described the project:

> *Chevron's original plan was conservative—drill about half of the planned 19 wells, begin producing oil, then study production data for 18 months before drilling the remaining wells. If the first wells didn't look good, the company could forgo drilling more and cut its losses. But as Chevron geologists counseled moving slowly, Mr. Moshiri [the Chevron executive in charge of the project] decided in the middle of 2005 to place a big bet on Frade. They would drill all the wells, one after another without a break. It was the oil industry equivalent of going all in with a poker hand. "That is our job, to take risks and push the envelope."*[3]

The original plan developed by Texaco called for the construction of a floating platform. This plan was scrapped in favor of an existing drilling rig. With the market for rigs extremely tight, the project team decided to retrofit a 30-year-old rig previously used offshore Eastern Canada. Chevron contracted with Transocean to employ the rig for $315,000 a day. Chevron then worked with SBM on the FPSO. Instead of a new build, they opted to convert a 30-year-old oil tanker to an FPSO. The conversion was done by Drydocks World-Dubai. Chevron owns FPSO Frade and SBM is the operator. In June 2008, the process of installing subsea pumps and valves on the seafloor began. Normally this installation would be done after all the wells were drilled. Because the specialized ship used to do the installation was only available for nine months and was booked years in advance, the project team opted to install the equipment before drilling was completed.

The project experienced a number of delays with both the drilling rig and the FPSO, but in the end, oil production began on June 20, 2009 only a few months behind schedule. As illustrated by table 4–1, Frade's delivered cost to Gulf Coast refineries is expected to be somewhere between $18/bbl and $20/bbl.[4] Clearly, combining the project's high capital cost with Frade's heavy crude selling at a discount to West Texas Intermediate (here assumed to be between 10% and 15%), the project as a whole is expected to probably take a relatively long payback period.

Table 4–1. Chevron's Frade Brazil economics

	Low End Estimate	High End Estimate
Reserves (bbls)	250,000,000	300,000,000
Production/day (bbls/day)	100,000	100,000
Production/yr (bbls)	36,500,000	36,500,000
Capital cost	$3,000,000,000	$3,000,000,000
Capital cost/bbl	$12.00	$10.00
Operating cost/bbl	10.00	8.00
Delivered cost/bbl	$22.00	$18.00
WTI Price ($/bbl)	$70.00	$70.00
Frade crude discount	15%	10%
Discount ($/bbl)	$10.50	$7.00
Frade price at refinery ($/bbl)	$59.50	$63.00
Less delivered cost	(22.00)	(18.00)
Possible gross margin ($/bbl)	$37.50	$45.00

Development Project Opportunities

The starting point in the upstream development process is a development opportunity that becomes a viable project. According to the SEC, "A development project is typically a single engineering activity with a distinct beginning and end, which, when completed, results in the production, processing or transportation of crude oil or natural gas. A project typically has a definite cost estimate, time schedule and investment decision; is approved for funding by management; may include all classifications of reserves; and will be fully operational after the completion of the initial construction or development. The scope and scale of a project are such that, if a project were terminated before completion, for whatever reason, a significant portion of the previously invested capital would be lost."[5]

Upstream development opportunities originate in many ways. The classic vertical integration opportunity comes from a firm's exploration activities in which the exploration arm of the firm identifies a potential resource and the development arm performs an investment appraisal that may or may not lead to project development. This is often the development process for IOCs and the large E&P independents. With this sequence, the firm controls all aspects of exploration, development appraisal, project development, and production management, and in turn, bears all the costs and risks of drilling. At any point in the development process the firm may decide to bring in partners.

Many alternatives besides the exploration-discovery-development option are available for the development of oil and gas acreage. Given the increasing technological complexity of projects, firms typically bring in partners to help in execution. This "help" may be restricted to financial investment or may involve technological and managerial support. Small companies may choose to farmout interest in their licenses to larger firms. Firms may invite other firms to be financial partners in their projects to offset the risks and costs of the project. National oil companies may invite bids for project participation in existing fields or new acreage. The recent deals in Iraq involve existing oil fields operating at far less than their potential. The project will be designed to upgrade production with better technology and improved management.

In today's oil industry, if firms were to restrict their development opportunities to their own exploration discoveries, IOCs in particular would have very limited reserve portfolios. With most of the world's oil and gas reserves controlled by national oil companies, oil companies need to be open to opportunities that come from acreage owned or controlled by states or other firms. Analysts assume that over the coming years, oil and gas firms will find new and innovative ways to gain access to development opportunities.

Project evaluation process

An upstream development project starts with a prospect or development opportunity. The next step is to perform an appraisal and analysis of the opportunity by evaluating risk, economic returns, project feasibility, and competitive challenges. If the project is deemed viable as it moves through the various review phases, a decision must be made about financing sources for the project. If financing is available, a final investment decision (FID) can be made. If the decision is to develop, the development design is done and reviewed, followed by the actual execution of the project and handover to the operators. In practice, for a major upstream development project, there will be many milestones and reviews that must be completed before final approval for capital spending.

The evaluation of a project requires a number of strategic, organizational, and financial determinations. Every E&P firm will have its own proprietary process for managing project development. As an example, table 4–2 shows a set of activities used by Shell's upstream development organization. Once the project, or *concept*, using Shell's terminology, has been approved for execution, a conceptual design is prepared. The conceptual design must then move through another approval process before detailed design, construction, and project execution can occur.

Table 4–2. Shell upstream project design, execution, and operation tasks

Concept Selection	
Formulate development concepts	Propose development concept
Evaluate concepts	Budget costs and economics
Conceptual Design	
Prepare basis of design	Prepare project execution plan
Execute conceptual design	Control budget and economics
Prepare project specification	Develop management systems
Develop contracting strategy	
Detailed Design and Construction	
Project management and reviews	Operational readiness
Design reviews	Technical support
Interface management	Commissioning support
Risk evaluation	Pre-start-up reviews
Contingency planning	
Operate	
Technical support	Facility-life extension
Debottlenecking studies	Production improvement
Postinvestment reviews	Abandonment planning studies

Source: Shell Global Solutions, Upstream Development Solutions, *http://www.static.shell.com/static/global_solutions/downloads/innovation/views/upstream_development_brochure.pdf.*

In practice, most E&P firms use a stage-gate project management process. Gates are the formal points in the development process where management approval (i.e., gatekeeping) must be given for continued work and resource allocation for the project. Gates provide the opportunity for quality checks and review of project acceptance criteria. Gate approval involves various discussions, such as:

- Review and agreement of details with key operational managers
- Agreement of support arrangement with executive sponsors
- Presentation to management teams for discussion and commitment
- Specific reviews
- Performance in areas such as technical, safety, and environment
- Economic viability
- HR resources necessary for completion
- Partner and contractor performance

- Procurement costs

- Legal and compliance issues

The gate may look in two directions: backward, to determine whether the previous phase of activity is satisfactorily completed, or forward, to determine whether the project is ready to continue and adds sufficient value. Figure 4–1 provides a simplified view of the stage gate process, breaking it up into four stages and gates that lead to actual project development.[6] At each gate a series of questions is addressed by the decision review board (DRB) and if not answered satisfactorily, gate approval will not occur. The four stages are feasibility, selection, definition, and execution. The first three stages are considered the front-end loading for the project, and at any one of these stages the project can be halted. Once the FID is made and the project execution begins, project reviews will continue, and projects may be halted during the execution phase.

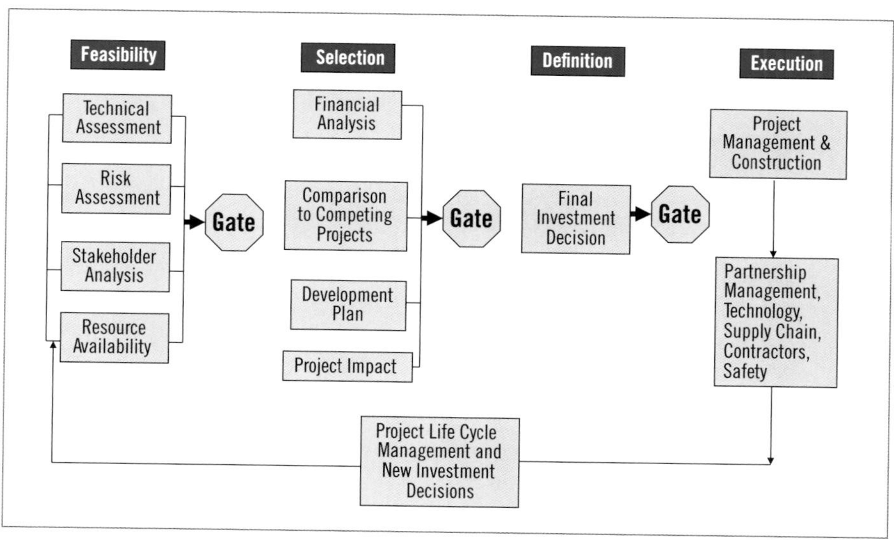

Figure 4–1. Project development stages

Key questions for stages 1 through 4 are as follows:

1. **Feasibility and appraisal**

   - Is the project technically feasible?

   - Is the project aligned with internal and external stakeholders?

- Are there other key stakeholders, and is the project aligned with their interests?
- Does the project require combined development (unitization) or cooperation across borders?
- What are the quantifiable and nonquantifiable full life-cycle risk factors associated with the project? Are they manageable?
- What are the key resources needed for the project, and are they available?
- Is the project aligned with the business strategy?

2. Selection

- Detailed financial analysis: does this project meet corporate hurdles?
- How does this project compare to other project opportunities?
- What are the key elements of the development plan?
- How will the project impact communities and stakeholders?

3. Definition

- Final investment decision (FID)—go or no go?
- Have all project issues been thoroughly examined?
- Are there stakeholder, partnership, or government issues that have not been resolved?
- Have there been material changes to the project economics?
- How will the project be funded?
- Has the development scope been finalized?
- Is the design sufficiently detailed?
- Are all cost estimates accurate?
- Are negotiations with contractors and suppliers concluded?
- Are negotiations with potential customers (especially for LNG) concluded?

Once the FID is made, the project is handed over to the development team and project execution begins. During the execution phase there will be regular reviews, questions asked, and occasionally, projects in mid-development will be shut down or postponed. In 2008 and 2009, oil prices dropped precipitously, and many projects approved for development were canceled. The International Energy Agency reported in 2009 that since October 2008, over 20 planned large-scale upstream oil and gas projects, involving around 2 mb/d of oil production capacity, were deferred indefinitely or cancelled. According to Daniel Yergin, "Prices have come down so far and so fast, it's become a shock to the supply system. . . . It's a classic—if extraordinarily dramatic—cycle."[7]

Project life-cycle reviews

After an oil field project is developed and handed over to the production operators, there will be regular reviews of the project life cycle. Figure 4–2 shows the life cycle of an oil field. Once the field is in production, further capital expenditures associated with the field may be required to extend the field's life. Depending on their size and the firm's threshold for review, these expenditures may go through a project evaluation process similar to the original process for the field.

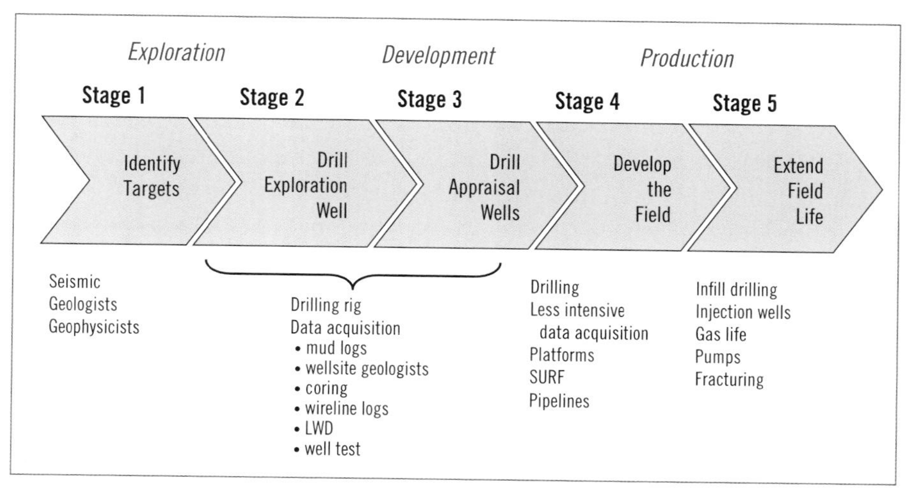

Figure 4–2. Life cycle of an oilfield

Source: "Oil & Gas for Beginners," Deutsche Bank, 2009, p. 49. Copyright © 2008 Deutsche Bank AG.

Joint Development/Unitization

Competing interests in the same field, either within one country or across boundaries, is an issue for considerable effort in the earliest states of development. In addition to the concerns of any individual producer (project size, production expectations, investment), societal interests are also intricately involved, as the maximization of ultimate oil recovery is important to society and the state, while the costs and returns of its production are obviously important to any private interests involved in development.

Rate of production and unitization

In the early days of the oil industry, the first principle of oil production was "faster is better." Since oil fields are the result of exploration and development investment, the sooner oil and gas is produced, the faster returns on investment are generated. But, the industry soon realized that an oil reservoir or field should be developed with its geophysical properties in mind, particularly the maintenance of its natural drives which force the oil and gas to the surface. Studies yielded the maximum efficient rate (MER) of field production, which was soon the standard used by both industry and regulators.

Another issue connected to "faster is better" is that oil reservoirs are common pools. If there are a number of different wells attempting to produce the same field, the wells are competing for migratory oil. In most countries this is not a problem, as the state owns and regulates the development and production of mineral and other subsurface resources. In the United States the situation is different because private parties are free to own and exploit mineral resources. A common oil field could lie underneath hundreds of different tracts of land and mineral leases. Each lease holder would then move as quickly as possible to produce the oil and gas within the reservoir before the next competing producer, the so-called *rule of capture.*

The US Bureau of Mines recognized the costs and risks associated with common pool overproduction as early as 1916, when it argued with a number of state authorities that common property resources like oil reservoirs need production regulation. The problems were confronted in East Texas in 1931 when it became clear that excessive production was rapidly destroying the reservoir dynamics of new oil field discoveries.[8]

Unitization. The arguments against competitive production were clear: excessive wells drilled, greater surface disturbance and excessive surface storage requirements. The longer term result is higher extraction costs because subsurface pressures are inefficiently depleted, which reduces overall oil and gas recovery.[9] The development of compulsory field unitization required private parties to coordinate reservoir production to minimize surface and production costs while managing reservoir pressure to maximize recovery. Unitization yielded reservoir development and fair distribution of the value.

> *The unitized area, usually a reservoir, is treated as a single unit for development purposes. It is as if the separate leases and licenses are merged into one single lease or license, with a single unit operator appointed to manage the development of the field, within the limits of the authority granted the unit operator by the unit operating agreement and the management committee composed of all the different lessees or licensees with interests in the unit.*[10]

In most countries today, the law allows first for the producing parties to create their own unitization production plan and only require a compulsory process if voluntary efforts fail. Most producing countries have unitization provisions in their petroleum development laws. However, with the exception of a few countries such as Peru or Angola, the laws are not as extensive and defined as those of the United States.

Cross-border unitization. At first glance unitization issues appear specific to the United States and a few other countries allowing private ownership. In those countries, unitization is not a major challenge when a reservoir lies completely within the individual state's boundaries. If production is undertaken by multiple licensees within the same country, it is well within the power of the single state to impose unitization, so-called *sole-country unitization.*

Unfortunately, oil and gas fields do not acknowledge geopolitical boundaries. Fields often extend across national boundaries. Cross-border unitization involves two or more different governments and two or more licensees. Although there are a variety of agreements used globally today, most cross-border unitization agreements—*international unitization agreements* (IUAs)—have a set of standard features: 1) unitization is required only after discovery; 2) the area of unitization is defined by the oil reservoir itself; 3) the participating countries collaborate via international agreement on all production issues related to the optimal development of the field; 4) all countries maintain their own sovereign rights related to activities within their borders; 5) the licensees agree to a single development plan and a unitized operating agreement; and 6) each licensee bears the proportion costs and receives proportion returns—the participation factor—of the field's oil and gas that lies underneath their physically defined borders and license.

The challenge grows when the field is in a geographic zone that is the subject of disputed sovereignty. This area, termed a *joint development zone* (JDZ), needs a set of unitization and joint operating agreements in which the surface rights (often water as much as land) are not well agreed upon. It is often best that an effective set of JDZ agreements be in place prior to discovery, including the establishment of a joint authority with overall production and management authority over the total field or reservoir.

The Nigeria–São Tomé & Príncipe Joint Development Zone (JDZ), illustrated in figure 4–3, is an area of overlapping maritime boundary claims that will be jointly developed by the two countries. In November 1999, the heads of state of Nigeria and of São Tomé and Príncipe provided the mandate for officials of both countries to commence negotiations on the territorial claims of the two countries. In December 1999, formal maritime boundary talks commenced, and in August 2000, the heads of state agreed on the joint development of resources in this region.

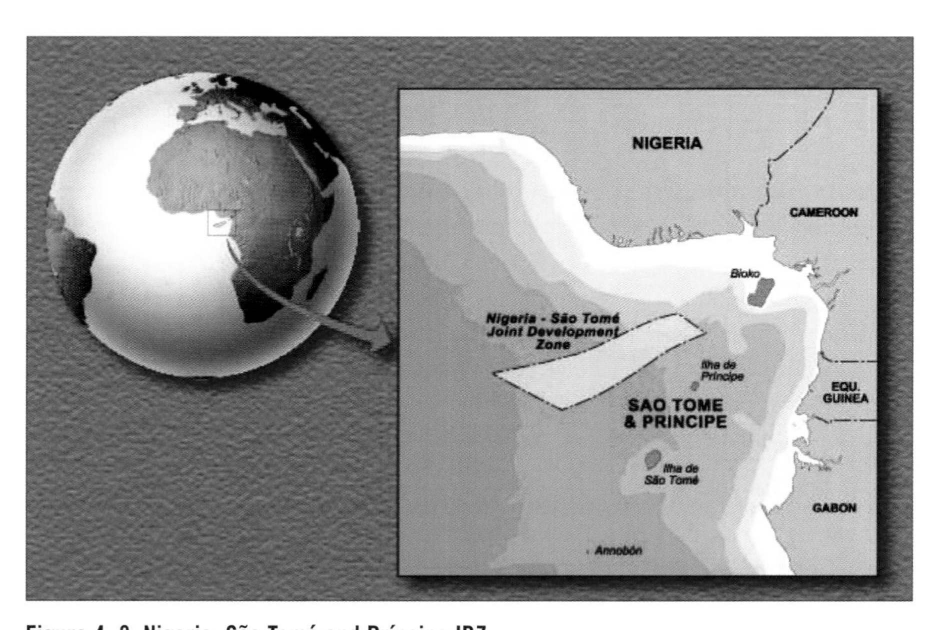

Figure 4–3. Nigeria–São Tomé and Príncipe JDZ.

Following a period of discussions, a formal treaty on the joint development of resources was developed. This was signed by the heads of state and subsequently ratified by respective legislatures in February 2001.

Because most JDZs originate from long-standing territorial sovereignty debates, the area in question is typically much larger than any specific oil field but may not contain the entire field. If a field lies within the zone, but then crosses over into specific known borders, the role of the single operating authority is even more complex. In addition to the Nigeria–São Tomé and Príncipe JDZ shown in figure 4–3, other joint development zones around the world include: Saudi Arabia, Bahrain, Kuwait and Abu Dhabi; Japan–South Korea; Sudan–Saudi Arabia; East Timor–Australia; Malaysia-Vietnam; Colombia-Jamaica; and Guinea Bisseau–Senegal. The East Timor–Australia JDZ serves as a case in point. Australia, Indonesia, and East Timor have been unable to agree on maritime boundaries. They have, however, concluded a joint petroleum development agreement (JPDA) splitting the petroleum product of the Timor Sea ("Timor Gap"), with 90% flowing to East Timor.[11]

Project Financial Analysis

As discussed, many aspects of projects must be analyzed and approved before project execution. From a business and shareholder value perspective, without financial viability the project cannot go forward. The financial analysis of projects in any industry follows traditional financial principles. These principles are adapted in a variety of ways for the realities of oil and gas. This section describes the financial analysis of a hypothetical oil development and pipeline project in West Africa.

Case example: West Africa

The West African Petroleum Development and Pipeline Project ("West Africa Project") combines the governmental interests of the host country with the private interests of an IOC. By early 2002, the financial analysis of the 100-million-barrel project was reaching final approval stages by the IOC.

Structure. The West Africa Project is a two-stage project to produce (the field system), transport, and export oil from 400 kilometers inland in a West African country. The field system encompasses the development of two main oil fields, which includes the drilling of 80 wells and the construction of all associated facilities and infrastructure. The pipeline/export system includes a 400-km pipeline running from the field system to the coast, where it then passes 8 km offshore to a floating storage and offloading vehicle (FSO). The pipeline is to be buried 2 meters below ground to provide both pipeline security and environmental protection. The pipeline will require two repressurization stations to assure proper flow. Once the oil is loaded on the permanently moored FSO, ocean-going crude tankers will be loaded from the FSO for transport to refineries anywhere in the world. The fields are expected to have a 20-year life span.

In early 2002, the company completed its project evaluation. Like any potential large-scale investment, a significant amount of capital had already been expended: $15 million in a signature bonus and $40 million in exploration and predevelopment expenses. Those were sunk costs (i.e., nonrecoverable). The analysis now underway will evaluate all incremental cash flows over time to decide whether to pursue development or walk away. As illustrated in table 4–3, after exploration and predevelopment, the company faces an investment over the next 20 years of $600 million. The investment will combine major upstream and pipeline/export facilities construction.

Table 4–3. West African oil development: Investment (millions US$)

Calendar Year	1999–2001	2002	2003	2004	2005	2006	2007	2008	2009	2002–21
Project Year	–1	0	1	2	3	4	5	6	7	Total
INVESTMENT CASH FLOWS										
Signature Bonus	15									
Exploration & Predevelopment	40	40	45							100
Upstream Development		30	85	90	20	15	5	2	2	273
Pipeline/FSO/Export	—	10	75	75	55	10	2	0	0	227
Total Investment	55	80	205	165	75	25	7	2	2	600

Table 4–4 presents the expected operating cash flows of the project over its 20-year life. These operating cash flows are estimated on the basis of both expected field production and expected crude price over time. For simplicity, the analysis is for the entire project and does not distinguish between IOC and state shares.

Volume. The two fields have been explored and analyzed in detail. Production volume is expected to rise rapidly, hitting a peak of 12.6 million barrels per year in the sixth year of operations and then declining gradually over time. Total recoverable oil is expected to be more than 100 million barrels.

Price. Projecting price is much more difficult than predicting production volume. The market price of oil has historically experienced extreme volatility. Financial evaluations of projects are usually extremely conservative in forecasting crude prices. They also employ various different scenarios and sensitivity analysis. In this case, since the crude will be exported out of West Africa and compete with

Atlantic Basin oils, its price will be pegged to Brent crude. The analysis in table 4–4 assumes a price of $20/bbl constant over time. This was a common, conservative price used in major African projects of the time.

Operating costs. Operating costs are estimated by the project sponsor on the basis of experience in various projects of this scale and complexity. The actual operating costs of both the fields and the pipeline are generally dictated by the scale and structure of the venture and will be largely fixed over the project life span.

Once the investment and operating cash flows have been estimated, the final net operating cash flows expected to flow from the project are subjected to a variety of financial metrics to determine both financial viability and the project's resilience to a multitude of potential risks to those cash flows.

Financial metrics

The net operating cash flows, both investment cash outflows and operating cash inflows, are analyzed using several financial metrics, the most common being net present value analysis and internal rate of return.

Net present value (NPV). This financial metric discounts all net operating cash flows generated over the lifespan of the project back to the present. The technique requires the identification of projected investment and operating cash inflows and outflows per year for discounting. The principle of discounted cash flow is based on the time value of money: a cash flow occurring at a future point in time is worth less today. For example, a $1,000 investment today for one year at 10% interest will yield $1,100 in one year. This in turn means that $1,100 received one year from today (the future value at time t, where t=1), if discounted at 10%, would be worth less—the present value, $1,000 in this case, today:

$$\textit{Present value} = \frac{\textit{Future value}}{(1+i)^t} = \frac{\$1{,}100}{(1.10)^1} = \$1{,}000$$

All identified net operating cash flows are then discounted back to the current period and summed for the calculation of net present value.

Table 4–4. A West African oil development project: Expected cash flows (millions of US$)

Calendar Year	2002	2003	2004	2005	2006	2007	2008	2009	2010	2011	2012	2013	2014	2015	2016	2017	2018	2019	2020	2021
Project Year	0	1	2	3	4	5	6	7	8	9	10	11	12	13	14	15	16	17	18	19
INVESTMENT CASH FLOWS																				
Signature Bonus	—	—																		
Exploration & Predevelopment	55.0	45.0																		
Upstream Development	30.0	85.0	90.0	20.0	15.0	5.0	2.0	2.0	2.0	2.0	2.0	2.0	2.0	2.0	2.0	2.0	2.0	2.0	2.0	2.0
Pipeline/FSO/Export	10.0	75.0	75.0	55.0	10.0	2.0	—	—	—	—	—	—	—	—	—	—	—	—	—	—
Total Investment	95.0	205.0	165.0	75.0	25.0	7.0	2.0	2.0	2.0	2.0	2.0	2.0	2.0	2.0	2.0	2.0	2.0	2.0	2.0	2.0
OPERATING CASH FLOWS																				
Production (millions of bbls)				0.57	6.10	9.42	12.40	10.80	9.49	8.30	7.26	6.36	5.56	4.86	4.26	3.70	3.26	2.85	2.49	2.00
Oil Price ($/bbl)	$20	20.0	20.0	20.0	20.0	20.0	20.0	20.0	20.0	20.0	20.0	20.0	20.0	20.0	20.0	20.0	20.0	20.0	20.0	20.0
Gross Revenue (millions of $)				11.4	122.0	188.4	248.0	216.0	189.8	166.0	145.2	127.2	111.2	97.2	85.2	74.0	65.2	57.0	49.8	40.0
Operating Cost ($/bbl)				6.20	2.60	2.40	2.30	2.36	2.40	2.46	2.54	2.64	2.72	2.82	2.94	3.08	3.24	3.40	3.60	3.40
Operating Expense (millions of $)				3.5	15.9	22.6	28.5	25.5	22.8	20.4	18.4	16.8	15.1	13.7	12.5	11.4	10.6	9.7	9.0	6.8
Net operating earnings				7.9	106.1	165.8	219.5	190.5	167.0	145.6	126.8	110.4	96.1	83.5	72.7	62.6	54.6	47.3	40.8	33.2
NET OPERATING CASH FLOW	(95.0)	(205.0)	(165.0)	(67.1)	81.1	158.8	217.5	188.5	165.0	143.6	124.8	108.4	94.1	81.5	70.7	60.6	52.6	45.3	38.8	31.2

Discount rate. A key component of NPV analysis is the selection of the discount rate, the *i* in the equation above. This discount rate used for the calculation of present value of project cash flows is defined as the "risk-adjusted cost of capital" for the specific project at hand. A company creates value for its shareholders when it invests in projects that yield results *above* the cost of the capital it utilizes for the project. But how much above? In practice, most companies will expect that any investment yield some minimum amount above the weighted average cost of capital (WACC), and this is called the *corporate hurdle rate.* In addition, companies also impose an additional risk premium for the specific project if the project risks differ from (exceed) the risks associated with the average activity of the company.

Consider the hypothetical calculation presented in figure 4–4. The company has a capital structure that is 25% debt and 75% equity. The company's average cost of debt is 8.00% (before taxes, noting that interest expenses are deductible expenses) and the average cost of equity capital is 12.00%. The risk-free cost of debt, for example the yield on a 10-year US Treasury bond, is assumed to be 6.00%. The company's bankers have assessed its credit quality as a 2.00% spread over treasuries, setting the cost of debt at 8.00%.

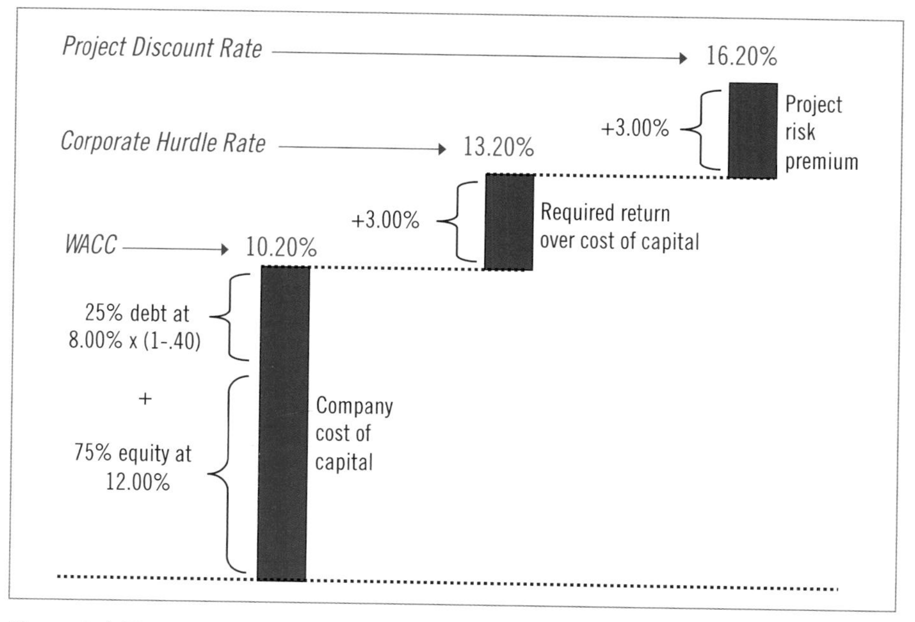

Figure 4–4. The cost of capital and discount rate

The company's cost of equity is calculated using the *capital asset pricing model* (CAPM), which adds a risk-adjusted market risk premium to the risk-free cost of debt. In this example, the cost of equity would be the risk-free rate of 6.00% (the treasury bond rate again) plus the company's *beta* of 1.2 (company returns are roughly 20% more volatile than the general market) multiplied by a market risk premium (the average return of equities over risk-free debt) of 5.00%: 6.00% + (1.2 × 5.00%) = 12.00%. Assuming the company has an effective tax rate of 40% and interest payments are indeed deductible towards corporate tax liabilities, the company's weighted average cost of capital (WACC) is:

$$\text{WACC} = [\ 0.25 \times 8.00\% \times (1 - 0.40)\ ] + [\ 0.75 \times 12.00\%\ ] = 10.20\%.$$

The company's WACC represents the average cost of raising capital for general company purposes at current market rates. The company then adds its minimum return above WACC, say 3.00%, to set its corporate hurdle rate. The company must determine whether the specific project poses additional risks beyond those typically undertaken. Some companies may assess an additional 3.00% for international investments. If this was an international project, the discount rate to be applied to the expected cash flows would be 10.20% plus 3.00% plus 3.00%, or 16.20%.

A positive NPV indicates that the project is expected to generate returns that exceed the company's costs of undertaking the project. The project is expected to create economic value, specifically for shareholders. Although NPV is widely considered the single best metric for determining a project's expected financial viability, a number of other metrics are also frequently used.

Internal rate of return (IRR). The internal rate of return is the discount rate that results in the NPV of the expected cash flow stream having a value of exactly zero. A project with an IRR greater than the sponsors' minimum rate of return (i.e., the *hurdle rate*) is an acceptable investment. IRR is often more intuitive for managers who are comfortable talking in terms of rates of return rather than the net present value concept. As a result, IRR is the second most commonly used financial metric for evaluating prospective projects.

As illustrated in figure 4–5 the West Africa Project is expected to have a positive NPV of $27.5 million and an IRR of 17.3%. Note that the NPV was calculated using the project discount rate calculated previously in figure 4–3 of 16.2%. With an IRR of 17.3%, it meets the required criteria of having a rate of return greater than the project hurdle rate. The project is financially viable according to both criteria. It should be noted that this evaluation is of the total project, field system and pipeline/export systems combined, and not from a single segment of the prospective investment.

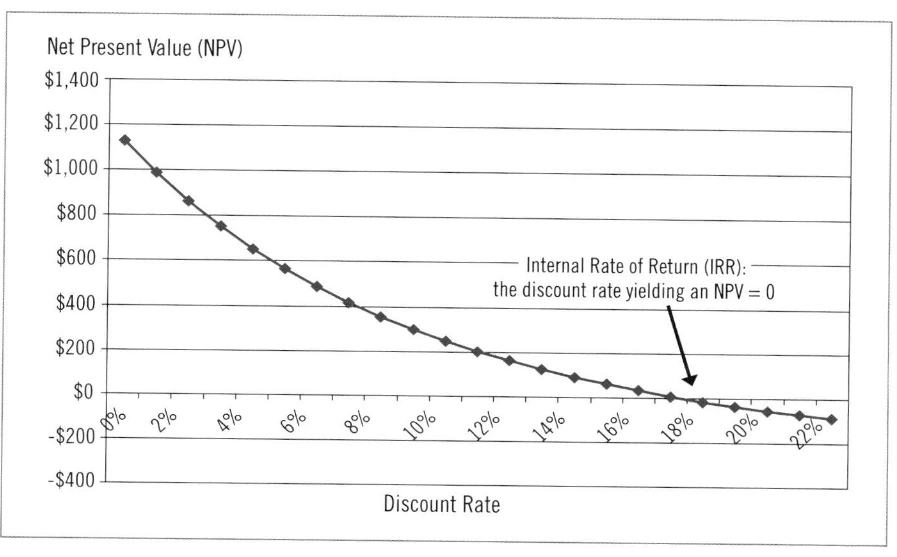

Figure 4–5. Project NPV and IRR. A project is deemed financially acceptable if it is expected to generate a positive net present value. This is technically equivalent to a project's expected internal rate of return exceeding the discount rate.

Risk analysis

Every project has its own inherent risks. A project evaluation must consider all risks and their likely impact on expected cash flows. For the West Africa Project, two primary risks are the future crude price and the actual volume and production rates from the oil fields.

Crude oil prices, as discussed previously, are highly unpredictable. Although every project involving oil must include a forecasted oil price, it is only an assumption for project evaluation purposes and not really an expectation of the future prices. What this means is that the project analyst does not hold any true expectation that the price of crude will be any specific value in the future—history has proven all the prognosticators wrong, over and over. But, the analyst also knows that some assumption about revenues must be made. Therefore, a very conservative estimate of price is "a necessary evil" in the words of some analysts.

Production volume is the second top-line risk that must be considered. Most oil and gas producers pride themselves in their ability to accurately explore and analyze fields and reservoirs. Like crude prices, the fields and reservoirs will likely have some surprises over time. A healthy sense of humility is called for when considering expected production volumes.

The Mongstad project evaluation

The primary cash outflows in many petroleum development projects are engineering and construction expenses, while the primary cash inflows arise from operating revenues after start-up. The challenge when forecasting operating revenues has always been the general inability to forecast future crude prices. The oil industry has always been much more "comfortable" estimating costs, both construction and operating. But there are some remarkable failures to project construction costs as well.

Statoil's Mongstad refinery complex 30 miles north of Bergen, Norway, was the subject of one very high-profile project evaluation mistake. In the late 1980s, Statoil undertook a highly controversial and expensive expansion of the complex. Estimated at NOK 8 billion, the expansion would increase refining capacity, and specifically allow much higher volumes of light petroleum distillates. These distillates were a higher valued-added crude derivative than the heavy fuel oils currently being produced.

The project evaluation proved a failure on two different levels. As a result of both engineering design and construction errors, the project suffered a NOK 6 billion cost overrun. This was highly unusual in a project involving proven mature technology. In addition to the cost cash flow underestimation, the project's revenue projections proved wildly overestimated. The original revenue projections were for added refining margins of $130 per tonne (metric ton) for the light distillates over heavy fuel oils. But before the project was less than halfway completed, the spread fell on the open market to roughly $50 per tonne.

As a result of the Mongstad project's cost overrun, many of Statoil's senior executives were eventually forced out, including the CEO Arne Johnsen, the man who for many years was called "Mr. Oil" in Norway. The "Mong" became a term of derision in Norway—a colloquial term for NOK 6 billion.

There are a number of additional risks that need to be considered in any prospective oil or gas development project, often separated into precompletion, postcompletion, and macroeconomic risk. Some examples of these risks are as follows:

Precompletion risks

- Resource base: Is the oil or gas resource base true?
- Technology: Does the project rely on existing or unproven technology?
- Construction: Are there substantial challenges to construction in a timely manner?
- Partners: Are there partner issues that could lead to conflict?

Postcompletion risks

- Markets: What are general and specific commodity price risks?
- Production and pipeline/transportation security: Are there material threats to the project?
- Environment: What will be the impact of major spills, explosions, pipeline failures, etc.?
- Geopolitical: Are there risks of government interference or expropriation?
- Stakeholder: Will stakeholders try to exert unexpected influence on the project?

Macroeconomic risks

- Inflation or economic disruptions: Is the local economy economically stable?
- Currency convertibility: Is the local government likely to restrict movements in or out of the local currency? Are there likely to be restrictions on moving in or out of the local currency to a more liquid global currency like the dollar or the yen?

As illustrated by the previous section about Mongstad Refinery's expansion, specific risks may interact in particularly disastrous ways in a project.

The analyst's conundrum

Although present value theory is the most common method used in the evaluation of investments across all industries, and is the theoretical core of most modern investment analysis, it does pose a major challenge for the oil and gas industry—financial time.

All major oil and gas investments, whether they are refinery retrofits in Aberdeen or new offshore drilling prospects in the Gulf of Mexico, are long-term investments. Present value theory tends to "punish" cash flows that occur further into the future and seemingly undervalue significant assets that produce cash flows for 20 or 30 years into the future. We witnessed an example of this conceptual challenge in practice. A senior VP for development at a major IOC once became so frustrated with the use of present value theory that he stood, kicked over the table in front of him, and proclaimed, "I don't want to hear the phrase 'discounted cash flow' ever again."

Table 4–5 provides a sample calculation to explain this challenge. Assuming a 10% discount rate, a $10 cash flow occurring five years into the future has only a $6.21 present value; a $10 cash flow seven years into the future only a $5.13 present value. As a result, many in the oil and gas industry have long believed that these techniques alone do not adequately capture many of the longer term strategic and organizational values that many projects include. This has led to a never-ending search for alternative valuation techniques.

Table 4–5. Present value factors and time

Year	Discount Factor (assuming 10%)	Future Value	Present Value
0	1.0000	$10	$10.00
1	0.9091	$10	$9.09
2	0.8264	$10	$8.26
3	0.7513	$10	$7.51
4	0.6830	$10	$6.83
5	0.6209	$10	$6.21
6	0.5645	$10	$5.64
7	0.5132	$10	$5.13

Project Execution

Once the FID is made, a project development team takes over and execution begins. The development team is tasked with getting the project built on time and on budget. This section considers some of the key business issues that must be managed by the development team. In doing so, we note that given the complexity of the project management field, it is not possible to present a complete discussion of project management (for the individual interested in the project management field there are many excellent books on the subject).[12]

Project management is a complex field and involves both art and science (see the upcoming section, "So you want to be a developer?"). The project leader must pull together a multidisciplinary team that includes managers with skills from the fields of engineering, construction, procurement, finance, legal, public relations, government affairs, safety, shipping and transportation, and community development. The project team will always be working toward tight timelines and often with conflicting stakeholder expectations. The size and complexity of today's oil and gas projects almost guarantee that unexpected events will occur despite the best efforts at planning for contingencies.

So you want to be a developer?

In the oil and gas industry, and especially in the upstream sector, successful project developers occupy an almost mythical position. All the major E&P firms have legendary developers in their corporate histories that broke new ground in complex projects. While enduring difficult living conditions in extreme heat or cold, hostile communities, as well as anxious shareholders, demanding bosses, and ridiculously tight deadlines, these developers managed to safely complete huge projects on time and on budget in areas like the Gulf of Mexico, West Africa, the Arctic, or the Middle East.

To be successful, lead project developers need a broad range of skills. Developers must:

- Be expert technical managers who also understand the broad scope of upstream activities and have experience in design, construction, and operation of facilities.

- Have outstanding communication and presentation skills because throughout the project stakeholders will expect regular updates and information.

- Build and manage teams composed of individuals who come from different backgrounds and parts of the organization. To build these teams they need to have strong personal networks that can find the right people for the team.

- Exert influence without authority in order to get the resources they need from disparate parts of the organization.

- Have strong finance skills because throughout the course of project development, as circumstances changes, real-time financial analysis must be done to ensure the project continues to meet corporate expectations.

- Be disciplined project managers because, as the old saying goes, "In failed projects, managers don't plan to fail, they fail to plan."

- Comfortably deal with ambiguity and uncertainty because no project unfolds exactly as planned, especially in the volatile and cyclical world of oil and gas.

- Manage the various stakeholders connected to the project, including partners, contractors, communities, and governments.

- Have the energy and the dogged determination to complete the job.

- Know how to manage the often conflicting goals of the project owners.

- Know how to emotionally detach from a completed project, hand it over to the operators, and move on to the next project.

Figure 4–6 shows an example of a project management plan for offshore well completions. As the figure illustrates, there are various distinct stages in project management. The various stages of this relatively straightforward project look quite complex. When the project is a mega project such as Kashagan in Kazakhstan or Gorgon in Western Australia, the complexity becomes enormous. In addition, as projects are developed in more remote locations, such as Papua New Guinea, Siberia, and interior Africa, logistics become critical. In remote locations there may be no roads, airports, telecommunications, potable water, electricity, housing, labor force, or medical care. For some of these projects, the technical challenges of the oil or gas field are dwarfed by the challenge of getting equipment, materials, and people to the job site. Not surprisingly, as we discuss, many mega projects experience significant time delays and cost overruns.

Although there is constant pressure to reduce the direct costs of development activities such as offshore wells, there is just as much pressure to reduce the completion time. As figure 4–6 details, the various stages of offshore well completion vary in time, and that time itself is a major source of cost. In recent years much of the pressure has fallen on the second- and third-tier suppliers of the pieces, parts, and systems to shorten their cycle time. They are increasingly expected to deliver yesterday on a design completed today.

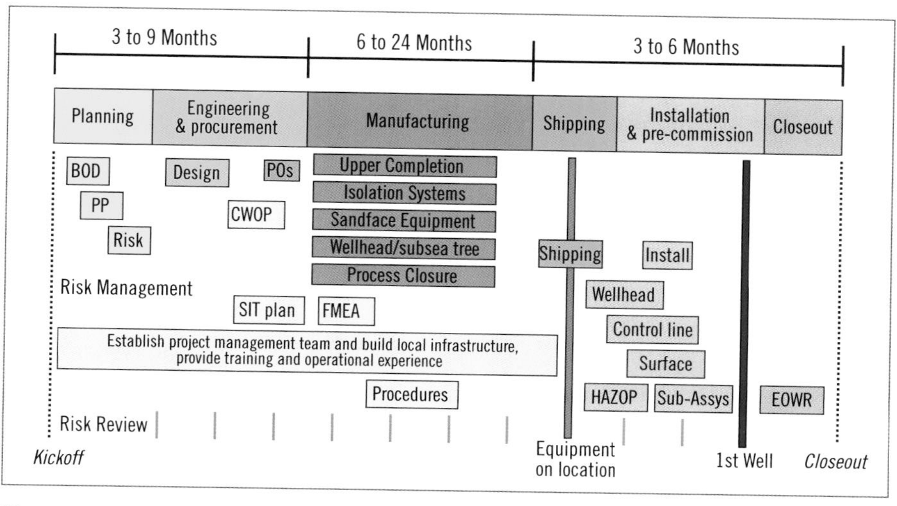

Figure 4–6. Project management of offshore well completions
Adapted from Caulfield, et al.,"Project Management of Offshore Well Completions," *Oilfield Review*, Spring 2007.

Contractor Relationships

The multitude of tasks, activities, processes, and technologies necessary for the execution of exploration, drilling, development, and production is daunting. No one firm can—or wishes—to do it all. There are thousands of firms worldwide of all shapes, sizes, ownerships, and interests that provide the many services and equipment needed in the field. For the E&P firm, successfully executing a project requires working with many contractors and building a reliable supply chain during project execution and into the production phase.

Contractor relationships and supply chains

Over the past few decades, the business of contracting and oil field service has changed significantly.[13] As the NOCs began to exert their authority and ownership over their reserves, they looked to contractors to do work previously done by IOCs. The large oil field service firms saw an opportunity to increase their activities. The legacy of the three biggest oil field service firms, Baker Hughes, Schlumberger, and Halliburton, goes back a century. The primary business for these companies used to be drilling and related services for the IOCs. The IOC did the exploration, the oil field service firm provided products and services up to the completion of the well, and then the IOC handled production.

In the mid-1980s, the oil field service firms began diversifying. Schlumberger acquired a geophysical exploration company in 1985 and formed an integrated project management group in 1995. In 1997, the firm won a contract from the Mexican oil company Pemex to redevelop the Burgos Basin by providing services ranging from geological analysis, road and tank farm construction, and postproduction reservoir management, as well as providing its traditional drilling service. In 1998, Baker Hughes merged with Western Atlas, an exploration services firm. The merged firm was able to provide integrated development services from exploration through crude production management, along with traditional drilling.

As the oil and gas industry expands into more challenging areas, and especially into deepwater areas, new technology is required. Some technology innovations will come from the E&P firms and some will come from contractors. The result in today's industry is a range of technologies and skills that is staggeringly complex. A set of large, well-capitalized contractors with technologies essential for IOC-led mega projects has evolved. For example, Technip, one of the largest contractors, has about 22,000 employees and offices across the globe. According to its website, "Technip executes infrastructure projects that are increasingly ambitious, complex and demanding: ultra-deep waters, extreme climates, mega-sized projects, nonconventional resources and optimization of environmental performance." Thus, to develop their mega projects, the IOCs have no choice but to rely heavily on a broad range of contractors. From a competitive viewpoint, there is a tug-of-war going on between IOCs, NOCs, and contractors. IOCs want to remain vital to the NOCs and maintain bargaining power with contractors; contractors want to occupy more of the value chain, and NOCs want to reduce their reliance on IOCs (see chapter 2 for more detail on NOCs and IOCs).

Although it is common to define contractors as companies that provide services but "do not produce petroleum," this is misleading. These companies provide the advanced products and services needed to drill, evaluate, appraise, complete, and produce oil and gas. Figure 4–7 shows the services and activities that contractors can provide in the exploration phase. Nevertheless, as one contractor recently commented, "The IOCs are the top of the food chain." Thus, while service firms and other contractors have increased their capabilities and scope, their business model has not changed. These firms work for a contracted fee, unlike the IOCs who seek to "own" the oil and gas. Because contractor business is project specific, their business results, level of investment, and employment ebbs and flows with the tides of crude oil prices and global industry activity. They are much more sensitive to the ups and downs of global oil activity than the IOCs themselves who have reserves on their books and can take a much longer term view of the industry. In response to the ups and downs, there has been a great deal of consolidation in the industry as contractors seek to diversify their portfolio of activities and increase their bargaining power with their clients.

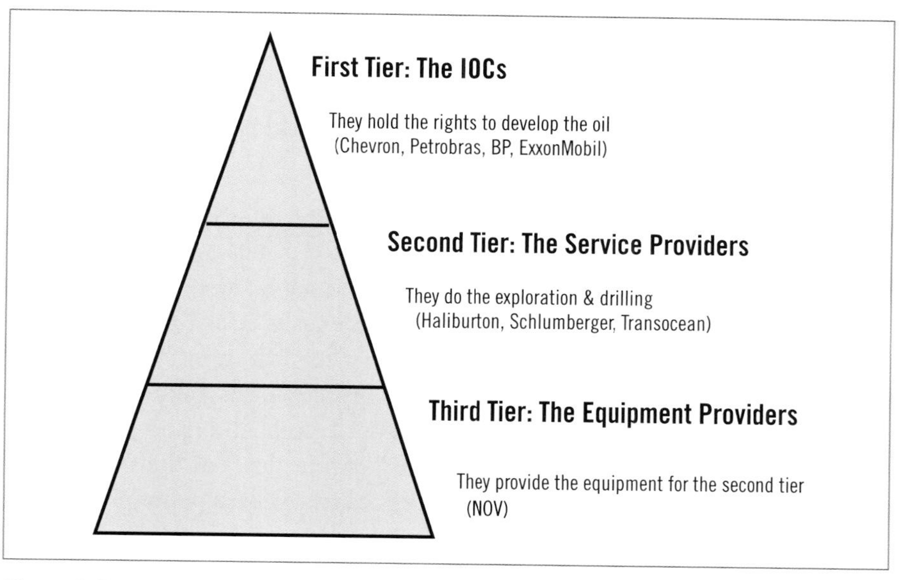

Figure 4–7. The tier structure of offshore services

Types of contractor relationships

Given the variety of possible projects, it is impossible to identify "one best way" to organize contractor supply chains. Two ends of a spectrum are: 1) the E&P firm does everything and uses no contractors, and 2) the E&P firm operates like a "virtual" firm and subcontracts every activity. Between these two ends are countless ownerships structures and contractor relationships. Using an offshore oil platform (with oil platform meaning the installations necessary to produce offshore oil) as an example, there are six possible contractor structures:[14]

1. Total integration—the IOC does the design, engineering, construction of the oil platform.

2. An independent engineering company designs the platform and delivers the service to the IOC; the IOC builds the platform.

3. An independent engineering company designs and builds the platform and sells it to the IOC.

4. An independent engineering company delivers services to a prime contractor that builds the platform and sells it to the IOC.

5. Engineers employed by the IOC provide services to the prime contractor that builds and delivers the platform to the IOC.

6. An independent engineering company provides drawings/solutions to the IOC. The IOC passes the drawings/solutions to a prime contractor who builds the platform and delivers it to the IOC.

These six simple examples grow much more complicated when the installation of the platform is considered. Does the IOC manage the installation or is a contractor used? If a contractor is used, is it the builder or a third party? Who manages platform maintenance? The more tasks added to the mix, the more complicated the supply chain becomes. Efficient management of contractor supply chains during development and into production is an area that firms use to establish a unique cost advantage relative to competitors. Problems with supply chain management can be very costly, and in times of tight contractor supply cycles, those firms with the best supply chains will be superior performers.

The 2010 BP *Deepwater Horizon* disaster in the Gulf of Mexico illustrates the use of contractors in the development process. As shown in figure 4–8, BP was the IOC with ownership to the oil-in-place, Transocean owned the *Deepwater Horizon* drilling rig and platform, and a variety of other contractors were involved in the various stages of drilling and cementing. One of the first organizational issues raised was whether all development parties were aligned in their strategic and business and technical objectives. One of BP's primary interests was completing and capping the well in a shorter period of time. Transocean, however, was compensated by the day of platform use. BP was paying Transocean a daily fee of $500,000 for the *Deepwater Horizon*. At the time of the explosion, the rig itself was already 43 days late in being moved to another BP site, a delay which had reportedly already cost BP $21 million. Although it may still be several years before a detailed analysis of the *Deepwater Horizon* disaster is available, it did raise a variety of questions regarding the structure and use of contractor relationships in edge-of-technology development projects.

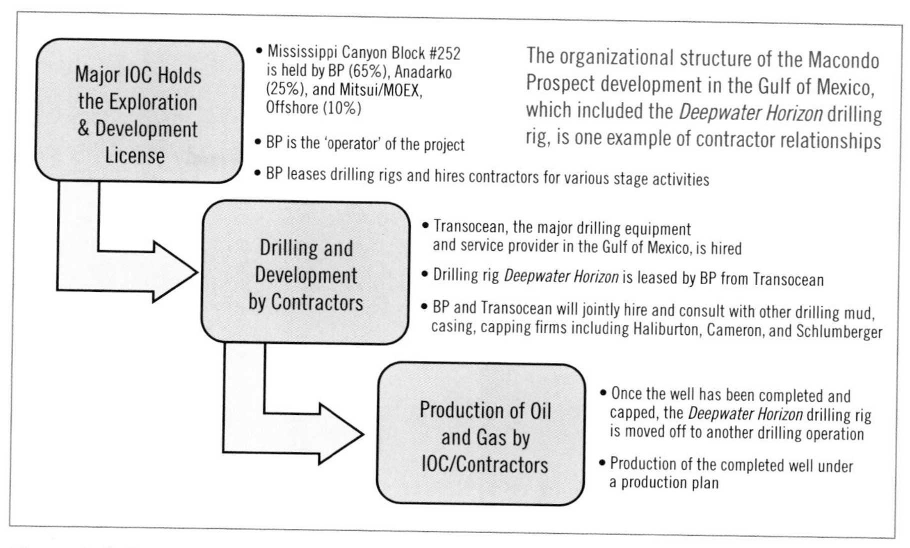

Figure 4–8. Organization of *Deepwater Horizon*

Successful contractor relationships

E&P companies increasingly prefer to deal with a limited number of suppliers and if possible, suppliers that can provide a turnkey solution. To get turnkey solutions there must be a collaborative relationship between the project operator, prime contractors, and subcontractors. Tasks must be closely integrated, increasing the scale of complexity for the project manager. As an example of how one firm approaches supplier relationships, see *Industry Insight: Becoming a Statoil Supplier*. The Insight lists the basic principles Statoil uses in selecting and working with suppliers.

Environmental impact

All oil and gas field developments will have environmental impacts, which means all projects must deal with environmental risk. Although environmental risk is inherent in the industry, one of the key objectives of the project development team is to mitigate the risks. There are many environmental risks that need to be considered. For example, a pipeline project must consider impacts to the soil and groundwater, air quality from burning debris and dust from vehicles, noise from the site, loss of vegetation and changes to the physical site, fish and animal habitats, and the risk of oil spills and accidents.[15]

Industry Insight: Becoming a Statoil Supplier

Basic Supplier Principles: Statoil's procurement process is based on competitive tendering and on the principles of transparency, nondiscrimination and equal treatment of tenderers.

- Statoil develops, integrates and implements procurement strategies to achieve the best possible agreements for the group. This is achieved through a category-based approach to goods and services, based on a coordinated control of demand, the global market situation and robust analyses in order to minimise risk in the execution phase.

- Suppliers must be prequalified in order to compete for tenders.

Integrity and Social responsibility: Statoil requires its suppliers to comply with the group's standards relating to group integrity and social responsibility, and to sign a declaration in confirmation of this.

- The declaration reflects Statoil's principles in areas such as ethics, corruption, human rights and labour standards. Our suppliers must be aware of the principles that form the basis for Statoil's activities.

- Statoil's strategy is to ensure social responsibility in the whole procurement process, from prequalification of tenderers to contract signing and finally the delivery of goods and services.

Procurement: Statoil has introduced the principle of division of roles/responsibilities between the line management and procurement function in order to meet the requirement for good control.

- The line management is responsible for planning demand, allocating technical and operational expertise and approving suppliers recommended by the relevant procurement entity.

- Only the procurement entities are authorised to sign binding agreements with suppliers on Statoil's behalf.

- The procurement entities are responsible for proactive coordination, planning and execution of procurements on behalf of the line management.

- Statoil has divided the products and services which the group has a verified requirement for into categories, and at present there is a total of 96 categories. Each category has one person who is responsible for having an overview of Statoil's total requirements in the relevant category.

- The categories are classified into 21 category groups, and the groups are put into one of five category areas (drilling and well, energy and retail, business support, operation and maintenance, projects). This category-based approach forms the basis for the organisation of the procurement function.

- There is a separate procurement entity for each category area.

Safety: Statoil cooperates closely with suppliers to ensure that our safety policy is clearly communicated and is understood and complied with.

Source: www.statoil.com.

Socioeconomic impact

Since most of the future major projects will be in emerging market countries, assessing the socioeconomic impact of projects has become an increasingly important part of the evaluation process. Upstream oil and gas projects are large and highly visible, which means E&P firms must assess and manage impacts very carefully. There are various impact categories: employment and economic development; demographic changes; infrastructure; individual, family, and community wellness; aboriginal use and culture; other economic sectors; and heritage and archaeological resources.

The Papua New Guinea (PNG) LNG project provides an interesting example of a socioeconomic impact study. The PNG project will develop gas fields in the interior of PNG. The gas will be transported by pipeline to an LNG facility 20 kilometers northwest of Port Moresby on the coast of the Gulf of Papua. There, the gas will be liquefied and shipped to gas markets overseas. The target is a first LNG cargo in late 2013 or early 2014. The project lifetime is expected to be around 30 years. Capital investment over the life of the project will be approximately $10 billion. The project is operated by Esso Highlands Limited, a subsidiary of ExxonMobil. The partners are Oil Search, Santos, Nippon Oil, Mineral Resource Development Company (MRDC), and Eda Oil.

PNG is a poor, undeveloped country with hundreds of tribes and languages. The economy is heavily based on agriculture, the majority of exports are mineral resources, and there is little manufacturing. A consulting firm, ACIL Tasman, conducted a study in 2008 to assess potential positive and negative impacts of the project on the economy of PNG. The firm reported impacts as follows:

- *Direct impacts:* Employment: 7,500 jobs in construction, about 20% PNG nationals; 850 FTE jobs in operational phase, about 75% to 90% PNG nationals; total payments of K3 billion to K4.5 billion, net of financing costs (kina is the currency of PNG) from company tax, personal income tax, royalties, and returns to equity.

- *Indirect impacts:* GDP doubles from K9 billion per year at present; government expenditure up 85%; household consumption up 105%; formal employment up 45%; and exchange rate appreciates about 11%.

- *Implications for business and commerce:* Increased demand for goods and services from the project; higher household disposable income and increased household consumption; competition for labor (especially skilled) pushes up labor cost; urban drift pressure on residential and commercial housing and rental costs; infrastructure of all kinds under pressure; strong kina—less competitive, lower yielding exports and higher costs of imported goods, in kina terms.

- *Potential winning economic sectors:* Quarrying, food processing, beverages and tobacco, transport, manufacturing, health, education, building and construction, commerce and finance, government administration and defense.

- *Potential losing economic sectors:* Agriculture, e.g., coffee, cocoa, palm oil, copra, especially smallholders; other mining and resources; metals and engineering.

- *Potential for classic "Dutch Disease" or "Resource Curse" situation:* Upward pressure on exchange rate and tendency for gas development to draw capital and labor away from traditional sectors of the economy (e.g., agriculture and forestry) has potential to reduce export competitiveness—classic "Dutch Disease" or "Resource Curse" situation.

Safety

A final business issue that must be at the forefront of project developers and the entire industry is safety. There are many different safety hazards in the oil and gas industry. For example, according to OSHA, hazards in the well drilling and servicing sector include being struck by various objects such as pipe, chain, whipping hose, tools, and debris dropped from elevated locations; being caught between spinning chains and pipe; and fires and explosion from well blowouts, releases of gas, poorly maintained electrical equipment, and aboveground detonation of perforating gun. Rigs can collapse, people can fall from elevated areas, and poisonous gases such hydrogen sulfide (H_2S) can be released. An additional hazard in the upstream in some areas is personal security, a reality of the industry's history and geographic scope.

All of the major E&P firms claim that safety of the workforce is their number one priority. Moreover, they would argue that safety is good for business because a safe operating environment creates a discipline that is transferable to other aspects of the project. Development teams will be staffed with safety professionals, and training of national workforces on safety will be an important area for project development, especially when projects are in countries that have had limited industrialization.

Problems in Project Development

Many things can go wrong with project development, leading to cost overruns and time delays, including: acceleration of projects to improve profitability, insufficient engineering and design, poor risk analysis, external stakeholder issues overlooked or discounted, overextended contractor supply chains, overly ambitious time targets, poor project leadership, and an overly rigid approach to project development. Companies involved in development project will build in contingency and response plans to deal with problems. As the BP *Deepwater Horizon* disaster shows, a failure to have an effective response plan in place can be devastating.

Many future oil and gas projects will be mega projects in challenging environments. Research has shown that many infrastructure mega projects have poor performance records in terms of economy, environment, and public support.[16] A recent study of mega oil sands projects in northern Alberta found that many projects suffered from huge cost overruns.[17] These projects require multibillion dollar investments and may require scheduling work for 7,000 to 8,000 people daily, with a personnel turnover as high as 300% annually. Some of the reasons for cost overruns and their underlying causes are:

1. Unrealistic or overly optimistic original cost estimate and schedules
    - Under-appreciation of project complexity and risks.

- Underestimating the cost to attract and maintain the labor (craft) workforce
- Underestimation of the labor productivity loss associated with working in cold weather climates and locations with severely shorter daylight hours in northern regions
- Shortages of skilled labor and lower-than-anticipated labor productivity due to mismanagement of the construction phase
- High labor turnover mainly due to the harsh working environment and competition between employers attracting labor

2. Because of the fast-tracking of mega projects, incomplete scope definition or inadequate front-end loading and poorly completed front-end deliverables.

3. Mismanagement of the construction phase
   - Inadequate field experience in home office engineering and procurement personnel
   - Later-than-anticipated engineering, vendor data, equipment, and material deliveries
   - Poor project controls
   - Inadequate plan of execution and poorly defined tasks and division of responsibility
   - Lack of knowledgeable leadership in the engineering, procurement, construction, and start-up of mega/major facilities
   - Inexperienced or poorly equipped project management personnel and supervisors
   - Lack of standardization and fit-for-purpose
   - Poor communication, teamwork, and alignment between the players, leading to adversarial relationships and protracted disputes
   - Poor site organization and layout leading to excessive time wastage and productivity loss during construction
   - Joint venture (JV) of project partners, contractors, and engineering firms that are not aligned

Project Development Case Example: Azurite Field in Congo

To conclude the chapter we provide another case example—the Azurite field in Congo, a project developed by Murphy Oil. In 2003, Murphy acquired an 85% working interest in two blocks—Mer Profunde Sud (MPS) and Mer Profunde Nord (MPN)—offshore of the Republic of Congo. The blocks, located in the Lower Congo Basin, were in water depths ranging from 450 to 6,900 feet and totaled approximately 1.8 million acres. At the time, the acreage had been only lightly explored by former operators. In early 2005, Murphy's first wildcat well discovered oil in the Azurite field. In 2005, the company successfully appraised the discovery and tested an appraisal well at 8,000 barrels of oil per day. Murphy drilled four unsuccessful exploratory wells on other parts of the MPS block in 2005. A third successful well in early 2006 further appraised the Azurite area. The company's board of directors approved the development of the Azurite field in late 2006. At the end of 2006, Murphy had $40,635,000 of exploratory well costs capitalized for the Azurite field.

Murphy proposed an ambitious development plan that would yield first oil by 2009. The initial plan called for a subsea infrastructure to link to an FPSO. Murphy submitted a bid package in June 2006 to vessel providers for the conversion and delivery of an FPSO. Murphy then changed the plan and decided to go with a new type of production system called an FDPSO (floating drilling production storage and offloading). The FDPSO concept had existed since the 1970s but had never been built. An FDPSO is in many ways its own vertically integrated value chain, capable of drilling the well(s), producing the oil, storing the oil, and finally offloading the oil to supertankers for transport. The Azurite FDPSO (figure 4–9) is equipped with a modular drilling package that can be removed and reused elsewhere when the drilling work is completed. It has a storage capacity of 1.4 million barrels of oil and a process capacity of 60,000 barrels of fluid and 40,000 b/d.

Figure 4–9. The world's first floating drilling production storage and offloading ship (FDPSO)

The decision to go with an innovative FDPSO system was motivated by the limited availability and high cost of mobile offshore drilling units. Murphy's field development manager for Azurite explained:[18]

> *The FPSO option did not solve our drilling-management challenges, and at the time rig rates were extremely high. . . . If you add all other costs associated with securing a rig capable of performing subsea drilling and completions in 5,000 ft of water, you're faced with costs in the range of USD 750,000–800,000 per day. We decided that from a cash-flow scenario, the FDPSO concept would allow us to avoid investing a great deal of money up front to drill wells and then wait on production until the FPSO is finished to obtain our returns. The FDPSO concept allows you to recoup your cash much more quickly, by drilling wells and bringing them onto production much sooner. . . . To drill 10 wells using a conventional rig at 50 days per well, you could easily be looking at costs of over USD 300 million. I anticipate that by project's end, we will have spent significantly less than that. By fitting a land rig onto a floating operation, our development costs may be half or even less than those from a separate drilling and production operation.*

The US$400 million contract for the conversion of the VLCC M/T *Europe* into an FDPSO was signed with Prosafe. The vessel underwent conversion between July 2007 and February 2009 at the Keppel Shipyard in Singapore. The operating deal with Prosafe was a seven-year contract with options for four additional two-year terms. The vessel was designed for a 15-year life. According to *E&P* magazine: "Murphy remained schedule-driven throughout the project and worked closely with the other companies involved to make sure the schedule was kept. One of the keys to success on the project was the focus on interfaces—interaction between operator and contractors and among the contractors working together on the project."[19] Murphy's project management team and subsea facilities team were based in Houston. The FDPSO team relocated to Singapore to oversee contractor engineering and hull conversion, outfitting, and integration of the drilling packages. A small facilities team relocated to Paris to work with Technip, the subsea installation contractor. Meetings took place on a quarterly basis in Houston between Murphy and all contractors,. There were also weekly interface meetings in Singapore between Prosafe and Murphy. Murphy had a dedicated, full-time group of 15 engineers working in Prosafe's offices.

Because the FDPSO was the first of its kind, no drilling rig had ever been installed in a floating production vessel. This created some unique construction challenges involving ship and drilling contractors who had never worked together. The existing ship deck was not strong enough to support the drilling rig, so a new support structure had to be built. According to the president of WJM Inc., the FDPSO design engineer, "It was a bit of trial and error" to get the rig designed and built. The rig was built in the PT Citra yard in Indonesia. Getting the vessel certified was also a challenge since there were no rules for FDPSOs and rules for drilling rigs and FPSOs were potentially in conflict. These unique engineering and construction issues resulted in a delay of four months from the original deployment plan.

In late 2007, Murphy sold down its interest in the MPS block (figure 4–10), including the Azurite field, from 85% to 50%. The 35% interest was sold to PA Resources, an independent E&P firm based in Sweden. The sales price was $110 million ($83.5 million with additional consideration of up to $26.5 million contingent upon achieving certain financial and operating goals for Azurite field development). Societe Nationale des Petroles du Congo holds the other 15%. In 2007, Murphy spent $129.3 million for development of the Azurite field, and $149.2 million for development of the Azurite field in the Republic of the Congo in 2008.

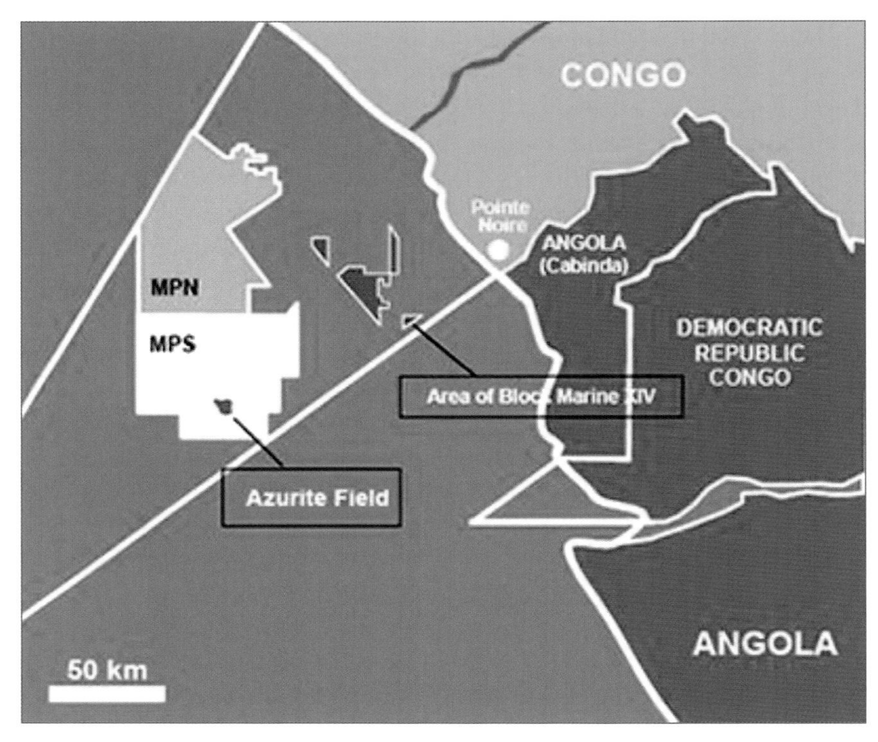

Figure 4–10. The MPS Block: Offshore Congo

Source: © Murphy Oil Corporation. All rights reserved.

In August 2009, production began from the Azurite field. The primary contractors for the project were:

1. Prosafe Production: conversion and operation of the FDPSO
2. Nabors Industries: drilling and rig fabrication
3. WJM: designed the rig decks and piping and oversaw their construction
4. Doris Engineering: facilities engineering
5. Technip: subsea equipment (flexible risers, production jumpers, umbilical, well jumpers)
6. FMC Technologies: 10 enhanced vertical deepwater trees
7. Framo: Multiphase pumps
8. DNV: classification

Development projects in the future

Four clear summary statements can be made about upstream development projects. First, flexibility in the development plan is necessary. In both cases the original development plan had to be reworked, which resulted in short delays in completion but a stronger final outcome. Second, technical challenges should be expected in upstream projects. These are some of the most sophisticated projects that exist, and it is unreasonable to expect all possible issues to be identified up front. Third, deep coordination and integration between contractors and project developers is critical for success. Finally, projects such as Azurite and Frade succeeded because project developers and firms were willing to take measured risks and think creatively.

As projects get larger and more complex, project risks increase. The consulting firm Booz Allen Hamilton identified several steps that E&P firms should take to improve the performance of development projects and accelerate project cycle time:[20] 1) Adopt more centralized and global management of their projects. Centralized management would help ensure a central depository of knowledge and standardized contractor management; 2) develop more sophisticated political risk management skills for doing work in "frontier" nations; 3) leverage design similarities across projects (i.e., "design one, build many"); and 4) create more integrated long-term contractor relationships to foster better advance planning, more effective use of resources, greater knowledge sharing, and more equitable sharing of risks and rewards.

Summary Points

- Project development is a unique part of the industry value chain because of its transitional nature. Unlike production, refining, and other activities, a project management team is created specifically to develop projects and then disband once the project is executed and handed over to operators.

- Cost overruns are a fact of life in project development; reducing the incidence and size of overruns is a huge opportunity for the industry.

- As mega projects get larger and more complex, project management skills will be a factor that distinguishes the most successful E&P firms from their rivals.

- With most mega projects located in emerging market countries, environmental and socioeconomic impacts of development projects will be at the forefront of many projects in the future.

- Competitive advantage in the development area is derived from a mix of different skills: technical, logistics, project leadership, financial, risk management, and partnering.

Notes

1. Mathew G. McKenna, Herve Wilczynski, and David VanderSchee, "Capital Project Execution in the Oil and Gas Industry: Increased Challenges, Increased Opportunities," Booz Allen Hamilton, 2006, http://www.boozallen.com/media/file/Capital_Project_Execution.pdf.
2. Russell Gold, "Chevron Project Offers Glimpse of Future: More Work, Less Oil," *Wall Street Journal*, October 30, 2008.
3. Russell Gold, 2008.
4. Michael Lynch, "Frade, Heavy Crude Oil Field Will Require Artificial Lift," http://www.rigzone.com, accessed, October 5, 2009.
5. http://www.sec.gov/divisions/corpfin/guidance/oilandgas-interp.htm.
6. G.W. Walkup and J.R. Ligon, "The Good, the Bad, and the Ugly of the Stage-Gate Project-Management Process in the Oil and Gas Industry," paper presented at the 2006 SPE Annual Technical Conference and Exhibition, San Antonio, Texas, September 24–27.
7. Jad Mouawad, "Big Oil Projects Put in Jeopardy by Fall in Prices," *New York Times*, December 15, 2008.
8. "Petroleum Institute to Study Oil Fields in East Texas to End Overproduction," *New York Times*, May 10, 1931.
9. Steven N. Wiggins and Gary D. Libecap, "Oil Field Unitization: Contractual Failure in the Presence of Imperfect Information," *American Economic Review*, Vol. 75, No. 3, June 1985, pp. 368–385.
10. Jacqueline Weaver and David F. Asmus, "Unitizing Oil and Gas Fields Around the World: A Comparative Analysis of National Laws and Private Contracts," *Houston Journal of International Law*, September 22, 2005.
11. Anthony Heiser, "East Timor and the Joint Petroleum Development Area," *MLAANZ Journal*, Vol. 17, 2003.

12. For example, see Robert T. Moran and William E. Youngdahl, *Leading Global Projects: For Professional and Accidental Project Leaders*, Burlington, MA: Butterworth-Heinemann, 2008.

13. http://www.ftc.gov/ftc/oilgas/gascolumn.html.

14. Ola Kvaløy, "Economic Organization of Specific Assets in the Offshore Industry," in Jerome Davis (ed.), *The Changing World of Oil: An Analysis of Corporate Change and Adaptation*, Burlington, VT: Ashgate Publishing, 2006, pp. 51–66.

15. Tim Van Hinte, Thomas I. Gunton, and J. C. Day, "Evaluation of the Assessment Process for Major Projects: A Case Study of Oil and Gas Pipelines in Canada," *Impact Assessment and Project Appraisal*, Vol. 25, No. 2, June 2007, pp. 123–137.

16. B. Flyvbjerg, N. Bruzelius, and W. Rothengatter, *Megaprojects and Risk: An Anatomy of Ambition*. Cambridge, UK: Cambridge University Press, 2003.

17. George Jergeas, "Analysis of the Front-End Loading of Alberta Mega Oil Sands Projects," *Project Management Journal*, Vol. 39, No. 4, pp. 95–104.

18. Ted Moon, "Congo FDPSO: An Industry First," *JPT*, January, 2010.

19. Judy Maksoud, "First Drilling FPSO Goes to Work Offshore Africa," *E&P*, May 1, 2009.

20. Mathew G. McKenna, Herve Wilczynski, and David VanderSchee, "Capital Project Execution in the Oil and Gas Industry: Increased Challenges, Increased Opportunities," Booz Allen Hamilton, http://www.boozallen.com/media/file/Capital_Project_Execution.pdf, 2006.

Chapter 5
PRODUCTION OF OIL AND GAS

No matter what the production was or what the hopes of the producer are, death and destruction surround that field, and it will only be a year or two at the most, when it will be numbered with last year's snows and forgotten.

—From a 1905 report to Standard Oil from Colonel Carter of the Carter Oil Company about the Kern River field; 105 years later (2010) the field's cumulative production has passed 2 billion barrels.

Competition is the keen cutting edge of business, always shaving away at costs.

—Henry Ford

After the development of an oil and gas field is complete, the extraction and recovery of oil and gas can begin. Because an oil and gas field can remain productive for 40 years or longer, the production activity must be managed very differently than the exploration and development areas. In many ways, the production of oil and gas is similar to manufacturing in that it is capital intensive and cost oriented, and lean thinking is the key to success.

The production of oil and gas has been the largest source of income for the oil majors over the past decade. Because profit margins in the production sector are directly tied to the price of crude oil, during times of rising crude prices, production income will typically be the highest in the energy value chain. As price takers, however, oil and gas producers can have limited impact on the price of crude. As a result, a focus on costs, efficiency, and productivity occupy much of the attention of production managers. Coupled with the cost focus is the use of innovative technology to lower the cost of production and to lengthen the life of producing fields.

In the future, most new projects will be in locations where producing oil and gas involves much more than just drilling wells and recovering the oil and gas. Consider the Shaybah Oil Field in Saudi Arabia, one of the largest oil fields in the world. Shaybah has over 100 wells, 600 km of pipelines, a residential camp for 700 people, an airfield and terminal, a fire station, recreation areas, maintenance and support workshops, power stations for generation and distribution, and a crude upgrading plant. Managing the production from an oil field like Shaybah involves tasks ranging from feeding and housing the workforce to generating electricity to lifting oil. This chapter examines the managerial challenges in the production area with a particular emphasis on costs and cost advantage as a strategic objective.

Defining Costs in the Upstream

Before we delve too deeply into the business of oil and gas production, we need to spend a moment detailing how the industry defines and accounts for various upstream costs. In the upstream, costs are divided between preproduction and production costs. Figure 5–1 provides an overview of this sometimes confusing terminology.

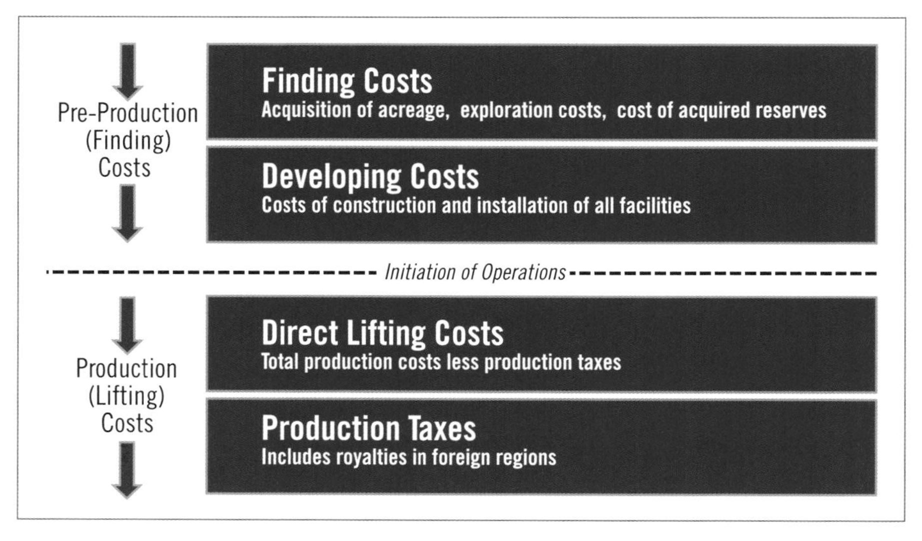

Figure 5–1. Upstream costs. Oil and gas costs in the upstream are divided between Finding the oil and gas and Lifting it. *Finding* is an industry term that includes both exploration and development. *Lifting* is the industry term for all operational production costs.
Source: Constructed by authors.

Preproduction costs

The Energy Information Administration (EIA) defines *preproduction costs* as "Costs of prospecting for, acquiring, exploring, and developing mineral reserves, incurred prior to the point when production of commercially recoverable quantities of minerals commences." This definition combines the traditional categories of exploration and development plus the costs of acquired or purchased reserves. Although exploration and access were covered in chapter 3, exploration costs are considered here because when a project is developed and put into production, the sum of preproduction and production costs is the real cost of producing oil or gas that must be reflected in the income statement.

Preproduction costs can be broken down into two main categories:

- *Finding costs* include the costs of finding the oil added to reserves. Finding costs include the costs of exploration and appraisal programs, which includes drilling, purchase of equipment, leasing of equipment, seismic analysis, and all labor costs in exploration. Finding costs also includes the cost of acquiring or purchasing reserves.

- *Development costs* are the costs of construction and installation of all facilities needed to produce and transport oil and natural gas. These costs are incurred after the oil and gas has been 'found' and a decision has been made by the company to proceed towards the commercial production of that hydrocarbon.

Somewhat confusingly, the terms *finding costs* and *preproduction costs* are often used synonymously. In addition, the term *discovery costs* is sometimes used instead of *exploration costs.* An alternative view of finding and development costs categorizes the costs incurred in preproduction as falling into four different categories or buckets: 1) acquisition of acreage; 2) exploration of that acreage; 3) purchase of existing reserves; and 4) development of successful exploration activities.[1]

Exploration costs and development costs are different in how they are planned for and budgeted. E&P firms have annual exploration budgets set at a specific amount for the year. In the words of one industry analyst, "Some years they find and others they don't." The exploration budget for an E&P firm is similar to the R&D budget for a pharmaceutical firm. For both types of firms, a fixed amount of capital is spent in a given year with the hope that commercial oil and gas fields or drugs will be the outcome.

Development costs are only incurred after exploration has been successful and the organization has committed to bringing the oil or gas to commercial production. Development costs are project-dependent and will vary significantly per project depending on field location, project size, technical difficulties, etc. The large E&P firms have a rolling portfolio of on-going project commitments and expenditure plans. Each project in development will have an annual budget, and the sum of the various project budgets will represent the development budget for the year. Although the variance between the E&P firms' development budgets and actual spending in a year is usually not large (at least for the majors), the same cannot be said for specific projects. In most projects, the actual amount spent on development will be different than budgeted—often more.

Production costs (lifting)

> *Those costs incurred to operate and maintain an enterprise's wells and related equipment and facilities, including depreciation and applicable operating costs of support equipment and facilities (paragraph 26) and other costs of operating and maintaining those facilities.*
>
> —Financial Accounting Standards Board (FASB), No. 19, paragraph 24.

Production costs, also called *lifting costs*, are the operating costs incurred after the oil and gas reserves have been found or acquired and developed for production. In other words, these are the costs incurred after preproduction.[2] According to the EIA, production costs are incurred to operate and maintain wells and related equipment and facilities, including depreciation and applicable operating costs of support equipment and facilities and other costs of operating and maintaining those wells and related equipment and facilities. The following are examples of production costs:

- Labor to operate the wells and related equipment and facilities
- Repair and maintenance
- Materials, supplies and fuel consumed, and services utilized in operating the wells and related equipment and facilities
- Property taxes and insurance applicable to wells and related equipment and facilities
- Severance taxes
- Well workovers
- Operating fluid injections and improved recovery programs
- Operating gas processing plants
- Ad valorem taxes

Depreciation, depletion, and amortization (DD&A) of capitalized acquisition, exploration, and development costs are not usually considered production costs. However, as mentioned above, DD&A becomes part of the total cost of oil and gas produced along with production (lifting) costs.

Accounting for upstream costs

E&P companies incur many costs before the actual production of oil and gas. The methods of accounting for those costs are extremely complex and beyond the scope of this book. In brief, companies have two main accounting options for exploration costs: the full-cost approach and the successful-efforts approach.[3] Under the *full-cost approach*, companies capitalize all exploration costs, including the cost of dry holes. With *successful-efforts*, only those costs associated with successful exploration efforts (i.e., that result in proved reserves) are capitalized. These costs include license and leasehold property acquisition costs and the cost of drilling exploration wells. Exploration costs that do not lead to new reserves are expensed. The successful efforts approach is the method used by most major oil and gas companies.

When a project is sanctioned for development, the capitalized exploration costs and the project developments costs are transferred to production assets and for accounting purposes are termed *property*, *plant* and *equipment*. The capitalized exploration and development costs are amortized on the basis of oil-equivalent barrels that are produced in a specific period as a percentage of the estimated proved reserves.

ED&P and the cost of business

In the end, it is also important to understand the organizational behavior of an E&P firm and the various factions and subcultures within the company itself. Consider the expenditures allocated to the three different activities—exploration, development, and production—and how they are viewed by leadership and management in the typical E&P firm.

- There is no 'optimal amount' of exploration expenditure. E&P firms have always struggled to determine whether they are spending/investing enough. Firms that do not find reserves to replace current production conclude they are not spending enough. Firms that do replace current production, consistently, often conclude they might be able to continue to do so with slightly less spending per year—at least until the results fall below current production levels. Then the cycle repeats.

- Once a development decision has been made, the expenditures required to bring a field to production are fairly well defined. Although there may be some ability to spend faster or slower, the spending necessary to bring a proved discovery to commercial production is known, although factors such as availability of materials and labor, competition for the services of contractors, and design changes will impact actual versus budgeted costs. Since development involves bringing proved reserves to commercial fruition, the activity is often considered critical within the organization. As a result, many organizations do not believe there is any real reason to constrain or limit spending in development.

- Production, the subject of this chapter, is in many ways a manufacturing process. That means oil and gas production, like shoe or personal computer manufacturing, will focus on efficiency, lowest per unit costs, high utilization rates for facilities, and low-cost competitiveness.

As a result, inside E&P companies there is often tension between the two sides—the exploration and development side that spends huge amounts of money to develop projects, and the production side, which is expected to be the low-cost operator, even though production had little control over project design.

Performance and Competitive Advantage

One of the major challenges in the upstream oil business is the inability to control the sales price of the core products (the natural gas industry operates somewhat differently from crude, as discussed in chapter 8). Since crude oil trades as a commodity, oil producers are price takers and crude oil as a product cannot be differentiated (we will discuss in depth the differing qualities of crude oil and the "Is crude oil a commodity?" debate in chapter 10). Thus, achieving a lower cost of production than competitors is the primary basis for competitive advantage. This section discusses the main elements of cost advantage in upstream production.

The primary driver of crude oil production cost is the nature and geography of the reservoir itself. All of the technology, efficiencies of processes, and organizational excellence in the world cannot make up for the massive cost advantages of $2/bbl or $3/bbl production costs enjoyed by producers operating in regions like Saudi Arabia or the United Arab Emirates. As illustrated by the analysis presented in figure 5–2, the major producing regions of the world (21 listed here) vary from as little as $1/bbl to $28/bbl[4] in the cash costs of production. This figure provides a clear illustration of how certain major producers in the world, for example Russia, may suffer massive pressures to shutin production if the dollar barrel price falls below $15 for an extended period of time.

Figure 5–2 also introduces yet another measure of cost—*cash cost*. Cash cost is actual cash expenses related to lifting, and does not include noncash expenses like depreciation and amortization of costs incurred in exploration and development. (Students of economics will recognize cash costs as the financial equivalent of variable costs associated with a specific oil field. These explicit costs that must be covered by oil revenues in order for operations to be sustainable over the long term.)

For perspective, keep in mind that cost pressures exist throughout the industry value chain. The discussion in this chapter is applicable wherever *low cost* is the basis for competitive advantage (which is the case for all commodity-based industries). Activities such as shipping, refining, and commodity chemicals are also sectors where competition is primarily cost-based. Besides crude oil, most refined products, petrol at the pump, and many basic chemicals have commodity-like characteristics.

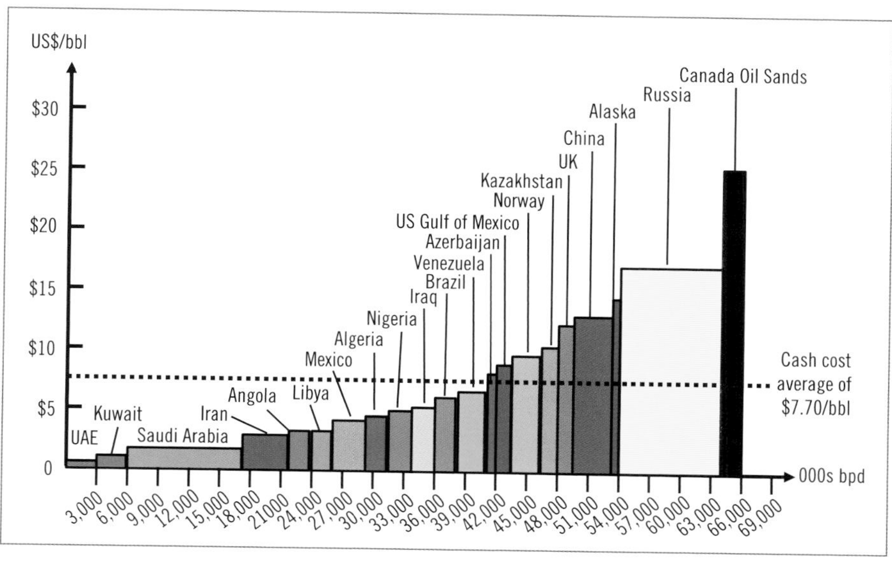

Figure 5–2. Production costs by producing region

Source: Constructed by authors using estimated cash costs per barrel from Deutsche Bank (2009) and region production levels for 2008 (BP Statistical Review). Cash costs = operating costs + royalties.

We have chosen to devote much of this chapter to cost management because first, the commodity nature of crude oil makes cost advantage the primary means for an oil producer to outperform a direct competitor. By direct competitor we mean firms producing in geographically similar areas, such as two E&P firms producing oil in offshore Angola, neutralizing the impact of cost differences highlighted in figure 5–2. Second, the production activity in the value chain is a long-term activity that could last decades for an oil field. As a result, continuous cost management and operational effectiveness are absolutely necessary, especially as oil production become a more challenging in a mature field. Third, decisions such as project size, location, partners, and major contractors are made during the exploration and development stages and before production starts. Once production starts, these decisions are "baked in," and project operators have limited scope for major change. As a result, their focus is on operating as efficiently and effectively as possible.

Fundamentals of business: Creating competitive advantage

The concept of competitive advantage is used to help explain the sustainability of business or firm performance. Firms that are able to outperform others in their industry are said to have a competitive advantage. Note that this concept applies to the business level and not the diversified firm level. Since an IOC has multiple businesses, we need to look at each business relative to the other competitive firms in that business. In other words, it is not logical to talk about firms like BP and Shell having a competitive advantage relative to each other. Competitive advantage gets created (or lost) at the business level, which in the case of an IOC could be the upstream production, fuels marketing, or lubricants businesses. Thus, BP could have a competitive advantage in fuels marketing and Shell could have an advantage in lubricants.

Figure 5–3, developed by BP, presents some upstream cost data for the six supermajors. The costs are equivalent to the cash costs shown previously in figure 5–2. The figure indicates that Total and BP have competitive cost advantages in oil and gas production at $6.00/bbl and $6.20/bbl, respectively, while ExxonMobil and Chevron are on the high end at roughly $9.80/bbl and $9.40/bbl, respectively. Note that this figure is based on barrel of oil equivalents, which means that oil and gas are combined into one unit cost. Also note that costs have risen significantly for all IOCs over the past decade.

In order to identify competitive advantage we start with the following questions: "What is a firm's relative position within its industry?" and "Is the firm's profitability above or below industry average?" If the firm has above-average performance based on a long-run view of the industry, the firm is said to have a sustainable competitive advantage. As an example, consider the US airline industry. For several decades, Southwest Airlines has outperformed all of its major competitors in the airline industry and has above-average performance based on various measures of financial profitability, such as return on sales and return on capital employed. Thus, we can say that Southwest Airlines has a competitive advantage relative to its US domestic competitors.

Drawing on the work of Michael Porter, competitive advantage is achieved when a company can offer a value proposition that customers will prefer and that competitors cannot easily match.[5] There are two basic strategic options for achieving an advantage: differentiation or low cost.[6] A differentiation strategy is based on the uniqueness of a firm's products in the eyes of the customer. The uniqueness will allow the firm to charge a premium price for its products. There are multiple sources of uniqueness, including technology, service, product features, and speed of delivery and service. An example of differentiation from

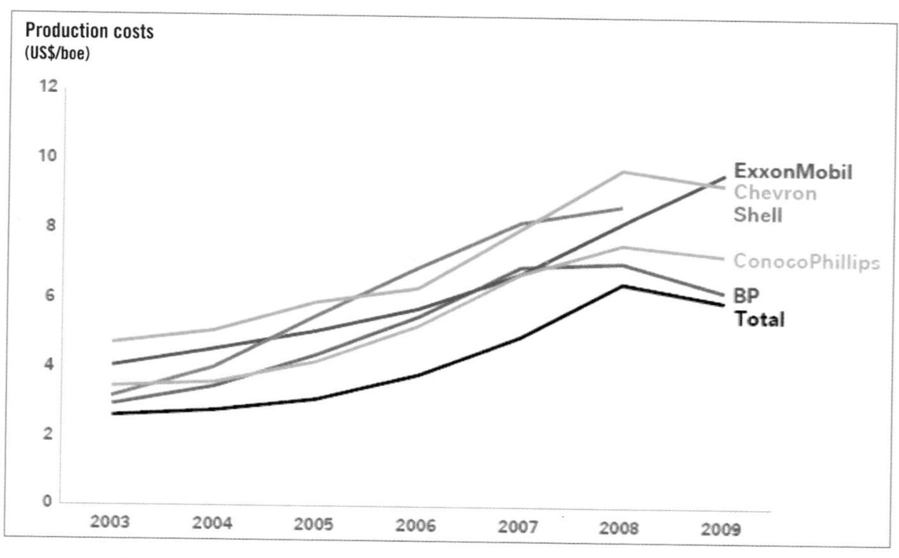

Figure 5–3. BP's estimates of IOC production costs

Source: BP Strategy Presentation, London, March 2, 2010, p. 41. Copyright © 2010 BP p.l.c. Reprinted with permission. Production costs and production from reserves per annual Supplemental Oil and Gas disclosure in 10-K/20-F. Consolidated subsidiaries only. Data prior to 2009 excludes mined oil sands. Total's 2009 production costs estimated based on disclosure from 4Q09 results presentation.

the car industry is Mercedes-Benz. The Mercedes brand is one of the best in the industry, and the firm is able to charge premium prices for its cars. Another example from the airline industry is Singapore Airlines. Because of its perceived higher quality service, Singapore Airlines is able to differentiate itself from its long-haul international competitors and charge premium prices.

Low cost is at the other end of the advantage spectrum. In some industries it is not possible to create a differentiation advantage, which means cost becomes the basis for competition. The extreme case for cost-based competition is a commodity industry where the firm must be a price taker and where customers are indifferent as to the products produced by one firm or another. Although there are chemical differences between crude oils, the product is traded on commodities markets and, therefore, oil producers must compete on cost. In a commodity market, the lowest cost producer will earn the highest profit. In some industries it is possible to have some firms with a differentiation advantage and others with a cost advantage. In the car industry, the luxury car producers such as Mercedes, BMW, Lexus, and Infiniti are seeking a differentiation advantage. They cannot compete on cost with firms like Hyundai or Honda and doing so would run the risk of tarnishing their premier image. Firms like Tata Motors in India, the producer of the Nano, the world's lowest price car, are clearly trying to establish a low-cost advantage (which can be translated into a low price for customers). To succeed and prosper at the low end of the automotive scale, low costs are essential, and per-unit margins will be slim relative to those earned by Mercedes-Benz. However, given the potential market size at the low end, total firm margins could be very large.

Managers are sometimes under the mistaken impression that having a differentiation advantage is better than trying to compete on cost. It is in fact quite possible to be very profitable in commodity-based industries, especially if a firm has the lowest cost. Likewise, it is possible to be spectacularly unsuccessful in industries where differentiation matters, as Ford's experience in luxury cars in the past few decades demonstrates. In commodity-based industries, firms that can achieve and sustain the lowest costs can be very successful. Those firms that have neither the lowest cost nor the ability to differentiate are termed *stuck-in-the-middle*, an undesirable state for any business. The reality in the oil business is that product differentiation is not, and never will be, possible.

Finally, if a firm has a true competitive advantage it will be difficult for competitors to imitate. If a competitor can easily imitate your firm's strategic choices and replicate your cost position or basis for differentiation, by definition you do not have a competitive advantage. We now turn our attention to a variety of factors that drive costs in the production of oil.

Oil Economics

The profitability of an individual well is core to an oil and gas firm. A few simple exercises highlighting the economic and business principles of the individual well are helpful in understanding many of the general business principles at work.

Assume a well has been drilled and brought to production. The *operating profitability* of an individual well of an operating lease can be simply stated as:

$$Profit = Revenue - Cost$$

Revenue is the price of the crude oil per barrel, P_{oil}, times the barrels produced for the operating period, *BBLs*, say one month. The *lease operating expense* (LOE) is composed of labor, equipment, ownership fees, power, and material expenses associated with an operating lease. LOE is stated here as a fixed currency amount, say $2,500 per month, and is not calculated directly on a per barrel produced basis.

$$Profit = (P_{oil} \times BBLs) - LOE$$

Break-even

This is a very useful little equation. To calculate the *break-even* production level for an individual well, where profit equals zero (a relevant question for small independent operators during every downturn in crude oil markets):[7]

$$Break\text{-}even = (P_{oil} \times BBLs) - LOE = 0$$

Rearranging the equation to solve for the barrels that need to be produced to break-even, *BBLs*:

$$BBLs_{be} = \frac{LOE}{P_{oil}}$$

Assuming the price of oil is $50/bbl, and the individual LOE is $2,500 per month:

$$BBLs_{be} = \frac{LOE}{P_{oil}} = \frac{\$2{,}500/month}{\$50/barrel} = 50 \text{ } barrels \text{ } per \text{ } month$$

The individual well would need to produce at least 50 barrels each month in order to cover the operating lease expenses, which is the break-even. Obviously, as the price of crude oil rises (or falls), the number of barrels for production break-even would decrease (or increase).

Multiple products

From this point, the equation can be expanded and built-out to include more realities of actual oil production. Frequently, crude oil production yields natural gas and condensates. Both have their own values and can be included in the revenue results on a per barrel basis. This is represented by P_{gas} and P_{con} as well as the relative quantities of natural gas, α, and condensate, β, produced per barrel.

$$Profit = [(P_{oil} + \alpha P_{gas} + \beta P_{con}) \times BBLs] - LOE$$

Although this is a simplistic representation of well revenue, the obvious implication is that if the well is producing associated gas and/or condensate with each barrel of crude, the value and profitability is greater. If the equation is rearranged to solve for break-even production levels, assuming the same lease operating expenses, the break-even production volumes—*BBLs*—should be lower.

Working interest

The next addition could be ownership interests. The *working interest* (WI) of an individual investor can be represented by the investor's respective share of operating expenses and the *net revenue interest* (NRI), the proportion of revenues from production after all burdens such as royalties have been deducted from working interest. If we return to the break-even analysis for an individual well producing only oil, the equation now captures more specifically the cash flows accruing to individual parties.

$$BBLs_{be} = \frac{WI \times LOE}{NRI \times P_{oil}}$$

Taxes

The last and possibly least pleasant element is taxes. Taxes enter the equation in the revenue component, reducing the available net cash flows to the investors and raising the break-even barrels production requirement.

$$BBLs_{be} = \frac{WI \times LOE}{NRI \times P_{oil} \times (1-T)}$$

The critical issue for many independent operators is the economic limit of their individual wells. Depending on the lease operating expense and the price of crude oil, the well's break-even productive capability will vary. Figure 5–4 provides a simple analysis of how lower cost operating leases have an appreciable financial impact during low market price environments—as one would expect.

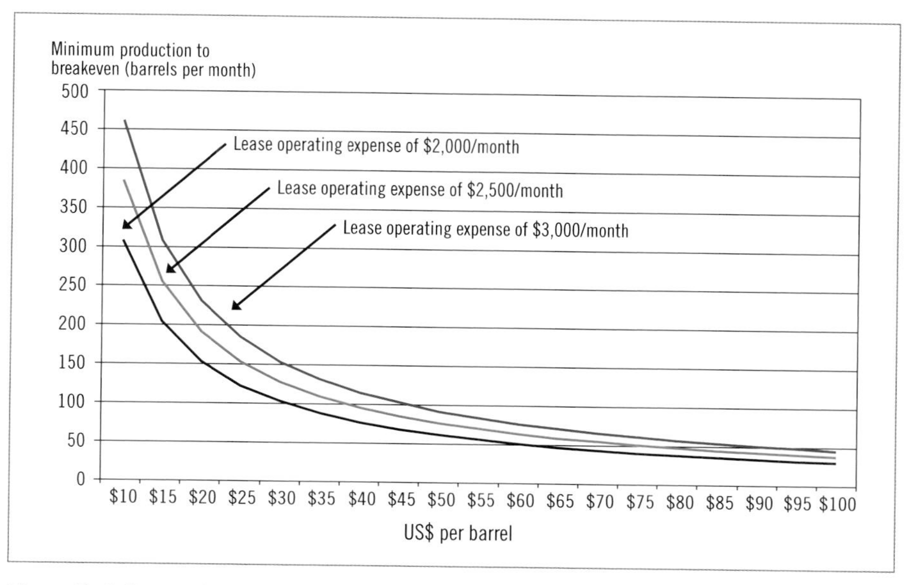

Figure 5–4. Economic limit: break-even production

Source: Author calculations. Assumes no gas content, 35% taxes, and 100% working interest.

Valuing production on a per barrel basis

When valuing producing oil fields, the industry uses a variety of calculations. One calculation is the US$/bbl value for a producing field, given production rates and patterns over time. This valuation is based on the present value (PV) of a field producing 1 b/d into the future. A simplified version of this calculation uses the following variables: profit per barrel, years of production, and discount rate. Assuming a profit of $10/bbl, a 15% discount rate, and 20 years of production, 1 barrel of oil production is worth about $23,000.

Companies actually doing a valuation per barrel for a producing field will have to incorporate many variables to do a proper analysis, including where the reserves are located (country, offshore or onshore, political regime and risk, etc.), the quality of oil (for example light sweet versus heavy oil), years of remaining production, additional capital investment required, operating expenses and lifting costs (discussed in further detail later in the chapter), and expectations of crude prices. In the not-too-distant past the approximate values used for high-quality producing fields was somewhere between $35,000 and $55,000 per barrel. After the run-up in crude prices in 2007 and 2008, $55,000/bbl is probably on the low end of the scale, while the high end may go up to $100,000/bbl.

Cost Management

The major upstream oil companies have learned from past experience that they need to focus on costs. Invariably, when boom cycles occur, less-experienced oil firms get carried away with rising prices and end up paying too much for acquisitions of new acreage, firms, partnership stakes, etc. When oil prices fall, the oil fields in the high-cost acreage will not be profitable, and the firms that paid the high prices will be forced to divest properties. The buyers are often the IOCs who sat out the bidding wars during the boom times. The IOCs take a very conservative approach when it comes to evaluating projects. Their cash flow projections are based on bottom-of-the cycle prices, and they focus very heavily on ensuring that costs are carefully controlled.

This is not to suggest that all costs are 100% controllable. Costs rise during boom times because more projects get started and there are constraints on drilling rigs, materials, technical personnel, etc. Figure 5–5 shows worldwide upstream costs over the past few decades, according to the EIA. The figure shows total upstream costs and a breakdown between finding costs (exploration and development) and lifting costs (the production costs associated with bringing a barrel of oil to the surface). Costs fell through the 1980s, remained flat from about 1993 to 2000, and then rose steadily as global demand jumped significantly in the 2002 to 2008 period. Although lifting costs have started trending upwards from a low of roughly $5/bbl to nearly $9/bbl, finding costs have jumped dramatically in the past 5 years, coming to rest at a near 30-year high of $18.50/bbl.

Strategic cost management

When a development project is completed, it is handed over to the operators. While the exact amount of recoverable reserves in any oil field is impossible to determine, once a field enters its productive life, many activities will be performed. Table 5–1 shows a diverse range of technical activities that must be performed to ensure that production continues on schedule. Each activity has its own set of costs to be analyzed and managed.

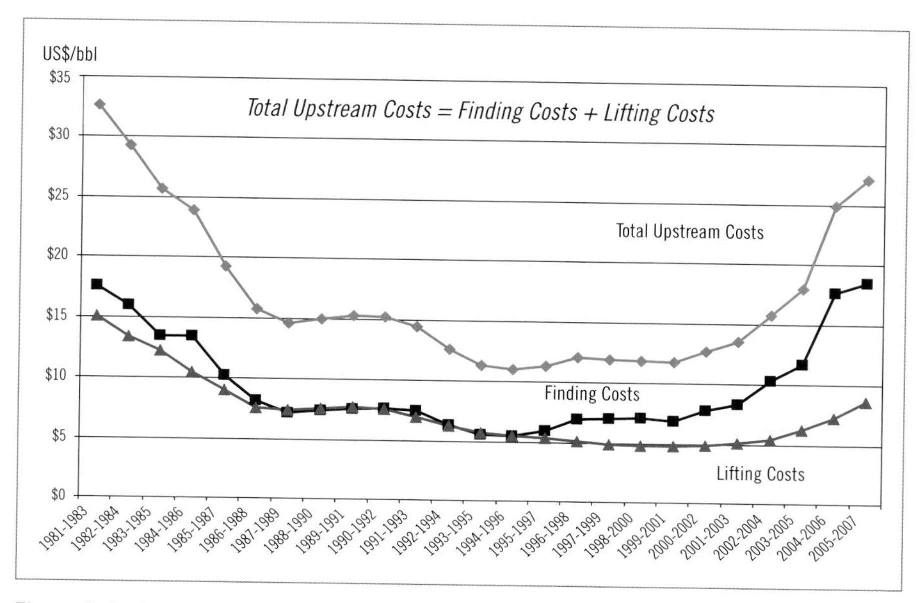

Figure 5–5. Worldwide crude oil production costs
Source: EIA. Constant 2007 dollars, US$ per barrel of oil equivalent.

Table 5–1 Technical activities during the production phase

Preparing Wells for Production	Extracting Oil & Gas from the Reservoir	Improving Flow and Stimulating Production
Completion and initial production	Primary production	Field abandonment
Completion and production equipment	Well servicing	Enhanced recovery and production termination
Sand control	Coiled tubing and nitrogen services	Pressure pumping
Pressure pumping	Production chemicals	Artificial lift
Fracturing and acidizing	Tubular inspection and coating	Production chemicals
Wellhead and valves	Compression services and equipment	Well servicing
Cementing equipment and services	Production platform services	Downhole monitoring
Casing services	Artificial lift	Reservoir stimulation
Casing and tubing	Field process equipment	Specialty chemicals
Tubular inspection	Downhole monitoring	4-D seismic
Marine construction	Reservoir performance monitoring	Reentry drilling
Platform fabrication	Corrosion monitoring and control	Plugging and abandonment
Subsea services and equipment		Site clearance and equipment removal
Well testing		Environmental remediation

Source: http://www.simmonsco-intl.com/energyindustry_universe2.asp.

There are several steps in strategically managing costs during the production phase. The first step is to clearly identify the specific production activities and their costs. Once the costs are identified, the specific cost drivers must be determined. A *cost driver* is any factor or activity that causes a cost to be incurred. Cost drivers can be classified in two categories: 1) structural drivers tied to strategic choices about the project's underlying economic structure, such as scale and complexity of operations, location, technology, and partnership structure; and 2) executional cost drivers dependent on the execution of the project, such as oil field layout, workforce involvement and productivity, work flow, operations integrity, safety, and managerial skill.[8] Each of the drivers in turn influences specific costs. For example, location of the oil field as a driver will influence logistics and transportation costs, the cost of the workforce, security, and the type of offices and housing required. The technology of the project will influence contractor relationships, purchases of materials such as steel and cement, and the ongoing research and technical support needed to keep the project operating.

Once the costs and their drivers are identified, the next step is to understand the cost dynamics and how they are impacted by scale, economies of scope, vertical scope, and learning.[9]

Economies of scale. The economy of scale concept is often described as "bigger is better because it will allow fixed costs to be spread over a greater volume." In reality, the concept is more complex. Greater scale may allow more specialized equipment to be purchased, reducing the labor requirements. Costs of designing, constructing, and administering an oil field should not increase proportionately with scale. After some point, scale economies become negligible and diseconomies of scale occur. In the oil and gas business, this could occur in an oil field of such enormous size and complexity that costs per barrel increase as the project gets bigger. In a world where mega projects seem to be increasing in number, oil companies must be particularly vigilant against a blanket assumption that a bigger project will mean lower production costs per barrel. It may well be that some of the mega projects currently under development end up with production costs that are higher than smaller projects.

Another area where scale can lead to cost efficiencies is *scale in basin,* which allows operators to share fixed costs and build an efficient supply chain around multiple assets or wells. For example, Paradox Basin Project in Utah includes two contiguous oil and gas prospect areas totaling 1,525,000 acres. Lynden Energy, an independent exploration and development company, has built several pipelines in the area and promotes them as a strategic advantage for potential investors.

Economies of scope. Economies of scope occur when there are benefits in having multiple production outputs. In manufacturing, including refining and petrochemicals, economies of scope are associated with having multiple products from the same site and being able to take advantage of distribution and other types of efficiencies. For example, a petrochemical plant might produce multiple products that can be sold into common sales channels. In upstream production, the most important scope issue is geographic scope and opportunities for reduced risk levels when there are multiple oil fields and in various locations. The largest E&P firms operate with a diversified portfolio of oil fields. Operating with a single oil field is obviously risky. For example, if a firm operates just one oil field in Venezuela and that field gets nationalized, that is the end of the firm. An extreme example, perhaps, but the concept of geographic scope is very important in the oil industry. A well-managed geographic scope strategy provides the E&P firm with several benefits:

- Hedging against political risk
- Opportunities to spread costs across multiple projects
- Opportunities to move down the learning curve using the concepts of "design one, build many" and "operate one, operate many"
- Opportunities to share people and resources across projects
- Opportunities to learn and transfer knowledge between projects

Vertical scope. What activities should be done in house and what activities should be contracted out? Once a contractor option is available, the analysis of cost dynamics becomes more complex. In an oil field and especially one in a remote site, there are a huge number of activities that must be performed on a daily basis. These activities range from the food and housing needs of the field workers to the most technical aspects of oil recovery. To decide on activities to contract out, firms must address a series of questions. For a given activity, if the answer is yes to any of the following questions, using a contractor is not advisable:

- Do you have a competitive advantage in the performance of the activity?
- Is there valuable IP associated with the activity?
- If you use a contractor, is there a risk of IP leaking to a competitor?
- Will using a contractor create new risks, such as corruption issues and the Foreign Corrupt Practices Act (for US firms)?

- Do you have the people and resources to perform the activity yourself?
- Do you have experience in performing the activity?
- Will you lose a significant amount of project control if you use a contractor?
- If you use a contractor, will there be issues of quality and delivery of the services, parts, or equipment?
- Will contractor costs be higher than the cost of doing it yourself?

The final question on the cost of using a contractor is probably the question that is the most heavily scrutinized. Unfortunately, focusing only on costs and ignoring the other questions can lead to some serious problems down the road. One of the questions raised after the BP *Horizon* disaster involved BP's use of contractors. Specifically, was BP too reliant on contractors and their safety and operational processes? The reality is that for a complex oil and gas project, no firm can perform every activity because no firm has the resources or the know-how to do so. As discussed in chapter 4, E&P firms must rely on many contractors to develop their projects. The most successful contracting strategies will be those that ensure project safety and integrity are maintained and the best skills are utilized on the project. Later in this chapter we discuss some of the issues associated with managing contractors.

Learning. Average costs per unit tend to fall as a firm gains experience in production. This phenomenon has resulted in the learning curve and experience curve concepts. The "design one, build many" concept is actively used by project developers in the upstream and is essentially an example of learning-by-doing. After a project is executed successfully, the next similar project is executed at a lower cost. When it comes to the production of oil and gas, the learning-by-doing benefits are more challenging. Research in the manufacturing area has shown that once production facilities are operating, learning spillover between plants does not always occur. In the oil and gas industry, the question is whether learning in one oil field is transferable to other fields, resulting in cost reductions. Given that the most valuable knowledge in any organization is tacit and resides in the heads of individuals, oil and gas firms that operate multiple fields need to create knowledge transfer strategies to ensure that the experience curve is created.

After cost drivers have been identified, the next step in cost management is controlling the drivers. For example, if location creates logistical challenges for spare parts, the firm could consider several alternatives for cost control, such as finding different sources for spare parts, increasing the inventory of parts, or developing a new system for ordering and receiving parts. The final, and by far most difficult, step in strategic cost management is reconfiguring the value chain. Reconfiguring the value chain means looking closely at the value chain and activities that generate the most value for the firm and then doing things differently. For example, over the past few decades, oil majors have increased the volume and type of activities outsourced to lower cost contractors. By outsourcing lower value added activities to oil field service firms, the majors can focus on their core strengths of technology, project management, and capital discipline. In doing so, the majors have protected their strategic uniqueness and made it difficult for NOCs to close the gap in terms of productivity and technological strength.

Maximizing value and minimizing costs

In all industries, costs change over time in response to industry growth, learning rates, the rate of technological change, competitive responses, and other factors. In the upstream oil and gas industry, finding costs and lifting costs are critical. Finding costs are, obviously, not part of the production activity. We have chosen to include finding costs here because a core theme of this chapter is cost management. Additionally, looking at the regional differences in finding costs and lifting costs is more useful when they are analyzed in tandem as we have done.

Table 5–2 shows changing finding and lifting costs for oil and gas across the globe on a dollar per barrel of oil equivalent basis through 2008 (latest available data).[10] Finding costs rose dramatically within the US during the period. Offshore finding costs are significantly higher (as would be expected) than onshore. For the other regions covered by the EIA's FRS system, it is hard to generalize about trends, as each producing region has its own complexity and geologic differences.

Lifting Costs, the second major component of E&P costs, have shown large but less dramatic changes in recent years. Lifting costs in the former Soviet Union states have suffered significant increases, and US lifting costs also show large increases. As opposed to what many people may believe, on a dollar per barrel basis, US production taxes are higher than the foreign average and worldwide total.

To reduce costs, firms must have a solid grasp of three elements: the workload of the oil field (such as number of wells drilled); workforce productivity, contractors and supply chains, equipment, power costs, and workover planning; and the cost drivers of the various factors of production (workforce, materials, equipment, services, etc.). Each of the three elements creates opportunities to reduce costs. For example, inefficient processes, lack of automation, and poor services support can lead to workforce inefficiency. The Kern River case at the end of the chapter provides an example of an oil field with consistent cost reduction.

Table 5–2. Finding and lifting costs globally for FRS companies

	Finding Costs			Lifting Costs						
				2007			2008			
Region	2005–2007	2006–2008	Percent Change	Lifting	Taxes	Total	Lifting	Taxes	Total	Percent Change
United States										
Onshore	$13.72	$24.31	77%	na	na	na	na	na	na	na
Offshore	$54.45	$63.89	17%	na	na	na	na	na	na	na
Total United States	$17.52	$29.11	66%	$8.53	$2.96	$11.49	$9.76	$4.86	$14.62	27%
Foreign										
Canada	$12.48	$27.80	123%	$10.24	$0.39	$10.63	$11.79	$0.42	$12.21	15%
Europe	$32.29	$61.37	90%	$8.66	$1.91	$10.57	$8.61	$2.99	$11.60	10%
Former Soviet Union	NM	$10.45	NM	$4.06	$0.50	$4.56	$6.78	$2.16	$8.94	96%
Africa	$39.11	$32.49	–17%	$5.78	$3.76	$9.54	$7.21	$4.08	$11.29	18%
Middle East	$4.88	$5.12	5%	$4.17	$4.62	$8.79	$5.48	$6.25	$11.73	33%
Other Eastern Hemisphere	$21.03	$12.45	–41%	$5.51	$2.89	$8.40	$6.15	$4.23	$10.38	24%
Other Western Hemisphere	$30.98	$27.36	–12%	$3.96	$2.09	$6.05	$4.29	$2.47	$6.76	12%
Total Foreign	$21.17	$18.75	–11%	$6.61	$2.46	$9.07	$7.44	$3.33	$10.77	19%
Worldwide	$18.98	$23.84	26%	$7.50	$2.69	$10.19	$8.54	$4.05	$12.59	24%

Notes: The above figures are three-year weighted averages of exploration and development expenditures, excluding expenditures for proven acreage, divided by reserve additions, excluding net purchases of reserves. Natural gas is converted to equivalent barrels of oil at 0.178 barrels per thousand cubic feet. Sum of elements may not add to total due to independent rounding. NM = not meaningful. Energy Information Administration, Form EIA-28 (Financial Reporting System).

Field Reinvestment and Renewal

All businesses with productive assets require reinvestment over time. This new investment in existing productive assets can be subdivided into 1) the repair and maintenance of existing equipment and operations and 2) additional capital investment for future returns. As a producing oil or gas field matures, it may be a potential candidate for either or both investments. As is always the case with business and accounting, the use of capital and cash flow may prove to be both an "investment" and an "expense."

Production enhancement

Oil and gas production investment does not stop with the initiation of production. A variety of different incremental activities, some larger than others, are used to enhance production. Figure 5–6 provides an overview of enhanced recovery, primary techniques, and the relative magnitudes of cost and investment required to increase the percentage of oil-in-place (OIP) recovery. Enhanced oil recovery (EOR) includes secondary and tertiary recovery techniques and involves the stimulation of the reservoir and wells with supplemental processes beyond the natural reservoir pressure.

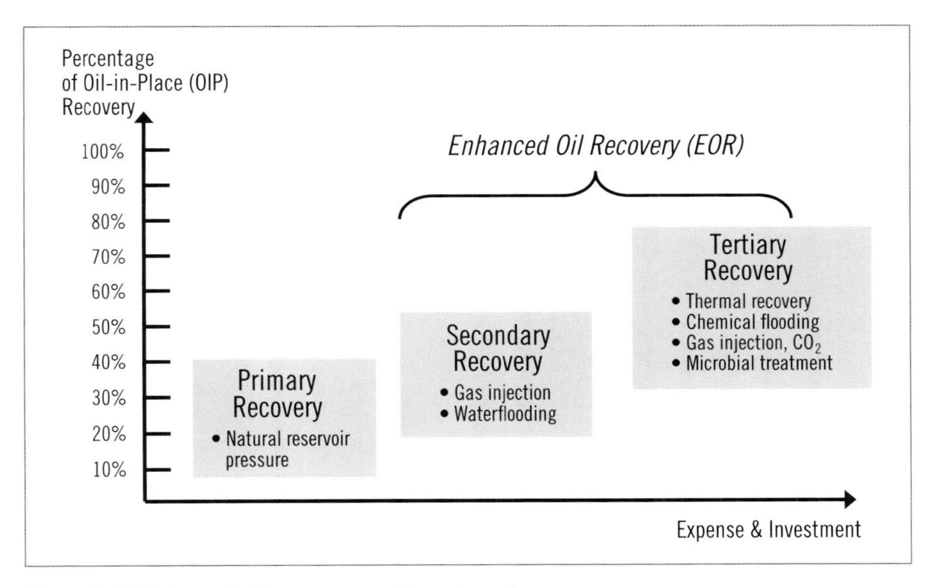

Figure 5–6. Enhanced oil recovery and investment

Secondary recovery is when an external fluid such as water or gas is injected into the reservoir through injection wells in order to preserve or increase reservoir pressure. In some cases the injection is designed to specifically drive the hydrocarbons to the well bore itself. The most common secondary recovery techniques are gas injection and waterflooding. *Gas injection* involves the injection of gas directly into the reservoir's gas cap in order to replace gases lost or dispersed through production. *Waterflooding* is more complicated, typically involving the injection of water into the rock strata holding the reservoir's hydrocarbons to push or *sweep* the oil toward the well bore.

A pressure maintenance program such as gas injection can begin during the primary recovery stage, but it is a form of enhanced recovery. Waterflooding and similar production enhancements are usually not required until well into a reservoir's productive life. A successful secondary recovery program may increase the total recoverable oil from a reservoir by an additional 5% to 20% of the oil in place.

Tertiary recovery is a third level of production enhancement. Whereas secondary recovery techniques try to maintain pressure and sweep the oil to the well bore, tertiary techniques try to change the chemical conditions of the hydrocarbons to allow greater movement and fluidity to the well bore. Three primary devices—thermal recovery, chemical flooding, and gas injection—alter the viscosity of the oil or alter the adhesive tensions between the hydrocarbons and subsoil sand and rock strata. Tertiary recovery can in some cases increase recovery by 15% to 50% of the oil-in-place, as shown in figure 5–7. Unfortunately, as shown, even after the application of enhanced recovery techniques, a large portion of the oil-in-place often remains.

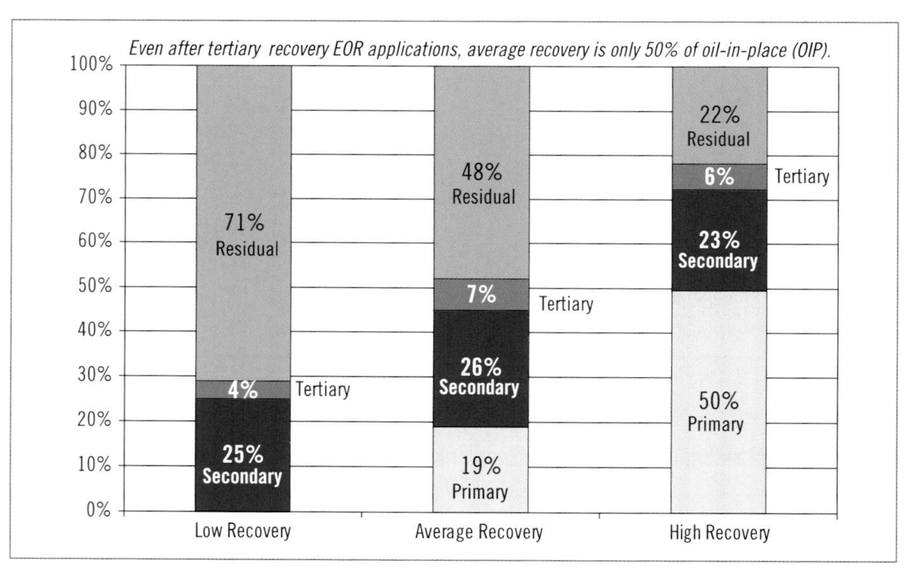

Figure 5–7. Ranges of Oil Recovery Factors

Source: "Oil & Gas for Beginners," Deutsche Bank, January 7, 2008, p. 70. Copyright © 2008 Deutsche Bank AG.

Enhanced recovery costs

Business reinvestment requires a disciplined approach to the use of capital and a careful evaluation of whether the higher investment and expenses incurred during secondary recovery and tertiary recovery are expected to yield positive financial results. The recovery techniques have traditionally been employed in older fields that were initially produced in the far past, when both recovery techniques and crude prices were considerably "less" (as in the Kern River example later in the chapter). And, both enhanced recovery methods are extremely expensive. For example, in 2010 ExxonMobil Corporation said its enhanced oil recovery project at Malaysia's Tapis field will start in 2013, with an estimated gross investment of more than $1 billion. The enhanced recovery project will be the first of its kind in Malaysia.[11]

Table 5–3 presents annual equipment and operating costs for a standardized 10-well oil lease in the United States. The analysis, conducted by the EIA, categorizes the costs by region and production depth. For equipment costs, there is little significant deviation across regions, with greater depths logically adding substantially to equipment costs.

Table 5–3. Equipment and operating costs for a 10-well oil lease

	Equipment Costs				Operating Costs			
	Producing Depth (feet)				Producing Depth (feet)			
Region	2,000	4,000	8,000	12,000	2,000	4,000	8,000	12,000
California	1,731,400	2,073,600	2,686,900	3,243,100	225,200	3,011,000	507,600	758,400
Mid-Continent	918,700	1,465,300	2,423,400	2,924,500	204,500	243,800	430,200	529,000
South Louisiana	1,116,600	1,545,100	2,013,300	3,288,300	244,100	350,900	419,700	626,900
South Texas	1,029,200	1,461,800	1,867,900	3,220,300	259,900	332,700	416,300	649,300
West Texas	938,700	1,438,800	2,348,700	2,950,200	186,400	2,235,000	326,200	497,100
Rocky Mountains	1,044,800	1,500,100	2,330,700	2,950,900	232,500	259,400	351,900	454,600
Lower 48 states (excludes off)	1,148,600	1,580,800	2,278,500	3,096,200	225,400	285,200	408,700	585,900
Additional cost for secondary recovery in West Texas	4,397,700	8,799,000	17,937,400	NA	434,900	593,700	827,900	NA

Source: "Oil and Gas Lease Equipment and Operating Cost, 1998–2006," Energy Information Administration, www.eia.org. Data is for 2006.

Secondary recovery costs are very expensive, especially for equipment. The secondary recovery costs presented (only those for West Texas), are roughly four times the equipment costs at 2,000 and 4,000 foot producing depths, and more than eight times similar costs at 8,000 foot depths. Operating cost impacts for secondary recovery in West Texas are about double the initial operating costs of primary production. Although this provides only a glimpse into the cost dynamics of enhanced recovery, these examples make it clear why operators and potential investors need to be convinced that secondary or tertiary recovery applications produce significant added resource recovery. Usually, recovery can be justified only when crude oil prices are trending higher.

Mature field valuation: An illustrative case in Brazil

A few years ago the Brazilian government decided to auction several older mature fields that had not been under production for a number of years.[12] The fields were thought to offer renewed profit potential because of advances in recovery technology and rising crude oil prices. Rising crude prices were thought to make the additional recovery investment feasible. The financial analysis of this opportunity is presented in table 5–4 and entails investment activities not seen in the original project evaluation.

Well intervention. Also referred to as *well work*, well intervention is any additional repair or reinvestment on a periodic basis in an operating well and field. For an older or mature field in which the wells have been either shut down or abandoned for a number of years, the cost of up-front restart may be substantial. One of the larger cost elements is the replacement of sucker rods and pumps, the basic aboveground infrastructure of older producing wells. The field will also likely require some additional reinvestment over the remaining productive years, though operators often minimize this in an attempt to reduce cost and maximize remaining returns.

When the intervention requires actually pulling and possibly replacing a well completion, it is termed a *workover*. This is an expensive and invasive process requiring the pulling of the Christmas tree and possibly wireline and tubing.

Divestment. States and governments want assurances that at the end of production fields are returned to preexploration and production states—*divestment* or *closeout costs*. Although there are significant regulatory differences across the globe, any potential field investor and operator needs to be well apprised of the exact nature of these expectations. As with all things regulatory, it is likely to be expensive! These costs vary dramatically by field depending on how the field has been operated over its lifespan. Regardless, the closeout costs must be included in any financial evaluation of a mature field.

In the analysis presented, costs are shown upfront as $800,000 in year 0 (equivalent to just over $2 million in year 10 using the company's discount rate of 10%). The reason for upfront costs is that in many countries, operators are asked to provide the funds at the reinitiation of field operations as a condition of their activity. This prevents undependable or unethical operators from walking away at the end of production and not fulfilling their closeout responsibilities.

Table 5–4 .Valuation of a mature field opportunity in Brazil

Item	Value	0	1	2	3	4	5	6	7	8	9	10
Reserves (barrels)			200,000	168,000	138,000	111,000	87,000	65,000	46,000	30,000	17,000	7,000
Production (barrels)			32,000	30,000	27,000	24,000	22,000	19,000	16,000	13,000	10,000	7,000
Capital Outlays												
Workover		(500,000)										
Sucker rods and pumps		(800,000)										
Average bid		(100,000)										
Divest (ending closeout)		(800,000)										
Project Income												
Price per barrel	$50.00		$51.00	$52.02	$53.06	$54.12	$55.20	$56.31	$57.43	$58.58	$59.75	$60.95
Price change per year	2.0%											
Revenue			1,632,000	1,560,600	1,432,631	1,298,919	1,214,489	1,069,854	918,949	761,579	597,546	426,648
Royalty	5%		(81,600)	(78,030)	(71,632)	(64,946)	(60,724)	(53,493)	(45,947)	(38,079)	(29,877)	(21,332)
Production cost	$8.00		$8.00	$8.40	$8.82	$9.26	$9.72	$10.21	$10.72	$11.26	$11.82	$12.41
Cost change per year	5.0%		(256,000)	(252,000)	(238,140)	(222,264)	(213,929)	(193,995)	(171,532)	(146,338)	(118,196)	(86,874)
Well interventions				(100,000)		(100,000)		(100,000)		(100,000)		
Maintenance			(50,000)	(50,000)	(50,000)	(50,000)	(50,000)	(50,000)	(50,000)	(50,000)	(50,000)	(50,000)
Income before taxes			1,244,408	1,080,578	1,072,868	861,718	889,845	672,377	651,480	427,172	399,484	268,454
Taxes	34%		(423,099)	(367,397)	(364,775)	(292,984)	(302,547)	(228,608)	(221,503)	(145,239)	(135,825)	(91,274)
Net income		(2,200,000)	821,309	713,182	708,093	568,734	587,298	443,769	429,977	281,934	263,660	177,179
Valuation												
Free Cash Flow (FCF)		(2,200,000)	821,309	713,182	708,093	568,734	587,298	443,769	429,977	281,934	263,660	177,179
Discount rate	10.00%											
NPV	1,203,964											
IRR	25.18%											

Source: Numerical analysis based on values presented in "Valuation of Onshore Mature Oil Fields: The New Bidding Rounds in Brazil," by Frederico Magahlhaes, Jr., Roberto Marcos da Silva Montezano, and Luiz Eduardo Teixeira Brandao, unpublished. Analysis is for a field of four wells.

Winning bid. The analysis presented in table 5–4 assumes that the operator will win the field on the basis of a $100,000 bid or bonus. This is included in the financial analysis to determine if, after paying this amount upfront, the firm can expect to earn an adequate return on the field.

As a final note for business valuation purposes, the top-line revenue forecast reflects the ever-confounding assumption of an oil price over time. All firms face the same basic challenge of "guestimating" this value. The top-line forecast also reflects what an individual operator believes the prospective field will yield. All operators' new and mature fields are not created equal. If the operator believes it has better competitive skills or knowledge of the field and believes it can produce greater volumes than competitors, this should be reflected in the production volume expectation. It may also appear in the operating costs (US dollars per barrel estimated) and in prospective workover or other replacement investments.

This hypothetical analysis produces a positive NPV, with an IRR of 25%. This return would be considered a highly profitable project of relatively low risk, given that the mature field's productive capabilities are known with a higher degree of confidence than a new field would present.

Continuous learning: Chad's production surprises

> *"The steps being taken to maintain oil production in Chad are designed to deal with a challenging and unusual combination of geologic issues in the Doba Basin oil producing area. The full extent of these geological challenges in Chad's Doba Basin—and the solutions to meet those challenges—could only be completely understood after oilfield development began and many wells had been drilled. Each new well yielded more data that helped engineers put together a more complete picture of the geological conditions a mile below the surface and what to do about them. So far, the array of solutions includes high pressure water injection, frequent well stimulations and drilling of additional wells."*[13]
>
> —Chad Cameroon Development Project, *Project Update No. 24*, Mid-Year Report 2008

It is not unusual for major challenges to be faced after production of a field has begun but some surprises are larger than others. The Chad-Cameroon Development Project provides a host of lessons as to how knowledge gained after production alters the producer's investment and activities over time.

After production in the Doba Basin in southern Chad had been underway for more than five years, the reservoir confronted the consortium with a series of challenges:

- **Declining reservoir pressure.** Although all reservoirs suffer declining pressure with production, the Doba Basin suffered declines much earlier in the production cycle than expected. To increase and maintain pressure the consortium invested in the construction of a high-pressure water injection system, in which water separated from the crude oil produced would be reinjected into the reservoir.

- **Pores in oil-bearing sands clog easily.** Crude oil is produced from wells as it flows through oil-bearing sand formations. The flow occurs through pores that in the Doba Basin were found to carry particles that combined with the crude to clog easily and rapidly during production. The project teams responded by initiating a number of well stimulations to back-wash the particles from around the well, in addition to adding in-fill wells to shorten the space and distance between wells.

- **Noncontinuous oil fields.** Once production commenced, the consortium found that many parts of the actual field were separate, and the oil sands formed in channels and layers, which limited actual reservoir yield. The complexity of the field was not known until after significant production results could be studied. The solution has been to increase the number of wells and density of drilling to increase access the individual pools.

In the Chad case, each of the solutions requires additional investment. The added investment and added operating activities will add substantially to ongoing operating expenses, expenses incurred in order to try to preserve and improve the operating results of this complex project. At the end of the chapter, we present a case study on Kern River. This case study shows how innovative approaches to oil recovery can significantly extend the life of an oil field.

In summary, E&P firms have several options for maximizing value from their oil fields. Reducing costs and improving and enhancing reserve recovery—altering both the denominator and the numerator of the dollar per barrel of oil equivalent—are the most important. Enhancing recovery is the most important lever for production value creation.[14] Companies should analyze the cost of recovery relative to the value of recovered reserves and ask two questions: when should increased investment be incurred to push for higher levels of recovery, and at what point will incremental value outweigh the value generated from incremental recovery? According to Boston Consulting Group, many companies struggle to find the appropriate level of investment in recovery. Companies often fall into the trap of focusing on the technical aspects of recovery instead of the financial outcomes of the chosen technique. Not surprisingly, oil field managers are usually strongest in technical areas and may have limited experience with financial analysis. Companies should apply business decision tools in order to choose the most financially beneficial recovery option. These tools will require the technical and financial aspects of recovery to be analyzed jointly.

Managing Contractor Supply Chains

In chapter 4, we discussed the importance of building contractor supply chains during project execution and construction. During production, building effective supply chains is also critical, and E&P firms are heavily reliant on a supporting cast of contractors. A key difference between development and production supply chains is that "on-time" (as in "get the project completed on time) is not as much of a factor during production. Given the long-term nature of oil production, production clients focus on contractor factors such as reliability, quality, delivery service, and cost. That said, time issues do not disappear because problems need to be fixed immediately to ensure production is not disrupted. Another difference between development and production is that production contractors might participate in an oil field for decades, whereas development contractors execute their part of the project and then the relationship with their client effectively ends until the next project begins.

Different stakeholders in the oil and gas industry have different contractor needs. E&P firms are looking for stable and long-term relationships. They are also looking for predictable and downward trending costs. To the extent possible, firms would like to establish relationships that cover multiple projects in order to push the contractor down the learning curve.[15] In recent years there have been constraints on the availability of high-quality contractors, although that has abated somewhat as the most recent recession ended.

One of the challenges faced by E&P firms is local content rules in countries where the development of local business is still in its infancy, such as Angola or Ghana. In these countries firms must abide by the rules and still ensure that they are able to get the desired quality and cost. Global suppliers are usually lower cost (initially) than local suppliers. However, if only global suppliers are used, there is no way for oil-producing nations to develop their own economy, increasing the risk that the curse of oil becomes reality. Thus, most E&P firms recognize that it is in their and the host country's best interests to develop local suppliers and train local employees.

Like their private sector competitors, NOCs also try to minimize contractor costs. However, NOCs have strong local content expectations, which as noted may conflict with cost management objectives. NOCs in the past have been notably more risk averse than IOCs, but as oil prices rose over recent years, they have been under pressure from their government owners to increase investment. The boom-and-bust nature of the industry and short-term expectations from governments is a particular challenge for NOCs. When prices fall, NOCs may struggle to weather the resulting drops in income, which puts pressure on contractor relationships.

Contractor bargaining power

In chapter 4, we discussed a tier structure for E&P contractors that is applicable in development and production (see table 4–4, chapter 4). Tier 1 suppliers have to work with their clients, the NOCs, IOCs, and independent E&P firms, and also deal with their own tier 2 contractors. As discussed in chapter 4, oil field service firms and other contractors are not equity oil owners and are paid for services rendered. Tier 1 contractors have several challenges. Their clients are expanding into ever more challenging technical and geographic areas and expect their contractors to support them over the life of the new fields. They have their own suppliers who are trying to move up the value chain and acquire a larger piece of the pie. And, they are under the never-ending pressure from their clients to drive prices down. In analyzing their bargaining power with clients, a contractor needs to consider various factors. In general, contractors have more power when:

- The client has few choices for the type of input provided by the contractor.
- The contractor inputs to the client are unique, making it hard for the client to switch contractors.
- Contractor inputs are critical to a high percentage of the client's business.
- The contractor can walk away from the client business with limited impact on its overall performance.
- The client cannot buy directly from the contractor's suppliers.
- Bringing the contractor business in house is not an option for the client.
- The main focus of the client-contractor relationship is more about quality, service, and technology than it is about price.
- The contractor has IP control over the inputs it sells to its clients.

- The contractor has full information about the client's oil and gas production costs.
- The contractor and the client have a working relationship across a portfolio of oil and gas projects.

Partnership Management

All large upstream projects involve partnerships. There are various reasons why partnerships are so prevalent: projects may be too large for a single company, no single company wants to have full risk exposure, and the resource owner/NOC partner does not have the technology to access its oil and gas. Chapter 3 discussed the role of partnerships during exploration, and chapter 4 suggested that partner conflict is one of the causes for cost overruns and other project development problems. In this chapter we discuss important managerial issues associated with oil and gas partnerships in the production activity.

As projects are developed and put into production, the project owners will need to work together to create value from the oil and gas field. In the US domestic oil and gas industry and in some international markets, E&P partnerships are usually governed by a legal structure called a joint operating agreement (JOA). A JOA spells out the participatory interest of each of the partners and designates one of the partners as the operator of the venture. A JOA governs the relationship between the parties, including budget approval and supervision, and crude oil lifting and sale in proportion to equity and funding by the partners. In addition to the JOA, there is often a memorandum of understanding (MOU) that governs the structure under which revenues from the venture are allocated between the partners, taxes are paid, and royalties and operating margins are distributed. The income derived from the operation is usually shared in proportion to the equity interests of the parties to the venture, with each party bearing the cost of its royalty and tax obligations in the same proportion. Allocations are also made from the revenue to take care of operating and technical costs.

A well-designed JOA identifies key areas of partner mutual interest and ensures that partnership responsibilities are clearly established. JOAs cover a variety of terms:[16]

1) Procedures necessary to operate the joint project:

  - Designation of the operator
  - Procedures for consenting to completion and subsequent wells

- Accounting records and audit mechanisms
- Voting procedures
- Default provisions
- Billing and payment timetables
- Budget provisions
- Provision of access to the well site and information

2) Future of the project

- Dispute settling mechanisms
- Future purchase rights
- Dissolution options
- Partner review process

3) Terms associated with operators and nonoperators

- Cost sharing
- Control over information

4) Rights and obligations of nonoperators

- Access to financial and scientific records of the operator
- Disclosure of financial performance of the entity
- Conditions for removal of the operator
- Access to operator proprietary know-how

One of the JOA challenges is funding issues due to the imbalance in financial capacity of the venture partners. If one partner has deep pockets and wants to expand the project and the other partner is broke, expansion plans could be held up for years. Financial capacity issues are particularly problematic when one of the partners is a government with other pressures on its resources. Venture financing disputes can lead to a reduction in production and losses in revenue. Another challenge of JOAs is allegations by the nonoperators of "gold plating" operating costs. Since nonoperators have limited control over operating costs, their concerns can lead to mutual suspicion between the parties and a perception that the operators are unfairly benefitting from the venture. JOAs must be very specific as to who will bear what costs if the operator or one of the JOA partners decides to undertake an activity when other partners choose not to.

Joint venture conflict

There are many areas of potential conflict between JV partners. Conflict can stem from disagreements about many areas, including venture strategy, venture management team, the size and extent of capital investments, the level of dividends, and the operation costs. When conflict does arise, the danger is that the conflict gets in the way of the bigger picture associated with mutual value creation. For an example of a JV with plenty of conflict but also a continuing life, see *Industry Insight: TNK-BP Joint Venture Issues.*

There is probably no other industry with partnerships as large and complex as those found in the upstream oil and gas sector. Mega projects in particular face a number of partnership challenges: technological uncertainty, massive capital expenditures, many partners, government and nongovernment partners with different objectives, private firms from different countries, uncertainty over price, long development and oil field life cycles, and difficult environmental conditions. Remarkably, the partnerships work most of the time, oil and gas are produced, and the world gets the energy it needs.

Managing operated-by-others projects

As indicated, ventures may run into problems when nonoperators and operators disagree over spending or other issues. From the perspective of nonoperators, one of the more challenging aspects of upstream projects is managing operated-by-others (OBO) projects. In an OBO project, the key managerial challenge for the nonoperators is influencing the operations of the project without having formal authority to do so. In a sense, managing an OBO project is a firm-level form of "influence without authority," the situation that exists for individuals attempting to gain influence over those they do not formally supervise.

When a nonoperating partner is dissatisfied with its level of influence and seeks changes in the partnership or greater influence over operating decisions, several options exist, all of which are difficult to implement retroactively—it is always better to design an effective partnership agreement up front:

1. Renegotiate the contract such that the nonoperators have more influence via contractual terms.

2. Establish a management services arrangement whereby the nonoperator contributes expertise and management to specific project areas.

3. Elevate the issue from the operational level to the management/joint venture board level.

4. Establish performance milestones for the operator; if the milestones are not reached, various actions could be taken.

5. Sole risk provisions that clarify who will bear what costs if a nonoperator partner decides to undertake an activity and the operator chooses not to participate. This option often occurs when one partner wants to drill additional wells and one or more partners decide not to participate. A party choosing not to participate is referred to as a *carried interest* or a *carried party.*[17]

6. Provide additional secondees to the project (*secondees* are individuals temporarily assigned to the project from one of the partners). Through their day-to-day interaction and relationship building with operating management, secondees can often play a key role in operations. In some cases, secondees can bring specific processes to the project.

7. Continue to "work the problem" through communications, board meetings, and other partner interactions.

Most of the above options could be termed "blunt" because they involve one partner exercising its legal authority. In most cases, the blunt option is not necessary because experienced partners know that all partnerships have their challenges and in the long run, resolving issues through discussion and communication will lead to a greater possibly of mutually beneficial outcomes. For an example of how a nonoperator increased its operating influence, see *Industry Insight: Influencing Partners—Syncrude Canada.* Imperial Oil, one of the partners, entered into an agreement to provide various services to the venture. A few years later Imperial convinced its partners to allow it to provide managers, including the CEO. By providing services and managers, Imperial increased its influence over operations.

Industry Insight: TNK-BP Joint Venture Issues

TNK-BP is a large integrated oil company in Russia owned 50% by BP and 50% by three Russian investors (oligarchs). The venture accounts for nearly one-quarter of BP's oil production and close to one-fifth of its reserves. In 2008 BP was embroiled in a highly public dispute with its Russian partners. The main areas of conflict and disagreement stemmed from various issues:

- The Russian shareholders wanted to expand outside Russia and BP wanted to restrict the JV to Russia and former Soviet states.
- The Russian shareholders wanted to replace the BP-appointed CEO.
- Russian immigration authorities approved only 71 visas for BP employees, far fewer than the 150 BP requested.
- The Russian shareholders wanted the 2008 investment plan of $4.4 billion cut by $900 million and the money paid as dividends. BP argued that TNK-BP was the best performing oil company in Russia and should reinvest for the future.
- BP tried to do a deal with Gazprom that involved TNK-BP without informing its TNK-BP partners.
- The partners accused BP of running TNK-BP like a BP subsidiary with too much bureaucracy.
- Both sides were threatening to sue each other. The Russian shareholders threatened legal action to strip BP-appointed directors of their powers. BP was contemplating a lawsuit against the partners to recover a portion of back taxes that were recently paid.

Despite all these issues, the JV was still operating two years later. A new CEO and board were selected, and BP made some concessions on the strategy. All indications were that, unless the Russian government intervened, the JV would continue to operate for many years to come.

Industry Insight: Influencing Partners—Syncrude Canada

Syncrude Canada Ltd, a project in northern Alberta, is a joint venture between Canadian Oil Sands Limited (36.74%), ConocoPhillips Oil Sand Partnership II (9.03%), Imperial Oil Resources (25%), Mocal Energy Limited (5%), Murphy Oil Company Ltd. (5%), Nexen Oil Sands Partnership (7.23%), and Petro-Canada Oil and Gas (12%). Syncrude Canada Ltd. is the operator, which means none of the partners is designated as project operator. In 2006 the joint venture approved its operator, Syncrude Canada Ltd., entering into a comprehensive management services agreement with Imperial Oil (majority owned by ExxonMobil) to provide operational, technical and business management services to Syncrude Canada Ltd.

Under the agreement, Imperial, with the support of ExxonMobil, would provide support in areas such as maintenance and reliability, energy management, procurement, safety, health, and environmental. Imperial Oil would be paid a fee for providing the service. In 2007 the partners agreed to let Imperial provide managers to the operation, including the CEO in 2007. In 2009 the joint venture reported, "The majority of Imperial Oil/ExxonMobil's systems have been introduced at Syncrude through the Management Services Agreement and the effort now is directed at fully integrating them into the operations as we drive to achieve Syncrude's design capacity of 350,000 barrels per day."

Learning and knowledge sharing in joint ventures

Knowledge management and the ability to create, transfer, assemble, integrate, and exploit knowledge assets are critical for firm success. Every organizational unit bears the responsibility for finding and transferring new knowledge. As knowledge gets transferred and put to use, its value increases and the network of knowledge also expands, creating further opportunities to exploit the value of the knowledge. The more knowledge can be transferred, the greater the opportunity to create knowledge-based economies of scale and innovations that underpin competitive advantage. John Browne, the former CEO of BP, said the following:

> *Learning is at the heart of a company's ability to adapt to a changing environment. . . . In order to generate extraordinary value for shareholders, a company has to learn better than its competitors and apply that knowledge throughout its businesses faster and more widely than they do. . . . No matter where the knowledge comes from, the key to reaping a big return is to leverage that knowledge by replicating it throughout the*

> *company so that each unit is not learning in isolation. . . . The wonderful thing about knowledge is that it is inexpensive to replicate if you can capture it.*[18]

Partnerships provide an excellent platform for learning. There is no question that many of the strongest NOCs, such as Petrobras and Petronas, have learned a tremendous amount from their IOC partners. Research has shown that if firms are committed to capturing knowledge from their partners, they must create innovative mechanisms to support learning.[19] In the joint venture area, these mechanisms include transfer and rotation of people from the parent to the venture, training programs, visits, networks of managers, and leadership intent to learn. Successful transfer of knowledge also requires consensus building about the value, relevance, and potential uses of the knowledge. Until sufficient individuals accept that knowledge has value and old knowledge must be modified, new knowledge will not have an impact on organizational action. Finally, the community of individuals interested in learning must have sufficient critical mass; a single individual is unlikely to have much impact on a large corporation's knowledge base.

Research has also shown that most firms underestimate the challenges of learning from a partner. Three barriers to learning can be identified. The first is the nature of knowledge itself. Valuable knowledge is highly dependent on the context in which it has been created. Knowledge created by any firm will be deeply ingrained in that firm's culture, systems, and processes. The successful transfer of context-dependent knowledge cannot occur until the knowledge source and recipient units are in alignment about the knowledge value and its potential impact on the organization. When knowledge is deeply embedded in a unique culture and context, its transfer becomes a very difficult and often costly proposition. The second barrier involves the displacement of old knowledge. To displace old knowledge requires a process of forgetting (i.e., discarding) existing knowledge and practices and adopting new ideas. Unfortunately, managers and organizations tend to be highly resistant to the notion of forgetting knowledge associated with their specific area of work. The third barrier to successful learning is identifying the wrong knowledge that should be discarded or ignored. Given the high cost of learning, focusing on the wrong knowledge can lead to poor outcomes. Organizations must ensure that efforts to transfer knowledge are focused on valuable knowledge that has the potential to be implemented and put to use in the targeted site.

Managing Political Risk

Chapter 2 examined political issues in the oil industry. One of the conclusions in the chapter was that politics and resource nationalism will play an increasingly important role in the future as demand increases and supplies are constrained. For oil and gas producers, in-country political risks are inevitable, and how they are managed will have a significant impact on business performance. At the firm level, political risk is defined as the unexpected impact on operating cash flows as result of political actions and events in the host country. The nature of political risk ranges from the extreme case of complete expropriation of assets with no compensation to minor day-to-day interference that increases the cost of doing business.

From a rational perspective, one would think that a government would resort to political interference only if the benefits from doing so exceeded its costs. However, the actions of governments in many places would suggest that rational behavior takes a back seat to political exigency. In other words, governments often do what they want with limited regard to the long-run consequences. In the past few years there have been many cases of countries politically interfering with the operations of oil and gas firms. Some recent examples include: PDVSA (Venezuela) seized the assets of oil service companies; Gazprom acquired a major stake in Sakhalin II project; Ecuador expropriated Occidental Petroleum's interest in the Block 15 Field; and a host of countries have unilaterally revised their fiscal terms with oil companies.

While one can argue about the legitimacy of these actions, E&P firms must put into place risk-mitigation strategies. Some would argue that the IOC's bargaining power is diminishing vis-à-vis that of host states and that there is an anti-Western IOC campaign.[20] If true, that could create opportunities for independent E&P firms, although unless these firms are quite small, it is likely they will face the same challenges.

Managing political risk should be viewed as a necessary cost for an international oil company. As one expert suggested, "While political risk can be managed through insurance, strategic alliances and partnering, it can also be minimized, by taking some actions, which may seem obvious, but are too often ignored. Effective techniques include keeping a low profile, maintaining close relationships with the host government, anticipating change and working with it, avoiding geographical concentration, being a good corporate citizen and utilizing local suppliers and personnel to the greatest extent."[21]

The goal in political risk management is to create an economic link with the host country. To be effective, the economic link should establish a constituency in the host country with a stake in the firm's continued political survival. Without such a constituency in the host country, the firm will be subject to short-term politically motivated whims of the government. For example, in 1995 the American firm Enron was building a power plant in the state of Maharashtra, India. During a state election campaign, the BJP, one of the parties contesting the election, had a slogan "Throw Enron into the Arabian Sea." The BJP told the voters that Enron and the power plant project were bad for India and, if elected, the BJP would shut down the project. To many people's surprise, including Enron's, the BJP won the election and promptly suspended the project. What ensued was a long negotiation period during which Enron had to rebuild its economic link with the government and establish a base of constituents within the BJP. Eventually, after various contractual terms and conditions were revised, Enron was allowed to resume construction. If Enron's ties to the BJP had been stronger, political interference might have been avoided (unfortunately, when Enron went bankrupt in 2001, the unfinished second phase of the project was abandoned).

Innovation and Technology

Throughout this book, the important role of technological innovation is emphasized. The oil and gas industry is highly dependent on new technology, without which the industry could not grow and the world's energy demand would not be met. Technology impacts all aspects of the oil and gas value chain. As discussed in chapter 3, innovations in exploration technology have allowed geologists to find new oil fields in more challenging areas. Innovations in deepwater and horizontal drilling, subsea equipment, and other areas have opened up many new fields for development. The following list examines one observer's assessment of the most important technology innovations in the upstream areas.

Technology innovations in the upstream oil and gas industry

1. Wireline—a cabling technology using a current sent to downhole logging tools in oil well exploration and completions.

2. Logging while drilling (LWD)—the measurement of the borehole rock properties during the excavation of the hole through the use of tools integrated into the bottomhole assembly.

3. Computers—an enabling technology for most of the other innovations.

4. Top drives—A device that turns the drillstring; modern top drives are a major improvement to drilling rig technology and are a large contributor to the ability to drill more difficult extended-reach wellbores.

5. Subsea equipment—a wide range of equipment that supports deepwater production.

6. Geophysical surveys—the systematic collection of geophysical data from the earth's subsurface. Improved geophysical survey data have greatly reduced the number of dry holes.

7. Drill bits—the tool used to crush or cut rock.

8. Reservoir modeling—the modeling of reservoir characteristics to optimize oil and gas production.

9. Enhanced oil recovery—a range of technologies to increase the amount of recoverable oil.

10. Drilling rigs—on land and especially on water, drilling rigs continue to evolve and allow for more advanced exploration and production.

Technology and cost control: Chevron's Kern River Field

In the production area, innovation is directly tied to two critical areas: cost reduction and increasing reserve recoveries. In the following case example of Chevron and Kern River, technology and cost control at a very old oil field are examined.

The Kern River oil field is one of the world's oldest oil fields. The field is located about 100 miles northwest of Los Angeles near Bakersfield and covers an area of about 15 square miles. Oil was discovered in 1899 from a well of about 70 feet, the first of more than 16,000 wells in the Kern River Field. By 2010 the field had produced more than 2 billion barrels and could hold another 1.5 billion barrels. The oil is trapped in sandy material, and it is a challenge to keep the well from becoming clogged with sand. The oil and water mixture extracted is about 90% water.[22]

By 1901, Kern River was producing 12,000 b/d. As the quote at the beginning of the chapter indicates, it was thought that the field would be dry within a few years. Although Kern River entered a period of maturity in 1910, with gradual production declines occurring over the next 50 years, the field is still active more than 100 years later. The field was unitized (i.e., the consolidation of all, or a high percentage of, the participating interests in an oil field) by Getty Oil in the 1960s. Getty was acquired by Texaco in 1984, and Chevron acquired Texaco in 2000.

Beginning in the 1960s, Getty began using steam flooding, a method of injecting steam into a reservoir to heat the oil and reduce its viscosity, allowing it to be pumped out of the ground. The field's output peaked in the 1980s at about 140,000 b/d. Although the field is in decline, Chevron's use of innovative technology has slowed the decline and keeps the field producing at a profitable level. Production is falling at a rate of about 2% per year, compared to an average of 7% per year from 1998 to 2005. The site currently produces about 80,000 b/d from 8,000 producing wells, which means on average, each well produces about 10 b/d.

Chevron uses Kern River like a real-world laboratory, and according to a Chevron engineer, "The thing about being in this old oil field . . . you can try stuff."[23] Chevron uses 3-D visualizations translate data from seismic surveys into a 3-D model that shows the subsurface structure of the field. The 3-D model helps determine the most efficient and cost-effective method to inject steam that will loosen the oil deposits that are more difficult to reach. A 3-D visualization facility also helps determine where to drill horizontal wells, which produce over 10 times more than typical vertical wells. The facility also allows older steam injection wells to be converted to production wells because oil in those areas is now thermally viable to pump out. Sensors inside 660 observation wells send real-time data on the soil's porosity, temperature, and makeup, to build these 3-D computer models.

Chevron cost control

As an older field in decline using enhanced recovery techniques, Kern River is a good example of the importance of cost control. An analysis by *Oil Drum* shows that Chevron does a number of things to keep costs down.[24] Chevron reuses equipment as much as possible. If a pump jack is not needed in one location, it is moved to another location. If an oil well is no longer producing oil, the borehole can still be used for surveillance of ground temperatures (to see how the steam is heating the area) and for other measurements that help determine whether oil remains in the area. Pump jacks are put on timers and turned on and off as needed to use as little electricity as possible.

Chevron uses natural gas to create the steam needed for injection into the wells. To save energy costs, Chevron uses a cogeneration electricity process. The waste steam generated in the steam injection process is used to provide power for the oil field, and the excess is sold to a local electricity utility. Through cogeneration, Chevron is able to attain very high efficiency in its use of natural gas.

Operating expenses are estimated by *Oil Drum* for the San Joaquin Valley to be $17.84 per barrel, and capital expenditures amount to $11.47 per barrel. The selling price for the heavy oil from Kern River is $6 to $7 below that of WTI, which means that as long as WTI is around $37/bbl, Chevron will earn a positive contribution margin from Kern River.

Chevron's conclusions from their experience with Kern River:[25]

- Continued innovation and application of technology has enabled the Kern River Field to continue as a viable field.
- Use of state-of-the art earth modeling techniques combined with dedicated reservoir surveillance programs provide optimal reservoir management.
- Innovation will continue to extend the field life after producing more than 60% of the oil originally in place.

When E&P companies talk about having a long-term time horizon, it is the production area where this is most apparent. As the Kern River example shows, an oil field can be productive for more than 100 years. About 20 to 30 years is the norm for productive life, but even that is a long time relative to the product life cycles found in most other industries. Given the long production periods and the fact that the product itself does not change over time, the emphasis of production managers is on driving the cost of production down and enhancing the size of recoverable reserves.

Summary Points

- The cost of producing a barrel of oil differs dramatically around the world, and those differences drive the competitive economics of the global industry.
- Competitive advantage in the production activity is directly related to cost management; for a given region, the lowest cost producer will have the highest income.

- Many factors impact the costs of production. Understanding these factors, such as workforce productivity, and their cost drivers is critical to success as an oil and gas producer.

- With many of the largest oil fields in decline, there are many opportunities for innovative techniques that can stem the decline rates. Enhanced recovery techniques, while costly, can extend the useful life of mature fields.

- All businesses with productive assets require reinvestment over time. This new investment in productive oil field assets can be sub-divided into 1) the repair and maintenance of existing equipment and operations and 2) additional capital investment for future returns.

- Technology plays a key role in driving down production costs.

- Since all major oil fields involving private sector firms are partnerships, partnership management skills are a key factor in successful production.

Notes

1. From "Oil & Gas for Beginners," Deutsche Bank, January 7, 2008.
2. *Oil and Natural Gas Production*, Energy Information Administration, 2008, p. 19.
3. For an in-depth analysis of accounting in the oil and gas field see Charlotte J. Wright and Rebecca A. Gallun, *Fundamentals of Oil & Gas Accounting*, 5th edition, Tulsa: PennWell, 2008.
4. "Deutsche Bank Analyzes Oil Production Costs," *Oil & Gas Journal,* March 16, 2009.
5. Michael E. Porter, *Competitive Advantage: Creating and Sustaining Superior Performance*, New York: Free Press, 1985.
6. Porter, 1985.
7. Mohammed A. Mian, *Petroleum Engineering Handbook for the Practicing Engineer*, Tulsa: PennWell, 1992, p. 447.
8. John K. Shank and Vijay Govindarajan, *Strategic Cost Management: The New Tool for Competitive Advantage*, New York: Free Press, 1993.
9. Richard P. Rumelt, *Note on Strategy Cost Dynamics*, Anderson School at UCLA, 2001.
10. http://www.eia.doe.gov/neic/infosheets/crudeproduction.html.

11. Eric Watkins, "ExxonMobil to Boost Tapis Field Production in Malaysia," *Oil & Gas Journal*, June 10, 2010.
12. This analysis is based on ideas presented in "Valuation of Onshore Mature Oil Fields: The New Bidding Rounds in Brazil," by Frederico Magahl-haes, Jr., Roberto Marcos da Silva Montezano, and Luiz Eduardo Teixeira Brandao, unpublished.
13. Chad Cameroon Development Project, Project Update No. 24, *Mid-Year Report 2008*, pp. 7–10.
14. Uwe Gunther, "Maximizing Value in Upstream Oil and Gas," Boston Consulting Group, 2007.
15. "Better Contracting in Oil and Gas Major Projects: Finding Mutually Acceptable Ways of Securing the Best Supply Chains in a Tight Market," CRA International-London, 2008.
16. J. B. McArthur and J. Leitzinger, "Balance Needed in Operating Agreements as Industry's Center of Gravity Shifts to State Oil Firms," *Oil & Gas Journal*, Vol. 98, Oct 23, 2000, pp. 74–81.
17. R. A. Gallun, C. Wright, L. M. Nichols, and J. W. Stevenson, *Fundamentals of Oil and Gas Accounting*, 4th edition, Tulsa: PennWell, 2001.
18. Steven E. Prokesch, "Unleashing the Power of Learning: An Interview with British Petroleum's John Browne," *Harvard Business Review*, Vol. 75, September–October, 1997, pp. 146–168.
19. Andrew C. Inkpen, "Learning Through Alliances: General Motors and NUMMI," *California Management Review*, 2005, Vol. 47, No. 4, pp. 114–136.
20. Vlado Vivoda, "Resource Nationalism, Bargaining and International Oil Companies: Challenges and Change in the New Millennium," *New Political Economy*, Vol. 14, No. 4, December 2009, p. 517–534.
21. Alan Berlin, "Managing Political Risk in the Oil and Gas industries," *Transnational Dispute Management*, Vol. 1, No. 1, 2004.
22. Steve Jacobs, "Breakthroughs: The Top 10 Oilfield Technologies of All Time," *Journal of Petroleum Technology*, January 2010, pp. 18–19.
23. Ben Casselman, "Chevron Engineers Squeeze New Oil From Old Wells," *Wall Street Journal*, October 9, 2009.
24. A Visit to Chevron's Kern River Heavy Oil Facility, http://www.theoildrum.com/node/5023.
25. Chevron North America Exploration & Production Company, "Kern River Field: Framework and Future of an Old Giant," 2007.

Chapter 6

FISCAL REGIMES

All rights to petroleum existing in its natural condition in strata in Great Britain or beneath the territorial sea adjacent to the UK are vested in the Crown. Title to petroleum produced passes to the license holder at the wellhead.

—PetroReports: Fiscal and Regulatory Guide—United Kingdom," Deloitte, September 2009, p. 77

All around the world, from Algeria to China, governments are changing the terms of investment in oil and gas, on the ground that they are not receiving their fair share of the profits. For critics of capitalism, last year's surge in oil and gas prices seemed to present a long-awaited chance to shift the global balance of power away from corporate behemoths and the governments that are closest to them.

—"Energy and Nationalism," *The Economist*, March 8, 2007

Oil and gas development involves risk and return. The development of a petroleum property requires the investment and allocation of expertise and capital—the *risk*. The proceeds of successful exploration and production can offer significant return. This two-edged sword has led to the development of a number of specific business forms and arrangements called fiscal regimes that are rarely seen or used in other industries. A fiscal regime establishes the relative proportions of both physical results (e.g., barrels of oil) and monetary results (shares or rates of return) from oil and gas production in a country.[1]

This chapter first describes the three most common fiscal regimes used in the development of oil resources: royalty/tax systems (typically described as concessions) and the two primary contractual systems used—service agreements and production sharing agreements. We then present a detailed example of a production sharing agreement (PSA) for a hypothetical sub-Saharan African project. We conclude our discussion with potential strategic positioning choices for governments and oil companies.

Development Agreements

Simplistically, the two parties of primary interest in any petroleum development project are the asset owner and the asset developer. Historically, traditionally, and globally, this translates into a sovereign state like Angola or Indonesia, and an international oil company like ExxonMobil or Shell.

Gaining the right to develop a project

The right to develop a project is gained through an individually negotiated agreement or through competitive bidding. Negotiated agreements were used throughout the Middle East and Asia in the first half of the 20th century. Since that time the majority of development rights have been obtained through competitive bidding, although under many systems even after a bid has been "won," many issues remain to be negotiated between the two parties. Regardless of how complex the agreement, they all still carry some of the fundamentals found in the original concession that Edwin Drake obtained in 1857, described in *Industry Insight: Excerpts from the Drake Lease of 1857.*

Industry Insight: Excerpts from the Drake Lease of 1857

"Demise and let" all the lands owned or held under lease by said company in the County of Vanango, state of Pennsylvania, To bore, dig, mine, search for and obtain oil, salt water, coal and all materials existing in and upon said lands, and take, remove and sell such, etc., for their own exclusive use and benefit, for the term of 15 years, with the privilege or renewal for same term.

Rental, one-eighth of all oil as collected from the springs in barrels furnished or paid for by lessees. Lessees may elect to purchase said one-eighth at 45 cents per gallon, but such election, when made, shall remain fixed. On all other minerals, 10% of net profits.

Lessees agree to prosecute operations as early in the spring of 1858 as the season will permit, and if they fail to work the property for an unreasonable length of time, or fail to pay rent for more than 60 days, the lease to be null and void.

Source: "International Petroleum Agreements –1: Politics, Oil Prices Steer Evolution of Deal Forms," *Oil & Gas Journal*, August 24, 2009, p. 20. Article is adapted from Owen L. Anderson, R. Doak Bishop, and John P. Bowman, *International Petroleum Exploration and Exploitation Agreements*, Second Edition, New York: Barrows Co., Inc., 2009.

The state owns the asset and has exclusive control over the *who*, *what*, *where*, and *how* of the production. In fact, Canada, the United States, and Norway are among the few countries in which a private individual, not just the state, can own the oil and gas resource. Although the state retains exclusive control, it typically has limited access to both development capability (competence and experience in the development of oil and gas) and capital (required in massive amounts for extended periods of time in order to develop petroleum deposits and distribute to markets). These are the competencies and resources brought to the table by the international oil company (IOC). (Note that throughout this book we use the term *IOC*. In this chapter, IOC refers to international oil company, which could be an integrated oil major or an independent E&P firm.) Although many countries have their own NOC to represent the state's interests, the NOC does not change the basic relationship and agreement needed between the two parties.

In the not-too-distant past a state often needed an IOC to not only develop a project but also to explore and identify the nature and economic promise of the resource. This informational advantage gained by the IOC through delivery has changed in recent years as states gained knowledge about their own oil and gas resources prior to the awarding of development contracts. This has shifted some of the informational advantages away from IOCs to NOCs, as discussed in chapter 2. Today most states have extensive knowledge of both their physical assets and their market prospects from development.

Interests, incentives, and behaviors

The state walks a fine line between two different interests when it constructs a fiscal regime. First, it wishes to promote investment and production by oil and gas development firms. This means that it must provide sufficient opportunities for the firm to recover its costs and make an appropriate rate of return on its investment. Second, the state must strike a balance between producers, consumers, and citizens (i.e., the mutual owners of the asset). As we discuss later in the chapter, a number of governments have periodically revised their fiscal regimes to either provide greater incentives for inward investment by IOCs, or to garner a greater share of returns when those returns start to rise in time.

Development agreements also include a number of features intended to encourage specific behaviors of the IOCs in their development of the resource.[2] Depending on the agreement, cost-saving efforts and incentives may be an important element depending on whether the IOC is entitled to recoup all expenses or whether all expense reductions alter the relative payout rates between the two parties. Many agreements now require the explicit development of national employees at all levels of the organization. This is often difficult for IOCs because technically qualified and experienced professionals are often in short supply in many developing countries. A third requirement frequently embedded in development agreements are requirements and incentives for use of nonlabor local content (such as materials, equipment, and construction firms). Although this requirement could have strong economic development benefits to the host country, it is often difficult to achieve beyond basic construction and operational support facilities and services because the developing country does not have the capacity to provide sophisticated equipment and services.

The interests and negotiating positions between the state and IOC will often change during project development. Predevelopment, the IOC often has more bargaining power as it is providing all of the expertise and capital but has not yet made its commitment. Once development is completed, the IOC carries nearly all the financial and business risks associated with the project. If the project fails, the state still has its asset (at least until depleted from production). Therefore, the IOC is motivated to negotiate a comprehensive development agreement prior to capital and operational commitment—prior to "lock-in." After the IOC has initiated investment, it is not surprising that the state often attempts to alter the agreement (renegotiate) to garner a greater proportion of the profits. The exposure the IOC experiences after lock-in is a serious dilemma that all parties to fiscal regimes must be ready to address.

Petroleum Fiscal Regimes

The two most widely used fiscal regimes are royalty/tax systems and contractual systems (see figure 6–1).[3] We begin with a discussion of concessions, the origin of royalty/tax systems, and then we discuss the two main fiscal regimes.

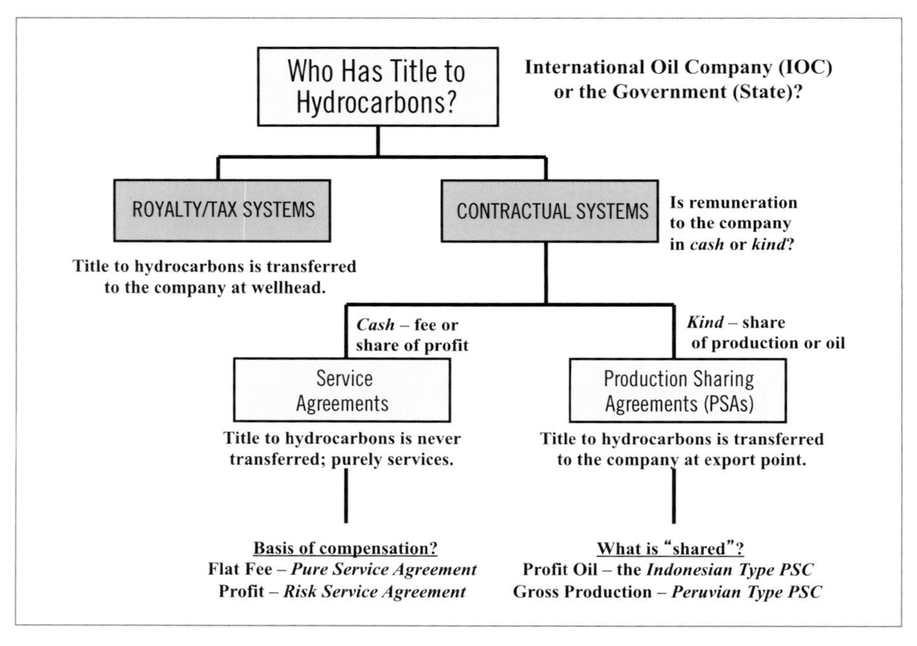

Figure 6–1. Petroleum fiscal regimes

Concessions

Historically termed concessions, now more commonly categorized as royalty/tax systems, these agreements grant exclusive rights to an individual firm, typically an IOC, with few requirements or limitations over its development activities. A traditional concession is a contract granting four principle rights to the IOC:

1. Rights for mineral development over a vast area

2. Rights for exclusive or near-exclusive development for a long period of time

3. Extensive control over the schedule and manner in which the resources are to be developed

4. Rights to all mineral development profits beyond the payment of a royalty to the state

The concession grants the IOC an extremely wide breadth of discretion in its exploration and development activities. In the past, the IOC could often explore for many years before identifying properties or projects thought to be truly economic. If a specific discovery was deemed commercially viable, the IOC had the right to develop and produce the oil and gas.

When hydrocarbons were produced, the IOC would take title to its share at the wellhead. This share was termed its *entitlement*. This entitlement equaled gross production less a royalty payment to the government. If the royalty were 10%, the international oil company could *lift* (take physical and legal possession of its entitlement of crude oil) 90% of production. The IOC also financed and owned all exploration and production equipment. The profits generated by IOCs on production were then taxed by the state. The effective tax rate was often the subject of considerable debate.

Most older concessions did not require the companies to drill on any of the lands or to release territory if exploration and drilling were not undertaken, often resulting in no effective development for decades. The states themselves had no right to participate in managerial decisions under the concession, and the state's sole reward was the royalty, paid in cash or kind. Many of the concessions also calculated state's royalty as a flat rate per ton, rather than as a percentage of the value of the sale price of the production. This meant that regardless of movement in crude price, royalty proceeds were set.

Concessions were widely used in the first half of the 20th century when many states had limited industrial development or had little interest in oil and gas development. Often an individual entrepreneur would approach the state and secure an exclusive agreement, leaving the state with few rights and fewer returns. Often little actual exploration or production occurred, yet the concession holder was under no obligation to develop on any specific time line, if ever. *Industry Insight: Agreement between Petroleum Concessions, Ltd, and Sultan of Muscar and Oman, 1937 (Excerpts)* provides excerpts of one such concession agreement with the sultan of Oman in 1937.

Industry Insight: Agreement between Petroleum Concessions, Ltd, and Sultan of Muscar and Oman, 1937 (Excerpts)

Article 2: The period of this agreement shall be 75 (Gregorian) Calendar years from the date of signature.

Article 3: The Sultan hereby grants to the company for a period of five years from the date of signature of this agreement (hereby referred to as "the option period") the exclusive right to explore, search for, drill for, produce and win natural gas, asphalt, ozokerite, crude petroleum and their products and cognate substances within the leased area.

Article 6: The Company may also at any time during the Option period declare in writing its intention to take up the concession over the lease area.

Article 9: The Company shall conduct its operations in a workmanlike manner and by appropriate scientific methods and shall take all reasonable measures to prevent the ingress of water to any petroleum bearing strata and shall duly close any unproductive holes drilled by it and subsequently abandoned. The Company shall keep the Sultan informed generally as to the progress and result of its drilling operations but such information shall be treated as confidential.

Article 19: a) The Company shall have the right at any time after the expiry of three years from the date of signature to give the Sultan notice in writing of its intention to terminate this Agreement and this Agreement shall absolutely determine on the date fixed for such termination in such notice. b) If such notice be given not later than thirty years after the date of such signature the Company shall be entitled to remove free of all taxes and duties all plant building, stored material and property provided that for a period of three months from the receipt of such notice the Sultan may purchase—should he so desire—the same at a price equal to the replacement value of that date less depreciation. c) If such notice be given later than thirty years after the date of such signature all the property aforesaid shall become the property of the Sultan free of all cost.

Article 24: Failure on the part of the Company to fulfill any of the conditions of this Agreement shall not give the Sultan any claim against the Company or be deemed a breach of this Agreement insofar as such failure results from force majeure. Force majeure as used in this Agreement includes the Act of God, war insurrection, riot, civil commotion, tide, storm, tidal wave, flood, lightening, explosion, fire, earthquakes an any other happening which the Company could not reasonable prevent or control.

Article 25: The Sultan shall not by general or special legislation or by administrative measures or by any act whatever annul this Agreement. No alterations shall be made in the terms of this Agreement except in the event of the Sultan and the Company jointly agreeing that it is desirable in the interest of both parties to make certain alterations.

Source: Smith, et al., *International Petroleum Transactions,* second edition, Rocky Mountain Mineral Law Foundation, 2000, pp. 413-414.

There are a number of examples of the one-sided nature of many early concessions. For example, William D'Arcy, a British entrepreneur backed by Burmah Oil (British Petroleum), obtained a concession with the Shah of Persia (Iran today) in 1901. Obtaining a concession was often contingent on proving that one was capable of effectively exploring and developing the concession property. In exchange for an initial consideration of $100,000 in cash, $100,000 in stock, and a 16% royalty rate, D'Arcy received exclusive development rights to 480,000 square miles of Persia for a 60-year period. Over the following years, a number of drilling attempts resulted in dry holes and increasing financial drains, finally forcing D'Arcy to combine his resources with the Burmah Oil Company. In 1908, the company finally struck oil at a depth of 1,180 feet. In 1909, the company was reorganized as the Anglo-Persian Oil Company (APOC), with D'Arcy as director. The company would later become British Petroleum.

Two other famous concessions followed the same basic structure. In 1933, Standard Oil of California obtained a concession from the king of Saudi Arabia. In exchange for 50,000 pounds of gold, Standard Oil received a 75-year concession covering roughly the same area as Persia, close to 500,000 square miles. Similarly, in 1939, five major foreign oil companies signed an agreement with the Ruler of Abu Dhabi giving them exclusive development rights over the entire country for 75 years.

Royalty/tax systems

The modern form of the concession, the *royalty/tax system*, is much more comprehensive in protecting the interests of the state. All four original Concession principles have been changed to balance, if not tip, the scales in favor of the state. The typical royalty/tax system structure today is for a much shorter period of time, a much smaller portion of a potential hydrocarbon deposit, and requires specific exploration and development efforts within a set period of time, or the rights expire.

The royalty/tax system's division of cash flows between the IOC and the state is detailed in figure 6–2. Since the IOC takes ownership of the hydrocarbons at the wellhead, the returns to the IOC are based on its profit from producing and selling the hydrocarbons, while the state's returns are confined to royalty and tax streams. Royalties are typically calculated as a percentage of revenues, and the stream of income to the state is more predictable and stable than income generated from profitability (such as taxes or dividend distributions). Although income will vary over time with the price of oil, royalties will generate an income to the state in each and every period of production, as opposed to the income accruing to the foreign oil company (FOC), which will occur only if the project is profitable in the period. Note that we have used a price of $50/bbl here to reflect what these calculations might look like in recent years, although as noted previously, tomorrow could see $20/bbl or $100/bbl just as easily.

In figure 6–2, the IOC pays all production costs and reaps all residual profits after paying state royalties and taxes. The state's earnings occur before and after production costs, assuring an income stream regardless of rising or uncontrolled production costs. This example reflects a modern form of royalty/tax system, as it contains a large draw on gross revenues in taxes. The earliest concessions often waived all IOC tax liabilities.

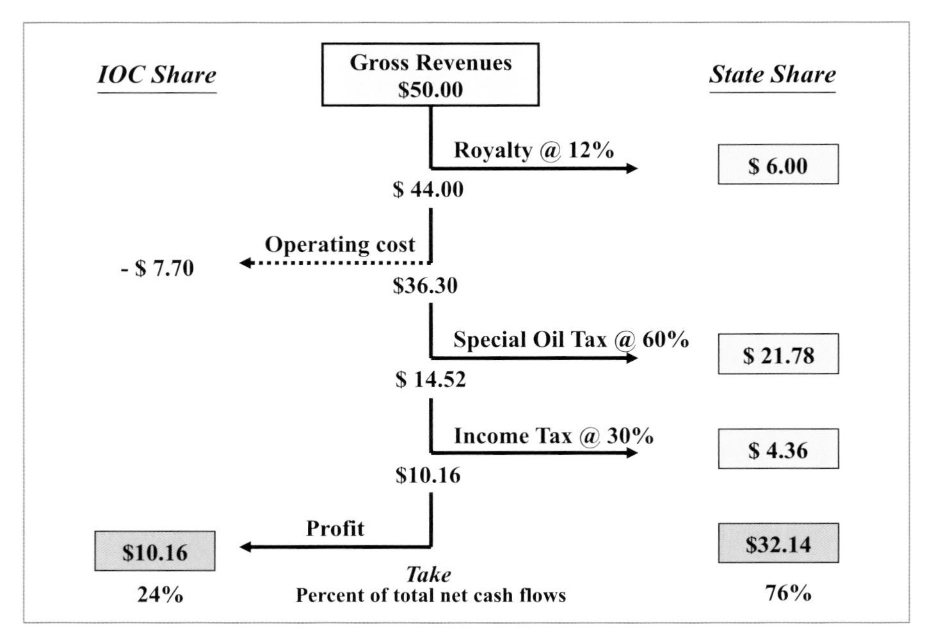

Figure 6–2. Royalty/tax system financial flows

Figure 6–2 also illustrates the individual takes arising from the agreement. The *take*—the percentage of cash flow accruing to each party of total net cash flows—is driven by the tax rates. The IOC's take is calculated as $10.16, or 24% of the net cash flows ($10.16/($10.16 + $32.14)). The state's take is $32.14 or 76% of total net cash flows ($32.14/($10.16 + $32.14)). The state's take is the sum of royalties, special oil taxes, and income taxes ($6.00 + $21.78 + $4.36 = $32.14). These takes are for one hypothetical year.

In this example any change in the gross revenue, royalty rate, operating cost, or income tax rate has relatively little impact on the takes of the individual parties. The single largest cash flow, the special oil tax, has a sizeable impact on the distribution of takes. Any increase (decrease) of the special tax increases (decreases) the state's take. It is important to note that the two operating or business risks, price and cost, result in little change in the relative takes between the two sponsors.

Royalty/tax systems still make up roughly half the fiscal systems used in petroleum development today. Tax rates vary across countries and usually range between 8% and 18%. Sliding-scale royalty systems are also frequently used. The so-called OPEC model used widely in the 1970s, which combined a 20% royalty with an 85% tax rate, resulted in little development interest from IOCs. The OPEC model shows the extreme pendulum swing from the lax concession systems of the 1930s.

Contractual systems

The experiences gained by a number of states under concessions led to the development of contractual systems (figure 6–1). The two most commonly used forms are the production sharing agreement (PSA) and the risk service contract.[4] The primary distinction between contractual systems and the traditional concession agreement is that the state maintains ownership of the hydrocarbons far beyond the wellhead. The IOC takes ownership of oil either allocated for cost recovery (those costs recognized by the state for contractual and tax purposes) or allocated from the profit split of volumes for distribution and sale.

Production sharing agreements. Under a PSA, the IOC is fully responsible for the development of the oil and gas. This includes all aspects of getting the oil and gas out of the ground and delivery to some point for transportation and sale. Under most PSAs, the state accrues cash flow returns via three primary channels: 1) royalties, 2) taxes, and 3) ownership interest. Figure 6–3 depicts a PSA structure with a 10% royalty, a 60/40 profit split (state/IOC), and a 40% tax rate. In principle, royalties should generally not exist in PSAs because there has been no transfer of asset ownership, just development. That said, royalties are often a component of PSAs. Many PSAs have cumulative production sliding scales for tax rates and profit oil splits. For example the state's profit oil split may rise as production volumes rise, either in an individual year or cumulatively over the production life of the reservoir.

The net revenues (revenues less the royalty) generated from production are unique to PSAs and are then reduced by the allowable cost recovery. *Cost recovery* in a PSA is, in principle, the deduction of a proportion of the oil—*cost oil*—to compensate for the capital and operating expenses related to the project. If, however, costs exceed the specified cost recovery limit, expenses are not deductible beyond this specified ceiling in the current year. The cost recovery limit is typically stated as a percentage of gross revenues earning during the period (40% in the above example).

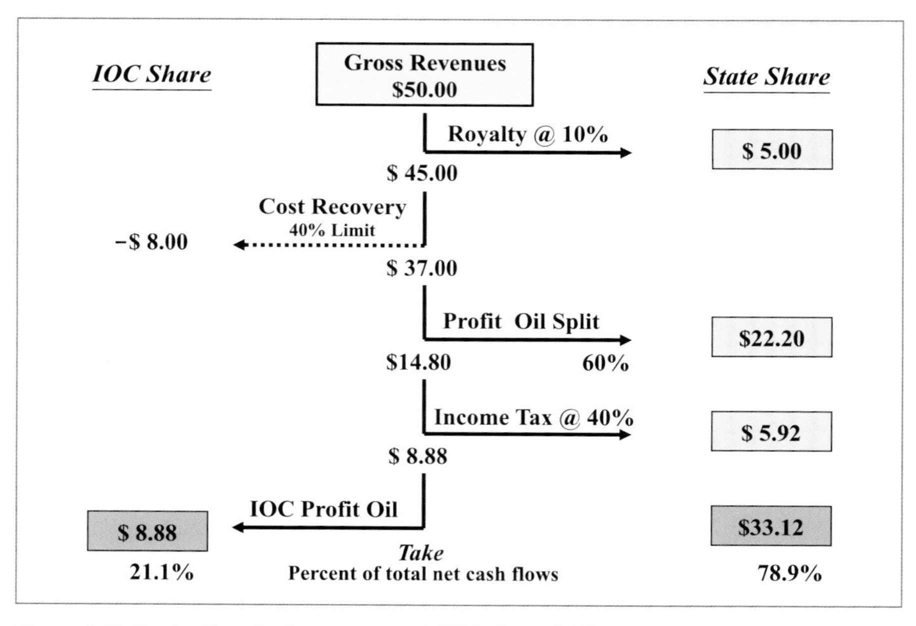

Figure 6–3. Production sharing agreement (PSA) financial flows

There are many different cost recovery systems and rates in use. Cost recovery limitations average about 65%, and approximately 75% fall between 40% and 70%. For example, Indonesia did not have a limitation for many years but now imposes a loose 80% value. Concession systems typically do not employ a cost recovery limit. Remaining costs (i.e., residual costs) would then be rolled forward in time until fully recovered. The obvious feature of this cost recovery limit is to ensure a minimum level of profitability for the state in any given year, regardless of what the IOC operator argued as legitimate deductible expenses.

All costs and expenses related to the development and operation of a single reservoir, field, or block, should be assigned and deducted from the revenues generated solely by that resource unit. This is the concept of *ring fencing*, which means separating a single reservoir or field from any others for clear identification of revenues and costs. Many governments, including Indonesia, require the establishment of a different and separable company for every field developed in order to eliminate confusion or controversy (from consolidation) related to the revenues and costs for a specific reservoir. Ring fencing increases the cost and complexity of managing the development.

The remaining oil after cost oil reductions, termed *profit oil*, is split between the IOC and the state. Only the profit accruing to the IOC is taxable, and the income tax represents a direct transfer of income from the IOC to the state. For many years, IOCs did not pay income taxes as part of their concession agreements. When the Venezuelan government pushed what was termed the *fifty fifty concept* in 1948 to gain ownership and profits of 50% of all oil developed, the principle of paying income taxes to the state spread rapidly. This has proven to be a major source of income—a major component of the state's take—in development of oil properties. The final IOC take is therefore 21.1%, and the state's take 78.9%. In both cases, for cost oil and profit oil, the payment may be defined in either cash (cash flow) or kind (oil).

Although at first sight the PSA appears to be a significantly different construction than a royalty/tax system, they are in effect nearly mathematically equivalent. The profit oil split used in the PSA is mathematically equivalent to the special oil tax levied in the royalty/tax system. If the PSA shown in figure 6–3 had used the same operating cost as in the previous royalty/tax example, $7.70 instead of $8.00, and the income tax rate of 30% instead of 40% above, the results would be identical in this simple numerical example. Both the IOC and the state would have the same takes.

Most IOCs were not anxious to enter into PSAs when they were first introduced in Indonesia in 1960. The prospect of investing significant amounts of capital in exploration, equipment, and development and not holding title to to the IOC's assets appeared too risky. The opposition of the global oil majors was broken when a number of smaller independent companies without established operations and concession agreements in other countries moved quickly to enter into PSAs and gain entry into markets where they might have previously been shut out.

Service agreements. Service agreements, either a pure service agreement or a risk service agreement, follows a simple premise: the IOC provides some or all capital for exploration and development and in turn is paid like any contractor or service provider. In a risk service agreement, if the IOC strikes oil, the state allows the IOC to recover all costs through the sale of the oil and gas. The state pays the IOC a service fee (hence the IOC's share is not called *profit oil*) calculated as a percent of the net proceeds (profit) after cost recovery. The state retains all ownership rights over the oil and gas while the contractor searches and potentially produces the oil at its own risk and expense. Importantly, and as discussed in more detail in chapter 7, the IOC would not be allowed to book the reserves under a service agreement.

A *service agreement*, as opposed to a risk service agreement, is one in which the IOC receives a flat fee for development services based on activities not profits. The Argentine Frondizi contracts of the late 1950s are examples of pure service contracts. Named after then-president of Argentina, Arturo Frondizi, IOCs were required to drill a specific number of wells per year per exploration block and in exchange received a set dollar amount based on a variety of metrics, including meters drilled, wells completed, and ultimately production volume per hour. The IOCs were pure service providers, as the resulting oil had to be sold to the Argentine national oil company, Yacimientos Petroliferos Fiscales (YPF).

The countries typically finding this structure most useful are those that import oil and gas to fulfill their total consumption needs. The risk service contract allows the state to maintain ownership and access to all oil and gas produced. As opposed to production sharing contracts, the IOC in a risk service contract never gains a share of the oil, so there is no "production sharing" or "profit oil split," although in the end this is more of a semantic distinction rather than an economic one.

Ecuador's service contract. Ecuador's version of the risk service contract is the prototype agreement structure that introduced the *R-factor* calculation or the average profit factor.[5] The example presented in table 6–1 assumes $50/bbl oil and an annual production level of 6,000,000 bbl, $3.0 billion in revenue.

In this case, the IOC's entitlement is based on cost recovery and a service fee that is taxed at a rate of 40%. This is basically a fee-based service agreement by a foreign corporation operating in a country market. It generates income in-country via the fee that in turn requires payment of corporate income taxes in-country. The fee calculation is based on a formula consisting of a sliding-scale R-factor, a profit rate (in effect a royalty rate) that changes with production volume measured in barrels of production per day (b/d). The Ecuadorian government has set the R factor rates by production range. For example as shown in table 6–1, the first 10,000 b/d of production carries an R factor of 0.30. As production on a b/d basis rises, the R factors for the subsequent increments decline to 0.25 then 0.23, then 0.20, etc. The final service fee paid to the IOC is the weighted average R factor on a b/d basis.

Table 6–1. The Ecuador risk service contract (R-factor)

Acronym	Description	Baseline	No Development	
Fee	Annual service fee			
Prime	Average prime rate of interest	10.0%	0.0%	
INA	Develop. & prod. costs less reimbursements	$25,000,000	$0	
P	Price of crude oil ($/bbl)	50.00	50.00	
C	Productions cost of crude oil ($/bbl)	1.67	1.67	
Q	Annual production (mmbbls)	6,000,000	6,000,000	
	which is equivalent in bpd	16,438	16,438	
R	Weighted average profit factor "R factor"	0.2804	0.2804	
Fee =	Prime × INA + R × (P – C) × Q	$83,815,225	$81,310,392	
Factor	**Range in bpd**	**R factor**	**bpd**	**Proportion R × bpd**
R1	From 0 to 10,000	0.30	10,000	0.1825
R2	10,000 to 30,000	0.25	6,438	0.0979
R3	30,000 to 50,000	0.23	0	—
R4	50,000 to 70,000	0.20	0	—
R5	Above 70,000	0.18	0	—
	Final R-Factor		16,438	0.2804

The annual service fee (Fee) formula is made up of two components: an interest return on investment and an annual service fee. The first, Prime × INA, is an interest return to the IOC for development and production costs that have not been reimbursed (a financial cost-of-carry payment). This is an interesting issue within this structure; the IOC operates as a contractor but is still investing capital without gaining an ownership stake. The second, the actual "service fee" component—the R-factor—is a percentage of operating profits, $P - C \times Q$ (net profit per barrel times barrels).

In the case of Ecuador, the R-factor service fee is calculated on a weighted average daily volume basis per the schedule shown. (According to the Ecuadorian sliding scale, the R factor would begin at 30% for very low daily production volumes, then decline toward 18% as daily production rose to above 70,000 b/d.) Using the values shown above, the formula would yield an R-factor of 0.2804 and therefore an annual service fee of $83,815,225. Table 6–1 also calculates a "no development" case in which the IOC does not incur unreimbursed development and production expenses.

There are three determinants of take in the case of the risk service contract: 1) the defined cost recovery component, 2) the capital costs, and 3) the associated R-factor scale components. A widely used version of the risk service contract today is one with a sliding-scale R-factor with both production volume and crude oil price. The result is a contractual form limiting the rate of return to the IOC and also limiting downside risk, while assuring the state most of the upside potential from production volumes and rising crude oil prices.

The R-factor approach is just one of many sliding-scale systems, generally referred to as *rate-of-return contracts* (ROR). Sliding scales based on rate of production, cumulative production, crude oil prices, or even variations in crude quality as in an oft-cited Guatemalan agreement are all designed to increase the state's take and put boundaries on the IOC's take as project life and profitability increase.[6]

Added Contractual Features

Regardless of whether the fiscal regime is a concession, a production sharing agreement, or a risk service contract, exploration and production agreements also include a variety of additional details. These details can have a substantial impact on the financial returns to the two primary parties and include signature bonuses, domestic market obligations, and investment uplifts.

Signature bonuses

Signature bonuses are up-front payments by the IOC to the state as part of the total fiscal agreement. The use of signing bonuses is common in many different industries to motivate contractors to explore, develop, and produce faster rather than slower (in order to start generating a return on the capital already expended and invested in the bonus). In the oil and gas industry, signing bonuses have become quite controversial in recent years. Some countries have also introduced production bonuses for either reaching specific production rates (such as barrels per day) or at cumulative levels of reservoir production). Although signature bonuses are generally not recoverable through cost oil or cost recovery, they are often deductible for both income and withholding tax calculation purposes.

As crude oil prices have risen, the size of bonuses as part of bids for rights to development blocks has skyrocketed. In the past, signature bonuses often fell between $1 million and $10 million. Recent exploration and development agreements have seen bonuses exceeding $100 million. In addition to signature bonuses, some concessions or PSAs may include discovery bonuses and production bonuses (payments triggered by specific volume levels of production) or, in some cases, even price-based bonuses tied to crude oil prices.

The controversy over signature bonuses is because of corruption and diversion of funds. For example in July 2000, Marathon Oil made a signature bonus payment of $13.7 million to Sonangol, the national oil company of Angola, as the first of three payments for the rights to develop an offshore block.[7] The payment was made to a bank in Jersey, one of the Channel Islands off the United Kingdom. Jersey is known for its bank secrecy laws and zero tax provisions on foreign source income (a so-called *offshore financial center*). A number of news reports argued that the funds were rewired in a matter of hours to a variety of other Sonangol accounts around the globe, some later proven to be the personal accounts of Angolan government ministers, the Angolan president, and a variety of senior government officials. The controversy, including repeated concerns by the International Monetary Fund (IMF) on lending to Angola given its poor record on transparency and oil signature payments, raged on for several years.

Alleged corruption like that of Angola has led to increasing demands by a number of organizations globally for transparency in payments made by IOCs to states. In February 2001, BP announced publicly that it would begin publishing an annual statement of payments made to Angola's national oil company Sonangol. BP paid a very large amount for the development rights to Angola's Block 31, a sum of $111 million. The Angolan government, in an attempt to appease many of its global critics, contracted with KPMG for a financial audit of payments received by the state from IOCs. KPMG later reported that the total signature bonus payments from all JV partners on Block 31 totaled $335 million.

Angola's Sonangol CEO, Manuel Vicente, responded harshly and quickly in a letter to BP's CEO and chairman, Lord Browne, that it considered BP to be "violating the conditions of legal contracts signed with Sonangol . . . and if confirmed, is a sufficient reason to apply measures established in Article 40 of the PSA [production sharing agreement] i.e., contract termination." The following letter is the complete text of the missive. Since that time no other major IOC has reported specific financial payments made in signature bonuses and associated revenues to states, preferring to conform to state requirements for confidentiality in order to gain access to ever more valuable petroleum reserves. There are a variety of competitive and contractual arguments in favor of the state's insistence over confidentiality. However, the loss of transparency on up-front cash payments to governments and NOCs and the inability to then track those payments is considered a major factor in the rise of corruption in many oil producing states.

The Sonangol Letter to BP

Dear Sir,

It was with great surprise, and some disbelief, that we found out through the press that your company has been disclosing information about oil-related activities in Angola, some of which have a strict confidential character. According to the media, your company promised to continue to supply further such information in a letter dated 06/02/01 and signed by Mr. Richard Oliver [sic], thereby seriously violating the conditions of legal contracts signed with Sonangol.

As a result, we are making enquiries to confirm the veracity of information that has been published which, if confirmed, is a sufficient reason to apply measures established in Article 40 of the PSA [production sharing agreement] i.e., contract termination.

We are aware that some oil companies have been under pressure by organized groups that use available means in an orchestrated campaign against some Angolan institutions by calling for "pseudo-transparency" of legitimate government actions. As the national authority that awards concessions, Sonangol is fully aware that its economic link with your company should not be mixed with other relationships that seriously violate existing contracts in order to attract bogus credibility. Given this situation, we highly recommend that your company scrupulously respect the agreements that it has signed with Sonangol, as well as Angolan legislation relating to the confidentiality of information. May we recall there are specific channels, which should be respected, to release any type of authorized information.

Given the seriousness of this situation, if the provision of information by your company is confirmed and we observe moral or material damage thereof, we reserve the right to take appropriate action. The same is valid if you repeat such practices in the future.

Finally, and in the hope of maintaining the good relations that we have always had with the oil companies that operate in Angola, we strongly discourage all our partners from similar attitudes in the future.

In closing, please accept our best wishes.

The President of the Administrative
Council
Manuel Vicente

Source: "Some Transparency, No Accountability," Human Rights Watch, Section IV. Government Attempts to Restrict Information, January 12, 2004, http://www.hrw.org/en/node/12195/section/7.

Domestic market obligations

Domestic market obligations are clauses in PSCs that require the IOC to sell a proportion of its profit oil back to the state at a stated discounted price. Because this results in an additional reallocation of oil at below-market rates, it effectively increases the state's share of the take. Although obviously within the state's rights, this method of redistribution of oil and value may overly complicate the IOC's ability to actually plan and value prospective properties.

Investment uplifts

Although commonly used in the oil and gas industry, the term *uplift* is used in a variety of different ways for fiscal regimes. Probably the most consistent definition of uplift is the ability of an IOC to recover some added percentage of investment as a deductible expense once operations begin. An investment of $10 million with an uplift of 10% would allow the eventual deduction of $11 million in recoverable expenses. A second often-used form of the term uplift is simply in the calculation of the rate of return target or limit in many ROR contracts.

Stabilization clauses

From the very beginning, one of the biggest risks to any concession holder was the possibility that the state would unilaterally change the agreement. As a result, many concessions contained *stabilization clauses*, which attempt to freeze the laws of the host state and prevent changes. The following clause from a concession in Kuwait is a classic example:

> *The Shaikh shall not by general or special legislation or by administrative measures or by any other act whatever annul this Agreement except as provided by Article 11. No alteration shall be made in terms of this agreement by either the Shaikh or the Company except in the event of the Shaikh and the Company jointly agreeing that it is desirable in the interest of both parties to make certain alterations, deletions or additions to this Agreement.*[8]

Unfortunately, regardless of the specific language used in the stabilization clause, the governing law of most concessions or PSAs has always been that of the host country. As a result, in instances of debate, many countries have argued that clauses such as these are unenforceable as they offend the host state's own (natural resource) sovereignty. Similarly, clauses and contractual features calling for dispute settlement and arbitration authorities outside the host country have seen limited success. Some concessions have attempted to "internationalize" the contract by calling for a conformity with the laws of the state as long as they do not diverge from generally accepted standards of international law or tribunals.

Top-Line Risks

All petroleum projects have two primary top-line risks: production and price. Market demand is not usually a risk; most oil and gas is saleable at some price. Market demand can become a risk when there is not a huge market for the product or the market is not very liquid (in terms of financial liquidity), such as the markets for LNG and extra-heavy oil.

Production risk is a function of petroleum geology and economic motivation. Price is driven by global supply and demand for oil and gas. Figure 6–4 illustrates a common oil reservoir's expected production profile. The reservoir's expected yield is 100 million barrels over its 20-year life. This production profile is customary in oil: a rapid ramp-up to peak production, followed by a decline over time, either gradual or rapid depending on the specific reservoir.

Although there is an absolute volume of total oil-in-place in any reservoir, the actual recoverable content is never clearly defined. Depending on rate of extraction combined with natural and supplemental recovery techniques (described previously in chapter 5 on production), the recoverable production will vary. Oil prices will also impact total recoverable production because at higher crude prices costly repressurization and injection techniques become more financially viable. This viability alters the planned production and development plan, changing projected operating costs over the long term.

The production decline depicted here is quite gradual. Many projects see a much more rapid decline in production, often falling off to residual amounts within five years of peak output. The rate of production and ultimate recovery for any given reservoir is dependent on many factors, including geologic complexity, facility developments, regulatory requirements, use of secondary or tertiary mechanisms, and other unforeseen or unique challenges. Chapter 5 discussed reservoir productivity challenges in the Chad-Cameroon project in Africa. The reservoir's annual production volumes have declined more rapidly than expected as a result of an extremely porous sand structure that envelops the oil, resulting in faster depressurization than predicted. As a result, total production is now expected to be roughly 80% of what was originally planned.

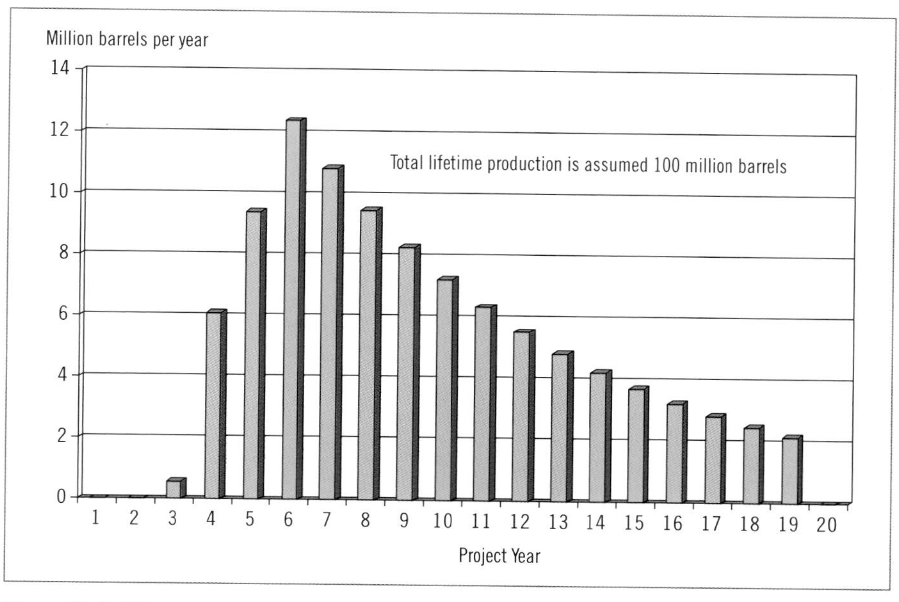

Figure 6–4. Oil reservoir production over time

Petroleum prices

The second major project driver—and primary value risk—is price.[9] There is probably no product price that has been the subject of more conjecture and predictive investment over time, with such poor results in prediction, than the price of oil. As illustrated in figure 6–5, the price of oil since 1970 has varied between a few dollars per barrel and the 2008 prices of $147 per barrel (slightly over $120 on average for the quarter).

The oil price presented in figure 6–5 is that of UK Brent crude. There are, however, a number of different types of crude oil, and their respective prices (crude prices and markets are discussed in detail in chapter 10). The obvious implication of figure 6–5 is the radical increase in the price of oil since 2003. Although oil prices have never been considered predictable, there was a long-standing belief/assumption that the price per barrel would in some way oscillate around $20/bbl over time. Many in the industry now believe that the long-term price may stay above $75 or $80/bbl for many years to come.

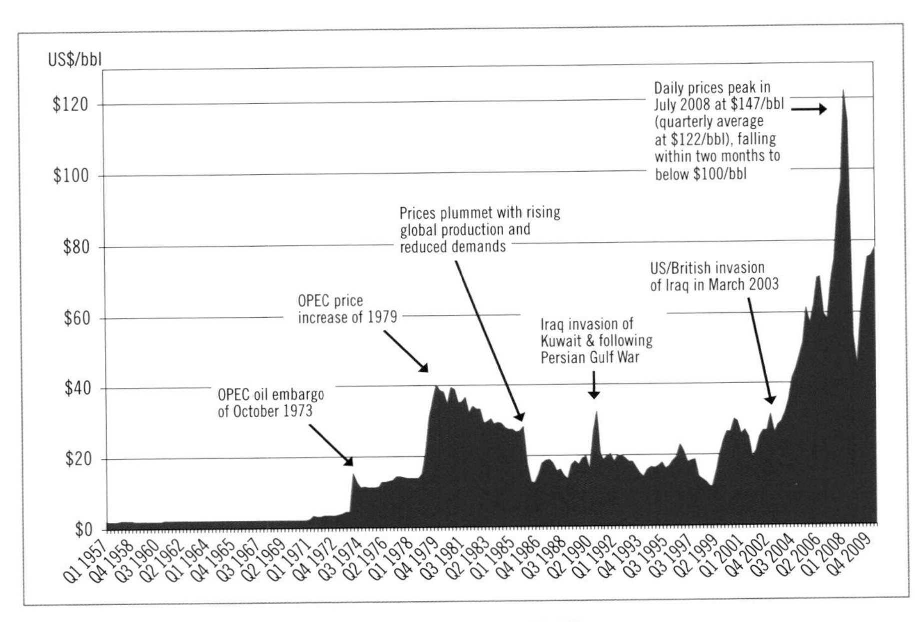

Figure 6–5. Brent crude oil prices (quarterly, 1957–2010)
Source: *International Financial Statistics*, monthly, International Monetary Fund, www.imf.org.

This price uncertainty has led to several consistent behaviors by both states and IOCs when planning the development of oil projects. First, developers in the most successful oil companies are conservative. Regardless of the current price of oil, there is a belief that the price of oil will always fall back to a level that has been observed over time. For example, until the 2006 to 2008 period, a typical 20- or 30-year forecast might assume $20/bbl. This price assumption was not based on a belief that $20 would actually occur at a specific point in time. Rather, it is a conservative behavior based on the reality that predicting prices is impossible, so the safest course of action is to assume prices will always return to a minimum sustainable level. The price of oil has never been found to be correlated with any other specific economic variable, whether it be the rate of inflation, economic growth, or even the price of gold. This lack of predictability has driven most oil industry companies to assume a long-term base price slightly above what is economically sustainable given baseline world oil demand and gradually rising production costs.

The rapid increase in crude oil prices between 2005 and 2008 had a strong impact on the behavior of many oil-rich states. Although the proportional division of profits and returns—the takes—are spelled out in detail in all concessions and PSAs, higher crude prices prompted many states and NOCs to argue for greater proportions of cash flow proceeds. The argument they have made is consistently the same:

> *Concession/PSAs spelled out a division of returns assuring IOCs of acceptable, appropriate, or even potentially highly profitable returns on their investments. Additional returns now being generated by oil prices above $100/barrel should go to the state.*

A Hypothetical Sub-Saharan PSA

Consider the potential development of a hypothetical petroleum project in sub-Saharan Africa. The project is expected to produce 100 million barrels of oil over a 20-year project life. The preliminary financial analysis is presented in table 6–2. Much of the final analysis depends on the negotiated components of the production sharing agreement (PSA).[10]

The production profile shows volumes peaking rapidly in year 6 and then declining gradually through years 19/20. The preliminary financial analysis assumes a $20/bbl share price over the 20-year life. Although neither party believes this to be a true forecast, for the sake of preliminary discussions, $20 is a cautious initial assumption. The $20 price is also a reasonable assumption for financiers to use as a basis for lending decisions on the project. Production costs are estimated over the life of the project, with a gradual increase of the cost per barrel produced.

The capital investment required by the project is $300 million and is spread over the first five years. This capital, all provided by the IOC, represents a significant investment. Depreciation expenses commence in the first year of operations and are based on a five-year straight line depreciation schedule, depending on the investment made by the respective operating dates.

Table 6–2. Sub-Saharan production sharing agreement (PSA): Cash-flow projection

Year	0	1	2	3	4	5	6	7	8
Production (1,000 bbl)		0	0	578	6,100	9,420	12,400	10,850	9,494
Oil Price ($/bbl)		$20	$20	$20	$20	$20	$20	$20	$20
Gross Revenue ($m)		0	0	11,560	122,000	188,400	248,000	217,000	189,880
Royalty (%)	10%	0	0	1,156	12,200	18,840	24,800	21,700	18,988
Net Revenue ($m)		0	0	10,404	109,800	169,560	223,200	195,300	170,892
Capital Costs ($m)		30,000	40,000	100,000	60,000	70,000	0	0	0
Depreciation ($m)				34,000	46,000	60,000	60,000	60,000	26,000
Operating Cost ($/1,000 bbl)				5.50	2.60	2.40	2.30	2.36	2.40
Operating Expense ($m)				3,179	15,860	22,608	28,520	25,606	22,786
Total Expenses ($m)		0	0	37,179	61,860	82,608	88,520	85,606	48,786
Cost Recovery Limit ($)	50%			5,780	61,000	94,200	124,000	108,500	94,940
C/R C/F ($m)		0	0	31,399	32,259	20,667	0	0	0
Cost Recovery ($m)		0	0	5,780	61,000	94,200	109,187	85,606	48,786
Total Profit Oil ($m)		0	0	4,624	48,800	75,360	114,013	109,694	122,106
State Profit Oil Share ($m)	60%	0	0	2,774	29,280	45,216	68,408	65,816	73,264
IOC Profit Oil Share ($m)	40%	0	0	1,850	19,520	30,144	45,605	43,878	48,843
Signature Bonus ($m)		10,000	0	0	0	0	0	0	0
Tax Loss Carry Forward ($m)			–10,000	–10,000	–39,549	–20,889	0	0	0
Taxable Income		–10,000	–10,000	–39,549	–20,889	20,847	66,272	43,878	48,843
Income Tax (30%)	30%	0	0	0	0	6,254	19,882	13,163	14,653
IOC Net Cash Flow ($m)		–40,000	–40,000	–95,549	4,660	25,482	106,391	90,714	60,190
DCF @ 12% (half-year)	12%	–37,796	–33,747	–71,975	3,134	15,302	57,043	43,427	25,727
Cumulative DCF		62,798	13%						
IOC cumulative take		23.6%							
Bonuses ($m)		10,000	0	0	0	0	0	0	0
Royalty ($m)		0	0	1,156	12,200	18,840	24,800	21,700	18,988
State Profit Oil ($m)		0	0	2,774	29,280	45,216	68,408	65,816	73,264
Income Tax		0	0	0	0	6,254	19,882	13,163	14,653
State Net Cash Flow ($m)		10,000	0	3,930	41,480	70,310	113,089	100,680	106,905
DCF @ 12% (half-year)	12%	9,449	0	2,961	27,898	42,222	60,635	48,198	45,694
Cumulative DCF		408,998	87%						
State cumulative take		76.4%							

9	10	11	12	13	14	15	16	17	18	19	Total
8,307	7,269	6,360	5,565	4,869	4,261	3,728	3,262	2,854	2,498	2,185	100,000
$20	$20	$20	$20	$20	$20	$20	$20	$20	$20	$20	
166,140	145,380	127,200	111,300	97,380	85,220	74,560	65,240	57,080	49,960	43,700	2,000,000
16,614	14,538	12,720	11,130	9,738	8,522	7,456	6,524	5,708	4,996	4,370	200,000
149,526	130,842	114,480	100,170	87,642	76,698	67,104	58,716	51,372	44,964	39,330	1,800,000
0	0	0	0	0	0	0	0	0	0	0	300,000
14,000											300,000
2.46	2.54	2.64	2.72	2.82	2.94	3.08	3.24	3.40	3.60	3.40	
20,435	18,463	16,790	15,137	13,731	12,527	11,482	10,569	9,704	8,993	7,429	263,819
34,435	18,463	16,790	15,137	13,731	12,527	11,482	10,569	9,704	8,993	7,429	
83,070	72,690	63,600	55,650	48,690	42,610	37,280	32,620	28,540	24,980	21,850	1,000,001
0	0	0	0	0	0	0	0	0	0	0	84,325
34,435	18,463	16,790	15,137	13,731	12,527	11,482	10,569	9,704	8,993	7,429	563,819
115,091	112,379	97,690	85,033	73,911	64,171	55,622	48,147	41,668	35,971	31,901	1,236,181
69,054	67,427	58,614	51,020	44,347	38,502	33,373	28,888	25,001	21,583	19,141	741,709
46,036	44,951	39,076	34,013	29,565	25,668	22,249	19,259	16,667	14,388	12,760	494,473
0	0	0	0	0	0	0	0	0	0	0	
0	0	0	0	0	0	0	0	0	0	0	
46,036	44,951	39,076	34,013	29,565	25,668	22,249	19,259	16,667	14,388	12,760	
13,811	13,485	11,723	10,204	8,869	7,700	6,675	5,778	5,000	4,317	3,828	145,342
46,225	31,466	27,353	23,809	20,695	17,968	15,574	13,481	11,667	10,072	8,932	339,131
17,641	10,722	8,322	6,468	5,019	3,891	3,011	2,327	1,798	1,386	1,098	62,798

9	10	11	12	13	14	15	16	17	18	19	Total
0	0	0	0	0	0	0	0	0	0	0	10,000
16,614	14,538	12,720	11,130	9,738	8,522	7,456	6,524	5,708	4,996	4,370	200,000
69,054	67,427	58,614	51,020	44,347	38,502	33,373	28,888	25,001	21,583	19,141	741,709
13,811	13,485	11,723	10,204	8,869	7,700	6,675	5,778	5,000	4,317	3,828	145,342
99,479	95,451	83,057	72,354	62,954	54,725	47,504	41,190	35,709	30,895	27,339	1,097,051
37,965	32,524	25,269	19,654	15,269	11,851	9,185	7,111	5,504	4,252	3,359	408,998

Negotiations continue between the state and the IOC over the following critical components of the production sharing agreement:

- **Royalty.** The IOC has proposed an 8% royalty rate as a percent of total gross revenue per year, while the state has argued consistently for 12% or more. The baseline analysis assumes 10%.

- **Cost recovery.** The cost recovery limit has been initially set at 50%, meaning that regardless of what the IOC suffers or claims in total operating expenses (including depreciation) in any one year, it will not be allowed to allocate more than 50% of gross production, either physical barrels or equivalent cash flow, towards cost recovery. This is a standard rate used in many PSAs today and will most likely be agreed to by both parties. Cost recovery carry forward (CRCF) is the amount of annual operating cost that exceeds the specified cost recovery limit, which is then carried forward to future years.

- **Profit oil split.** The profit oil split (net revenues less cost recovery) is to be 60/40, 60% going to the state and 40% to the FOC. The IOC has continually argued for an equal split of 50/50, but the state has been adamant over having the majority share. Most PSAs signed in recent years have tended to fall near the 60/40 split.

- **Tax rate.** The IOC will pay 30% corporate income taxes to the state on its profit oil earnings. As noted previously, income taxes paid in the country by IOCs is a large source of income in many oil projects.

- **Signature bonus.** A $10 million signature bonus payment by the IOC to the state on the signing of the development agreement has been agreed. Although most signature bonuses were $10 million or less for many years, bonuses have risen into the hundreds of millions for many of the most promising developments in recent years.

The final net cash flows to each party are then collected and discounted to determine the net present value of the proposed petroleum project. The IOC's take, its share of net cash flows, comes exclusively from its profit oil split. The state's take arises from four different cash flow components—royalties, profit oil split, income taxes, and the signature bonus. As shown in table 6–2, using a discount rate of 12%, the proposed project and estimated cash flows produce a positive net present value (NPV) for both parties. The discounted cash flow analysis presented in table 6–2 uses the half-year convention often used in the oil industry, rather than the typical textbook assumption of end-of-year occurrence. The discount rate itself is an item of individual choice. The state and the IOC will choose their rate for their own purposes of evaluation.

In many ways a fiscal regime analysis of a project is a zero-sum game. Alterations in royalty rates, cost recovery limits, profit oil splits, and the various taxes imposed on IOCs by states move cash flow between the two parties. As is usually the case with present value analysis, large cash inflows or outflows that occur early in the life of the project have the largest impact on NPV. For example, if the signature bonus were increased significantly, it would damage the project NPV as a result of its falling so early in the project. (Obviously the signature bonus is a major value-add for the state, as it occurs immediately and is paid regardless of eventual oil produced or prices received.)

The cash flow analysis presented in table 6–2 is actually quite simplistic compared to many agreements used globally. Royalties, profit splits, and tax rates are increasingly executed according to a variety of sliding scales, most of which are constructed to increase the take of the state under increasing production and profitability conditions and to protect the IOC against low crude prices over the project life.

The operating margin for the reservoir is the difference between price per barrel and cost per barrel. This margin drives results. Although the other factors redistribute earnings between the state and the IOC, the operating margin combined with acceptable capital outlays determine the project acceptability. A large operating margin in which the price per barrel far exceeds the production cost per barrel can also be misleading if the project's capital costs are not estimated correctly (though cost recovery for the IOC serves as a buffer against such errors), or are difficult to truly estimate upfront given the physical or technological challenges posed by the project.

One example of technological risk was the proposed Camisea gas field in eastern Peru studied by Mobil and Shell in the late 1990s. The proposed field would be developed in the middle of the Peruvian cloud forest, a socially and environmentally sensitive region, increasing site infrastructure and operating costs radically. It would then have required the construction of a dual pipe over the Andes to the Pacific Coast near Lima, dropping 12,000 feet in altitude in a little more than 20 miles. Mobil and Shell walked away because this was a challenge thought to be too technically risky given expected returns at the time. The project was later undertaken and completed, on a smaller scale and alternative structure, by a consortium of companies including Hunt Brothers and PeruPetrol.

Figure 6–6 illustrates the net cash flows—the takes—accruing to the state and IOC over the 20-year project life. Note that the net negative cash flows occurring in years one through three are the result of the investment required to develop the reservoir. This negative cash flow is borne completely by the IOC. The state in this example provided none of the investment capital. The 60/40 profit oil split is the primary parameter that essentially guarantees that the state's take will always exceed the IOC's take.

The state's take is always positive, showing that all net investment is made by the IOC and not the state. This is integral to the construction of the modern PSA because the combined cash flow earnings assure the state of reaping the majority of benefits. The IOC bears nearly all cost/investment and risk. The cumulative cash flows accruing to the two parties over the life of the project are $1,010.518 million to the state and $425.663 million to the IOC. This represents a final take of 76.4% to the state and 23.6% to the IOC. Of the state's total take under the baseline analysis, 18.2% is in royalties, 67.6% in profit oil split, 13.2% in taxes, and less than 1% (0.9%) in the up-front signature bonus.

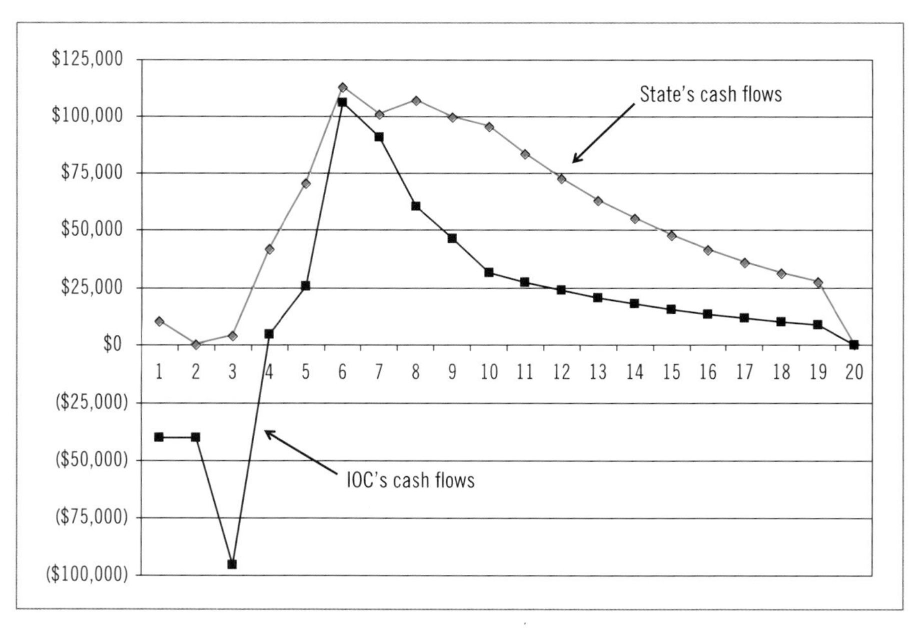

Figure 6–6. Sub-Saharan sample cash flow takes

Table 6–3 provides a comparison of outcomes over a variety of different sensitivities and scenarios. The baseline analysis is one of the more favorable outcomes for the IOC. As royalty rates and taxes increase, the state's share of final cumulative take increases past 80%. Changes in royalties and taxes are also clearly seen to redistribute earnings as total cash flow take does not change. Higher oil prices have a slightly counterintuitive impact, as the state's share falls slightly if the project enjoys higher crude oil prices over its lifetime. Total cash flow earnings increase to both parties with higher per-barrel prices. Although not included in the model presented in tables 6–2 and 6–3, a growing number of PSAs around the globe have adopted increasingly sophisticated profit oil splits that increase the state's share as prices rise, sometimes approaching 100% of profits above a specified crude oil price. These PSAs have obviously eliminated much of the upside to oil and gas developments for the IOCs.

Table 6–3. Comparison of sub-Saharan PSA assumptions and project proceeds

PSA Component	Baseline	High Signature	High Royalty	Higher Price	Higher Tax	High State
Petroleum price ($/bbl)	$20.00	$20.00	$20.00	$30.00	$20.00	$30.00
Royalty rate (%)	10.0%	10.0%	20.0%	10.0%	10.0%	20.0%
Cost recovery limit (%)	50.0%	50.0%	50.0%	50.0%	50.0%	50.0%
Profit oil split:						
State split (%)	60.0%	60.0%	60.0%	60.0%	60.0%	60.0%
IOC split (%)	40.0%	40.0%	40.0%	40.0%	40.0%	40.0%
Income tax rate (%)	30.0%	30.0%	30.0%	30.0%	50.0%	50.0%
Signature bonus ($m)	10,000	20,000	10,000	10,000	10,000	20,000
Cumulative net cash flows (%)						
State take	76.4%	78.3%	80.3%	75.7%	83.1%	85.3%
IOC take	23.6%	21.7%	19.7%	24.3%	16.9%	14.7%
Total take	100.0%	100.0%	100.0%	100.0%	100.0%	100.0%

PSA Evolution

The first PSA was introduced in Indonesia in 1966. Since that time, PSAs have increased in complexity and sophistication as state and NOC knowledge of oil and gas development have risen at the expense of IOCs. The evolution of PSAs helps in understanding how changing petroleum market conditions and the interpretation of laws and incentives for states and IOCs combined to alter many PSA conditions.

First generation (1966–1975). The state retained ownership of all oil and gas produced, including that stored at export terminals. Although there were no royalty rates and taxes applied, the state was guaranteed revenue as a result of a specified profit split, regardless of cost recovery.

Second generation (1976–1983). Starting around 1976 and following the OPEC market interventions in 1973, all petroleum states were now aware of the growing value of their assets and their commensurate increase in relative bargaining strength. More flexible cost recovery limits were now more widely used (Indonesia eliminated them) as more difficult production areas came into consideration, while the profit oil split was pushed upwards to 85/15 in many markets to reap the higher crude oil price.

There was an additional more subtle change that was critically important for all US-based majors—the requirement of IOCs to pay a tax *directly* to the state, rather than to NOCs. Under US tax law, payments made by IOCs to NOCs under PSAs were not corporate income taxes. This meant the payments could not be credited with foreign tax credits when profits were remitted to the US parent company, reducing the after-tax profit for US majors dramatically. By changing the tax payment to the state, the US tax authorities now classified the tax paid a deemed-paid foreign tax payment and therefore eligible for foreign tax credit classification. This removed a major impediment for US majors to enter many foreign markets.

Third generation (1984–1987). A combination of slight adjustments to investment tax credits, corporate tax obligations, and corporate tax rates combined to increase the general sophistication of PSAs but not drastically alter its incentives or disincentives to IOCs.

Fourth generation (1988–present). With the collapse of oil prices in the mid-1980s, the new PSAs signed, beginning in 1988, showed increased flexibility in terms and conditions in order to provide additional negotiating space for states to attract IOCs (table 6–4). Many new major investments in exploration and development activity were now seen as marginal, others as declining profit opportunities.

Table 6–4. Key components across production sharing agreements

Component	Indonesia (1966)	Indonesia (post-1988)	Angola (1989)
Area	Designated blocks	Designated blocks	Designated blocks
Duration: Exploration	5 years	3 years	5 years (3 + 2 ext)
Production	20 years	20 years	20 years
Signature Bonus	Yes	Yes	Yes
Production Bonus	None	Yes	Yes
Royalty Rate	None	None	None
Cost Recovery Limit	40%	80%	50%
Depreciation	5 years straight line	25% decl 5 years	5 years straight line
Profit Oil Split (State / IOC)	65 / 35	71.2 / 28.8	Sliding scale up 45/55 to 90/10
Tax Rate	None	48% income tax	50% income tax
Domestic Market Obligation	25% (15% of mkt price)	25% (10% of mkt price)	None
Investment Uplift	None	None	40% tangible capital
Ringfencing	All	All	None
Price Cap ($/bbl)	None	None	State takes > $20/bbl
State Participation	None	Up to 50% in joint operating	Yes, back-in option for development (to 51%)

Source: Compiled by authors from Johnston (1994) and Bindemann (1999).

Indonesia's fourth generation also introduced *first tranche petroleum*, which split the first 20% of production equally between IOC and state (the 71.2/28.2 split shown in table 6–2) before cost recovery. This has been viewed by many industry analysts as an effective royalty, as the state is guaranteed a return prior to any cost recovery. The remaining 80% of production was available for cost recovery (no limit). Although on the surface this appeared to be a less restrictive cost recovery structure, it was an exact cost recovery limit as seen previously. Marginal fields now had a slightly more attractive profit split (for the IOC) of 75/25 rather than the standardized 80/20 under second generation PSAs (this was revised to 65/35 in 1994 as a result of declining oil prices).

The Angolan PSAs signed in the mid- to late 1990s were some of the most demanding, as rising oil prices were met with sliding production volume profit oil splits and price caps, where the state took nearly 100% of oil above $20/bbl prices. Regardless of oil prices and rising production costs, all states now reserve the right to take equity positions in many development projects—state participation—allowing them to become equity partners after exploration and development efforts and investments have been made.

Fiscal Regimes Today

Fiscal regimes in effect today cover a very wide range of markets and rates. Figure 6–7, using analysis from Daniel Johnston, provides an overview of countries, agreements, and their estimated takes. From this spatial array, it is clear that the sharing agreements average the highest government takes, the royalty/tax regimes the lowest, and the PSA/PSCs the middle ground.

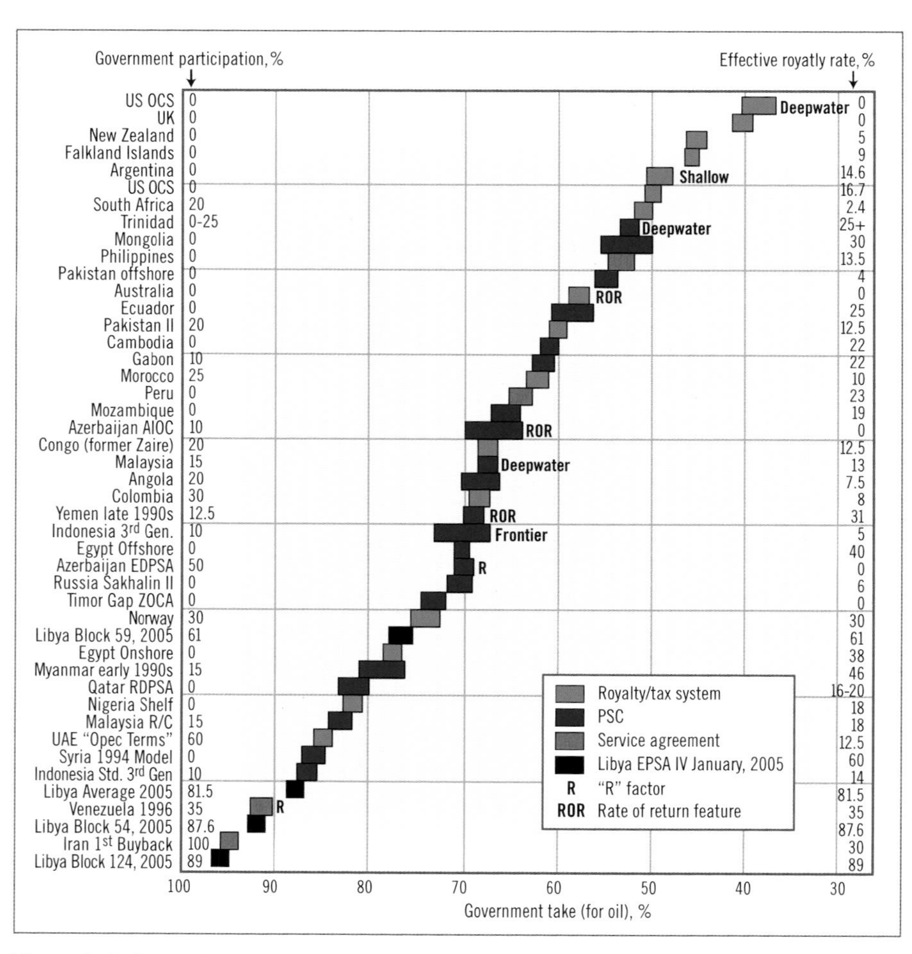

Figure 6–7. Global fiscal regimes and takes. Although every fiscal regime is slightly different, the royalty and tax systems average the lowest government takes, while service agreements average some of the highest government takes.

Source: Daniel Johnston, "Impressive Libya Licensing Round," *Oil & Gas Journal*, April 18, 2005, Figure 4.

While figure 6–7 focuses on the takes of different countries, table 6–5 summarizes the three major fiscal regime classes and their basic business features.

Table 6–5. Comparison of common fiscal regimes

Feature	Royalty/Tax	Production Sharing Agreement	Risk Sharing Agreement
Ownership of Hydrocarbons	Transferred to IOC at the wellhead	Transferred to IOC at delivery point	Stays with the State or NOC
IOC Control	High	Moderate to low	Low
Government Control	Low	Moderate to high	High
IOC Lifting Entitlement	Typically around 90%	Typically 50% to 60%	None
Cost Recovery Limit	None	Frequently	Rarely
Ownership of Production Facilities	IOC	Typically state/NOC	NOC/state
Limits to IOC Profitability	Few	Significant	Absolute

Royalty/tax agreements generate on average the greater takes for the E&P firms and the lowest government takes. Because many of these regimes limit government returns to a royalty rate that is often less than 20% of the value of the oil, they represent the greatest potential returns to private firms. Service agreements generate the lowest returns for E&P firms and reserve all true price or other market-based returns to the state. The E&P firms here are in essence nothing more than contractors and are paid for services rendered on a limited basis. PSAs, as opposed to common misconception, are not necessarily instruments of maximum government take and often fall into a mid-range of balance between the IOC and the state. Where any individual PSA/PSC falls is in the details of the individual agreement.

IOC operating strategies are heavily influenced by the type of fiscal regime. Although the reservoir or field itself is the primary determinant of rate of extraction, the IOC is motivated to produce faster under most royalty/tax systems. If the IOC is not constrained by cost recovery limits, owns the oil, and pays the state a royalty and set of taxes, it recovers its investment faster and increases the project's net present value by producing rapidly. In a favorable price environment, the value capture is even faster.

An IOC operating under a typical PSA will be quite cognizant of cost recovery limits and the ability to recover investment outlays. A more difficult dimension is the increasing number of PSAs that alter IOC profit oil split on the basis of the IOC's investment recovery or rate of return. This contractual feature, if combined with higher crude oil prices, can lead to some counterintuitive outcomes. For example, one offshore Angolan PSA in effect limits the physical profit oil split of the IOC to its recovery of initial investment. When crude oil prices skyrocketed in 2007 and 2008, the IOC recovered its investment much more rapidly than had been expected and found itself with decreasing profit oil volumes much earlier in the project life than expected. Since those volumes were important for the company's general business portfolio, it was a significant hit to the IOC's strategy.

Government and Policy Change

Governments compete to attract investments. But the competition is primarily regional, and as a result some countries or areas are not competitive at a global level. On average, fiscal systems make small but potentially profitable oil fields uneconomic. Although this approach might be expected from oil-producing and oil-exporting countries, it is not in the best interests of oil-importing and self sufficient countries. Moreover, world oil production could be increased significantly if importing and self sufficient countries offered better terms for such fields.

—Chakib Khelil, "Fiscal Systems for Oil,"
Note 46, The World Bank, May 1995, p. 4.

Unlike some of the earliest concessions signed in the Middle East in the 1930s, which offered 70-year exclusivity and $0.35/bbl royalty rates, fiscal regimes today are complex and constantly changing. As shown previously, the state now enjoys relatively more power and more of the take of cash flows from the development of its oil.

But calculated take statistics do not tell the whole story and may not prove to be very good indicators of who benefits in what ways over time. Fiscal regimes have often reflected the short-term business environment and are particularly tied to the price of oil. Although the major oil and gas developers have long argued that it is a long-term industry of long-term capital investments and returns, global political environments remain decidedly short-term in their outlook.

In the late 1990s, when oil prices dropped to $10/bbl, countries like Venezuela offered very attractive fiscal regimes with low tax and royalty rates to attract domestic and international oil companies.[11] Those same regimes became the subject of effective nationalization a decade later when the price of oil went above $100/bbl. Similarly, Russia in the early 1990s was effectively bankrupt and desperate to attract foreign E&P firms to develop the oil and gas in the bitterly cold waters off Sakhalin Island. The Sakhalin II PSC guaranteed the IOCs a full return of their investment before the state enjoyed any returns. Not surprisingly, a resurgent and more aggressive Russian state opted later for the near-nationalization of the project and an alteration of the ground rules.

When governments modify terms of existing fiscal regimes, is it because the original contracts did not adequately incorporate change expectations? Possibly, especially when looking at some recent fiscal regimes that were constructed with expectations of a changing environment. For example, the fiscal regimes in Malaysia and Azerbaijan have detailed and strict price escalator clauses increasing the state's take when certain price thresholds are passed. Angola's fiscal regimes have gone even further, setting a limit on the rate of return that an IOC investor can reap from offshore developments. At the same time, the extreme contracts in Iran require all cost overruns to be incurred in full by IOCs, resulting in E&P firms motivated to work slower, cheaper, and sometimes with out-of-date technology.

Fiscal regime changes affect all private parties and not just foreign companies and oil majors. Various governments considered tax law changes—so-called *windfall profits taxes*—during the recent run up in oil prices.[12] Those debates, heard loudly in both the US and UK, would alter the business returns and incentives for future oil and gas exploration and development.[13] It is increasingly clear that a private company conducting E&P anywhere in the world must trade off the terms of a regime for expectations of stability. What once appeared so attractive in Venezuela proved unsustainable. Other countries such as Qatar, though bargaining hard and imposing significant restrictions on IOC behavior, have in the end offered the IOCs a higher rate of return from a stable political and business environment and fiscal regime. Ironically, even in postwar Iraq where the United States has been accused of intervention and even war for oil, the resulting service agreements are hardly attractive to the IOCs, as described in the *Industry Insight on Postwar Iraq and Oil Sharing Agreements*. Change is also not necessarily whimsical or even bad simply because of government policy. As the following case example shows, the UK serves as something of a test case.

Industry Insight: Postwar Iraq and Oil Sharing Agreements

A consortium led by ExxonMobil Corp. and Royal Dutch Shell PLC accepted contract amendments made by the Iraqi government to develop an oil field in southern Iraq, a senior Ministry of Oil official said. Exxon Mobil, the first US company to gain access to Iraq's oil reserves, and Shell won the right to develop the West Qurna Phase 1 oil field following the country's first postwar round of oil bidding in June 2009. The field holds 8.7 billion barrels in proven oil reserves.

The Exxon-Shell consortium signed the deal in November but was awaiting final approval from the Iraqi government. The government proposed amendments to all the contracts recently awarded to foreign oil companies, though Iraqi oil officials declined to speak about the changes. However, an oil-industry executive said one amendment gives the government the right to change production levels in order to comply with quotas that the Organization of Petroleum Exporting Countries might impose on Iraq. Also, all companies need to insure workers and equipment, the executive said. He said Iraq will impose taxes on imported materials and equipment used to develop the fields.

ExxonMobil and its partner will be paid $1.90 for each extra barrel of oil the companies extract on top of current production at the field. They pledged to increase output to 2.325 million barrels a day from 279,000 barrels a day. ExxonMobil has 80% of the venture, with Shell holding the remaining 20%. The Exxon-Shell team beat out two other consortia, one led by Russian giant OAO Lukoil that includes ConocoPhillips, and the other led by China National Petroleum Corp.

Source: Excerpted by authors from Hassan Hafidh, "Iraq Says Exxon, Shell Accept Contract Changes," *Wall Street Journal*, January 19, 2010.

Case example: Fiscal regime change in the UK

> *By that measure (volatility), the worst place to produce oil is not Russia or Venezuela, but Britain which is constantly tinkering with its tax rates.*
>
> —"Barking louder, biting less," *The Economist*, March 8, 2007

The United Kingdom's fiscal regime for its primary production area, the United Kingdom Continental Shelf (UKCS), is an example of a regime evolving with both the production and the global market.[14] The complexity of the UK political and legal environment is the starting point for understanding the fiscal regime. The UK incorporates four distinct legal systems: Northern Ireland, Scotland, England, and Wales. Unifying the various laws is extremely difficult.

The 1934 Petroleum Act confirmed that legal title to petroleum found in its natural state belongs to the Crown. The act also instructed the government to provide licenses for the exploration and development of the petroleum. Since 1964, 26 "seaward licensing" rounds have been completed. "A key objective of the government is to ensure that an appropriate share of profits generated from the production of oil and gas in the UK and on the UKCS is taxed, while maintaining industry interest in the oil and gas sector."[15]

Fiscal terms in the mid-1970s were focused on generating greater tax revenues. The focus shifted in the early 1980s to encourage investment in exploration and development. This focus was heightened by the realization that North Sea production was mature and on the decline. The UK is the only country within the European Union with any significant energy exports, and that is now changing. Oil production in the UKCS is estimated to have fallen by 50% since 1999. In 2008 production was approximately 1.5 mbpd per day. In that same year the UK consumed 1.7 million b/d, making the UK a net importer of crude oil. Thus, if E&P firms were to commit more capital and risk in pursuing smaller opportunities in the UKCS, when other more attractive opportunities were opening up across the globe, it was clear that the fiscal regime needed to adapt to changing times. Effective January 1, 2003, royalties were abolished on all onshore and offshore fields. In previous years in specific fields the royalty rate had been 12.5%.

The fiscal regime in effect today is made up of three tax components: 1) ring fence corporation tax of 30% on upstream profits; 2) a supplementary charge of 20% to the above ring fence tax; and 3) the petroleum revenue tax (PRT) of 50%. But each of these components has been altered over time to reduce the burden on private enterprise. The ring fence corporation tax, as shown in figure 6–8, has been whittled downward consistently over time. The supplementary charge, first introduced in 2002 at 10% and then increased to 20% in 2006, has been amended to provide additional credits if applied to PRT-taxable properties. The petroleum revenue tax (PRT) was eliminated for fields licensed after March 1993 and is now the subject of increasing reduction on those continuing fields licensed in the years previous to its elimination.

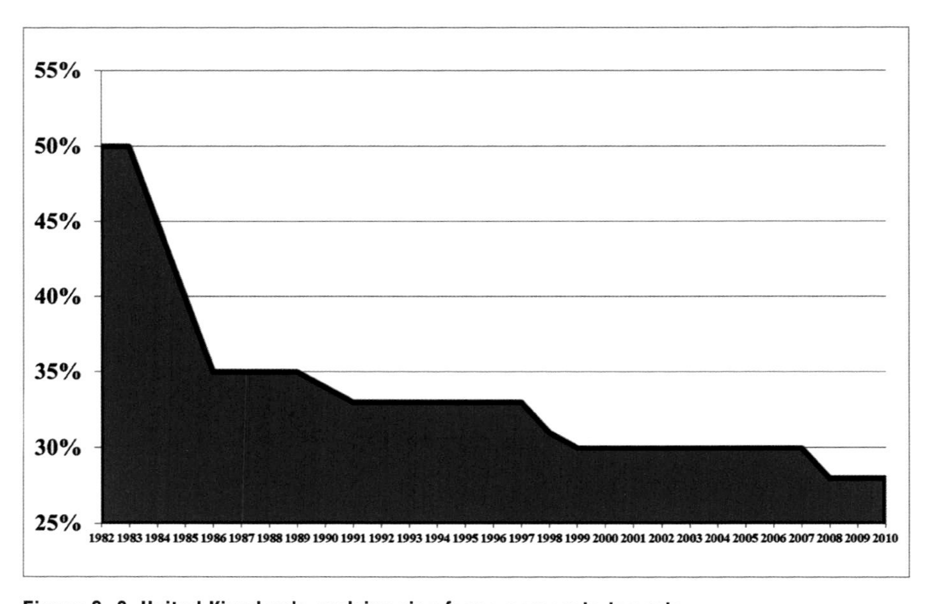

Figure 6–8. United Kingdom's evolving ring fence corporate tax rate

Source: Constructed by authors from *PetroReports: Fiscal and Regulatory Guide: United Kingdom*, Petroleum Service Group, Deloitte, September 2009, p. 63.

And the constant change has not stopped. From 2005 to 2008, the UK conducted an ongoing assessment of the UKCS fiscal regime to determine whether the current regime was undermining ongoing stability and whether it would in any way hamper the UK government's objective of maximizing the economic recovery of the oil and gas. That review has led to a more rapid rate of PRT phase-out, the expansion of tax carry forwards and carry backs on decommissioning costs (more important in a declining UKCS), and a number of accounting and cost changes aiding the smaller E&P operators in the North Sea. As opposed to what the earlier *Economist* quote implied, change or volatility is not necessarily bad.

Concessions in the United States

The United States is one of the few countries in which the government does not assume full ownership of all mineral rights within its geographic boundaries. Private owners of land and subsurface rights have the right to develop or transfer those rights to other private parties. The US government does, however, hold the development rights over those lands and offshore resources not held by private parties. For those development rights, the US uses a royalty regime. Over the past century, the US has charged oil and gas developers between 12.5% and 16.67% of all subsequent revenues generated from properties for the rights to explore and develop.[16] Payments by developers have gone primarily to the US Treasury and the management and monitoring authority for royalty payments, the Minerals Management Service of the US Department of the Interior. Cumulatively, more than $100 billion in revenues has been generated for the US government, the second largest contributor to US Treasury revenues behind personal income taxes.

The US government has passed and applied a number of changes to its royalty system in the past 15 years in an effort to stimulate more oil and gas exploration and development. For example, in 1995 the US passed the Deep Water Royalty Relief Act granting a royalty holiday to oil and gas companies drilling in government-owned deep waters in the Gulf of Mexico for leases sold between 1996 and 2000. Specifically, under the program, companies would not have to pay the normal royalties except when market prices reached $34 a barrel for oil and $4 per thousand cubic feet (mcf) for natural gas. Because oil and gas prices were quite low at that time, it was thought to be an incentive for increased oil and gas exploration and development. In 2004, the government estimated that the act would eventually cost the US nearly $80 billion in lost revenue on royalties.[17]

In 2004, as oil and gas prices began to climb, the US government offered new royalty incentives to companies exploring shallow waters, raising the threshold price at which producers had to begin paying royalties to $9.34/mcf. In 2006, the US passed the Gulf of Mexico Energy Security Act, opening an additional 8.3 million acres in the Gulf of Mexico to exploration and development. The US government has also expanded the Royalty-in-Kind (instead of cash payment) program in recent years. Under this new policy, the Minerals Management Service (MMS) dramatically expanded its program to take oil and gas royalties-in-kind (RIK), meaning the industry gives the government a portion of the oil and gas it takes from federal lands rather than paying royalties in cash.[18] Much of the oil taken under this program has been used to fill the federal government's Strategic Petroleum Reserve. The in-kind program has since been discontinued.

Change is also clearly in the cards for the US regulatory frameworks. The *Deepwater Horizon* disaster of 2010 resulted in the government trying to impose a six-month moratorium on all offshore drilling. The moratorium was overturned in a US District Court, leading to uncertainty as to what would happen. Other changes to offshore (particularly deepwater) exploration and development in the Gulf of Mexico are expected as a result of the disaster.

The Future

Figure 6–9 provides a snapshot of the ongoing changes among fiscal regimes around the world. Although clear trends cannot be discerned, we expect governments to continue to compete for exploration and development investments and IOCs to search out regimes allowing them adequate returns for ever-riskier investments. We also expect change to become a mainstay of fiscal regimes. As noted by Nana Asafo Adjaye, managing director of the state Ghana National Petroleum Corporation (GNPC), when commenting on the ongoing debate over one company's sale of development rights to a major IOC:

> *"It is in our interest that we have a system where the rules are clear. The fiscal regime is dynamic and already allows for continuous change."*[19]

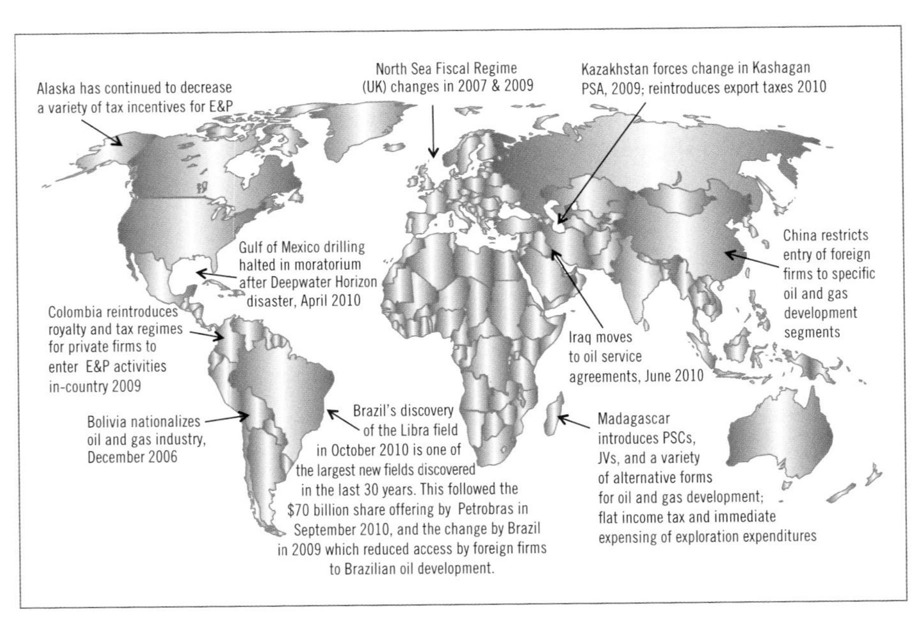

Figure 6–9. Recent changes in E&P fiscal terms

Summary Points

- Exploration, discovery, and production of oil and gas is only achievable after access to the prospective properties is gained through a contractual arrangement called a *fiscal regime*.

- Exploration and development contracts, usually between an international individual, organization, or company and a sovereign state, specify the timing, degree of commitment, and structure and distribution of investment and cash proceeds of the oil and gas venture.

- Most of the oil and gas development made in the first half of the 20th century was made under the auspices of concessions, in which an IOC made all of the investment. In return, the IOC paid a percentage of the revenues generated—the royalty—and possibly a portion of the profits to the sovereign state. The IOCs were largely believed to have the majority of the power under these long-term development agreements.

- Beginning in the middle of the 20th century, the power and returns of oil and gas development shifted from IOCs to states. The development of the PSA marked an important structural change in the oil and gas industry. With the development of PSAs, the government take has clearly increased.

- Recent years have seen the growth of the risk sharing agreement, a third category of fiscal regime in which the E&P company is relegated to contractor status and where the majority of returns accrue to the state. With risk sharing agreements private firms are often expected to cover substantial levels of investment and expense.

- Fiscal regimes create the initial rules for long-term oil and gas development. Over time the market, the hydrocarbons, and the political fortunes of many will dictate realized outcomes. In some cases the most restrictive fiscal regimes have resulted in the greatest IOC return as a result of good fields, good prices, and stable and sustainable development policies.

Notes

1. This chapter draws from a variety of sources including Daniel Johnston, *International Petroleum Fiscal Systems and Production Sharing Contracts*, Tulsa: PennWell, 1994; Smith, et al., *International Petroleum Transactions*, Second Edition, Rocky Mountain Mineral Law Foundation, 2000; Daniel

Johnston, "How to Evaluate the Fiscal Terms of Oil Contracts;" and *Escaping the Resource Curse*, M. Humphreys, J. Sachs, and J. Stiglitz, editors, New York: Columbia University Press, 2007. The authors would like to thank Mark Cooper for detailed comments as well.

2. Alfred Kjemperud, "General Aspects of Petroleum Fiscal Regimes," The Bridge Group A/S, Cambodia, December 2004.
3. David Johnston, Daniel Johnston, and Tony Rogers, "International Petroleum Taxation, for the Independent Petroleum Association," July 4, 2008.
4. We will use the term *production sharing agreement* throughout. It is also commonly referred to as a *production sharing contract* (PSC). The two terms are generally interchangeable.
5. Daniel Johnston, *International Petroleum Fiscal Systems and Production Sharing Contracts*, Tulsa: PennWell, 1994, pp. 90–92.
6. Johnston, 1994, pp. 92–96.
7. David Leigh, "Angolan Oil Millions Paid Into Jersey Accounts," *The Guardian*, November 4, 2002; Henrie E. Cauvin, "I.M.F. Skewers Corruption in Angola," *The New York Times*, November 30, 2002; Justin Pearce, "IMF: Angola's 'missing millions': The sum is three times the amount paid in aid," BBC, Friday October 18, 2002; "Angola battles with IMF accusation of corruption," *Alexander's Gas and Oil Connections*, News and Trends Africa, Vol. 7, Issue #22, November 13, 2002.
8. M. Sornarajah, *The International Law of Foreign Investment*, second edition, 2004, p. 409, as described by Simon Brinsmead in "Oil Concession Contracts and the Problem of Hold-Up," 2008.
9. Bernard Mommer, "Oil Prices and Fiscal Regimes," Oxford Institute for Energy Studies, WPM 24, May 1999.
10. This numerical example is based on one presented by Kirsten Bindermann in "*Production Sharing Agreements: An Economic Analysis*," Oxford Institute for Energy Studies, October 1999.
11. "International: Barking Louder, Biting Less; Energy and Nationalism," *The Economist*, March 10, 2007, p. 71.
12. Steve Hargreaves, "Taxing Oil Profits: Proceed with Caution," *CNNMONEY.COM*, May 6, 2008.
13. "Editorial: Royalty Rip-Off, *The New York Times*, April 11, 2010.
14. "The North Sea Fiscal Regime: A Discussion Paper," HM Treasury, March 2007.
15. *PetroReports: Fiscal and Regulatory Guide: United Kingdom*, Petroleum Service Group, Deloitte, September 2009.
16. "FAQ: Oil and Gas Royalty Relief," PBS, June 16, 2006.

17. "Audit Report: Minerals Management Service's Compliance Review Process," Department of the Interior Office of Inspector General, December 2006; "Oil and Gas Royalties: Royalty Relief Will Likely Cost the Government Billions, but the Final Costs Have Yet to Be Determined," General Accounting Office, Testimony before the Committee on Energy and Natural Resources, United states Senate, January 18, 2007.
18. Edward T. Pound, "A Billion Here, A Billion There . . . How One Big Goof in a Little-Known Federal Agency Gave Taxpayers a Big Black Eye," *US News & World Report*, October 12, 2006.
19. William Wallis, "Finds Put Ghanaians on a Roller Coaster of Expectations," *Financial Times*, December 4, 2009, p. 3.

Chapter 7

FINANCING AND FINANCIAL PERFORMANCE

Do what you will, the capital is at hazard . . . All that can be required of a trustee to invest, is, that he shall conduct himself faithfully and exercise a sound discretion. He is to observe how men of prudence, discretion, and intelligence manage their own affairs, not in regard to speculation, but in regard to the permanent disposition of their funds, considering the probable income, as well as the probable safety of the capital to be invested.

—Prudent Man Rule,
Justice Samuel Putnam, 1830

The petroleum industry is also highly capital-intensive, so strong returns are critical to attracting low-cost debt and equity capital. In fact, while many of the integrated companies have the cash flow and financial wherewithal to fund capital spending internally, they frequently rely on external debt and new equity capital, particularly to finance larger acquisitions and mergers.

—"Rating Methodology: Global Integrated Oil & Gas Industry," Moody's Investor Services, October 2005, p. 12

All business owners know that having access to capital, at acceptable costs, is critical for the conduct and growth of the business. The global oil and gas industry is no different, although its capital needs are growing more rapidly than most other industry sectors. A number of different forces have driven many changes to the financing of the oil and gas industry in the past 30 years.

- First, as the upstream global oil and gas industry has had to reach ever farther and deeper to find new reserves, the capital required has increased dramatically. New developments like those in the Caspian, offshore Brazil, sub-Saharan Africa, or even shale gas deposits in the US Midwest, all require more capital per well and per barrel than ever before. And they all are taking longer to initiate and execute.

- Second, these remote and difficult projects carry significantly more risk than ever before. Risk, when associated with access to capital, means less capital available and higher cost for that which is available. This risk has forced many players in the oil and gas value chain to get more creative in how they organize and fund projects, including who they take on as partners.

- Third is the growing power of the state. The fiscal regimes described in chapter 6 implicitly assumed that the international oil companies provided all of the capital for exploration and development. Although this was often the case historically, in many countries the state has taken a larger role in the development of petroleum projects—and its returns. As a result, financing is now coming from a number of different sources as government entities are increasingly active participants.

- Fourth, as the basic costs of crude oil and natural gas feedstock have risen in recent years, the downstream sectors of the industry have worked to increase their efficiency in yields, requiring investment in new and more expensive technology. Although the technology will yield its own return, that return is often spread further out over many years.

In short, oil and gas firms need more capital for longer periods of time to develop projects of ever-increasing risk—a daunting task to say the least.

This chapter provides an overview to the financing and profitability of the oil and gas industry. We begin with a primer on business finance, focusing on funding of assets and net working capital for firms in the private sector. We then describe management concerns over the use of debt and new forms of equity participation. The third section analyzes the primary financing sources—

corporate finance, project finance, and institutional lending—as well as some newly emerging financing innovations. The fourth section describes financial performance of the industry in general, noting a number of the most frequently used measures of profitability, including the industry's preferred metric, return on capital employed (ROCE). We conclude the chapter with some thoughts on the integration of corporate strategy and financial performance.

Business Financing: A Primer

Financial statements are like fine perfume: to be sniffed but not swallowed.

—Abraham Brilloff

All businesses, whether product or service providers, require capital. The amount of capital needed differs greatly across businesses depending on the type of company and its products or services. The three major financial statements that characterize the financial and operating results of any company are the income statement, the balance sheet, and the statement of cash flows. We begin with the balance sheet, the key to understanding the financing needs of the firm.

Basic balance sheet

Figure 7–1 provides a simple view of how a firm needs to raise capital to fund its activities as reflected on a simple balance sheet. Simplistically, the left-hand side of the balance sheet is the business, and the right-hand side is the funding of the business. Different businesses and enterprises will then have very different needs depending on what they actually have on their left-hand side. Funding is obtained from both equity and debt. Equity is the business owner's own money, while debt is, as they say, "other people's money."

The left-hand side of the balance sheet is composed of both current assets and fixed assets. *Current assets* include all items or activities that come and go in the firm in less than a year, like the purchase of an input for the construction of a piece of equipment. Although these items individually do not "last" more than a year on the firm's balance sheet, they will most likely be continually replaced—new ones purchased for new uses—so that the firm's current asset levels may remain relatively constant over time. *Fixed assets* are trucks, ships, machinery, and buildings, anything that will have an economic life extending over multiple years.

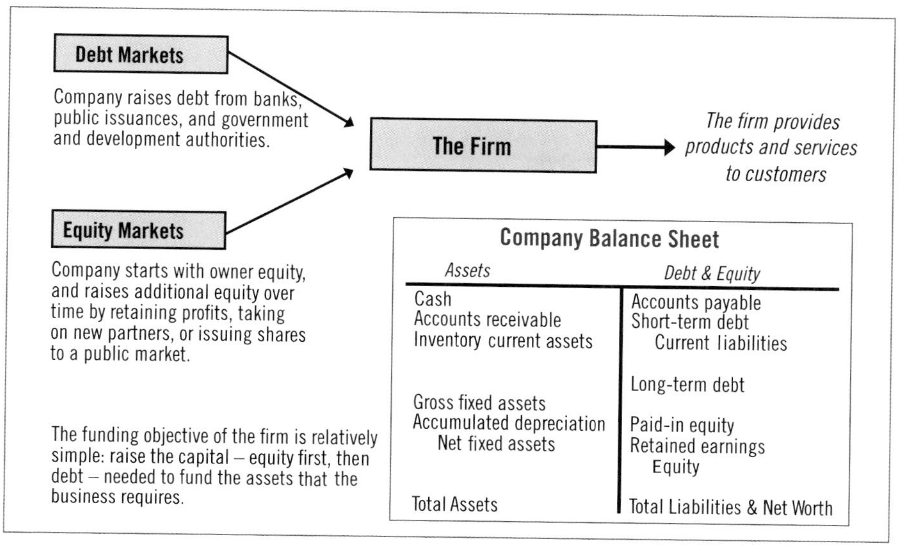

Figure 7–1. Funding the firm

The right-hand side of the balance sheet is composed of *current liabilities* (accounts payable, accrued expenses, and short-term debt), *long-term liabilities* (long-term debt such as bank loans or bonds issued to the market), and equity. Equity capital is the money that ownership itself has put into the business. This occurs at both startup and on a regular basis over time if the company makes a profit and reinvests some of those profits in the company (equivalent to making a profit and not paying all of the profits out to the owners). Long-term liabilities are essentially long-term (longer than one year in maturity) debt obligations which the firm has acquired. Current liabilities are comparable to current obligations of the company that will come and go within a year.

Managerial balance sheet

Using the line items defined above, we can now reduce the balance sheet down to what is termed a *managerial balance sheet*, one that reflects management activities rather than accounting distinctions. The left-hand side of the managerial balance sheet consists of only three items: cash, net working capital, and net fixed assets. As illustrated in figure 7–2, these three reduced items then make up invested capital.

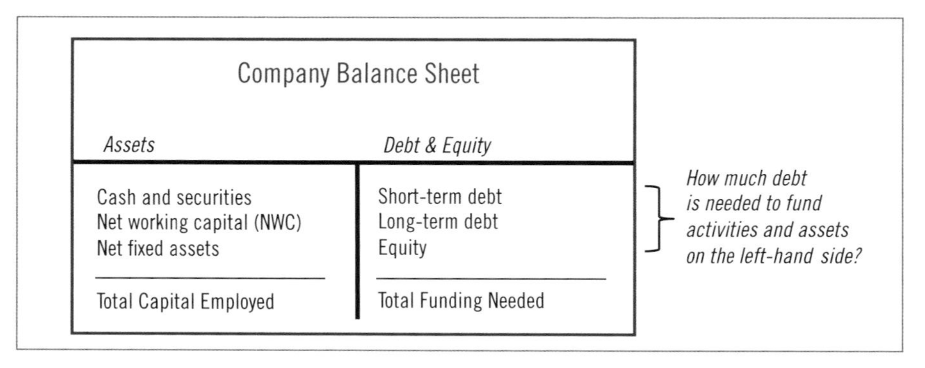

Figure 7–2. The managerial balance sheet reduced the components of the balance sheet down to three categories of assets: cash, net working capital, and net fixed assets. These three subcategories make up total capital employed. The reduction of assets also reduces the right-hand side of the balance sheet down to pure funding sources—debt (short-term and long-term) and equity. Many analysts and analyses will further reduce the firm's measure of capital or total funding to "net debt," where the total debt from the right-hand side will be reduced by the amount of cash on the left-hand side. Total debt could always be theoretically reduced by the use of this cash.

Cash. *Cash*—cash balance, not cash flow—is separated for one very important reason: management of the firm chooses what level of cash to hold. Made up of both cash and short-term marketable securities (e.g., treasury bills), this cash balance results from raising capital or retaining cash flow from operations. Management of the organization may choose to accumulate a large cash balance for a variety of reasons: for a buffer stock against possible cyclical declines; for strategic opportunities that may arise in the future, allowing a quick response; or for a specific acquisition at a future time of the buyer's choosing. Alternatively, management may decide to pay out large cash balances to ownership, stockholders, through an extraordinary dividend, a one-time major payment of returns to owners, as done by Microsoft and Porsche in recent years.

Net working capital. *Net working capital*, the second asset category of the managerial balance sheet, is a composite concept combining selected current assets and current liabilities. The simplest and most widely calculated version is:

Net working capital (NWC) = (Accounts receivable + Inventory) – (Accounts payable)

NWC is in many ways a measure of activities which spontaneously change on the balance sheet from executing the business. The firm purchases inputs from suppliers (accounts payable combines those inputs for the creation of unfinished and finished product—inventory), and then sells final product to customers, creating accounts receivable.[1] Some readers may recognize this from previous financial training as the *operating cycle*.

Net working capital is one of the oldest and most fundamental concepts in business finance. It is in principle the ongoing activities of the firm to conduct sales: a sale creates a receivable, but a sale is only created by having a product in inventory, and in turn owing suppliers for inputs purchased and activities conducted to create the product or provide the service. This is also why cash is not included in net working capital; cash balances are dictated by management, not spontaneously created by the business. It is also important to note that even a service firm, for example an oil services firm, will still have significant NWC management issues, primarily driven by receivables and payables, but not inventory.

The management of any firm will wish to minimize net working capital because funding net working capital is expensive—the cost of debt and equity. It does so by collecting receivables faster (cash is good, prepayment even better), reducing inventory or product/service development and delivery times (cycle time, and concepts such as just-in-time (JIT) inventory and lean manufacturing), and by paying suppliers ever later. Net working capital can actually be zero or even negative (e.g., Dell Computer and WalMart are famous for having negative net working capital levels). The smaller the net working capital, the smaller the total balance sheet, and therefore the fewer funds needed to support the business.

Net fixed assets. Net fixed assets are the total fixed assets of the firm reduced by the accumulated depreciation of those assets as they age according to accounting principles. The value on a balance sheet for net fixed assets does not represent replacement cost or fair market value, only its historical cost. Traditionally a fixed asset was always considered value in itself. However, with increasing focus on actual cash flow, the focus has moved towards the cash flows generated by fixed assets, not simply owning the asset itself. This has been one of the primary drivers for the hiring of outside contractors who own and maintain their own capital equipment, as we discussed in previous chapters.

The scale of assets required in the global oil industry has risen dramatically with the innovation and application of technology to an expanding set of activities and services necessitated by growing industry complexity. These assets are increasingly expensive and have resulted in many firms exploring a multitude of ways to acquire the services of those assets without actually purchasing them—holding them on their own balance sheet. Other sections of this book highlight these methods such as joint operating agreements and joint ventures, which attempt to spread the cost and risk and assets out over a number of different partners.

Invested capital. These three asset categories, cash, net working capital, and net fixed assets, sum to what is often termed *invested capital.* Note that this is not the same as total assets, because invested capital should be significantly smaller as a result of reducing net working capital by deducting payables (and any other nonfinancial current liabilities). Invested capital is our objective; it is the total net assets that the firm itself must fund. Since debt and equity capital costs money, management will attempt to minimize the total capital employed for any specific level of sales or profits, or alternatively, maximize the returns on the capital employed. We will return to this very concept, return on capital employed (ROCE), in a later section.

Funding the firm: Debt and equity

In chapter 4, we described in detail the costs and characteristics of both debt and equity. A few characteristics of each are worth repeating:

Equity. Equity is the capital the owners of the firm have in the business and is capital at risk. In addition to the capital initially invested, as the firm operates over time, profits earned and not paid out to the owners are retained; these retained earnings add to the equity base of the firm. The returns to equity investment in any firm are profits in private firms, and possible share price appreciation (capital gains) in publicly traded firms. If the owners of a firm wish to attract other equity investors, they typically must give up some degree of control over the organization.

Debt. Debt is capital gained from outside parties for a limited period of time at a relatively certain cost. Although cheaper than equity, *debt service* (the repayment of principal and interest) is made before returns to owners. Interest expenses on debt are generally deductible towards company tax obligations, globally, subject to specific government requirements. Debt is often seen as a burden on the firm and its cash flows, and generally is to be used with prudence.

The relative amounts of debt and equity that should be used to fund a firm, termed *capital structure,* is a subject that the field of finance terms "complex," meaning it finds it nearly impossible to explain. Without delving into the academic debate, we will summarize the capital structure of a specific company as follows: first, firms operating in the same segments of the value chain typically possess similar capital structures (but there may still be significant differences between firms); second, management of firms, like homeowners, generally take on debt levels that they believe to be supportable; and third, firms always attempt to use debt judiciously, but when confronted with a choice between taking on burdensome debt or missing what they consider important business opportunities, they often choose taking on the debt.

This second point is critical to understanding business finance. In general, firms that nearly fail from overly burdensome debt were not intentionally being foolish, nor were their lenders acting recklessly. Both parties evaluated the risks associated with the debt as being acceptable for the potential returns from the use of the capital. It is only in hindsight, with its crystal clear lens, that they often appear foolish. (It is also why many people believe that business entrepreneurs must be risk takers, as most accountants would likely shut down the business in order to avoid losing money!)

Figures 7–3 and 7–4 provide some insight into how debt usage differs across oil companies. Figure 7–3 shows the debt-to-capital ratios for 14 companies in 2009. ExxonMobil has negative net debt of –1% for 2009 (meaning that its cash balance exceeded its total debt obligations) and Repsol YPF has a relatively high level of 38%. Figure 7–4, also based on Credit Suisse analysis, tracks the average debt-to-capital percentage over the past decade for this same set of companies. Based on these two figures, it is fairly obvious that most of the major companies have chosen to carry debt levels that range between 18% and 22% of capital for the last decade. ExxonMobil and Chevron are clearly unusual in their extreme levels of negligible debt.

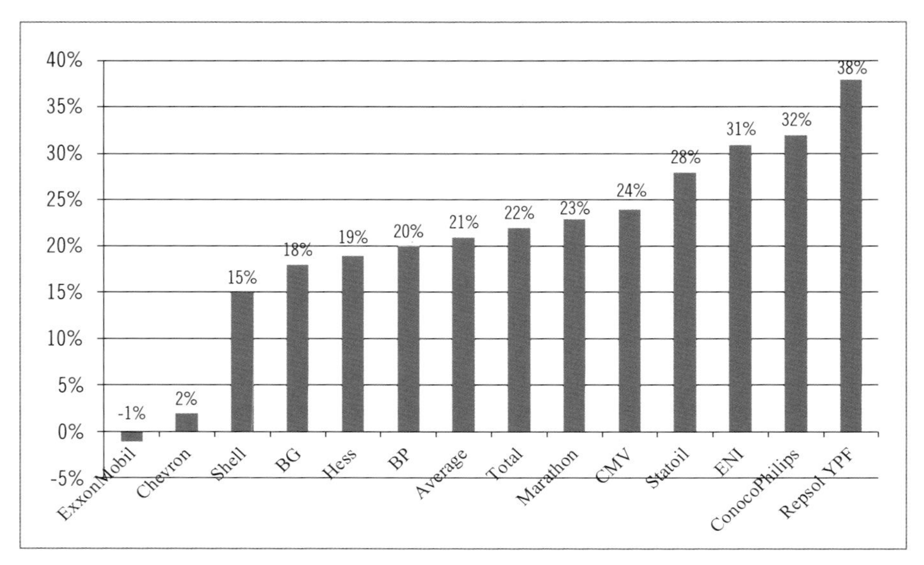

Figure 7–3. International oil companies: net debt to capital 2009

Source: "Global Integrated Oils: Sector Review," Credit Suisse, July 23, 2010, p. 40, Exhibit 100. Net debt is total debt less cash as a percentage of debt + equity – cash. Debt levels are for end-of-year 2009.

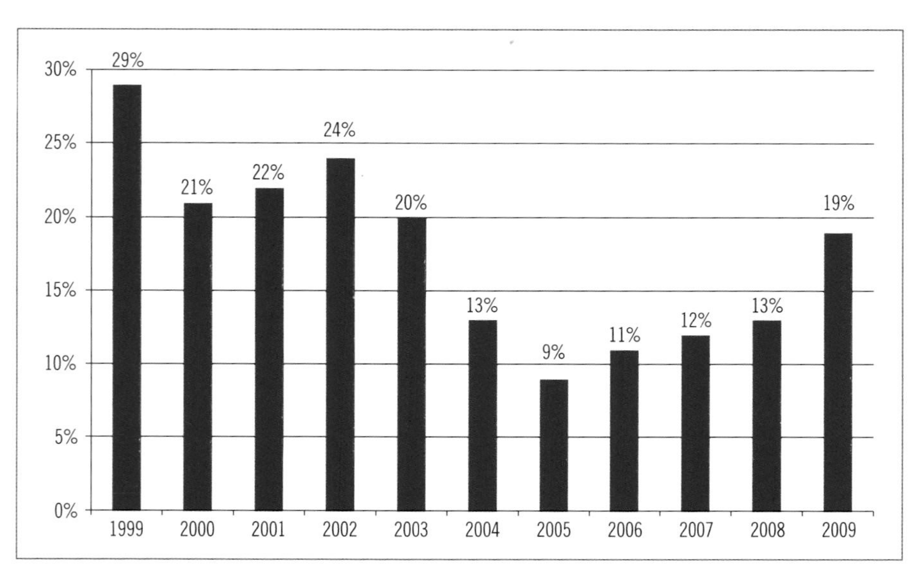

Figure 7–4. International oil companies: net debt to total capital 1999–2009

Source: "Global Integrated Oils: Sector Review," Credit Suisse, July 23, 2010, p. 40, Exhibit 100. Net debt is total debt less cash.

Revenue and earnings

The second major financial statement, the *income statement* (often also termed the *profit and loss statement*), records the revenues (sales), operating expenses, and net income (also called profits or earnings) for a specific period of time such as a quarter or year. Figure 7–5 provides a general simplified income statement for a firm.

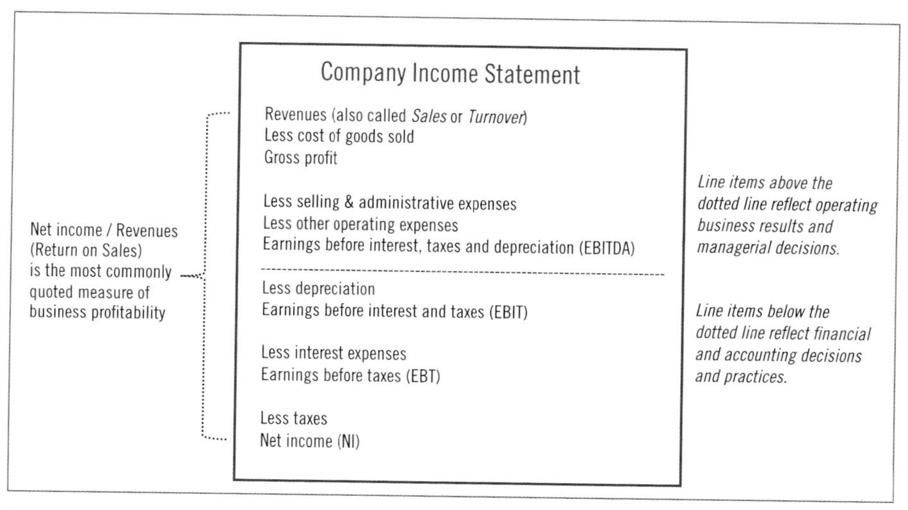

Figure 7–5. The income statement (P&L). The income statement or P&L for a business or business unit is by far the most frequently used financial measure of business results. One of the most frequently heard measures of financial accountability is the phrase "I have P&L responsibility for my unit." Internally, management is often held accountable for the results above the dotted line because many of the decisions and practices that determine the line items below the dotted line are driven by corporate level executive decisions, financial strategies, and accounting practices.

The income statement is by far the most widely used measure of firm financial results. The company's total *sales*, *revenues*, or *turnover*, whichever term is used, is the top line. The firm's after tax *profits*, *net income*, or *earnings*, whichever term you prefer, is the bottom line. And although the focal point of leadership in business is to ultimately increase the profitability of the business—the bottom line—it is in the long-run nearly impossible to sustain increased profitability without growing the top line.

Many firms also separate the top half from the bottom half of the income statement for managerial performance purposes. Line items above the horizontal dotted line in figure 7–5 reflect operating results from management decisions, whereas line items below the dotted line reflect financial and accounting decisions and practices (which are often centralized in multinational companies). This is also why earnings before interest, taxes, depreciation, and amortization—EBITDA, is so widely used as a key financial metric.

The most commonly used measure of profitability in business, return on sales, draws both items from this one financial statement. It is simply the bottom line, net income, as a percentage of the top line, revenues. Figure 7–6 presents a graphical analysis constructed by the American Petroleum Institute (API) of return on sales across industry groups for the first quarter of 2010. The message it conveys is very clear: ROS profitability differs dramatically across industries; the profitability of the oil and gas industry was relatively "average" during this period, and a number of major industries such as pharmaceuticals were much more profitable than oil and gas during this period.

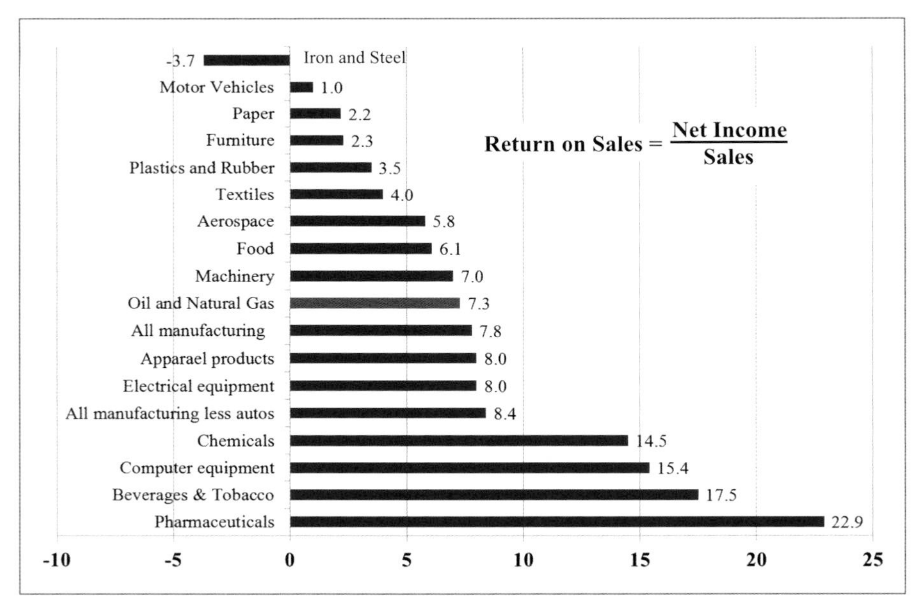

Figure 7–6. Return on sales across industries (%)

Source: "Putting Earnings Into Perspective," American Petroleum Institute, July 2, 2010, p. 2. Original data drawn from U.S. Census Bureau and *Oil Daily*. Data is for the first quarter of 2010.

Financial performance

Now that we have reviewed the basic financial statements we can turn to financial performance and ask the question, "What is the best measure of financial performance to judge a business, more specifically, an oil and gas business?" To be honest, we don't know, but offer up an alternative approach to the debate. A short list of the most commonly used measures of financial profitability include the following, beginning with return on sales introduced previously.

Return on sales. *Return on sales* (ROS), is calculated by dividing net income (profit) by total sales in the same period.

$$\text{Return on sales (ROS)} = \frac{\text{Net income}}{\text{Sales}}$$

Both financial values are drawn from the company's income statement. Income statement items (also called the *profit and loss statement* or *P&L*) are driven by accounting practices and do not necessarily indicate actual cash flow results. For example, a product or service will be booked on the income statement as a sale when the product is shipped or the service initiated, yet payment from the customer may not yet be received. ROS is one of the oldest measures of profitability and is still used internally within many companies on a business unit level because of its ease of calculation and ready access to data (most firms track sales and profits over very short intervals of time).

Although there is no specific that which is typical or best (biggest is clearly better), the S&P 500 companies, in some of the better years for business results, may average a return on sales of 7% to 8%. But that is only an average, and in any year, good or bad, different industries will show very different results. For example, during the 2005 to 2009 period, the automobile industry as a whole experienced negative return on sales, while the pharmaceutical industry produced double-digit returns.

Gross profit margin. *Gross profit margin* is calculated by dividing the gross margin, sales less cost of goods sold, by total sales.

$$\text{Gross profit margin} = \frac{\text{Sales} - \text{Cost of goods sold}}{\text{Sales}}$$

Again, both financial values are drawn from the company's income statement and are driven by accounting practices not necessarily indicating actual cash flow results. By focusing on direct costs of goods sold, gross profit does not address other general and administrative costs that often differ drastically across different industries and firms.

Return on assets. *Return on assets* (ROA) is calculated by dividing net income by the total assets of the business in the same period.

$$\text{Return on assets (ROA)} = \frac{\text{Net income}}{\text{Total assets}}$$

The calculation of ROA draws from two different financial statements. The net income value is drawn from the income statement, while the total asset value is drawn from the balance sheet.

Again, like ROS, the bigger the value for ROA the better. And like ROS, different industries will have very different typical performance results. ROA is particularly driven by the capital intensity of the industry or business line, with service firms having few assets compared to manufacturing firms. Although it is possible to use ROA for measuring service firm performance, it has not been the norm historically. One significant advantage of ROA is that profits are evaluated on a per-unit-of-asset basis, which is important given that assets represent capital invested, at least on a gross basis, and as we have noted previously, capital is expensive.

Return on equity. *Return on equity* (ROE) is calculated by dividing net income by the total equity capital invested in the firm.

$$\text{Return on equity (ROE)} = \frac{\text{Net income}}{\text{Equity}}$$

Like ROA, ROE draws its data from two different financial statements, the income statement and the balance sheet. Neither ROA nor ROE, however, draw from the third financial statement, the statement of cash flows. If the primary objective of the firm is to generate returns for its owners, its equity holders, then the ROE is the more focused measure of financial performance for ownership. But, given that many firms in the oil and gas industry are publicly traded, including all the supermajors and many NOCs, ROE does not include share price, the primary financial value for determining shareholder return.

Table 7–1 shows the financial performance and metrics chosen by the Congressional Research Service when reviewing oil and gas industry profitability in 2007. The two metrics presented, ROS and ROE, show relatively solid but unexceptional financial performance for major IOCs at that time (although ExxonMobil's ROS and ROE values are quite good, as is Occidental's ROS that period).

Table 7–1. Financial performance of international oil and gas companies, 2007 by CRS

Company	Revenue	Percent Change	Net Income	Percent Change	Return on Sales	Return on Equity
ExxonMobil	404,552	7.1%	40,610	2.8%	10.0%	33.4%
Royal Dutch Shell	355,782	11.6%	27,564	8.7%	7.7%	22.2%
BP	291,438	6.2%	17,287	–22.3%	5.9%	18.5%
Chevron	220,904	5.1%	18,688	9.0%	8.5%	24.2%
ConocoPhillips	194,495	3.2%	11,891	–23.5%	6.1%	13.4%
Marathon	65,207	–0.4%	3,956	–24.4%	6.1%	20.6%
Amerada Hess	31,924	11.2%	1,832	–4.6%	5.7%	18.8%
Occidental	18,784	9.4%	5,400	28.8%	28.7%	23.7%
Murphy	18,438	28.9%	766	18.8%	4.2%	15.1%
Total	1,601,524	7.1%	127,994	–2.9%	8.0%	22.7%

Source: Robert Pirog, "Oil Industry Profit Review 2007," CRS Report for Congress, Resources, Science, and Industry Division, April 4, 2008. This is Table 2 in this report, and the data is drawn from *Oil Daily, Profit Profile Supplement*, Vol. 58, No. 52, March 17, 2008, p. 6, and company annual reports.

Return on invested capital. Return on invested capital (ROIC) is calculated by dividing current period earnings before interest and taxes by the invested capital in the firm.

$$ROIC = \frac{EBIT\ after\text{-}tax}{Invested\ capital} = \frac{EBIT\ after\text{-}tax}{Sales} \times \frac{Sales}{Invested\ capital}$$

As illustrated, it is often decomposed one step further into two separate ratios, the first measuring profitability on sales, the second a ratio of sales to invested capital. As with ROA, the calculation draws values from both the income statement and balance sheet.

ROIC refines both the measure of profit and capital from the previously described ROA. By using EBIT after-tax (earnings before interest and taxes, after-tax), ROIC eliminates the deduction of interest expenses from profits. Second, by using invested capital rather than total assets, the calculation reduces the total capital value by the extensions of credit by suppliers, the accounts payable deduction we made in the calculation of net working capital earlier in this chapter.

The further decomposition of the measure into two separate ratios allows further analysis for profit performance evaluation. The first ratio, EBIT after-tax divided by sales, often termed the *operating margin* of the business, allows evaluation of current period expenses versus sales. The second ratio, sales divided by invested capital, is the lesser known ratio, termed *capital turnover ratio*, showing a gross relationship between the top-line sales of the firm and capital used to generate those sales.

Return on capital employed. Return on capital employed (ROCE) is very similar to ROIC, but is calculated by dividing EBIT after-tax by *capital employed*, not *invested capital*. Depending on who is using it, *capital employed* is defined in several ways and is typically much smaller than invested capital as it reduces the total capital value by nonfinancial long-term liabilities (large items like deferred taxes and postretirement pension reserves):

$$ROCE = \frac{EBIT\ after\text{-}tax}{Capital\ employed}$$

An additional distinction is that to calculate their metric, many firms use an average base, essentially a return on *average* capital employed. As with ROIC, ROCE can potentially be broken down into two separate ratios for further managerial and operational focus.

One of the champions of ROCE and capital employed is ExxonMobil, which defines it as follows:

> *Capital employed is a measure of net investment. When viewed from the perspective of how the capital is used by the businesses, it includes ExxonMobil's net share of property, plant, and equipment and other assets less liabilities, excluding both short-term and long-term debt. When viewed from the perspective of the sources of capital employed in total for the Corporation, it includes ExxonMobil's share of total debt and shareholders' equity. Both of these views include ExxonMobil's share of amounts applicable to equity companies, which the Corporation believes should be included to provide a more comprehensive measure of capital employed.*[2]

It then defines how ROCE is calculated for the total corporation:

> *The Corporation's total ROCE is net income attributable to ExxonMobil excluding the after-tax cost of financing, divided by total corporate average capital employed. The Corporation has consistently applied its ROCE definition for many years and views it as the best measure of historical capital productivity in our capital-intensive, long-term industry, both to evaluate management's performance and to demonstrate to shareholders that capital has been used wisely over the long term. Additional measures, which are more cash-flow based, are used to make investment decisions.*[3]

Note that the numerator of the calculation, "net income excluding after-tax cost of financing," is the same [EBIT–taxes] term used in the numerator of ROIC. Table 7–2 provides a brief snapshot of how ExxonMobil's balance sheet is converted to its measure of capital employed.

ExxonMobil is not alone in its use of ROCE. Most of the major IOCs now use it as one of their primary measures of financial performance. The investment banking profession has followed and, as illustrated in figure 7–7, calculates their own averages for the industry's performance. Despite its popularity, many academic studies have concluded that although ROCE may be a good measure for aiding management, and possibly investors, in determining the capital efficiency of an oil and gas firm, ROCE values have shown little ability to predict or even correlate with share price.[4]

Although there is no assurance that all companies and analysts define and calculate their ROCE metrics identically, the general acceptance of the measure for much of the oil and gas industry's performance aids the evaluation and comparison of different firms. Not surprisingly, as seen in figure 7–7, which presents the ROCE calculations of Credit Suisse for the 1999 to 2009 period for major IOCs, ROCE also shows the general increase in returns on capital during the mid-2000s when crude oil and natural gas prices reached their peaks in 2008, plummeting in 2009.

Table 7–2. ExxonMobil's balance sheet and calculation of capital employed, 2009

Assets		Liabilities	
Current assets		Current liabilities	
Cash and cash equivalents	10,693	Notes and loans payable	2,476
Marketable securities	169	Accounts payable and accrued liabilities	41,275
Notes and accounts receivable, less estimated doubtful	27,645	Incomes taxes payable	8,310
		Total current liabilities	52,061
Inventories			
Crude oil, products and merchandise	8,718	Long-term debt	7,129
Materials and supplies	2,835	Postretirement benefits reserves	17,942
Other current assets	5,175	Deferred income tax liabilities	23,148
Total current assets	55,235	Other long-term obligations	17,651
Investments, advances and long-term receivables	31,665	Total liabilities	117,931
Property, plant and equipment, at cost, less accum. depreciation	139,116	Equity	
Other assets, including intangibles, net	7,307	Common stock without par value	5,503
		Earnings reinvested	276,937
		Accumulated other comprehensive income	(5,461)
		Common stock held in treasury	(166,410)
		ExxonMobil share of equity	110,569
		Noncontrolling interests	4,823
		Total equity	115,392
Total assets	233,323	Total liabilities and equity	233,323

Capital Employed	2009	Description
Total assets	233,323	
Less current liabilities excluding notes and loans	(49,585)	Subtracting A/P and taxes due
Less long-term liabilities excluding long-term debt and equity	(58,741)	Subtracting deferred taxes and postretirement pension reserves
Less minority share of assets and liabilities	(5,642)	In the notes to ExxonMobil's financial statements
Add XOM's share of debt-financed equity-company net assets	5,043	In the notes to ExxonMobil's financial statements
Total capital employed	124,398	

Source: Derived by authors from ExxonMobil, *Summary Annual Report 2009.*

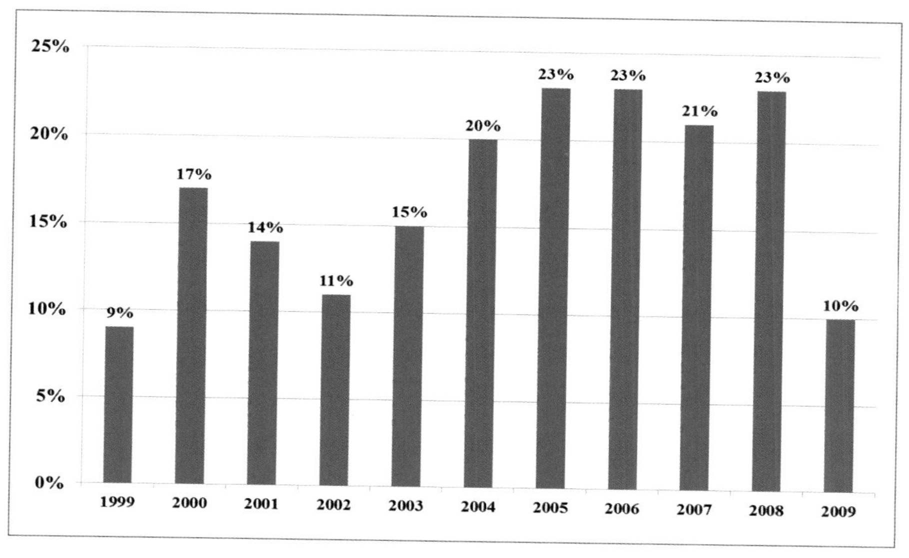

Figure 7–7. International oil companies: ROCE, 1999–2009

Source "Global Integrated Oils: Sector Review," Credit Suisse, July 23, 2010, p. 40, Exhibit 100. ROCE is return on capital employed and is calculated as EBIT after taxes as a percentage of capital employed. Companies are the same as in figure 7–3.

Performance and creditworthiness

Regardless of how well-designed and precisely calculated any individual financial performance metric may be, no one metric can tell all, any more than one gauge tell the pilot of a jet aircraft everything he or she needs to know. The pilots—management and leadership of the oil and gas business—need all the help they can get from as many gauges on the instrument panel as possible.

Metrics also differ by purpose. For example, ExxonMobil in the previous section defined ROCE as the measure "both to evaluate management's performance and to demonstrate to shareholders that capital has been used wisely." But what if you were a creditor, one of the many financial services firms that provides loans and debt capital to the oil and gas industry? The primary concern of creditors is not the general profitability of the company, or even how efficiently it uses capital, but the company's ability to repay debt.

Table 7–3 provides some insight into the metric set that the credit rating agencies, in this case Moody's Investor Services, base their credit ratings of major IOCs.[5] Moody's uses five factor categories, including financial issues like ROCE, finding costs, debt, and income ratios, as well as proved reserves, and reserve replacement.[6] Although these five factor categories include financial variables, it is still the underlying drivers of company competitiveness and production capability—oil and gas reserves—that dictate a large portion of the rating. Reserves matter. As noted by Moody's:

> *There is a high degree of positive correlation on the mapped ratings for reserves and production and the credit ratings of the integrated oil peer group. A substantial portion of the industry's reserves are concentrated with the majors and national oil companies (NOCs), reinforcing all the benefits of scale, asset and cash flow durability, and diversification cited. This trend has only intensified with the industry consolidation that has taken place over the past decade.*[7]

Table 7–3. Mapping the Moody's ratings methodology criteria

Rating Factor (and Relevant Sub-factor)	Weighting	Aaa
Reserve & Production Characteristics	25%	
Average Daily Production (Mboe/d)	8%	> 2,750
Proved Reserves (Million boe)	12%	> 10,000
Total Proved Reserve Life (yrs)	5%	> 13
Reinvestment Risk	10%	
3-Year All-Sources Reserve Replacement	5%	> 150%
3-Year All-Sources F&D Cost ($/boe)	5%	< $9
Operating & Capital Efficiency	10%	
Return on Capital Employed (ROCE 3-year average)*	5%	> 30%
Leveraged full-cyle ratio	5%	> 4x
Downstream Rating Factors	15%	
Total Crude Distillation Capacity (1,000 bpd)	5%	> 3,000
# of Refineries with Capacity > 100 M bpd	5%	> 15
**Segment ROCE (3-year average)	5%	> 25%
Financial Metrics	15%	
Retained Cash Flow / Net Debt (3-year average)	10%	> 60%
EBIT / Interest Expense (3-year average)	10%	> 20 x
Gross Debt / Total Proved Reserves	10%	< $3.50
Gross Debt / Capital	10%	< 25%
Total	100%	

* EBIT/Average Capitalization (including debt); ** Downstream EBIT/Average Downstream Capital Employed. Source: "Rating Methodology: Global Integrated Oil & Gas Industry," Moody's Global Corporate Finance, Moody's Investors Service, November 2009, pp. 4–14.

Much of this type of analysis must be forward-looking, in that the challenge is to forecast the company's repayment capability tomorrow, not today. As a result, it must make some serious assumptions about the sustainability of oil and gas prices, the ability to replace reserves and grow production, all within an industry with rising costs and consolidation.[8] Moody's also notes that in many cases the firm weights the six different categories slightly differently depending on the peculiarities and specifics of the firm in question. One of the factors it points to is when a company is a "single-country player," when a large portion of its reserves, production, and results are dependent on an individual country.

In some ways table 7–3 is an instrument panel, a collection of gauges and metrics that attempt to provide an overview of the current and possible future performance of the firm. But again, this is the panel of a creditor who is most concerned with the security of their capital and the repayment capability of the borrower.

Aa	A	Baa	Ba	B	Caa
1,100 to 2,750	550 to 1,100	140 to 550	55 to 140	27 to 55	< 27
5,000 to 10,000	2,000 to 5,000	500 to 2,000	100 to 500	30 to 100	< 30
11 to 13	9 to 11	7 to 9	5 to 7	3 to 5	0 to 3
130% to 150%	110% to 130%	100% to 110%	80% to 100%	60% to 80%	< 60%
$9 to $11	$11 to $13	$13 to $15	$15 to $18	$18 to $22	> $22
25% to 30%	20% to 25%	15% to 20%	15% to 10%	5% to 10%	< 5%
2.5x to 4.0x	1.75x to 2.5x	1.0x to 1.75x	0.5x to 1.0x	0.25x to 0.5x	< 0.25x
2,000 to 3,000	1,000 to 2,000	500 to 1,000	250 to 500	50 to 250	< 50
9 to 15	6 to 8	3 to 5	2	1	0
20% to 25%	15% to 20%	12% to 15%	7% to 12%	5% to 7%	< 5%
40% to 60%	30% to 40%	20% to 30%	10% to 20%	5% to 10%	< 5%
15x to 20x	8x to 15x	3x to 8x	2x to 3x	1x to 2x	< 1x
$3.50 to $4.50	$4.50 to $5.50	$5.50 to $6.50	$6.50 to $7.50	$7.50 to $10.00	> $10.00
25% to 35%	35% to 45%	45% to 55%	55% to 65%	65% to 75%	> 75%

Windfall profit

> *Being largely price-driven, with no increase in output, and with little new production resulting from increased oil industry investment, many believe that a portion of the increased oil industry income over this period represents a windfall and unearned gain, i.e., income not earned by any additional effort on the part of the firms, but due primarily to record crude oil prices, which are set in the world oil marketplace.*
>
> — Salvatore Lazzari and Robert Pirog, "Oil Industry Financial Performance and the Windfall Profits Tax," Congressional Research Service, September 30, 2008, p. CRS-1.

We would be remiss in our discussions of the oil industry's financial performance if we did not address the debate about so-called windfall profits. Periodically, the question arises as to whether the industry makes "too much profit" and whether those profits should be taxed away by government. In 2008, when crude oil prices hit record highs and the oil industry had record profits, the debate was quite vigorous.

There are a number of dimensions to this debate. For example, the United States Congressional Research Service's report to Congress in 2008 focused on the following issues:[9]

- First, were profits in the oil and gas industry high relative to other standards? When measuring profitability by return on sales (ROS), their conclusion was that "this indicator of industry performance is not out of line with the rate in the manufacturing industry generally." When measuring profitability by return on equity (ROE), the study concluded that "the oil and gas industry's ROE was, on average, significantly greater than the ROE for the manufacturing industry generally."[10]

- Secondly, the CRS concluded that the data, for the 2003 to 2008 period, "characterize an industry that was unable to respond to the market signal of higher price by increasing output as predicted by economic theory." Although it is unclear what is expressly meant by this conclusion, the implication from the report is that the industry was not responding to market forces and that it preferred to garner the greater profits.

- Third, the report argues that, according to both economic and financial theory, "when firms earn returns in excess of market rates of return they are likely to reinvest in their businesses to expand output and improve their technologies." The CRS analysis concluded that although the big five increased capital expenditures during this interval of time, "those investments have not met with noticeable success." The report also notes that these same five companies have not constructed a new refinery in the US since the 1970s, but acknowledges that existing refineries have been expanded and upgraded. It goes on to imply that the accumulation of cash and the increased return of profits to shareholders during this period were both at higher rates than would be expected according to traditional theory.

- Fourth, since most of the higher profits were a result of higher market prices for oil and gas, and not higher levels of production volumes, the report also implies that the industry may not *deserve* these profits. This argument rests on the premise that this income was "not earned by any additional effort on the part of the firms."

- Fifth, the big five firms paid corporate income taxes on the increased profits during this interval of time. The report notes that any windfall profit tax initiated would be above and beyond those considerable taxes already being paid.

We will not spend an inordinate amount of time and space here analyzing and debating these issues, but confine our thoughts and observations to the following basic points.

Industry profitability. The general profitability of the oil and gas industry has not been exceptional over the long term. As we described in the previous section, there are many different measures of financial performance, and the most commonly used, ROS, has shown the industry to be about average for US industries.

Industry response to higher prices. Given the time it takes to find, develop, and produce additional volumes of oil and gas, and the increasing difficulty in both finding and gaining access to new reserves, it is not surprising that volumes did not grow during this arguably short period of time. The industry has long noted that it must invest and operate for the long term and that would seemingly apply to when prices are high as well as when prices are low.

Industry reinvestment. The subject of reinvestment is very complex and only superficially addressed by most public policy studies such as that of the CRS. According to economic and financial theory, firms need to both reinvest in their businesses and provide returns to ownership. Reinvestment in the business includes capital expenditures, acquisitions, and paying down debt. Providing returns to ownership includes dividend distributions and share repurchases. All five major firms studied by the CRS undertook dividend distributions and share repurchases during the time period noted, but not always in the proportions that different stakeholders thought appropriate. This is a very difficult area of debate.

Deserving of profits. This is a rather unusual area of debate about which most economic and financial theory has little to say. It is not clear that higher levels of profitability are more deserved from expanding production or sales volume by gaining competitive advantage from unique customer value propositions (such as Apple or Southwest), creating patented technology (such as in the pharmaceutical industry), or finding and developing scarce mineral resources (such as in the oil and gas industry).

Figure 7–8 provides a bit of perspective on relative profitability across companies for a longer period of time, from 1991 through 2009. The four selected companies—ExxonMobil, McDonald's, Pfizer, and WalMart—are some of the largest multinational companies in the world in four different industries. At least with this most common measure, return on sales, ExxonMobil—the most profitable global IOC over the past decade, is consistently less than the Golden Arches people, and clearly below the record profit levels seen by Pfizer. Evidently "windfall" is like beauty, in the eyes of the beholder.

Although the debate over windfall profits has subsided in the post-2008 environment of lower oil and gas prices, it will inevitably resurface as soon as these same prices rise again in the future—which they will. Interestingly, during prolonged periods of low profitability, such as when crude oil prices were at $10 per barrel in 1998, one does not hear arguments on Capitol Hill that these same companies deserve some type of tax break as a result of their performance.

Finally, windfall profits is not just an issue in the United States. In 2006, China introduced a windfall profits tax on the sales of crude oil produced in China. The windfall profits tax is triggered when crude oil prices reach $40 a barrel. In theory, revenue from the special profits tax is to be distributed to farmers, fishermen, taxi drivers, and others adversely affected by rising oil prices. The impact on Chinese oil companies has been significant. For example, in the first half of 2010, CNOOC paid the government 7.98 billion yuan in windfall taxes, an amount equal to 31% of net income for the period.[11]

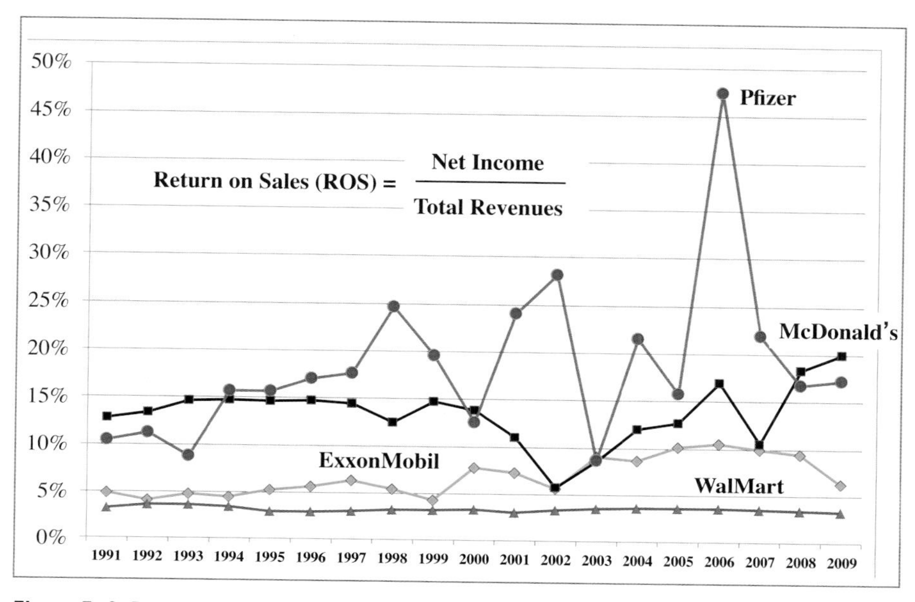

Figure 7–8. Return on sales for selected companies
Source: Derived by authors from individual company financial reports.

Capital Sourcing for the Oil and Gas Industry

The oil and gas industry, as a result of its nearly insatiable need for capital and its growing complexity and risk, uses basically any possible source of capital it can find. Like most industries, however, smaller firms are typically limited to their sources of capital, many of the smallest being private firms with only bank loans available for debt financing. Larger firms, particularly the publicly traded ones, have access to nearly every form of securitized debt or venture capital source that exists.

Corporate finance

Corporate finance deals with the traditional financial structure used by a company to raise capital, both debt and equity, for its general portfolio of businesses. There is no separation of general funding and the various businesses and activities in which the company is engaged. However, this is not a very attractive approach to financing many oil and gas projects around the world, as an individual project may require an amount of capital that is as large or larger than the existing company. With that level of capital at risk, and given the high levels of both project and country risk carried by many petroleum projects, corporate finance may be considered unattractive by oil and gas companies for the largest projects.

Raising capital for upstream or downstream oil and gas development is obviously critical. Going outside the firm to raise capital in the capital markets—debt and equity—is costly and time-consuming. Most companies will seek to raise both debt and equity in relatively set proportions over time if the business is growing. Although debt is cheaper than equity, it is considered a drag on the firm, a burden, and increases the risk of failure. Therefore, equity must be raised in any successful firm over time if a balance is to be maintained. A profitable operating firm that retains profits in the firm year after year (rather than paying the profits out to stockholders) is gradually adding equity capital (retained earnings). If this is not at a rate sufficient to fund the growing capital needs of the business in line with corporate strategy, ownership will have to go outside the firm for additional capital (which may be debt or equity—private or public).

One way in which oil companies can manage this size and risk profile is in the use of joint ventures, which have been discussed in chapters 3 and 5. Joint venture structures offer an attractive alternative to sole development since the capital investment and risk can be shared across companies. For megaprojects, there are frequently many partners, such as in the project structures used in the development of the North Sea. In the case of joint ventures, a separate operating agreement will then specify which company or companies are responsible for actual project development and operation. By partnering, companies gain greater diversification and can afford to be investors in more projects globally.

Public equity

Investors can invest in and own equity in companies two ways—publicly or privately. Publicly traded companies, with shares traded on stock exchanges all over the world, have shares that essentially can be purchased by anyone anytime. When a private company first goes public, issuing its shares to the general public, it is termed an *initial public offering* (IPO), the company raising capital for general business purposes. Companies that grow and mature often issue additional shares to the market over time.

The public share markets have been home to the oil and gas industry (upstream and downstream) for many years. The world's IOCs and many of the major NOCs are all publicly traded, either in New York, London, Tokyo, or Hong Kong. Although there are dozens of other stock exchanges around the world, these markets have long dominated equity trading. In addition to the majors, many of the medium-capitalization companies in the industry are also traded publicly, and even many of the so-called juniors.

The interest the public equity markets have shown in the oil industry historically has been based on crude oil and gasoline prices. When prices rise as in 1973, 1979, or 2003 to 2008, the market is very receptive to new offerings of publicly traded shares in firms attempting to profit from these rising prices. Of course when prices fall, equity market interest fades. The most recent fall in crude oil prices in the second half of 2008 was also at the same time as the global financial markets went into an even larger collapse of their own. The result was a public marketplace not very receptive to major oil and gas capital raising efforts.

Private equity

An investor wishing to become an equity owner in a private company may only do so with the current owner's permission. The potential investor must reach agreement with current ownership on the amount of capital and the resulting proportional ownership and control. This private equity can then be used to buy into an existing operating company or for outright acquisition of an existing company. The capital that private equity investors use is raised from many different sources, including individual wealth, institutional investors, or even funds.

The term *private equity* is used in many different and confusing ways, mixing its basic motivation with its specific investment strategy:

- *Venture capital*—often for investments in technological startups
- *Leveraged buyouts*—borrowing heavily to gain control, with selected management, to gain control of a firm
- *Growth capital*—investing in high-risk small company startups, but not obtaining control
- *Distressed assets*—investing in companies on the verge of failure or bankruptcy
- *Mezzanine investments*—investing for a share of ownership, but not control

All are forms of private equity investing.

It is also possible for a private investor to buy out in total a publicly traded company, de-listing it from its public exchanges, and taking it private.

Depending on the private equity investor and associated strategy, the investor's holding period can be anywhere from one to two years to five years or more. The private equity investor never truly wishes to be the owner of the asset or company in perpetuity, but only to hold it until it can build significant value. The private equity investor then usually sells the company to another investor or investment group or takes it to the public market, to reap its return and exit the investment.

The oil and gas industry has been fertile ground for private equity. The hundreds or thousands of small firms in the industry, particularly in the upstream, represent a sector of firms with significant promise or potential, but that suffer from limited capital. The owners of these firms, say a wildcat exploration firm, may have little in the way of current operating cash flows or earnings to show a banker and gain debt. Instead, they take on partners, who may play either passive (most common, a type of *grub stake*) or active (more rare) roles in the running of the company.

Consider the following three examples of the many private equity investors in the oil and gas industry.

- **Energy Trust Partners.** Funds from Energy Trust Partners make private equity investments in the upstream sector of the oil and gas industry. ETP collects capital in creating funds, which then selectively invest in upstream ventures. ETP's investment philosophy is to fund experienced management teams and their efforts to secure oil and gas reserves either organically (through exploration and development) or inorganically (through acquisition).[12]

- **3i.** 3i is a global partner for companies in the upstream oil and gas sector, particularly in some of the leading companies in the oil service and the exploration and production sectors. 3i started with investments in the North Sea in the early 1970s, and currently has a portfolio of 17 companies with a total equity commitment of more than $1.4 billion. It would be considered a growth capital firm.[13]

- **HitecVision.** HitecVision Asset Solutions is a European-based private equity fund that invests exclusively in the asset side of the offshore oil and gas industry. This includes assets such as drilling rigs, supply vessels, and subsea construction vessels. The fund is one of several HitecVision Funds related to the oil and gas industry.[14]

The oil and gas industry has relied heavily on private equity in its multitude of forms for many years. As noted by the *Oil & Gas Financial Journal*, "90% of petroleum produced in North America is from independent producers, many of them private companies that rely at least in part on equity investment in order to grow."[15] During boom periods, for example the run-up in crude oil and natural gas prices between 2003 and 2008, private equity provided the capital for hundreds of businesses of all shapes and sizes in the industry to grow and compete. Private equity firms were very important in all stages of the upstream and midstream, including oil services and new technology sectors. (*Industry Insight: Wildcatters, Private Equity, and the Boom* highlights the role of private equity.) But with the fall of crude prices in the second half of 2008, as well as the global financial crisis at nearly the exact same time (basically beginning in September 2008), its support for the industry has obviously waned. Later in the chapter, the strategy of Kosmos Energy, an independent E&P firm funded by private equity, is discussed.

Industry Insight: Wildcatters, Private Equity, and the Boom

The American oil patch, once left to languish during an extended period of low oil prices, is once again on the rebound, and would-be wildcatters like Bryant are ready to pounce. With oil prices hovering around $60 a barrel—three times as high as they were throughout the 1990s—the industry is expanding at a pace last seen decades ago. This revival of small exploration companies has been helped by the abundance of private equity funding. Dozens of wildcatters have been popping up across the United States in recent years, mostly to squeeze more oil out of existing fields.

"The oil industry has changed dramatically in the last 20 years," Bryant said. "Barriers to entry have dropped significantly. It doesn't matter if you've been in the business 100 years or 100 days." Easily available capital and technology, once the preserve of traditional oil companies, are reordering the business. Investors are lining up to fund energy projects, while leaps in computing power, imaging technology, and collaborative online networks now allow the smallest entities to compete on an almost equal footing with the biggest players. "There's a lot of money out there looking for opportunities," said John Schaeffer, the head of the oil and gas unit at GE Energy Financial Services. "It seems like everyone wants to own an oil well now."

Still, oil exploration remains a costly business fraught with peril. While the odds have improved, success is mostly elusive; three-quarters of all exploration wells come up dry, either because there is no oil or because geologists cannot pinpoint exactly where hydrocarbon deposits are located. All of which means that Bryant's start-up, Cobalt International Energy, based in Houston, which plans to begin drilling next year, faces formidable hurdles.

"There's no sugar-coating this—at the end of the day, it's a high-risk venture," Bryant said. "Financially, we're definitely wildcatting. It's either all or nothing."

Source: Jad Mouawad, "Wildcatters find their niche in the oil industry," *The New York Times*, May 19, 2007.

Venture capital

Venture capital (VC), despite conventional wisdom, is not just for the development of advance technologies.[16] *Venture capital* is defined as investment focused on the development of private, young, and [hopefully] fast-growing companies. The global oil and gas industry has thousands of such companies, but their access to venture capital has not always been strong.

In the early 1980s, VC investments in the energy and industrial-energy field made up more than 20% of all venture capital; by 2000, however, this had fallen to only 1%. The growth of a number of startup sectors was instrumental in both growing the pool of venture capital funds and in attracting their interest. These startup sectors included communications equipment, computer hardware, computer software, biotechnology, medical devices, and Internet-specific investments. The Internet and dot.com boom of the late 1990s clearly garnered a dominant share of the capital. When the dot.com bubble burst, and with energy prices rising in the 2002 to 2008 period, the energy sector gained a bit of that interest back, rising to roughly 3% of all venture capital in 2007. However, it appears that most of this investment is focused on so-called clean tech (clean technology), rather than the traditional fields of oil and gas development.

The oil and gas industry does not represent great promise for venture capitalists. Most new energy technologies take extremely large quantities of capital (often averaging more than $100 million), and typically more than 10 years to full-scale commercialization and profitability.[17] It is also an industry perceived as mature, and not particularly fruitful for the development of ground-breaking technological investments of profit.[18] This does not fit well given the typical short-term hit-or-miss strategies applied in most venture capital undertakings. That said, in recent years there have been many venture capital investments into nonpetroleum types of energy, such as algae (discussed in chapter 15) and biomass.

Debt

Raising debt in the oil and gas industry is problematic. Most of the smaller and medium-sized firms in the global oil industry rely upon commercial banks for short- to medium-term loans for part of their funding. However, debt has traditionally been somewhat limited in the upstream sector, depending on the specific activities of the firm. Small firms expressly focused on exploration often have major capital needs, few existing operating cash flows, and carry high risk—exploration risk.[19] Lenders are looking for collateral to lend against, possibly producing assets or proven reserves. As noted by Taylor-DeJongh, a financial advisory firm in the oil and gas industry:

> *A company's ability to monetize its assets or to put forward significant collateral in the form of P1 reserves can greatly improve the risk profile and therefore improve the terms and availability of finance. Graduation from exploration-only to exploration and production is therefore highly advantageous not only from the public equity perspective, but also for improving access to debt.*[20]

In principle, the more diversified the business lines and the longer the history of business performance, the better risk the business looks in the eyes of creditors. And, although short-term financing may be available to a great number of firms, companies wishing to purchase major fixed assets and technology need longer-term capital and longer-term loans, and they may find the debt markets unresponsive during periods of low and flat crude oil or natural gas prices. The super majors, by their very definition of diversification across the value chain, and in their size, depth, and diversity of risks (described in more detail in the previous section on mapping creditworthiness), have long enjoyed ready access to debt and debt markets.

Project finance

> *The only people who prefer project finance are the ones who've never done a deal using project finance.*
>
> —Senior executive, BP

Project finance is the financing of a project arranged in such a way that lenders rely solely on the assets and cash flows of the project for interest and loan repayment.[21] This is fundamentally different from *corporate finance*, where lenders rely on the cash flows and financial strength of the entire corporate entity for debt service. Project finance separates the project and its funding from the rest of the corporation—the *sponsor*. It is usually used for the financing of long-term capital projects, large in scale, long in life, and generally high—but manageable—in risk.

Project finance has also been widely used in the oil and gas industry for decades. It is a financial structure that lends itself to the various elements inherent in many petroleum development projects, including the creation of a high-debt financing structure that lowers the cost of total capital for development, while separating the sponsor's financial risks—their specific additional capital or legal liabilities associated with production cost overruns or political risks associated with specific projects.

Project finance is not new. Examples go back centuries and include many famous early international businesses such as the Dutch East India Company and the British East India Company. These entrepreneurial traders financed their ventures to Asia on a voyage-by-voyage basis, with each voyage's financing being like venture capital; investors would be repaid when the shipper returned and the fruits of the Asian marketplace were sold at the docks to Mediterranean and European merchants. If all went well, the individual shareholders of the voyage were paid in full. But if the specific voyage met disaster, or if the goods yielded less than what had been expected and invested, shareholders lost. There was no other sponsor to apply to.

Although each individual project has unique characteristics, most project financings are highly leveraged transactions, with debt making up more than 60% of the total financing. Equity is a small component of project financing for two reasons: first, the simple scale of the investment project often precludes a single investor or even a collection of private investors from being able to fund it; second, many of these projects involve subjects traditionally funded by governments—such as electrical power generation, dam building, highway construction, energy exploration, production, and distribution.

Debt levels of 60% or more place an enormous burden on cash flow for debt service. Therefore, the structure usually requires a number of additional levels of risk reduction. The lenders involved in these investments must feel secure that they will be repaid (the multitude of contracts and commitments behind the projects effectively reduce the business risk of the project). Commercial bankers are not by nature entrepreneurs and do not enjoy entrepreneurial returns from project finance.

Project characteristics. A project must possess four basic properties in order to be a good prospect for project financing.

1. **Separability of the project from its investors.** The project is established as an individual legal entity, separate from the legal and financial responsibilities of its individual investors. This not only protects the assets of equity investors, it provides a controlled platform upon which creditors can evaluate the risks associated with the singular project. The ability of the project's cash flows to service all the debt itself assures that the debt service payments will be automatically allocated by and from the project itself (and not from a decision by management within a multinational enterprise). Figure 7–9 highlights project separability.

2. **Long-lived and capital-intensive singular projects.** Not only must the individual project be separable and large in proportion to the financial resources of its owners, its business line must be singular in its construction, operation, and size (capacity). The size is set at inception and is seldom changed.

3. **Cash flow predictability from third-party commitments.** An oil field or electric power plant produces a homogeneous commodity product that can yield predictable cash flows if third-party commitments to take or pay can be established. Nonfinancial costs of production need to be controlled over time, usually through long-term supplier contracts with price adjustment clauses based on inflation. The predictability of net cash inflows to long-term contracts eliminates much of the individual project's business risk, allowing the financial structure to be heavily debt-financed and still safe from financial distress.

   The predictability of the project's revenue stream is essential in securing project financing. Typical contract provisions intended to ensure adequate cash flow normally include the following clauses: quantity and quality of the project's output; a pricing formula that enhances the predictability of adequate margin to cover operating costs and debt service payments; and a clear statement of the circumstances that permit significant changes in the contract, such as force majeure or adverse business conditions.

4. **Finite projects with finite lives.** Even with a longer-term investment, it is critical that the project have a definite ending point at which all debt and equity has been repaid. Because the project is a standalone investment whose cash flows go directly to the servicing of its capital structure, and not to reinvestment for growth or other investment alternatives, investors of all kinds need assurances that the project's returns will be attained in a finite period. There is no capital appreciation, only cash flow.

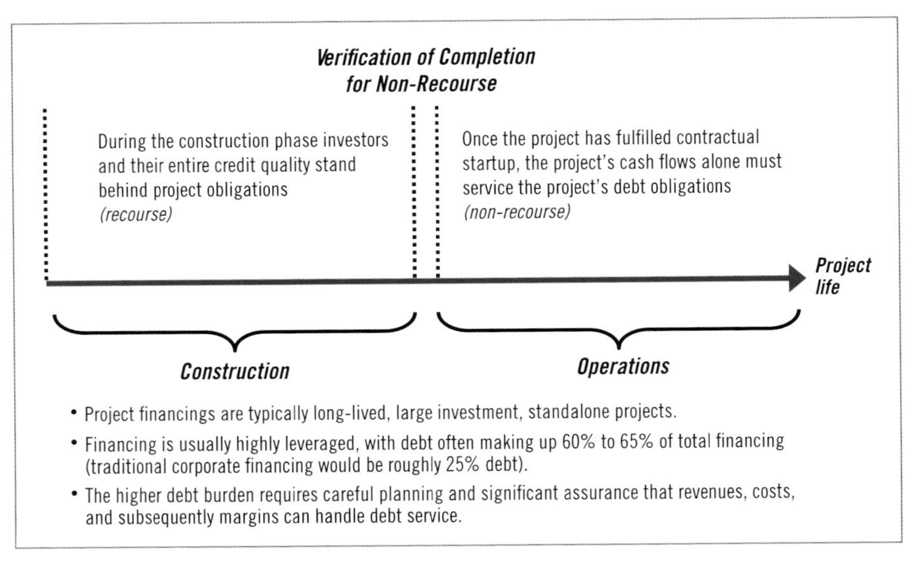

Figure 7–9. Conditions for separability

Examples of project finance include major LNG projects such as the QatarGas II project, as well as many of the largest individual pipeline investments undertaken in the past three decades, such as the Trans-Alaska Pipeline, Chad-Cameroon Pipeline, and the Baku-Tbilisi-Ceyhan Pipeline, and often combine project finance with equity joint ventures. For example, the Trans-Alaska Pipeline was a project finance joint venture between Standard Oil of Ohio, Atlantic Richfield, Exxon, British Petroleum, Mobil Oil, Phillips Petroleum, Union Oil, and Amerada Hess. A project of this scale and scope represents capital expenditures that no single firm would or could attempt to finance on its own. Yet, through a project finance joint venture arrangement, the higher-than-normal risk absorbed by the capital employed can be managed.

Many project financings in the petroleum industry are actually structured through a *special purpose entity* (SPE), an off-balance-sheet partnership set up by the company to separate and segment the risk of the project investment from the rest of the corporation and to allow other investors to potentially participate in the investment.

Equator Principles. Because project financing is used in many development projects around the world, and often in emerging markets lacking many of the traditional structures and regulatory environments for social and environmental protection, a set of private sector banks developed the Equator Principles. They are "a financial industry benchmark for determining, assessing and managing social and environmental risk in project financing."[22] These principles, based on the environmental standards and social policies long used by the World Bank and the International Finance Corporation, have become the accepted standard for assessing major development projects, including onshore and offshore oil and gas, around the world. First launched in 2003, they were expanded and amended in 2006 and 2010. As of August 2010, there were 67 signatory financial institutions and advisories globally.

Multilateral lending

Multilateral lending, loans provided by multilateral organizations like the World Bank to both public and private enterprises, has evolved over time. The World Bank has separated the agencies (table 7–4) that lend to public enterprises (government and semi-government organizations) versus those that lend to private enterprises (for-profit enterprise).

The International Bank for Reconstruction and Development (IBRD) and the International Development Agency (IDA) provide debt to public enterprises, while the International Financing Corporation (IFC) lends to private enterprises. *Private enterprises* are typically defined as any organization that operates with the objective of making profits. Originally created to support the reconstruction and maintenance of the infrastructure of economies torn by the Second World War, they have evolved into institutions whose primary goal is economic development. Conducted by multilateral lending institutions such as the World Bank, the InterAmerican Development Bank, the European Bank for Reconstruction and Development, and the African Development Bank, to name but a few, these loans provide external capital for the purposes of economic development of business, infrastructure, and industry in middle- to low-income countries.

Table 7–4. Multilateral lending organizations for the oil and gas industry

International Bank Facilities	Founded	Primary Lending and Funding Activities
International Monetary Fund (IMF)	1944	Provides financial assistance to countries, governments, to help them through serious periods of economic adjustment
International Bank for Reconstruction and Development	1944	Also referred to generally as the *World Bank*, it was founded to reconstruct post-War Europe, and now supports loans to governments worldwide to support economic and social development
International Development Association (IDA)	1960	Interest-free loans, credits, to finance projects that reduce poverty
International Finance Corporation (IFC)	1956	Lends directly to private companies, without governmental guarantees, to promote private enterprise
Multilateral Investment Guarantee Agency (MIGA)	1988	Provides investment guarantees (currency, war, expropriation, breach of contract) to private companies investing in developing countries
Energy Sector Management Assistance Program (ESMAP)	1974	Provides advice and analysis, but not funds, for shaping energy sector development and policy (a JV of the United Nations and World Bank)
Global Environment Facility (GEF)	1991	Provides grants for studies and projects involving national and regional environmental benefits
Carbon Finance Facility	2004	Supports carbon reduction policies and programs in OECD and non-OECD countries who are aligned, using the World Bank Carbon Finance Unit (CFU) as structure and manager

Regional Development Bank Facilities	
Inter-American Development Bank (IDB)	Asian Development Bank (ADB)
African Development Bank (AfDB)	European Union (EU)
European Bank for Reconstruction and Development (EBRD)	European Investment Bank (EIB)
Nordic Finance Group	Islamic Development Bank
OPEC Fund for International Development	Arab Fund for Economic & Social Development
Arab Bank for Economic Development in Africa (BADEA)	

Source: Hossein Razavvi, *Financing Energy Projects in Developing Countries*, PennWell, 2007.

Multilateral lending organizations have often been criticized as no longer necessary given the growth and development of international financial markets. As a result, they are often seen as impediments to the free marketplace for capital, and are often criticized as being highly politicized in their lending policies. Proponents of multilateral lending agencies argue that these institutions provide affordable capital in high-value-high-risk countries when the international capital markets may not be willing to lend at affordable rates. The projects that benefit are also not necessarily private enterprises, but public enterprise projects that yield benefits to the people of these countries.

State interests

The state's motivations and interests regarding capital investment and net cash flows is always the same: to reap the positive net cash flows generated during the production phases of a petroleum project based on state-owned assets and not to bear the negative cash flow investments required in the other phases.

As illustrated in figure 7–10, investment is required in the prelicense phase, in which very early geology and petroleum studies are conducted at the expense of the state or the E&P company. If the results of prelicense work are promising, a development contract is signed (time, t_1), the E&P company may provide most of the capital for exploration, development, and production (t_1 to t_4). Some properties, depending on market conditions and crude oil prices, may undergo production rehabilitation (for example reinjection of gas or water to increase total production). Regardless of rehabilitation, abandonment will require additional use of capital to plug and rehabilitate properties post-depletion, an investment which has frequently been borne by the state alone.

The state has varying degrees of access to affordable capital for petroleum development. Many of the world's petroleum properties are located within the political boundaries, both land and water, of emerging market countries with limited access to capital for petroleum development. At the same time, the state frequently does not want to bear the risk of development failure but does wish to reap the gains associated with successful developments.

E&P companies usually have greater access to capital than states, given both their home country of incorporation (which is frequently a large highly industrialized economy with well-established access to the global capital markets) and their potentially diversified multinational business structure. The problem is that most petroleum developments combine two very unattractive components when it comes to financing: they are very large and very long in capital requirements, and they often occur in countries considered highly risky. The capital requirements of petroleum development have long been known and have driven many of the world's major oil companies to use alternative financing structures such as joint ventures and project finance over traditional corporate finance.

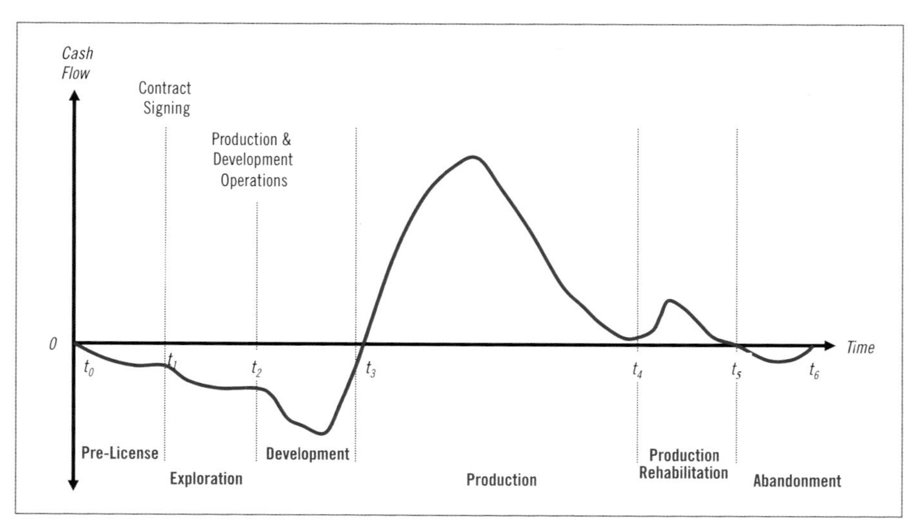

Figure 7–10. Petroleum development cash flows. Most petroleum developments follow a net cash flow pattern similar to that shown here. The periods of net negative cash flow (when the cash flow falls below the horizontal axis) are predominantly borne by the IOC, while the majority of the net positive cash flow proceeds (above the horizontal axis) accrue to the state.

Oil loans

Completing the circle of development and evolution of global petroleum assets is the concept of oil loans. Similar in concept to the early fiscal regimes, states have always suffered either a lack of interest in financing petroleum themselves, or the inability to gain ready access to sufficient affordable capital for development. But like fiscal regimes and their evolution towards the modern PSA, financing has now come full circle to allow many states to gain access to capital of their own for petroleum development—the use of the hydrocarbons in place as collateral for borrowing from the commercial banking sector.

Many of the countries that, for better or worse, are the domains of the hydrocarbon deposits to be developed have historically been plagued by poor economic development and high foreign debt. The debt, much of which was provided by the IMF and World Bank and other multilateral lending organizations over the past 40 years, has created a borrowing profile that is anathema to commercial banks the world over. Bad borrowers by track record, they possess billions of barrels of collateral to gain access to debt.

Angola is one of the more aggressive borrowers in recent years. In March 2004, China's Eximbank extended Angola an oil-backed credit line of $2 billion.[23] The credit was backed, and is to be repaid, in Angolan oil. The massive credit and later ones have come under much international scrutiny as questions over the allowed use of the credit to fund major infrastructure projects (70% of the funds have to be expended with Chinese companies) that may end up being white elephants when completed, a problem seen in much of Africa in the past 40 years. (A *white elephant* is any asset or object that the owner cannot dispose of, but whose operating and maintenance costs exceed its purported use or value. The term is derived from the ancient practice in some monarchies in Burma, Cambodia, Laos, and Thailand of owning a sacred white elephant.) Little of the capital is being spent on establishing the operational and maintenance skills and competencies needed to sustain the facilities once completed, many of which are for oil development.

Angola's oil-backed loan program expanded again in 2005, when Standard Chartered Bank (UK) led a consortium of lenders in extending a $2.35 billion, five-year loan.[24] Priced relatively high at LIBOR plus 2%, the loan has fostered a heated debate over whether it represents the best or the worst of the continually developing international capital markets. The lack of transparency, the lack of conditionality, and the pricing have all been arguments posed against the credit. Angola used a portion of the proceeds to immediately pay off the balance of a $750 million loan to its European "parent," Portugal. The fact that the international commercial banking sector is now willing to accept the high political risks associated with Angola's regime, as long as those risks are backed by petroleum assets, is the argument used by the proponents of the deal. Angola, for one, seems to have found great new financial opportunities for oil and gas still in the ground.

> *'Angola has no interest in transparency and there is no source of external leverage on the government right now,' said Monica Enfield, an analyst at PFC Energy, an energy consulting firm in Washington. 'With all their oil revenue, they don't need the IMF or the World Bank. They can play the Chinese off the Americans.'*[25]

The oil asset value as collateral for debt is not limited to new and emerging players like Angola. It is also an increasing conduit for cash-rich but oil-poor country governments to obtain the capital they need for infrastructure and industrialization. In some cases, such as in the *Industry Insight: China Bails Out Russian Debt for Russian Oil*, Russian oil has served as one critical way out from under rising debt loads to the West. Russia has also concluded similar deals in recent years with President Hugo Chavez of Venezuela.

Industry Insight: China Bails Out Russian Debt for Russian Oil

MOSCOW—As chances of rolling over loans with troubled Western banks dwindle in the financial crisis, Russian oil companies are negotiating multibillion dollar lines of credit from a more reliable source: the cash-rich Chinese government. Prime Minister Wen Jiabao of China was in Moscow on Tuesday for talks with his Russian counterpart, Vladimir Putin, about a potential loan-for-oil deal that would be backed by future exports to China, Russian officials said.

It is unclear how much oil China might receive in exchange for the credit. Plans for a pipeline to China, a spur off a trans-Siberian pipeline that is currently under construction, specify that it could carry about 300,000 barrels of oil per day. Reuters, which first reported the Russia-China talks on Monday, cited industry sources as saying that Russian oil companies could borrow between $20 billion and $30 billion and pledge to export about 2 billion barrels of oil to China over the next 20 years. That would cover about 4% of China's oil demand.

A crushing need for cash by Russian oil companies that are no longer able to draw credit from Western banks seems to have brought the two countries closer to agreement. The state oil company Rosneft, for example, has about $21 billion in debt and some of its creditors are demanding early repayment.

Source: Andrew Kramer, "Russia seeks to negotiate loans-for-oil deal with China," *The New York Times*, October 28, 2008.

Petroleum finance today

As the search for new oil has caused the industry to go further, deeper, and to more difficult parts of the earth, the size of projects—in both scale and capital—has continued to grow. The massive amounts of capital now required, combined with the growing political, technological, and business risk, have caused more adaptation in the financial structures of new projects.

Are E&P Firms Better Off If They Compete in the Downstream?

Chapters 2, 3, 4, and 5 covered the value chain activities from exploration to production that collectively represent the activities of an E&P firm. This chapter has tried to provide insight into the financial needs, sources, and management of the oil and gas industry. In this final section, we consider an important corporate strategy question, with major financial implications, for E&P firms: Is an E&P firm better off if it also competes in downstream refining and marketing?

Firms can diversify across the value chain or they can narrowly focus on a limited span of activities. Chapter 10 discusses the diversification of IOCs and their natural hedge against adverse price movements of refining spread components because they have some control over their entire supply chain. Chapter 13 examines the fuels marketing business and its role in IOCs' diversification across upstream and downstream activities. Some analysts argue that independent E&P companies are at a disadvantage because E&P companies need to fund their growth and will generate the greatest cash flows when demand and price for derivative products like gasoline and diesel are at their highest. When the demand for gasoline and diesel is strong and crude prices go up, E&P companies earn more profit. Unfortunately, when crude prices are rising, costs for inputs like rigs, steel, cement, and people are also rising. If firms try to wait out the cycles to lower their development costs, they may not have sufficient cash flow when oil prices drop.

In contrast to the stand-alone E&P firm, IOCs can fund their upstream growth from their diversified portfolio and can often wait out the high-cost cycles. When costs go up, IOCs can sit out the inevitable speculative frenzy and wait until "normalcy" returns. Independents cannot take such a conservative approach because they do not have a diversified profit stream and need to maintain their investments regardless of the cycle stage. During down cycles, IOCs can use the downstream and/or chemicals cash flows to maintain capital investment and keep their shareholders satisfied. As the argument goes, independents have no other profit streams and end up suffering more during down cycles.

A counterargument is that two decades of low investment by the IOCs in the down cycle period of the mid- to late 1980s and 1990s created conservative and risk-averse cultures. During the boom years of 2004 through 2007, the large IOCs were accused of being too cautious with their upstream capital investments. Because the independents could not wait out the cycles and hope for "golden ages" in refining or chemicals, they had to invest and take risks that the IOCs shied away from. Cautious IOC behavior opened the door for aggressive independents like Devon Energy in deep water, Kosmos Energy in offshore Africa, and XTO Energy in shale gas. By the end of the decade, the cycle had turned and oil and gas prices had fallen precipitously. Moreover, many analysts were saying that refining and downstream in general were entering a long period of steady decline, raising questions about IOC commitment to diversification.

So, although a definitive conclusion on the value of integration is impossible, a few points can be noted. One, the IOCs have more than a century of dealing with oil and gas cycles and diversified business portfolios. Two, there are many successful and well-managed independent E&P firms. Although they may not have a century of history, the independents provide strong competition to the IOCs. Three, there will always be boom-and-bust cycles in oil and gas, and the best-managed companies will do well regardless of the cycle. Finally, independent E&P companies often make more risky bets than the IOCs. *Industry Insight: Kosmos Energy* provides an overview of the company's strategy. Kosmos's business model is to search for oil and gas in exploration areas that are not of interest to the majors. Once oil and gas are found, Kosmos hands off the project for development and production to a larger firm. Independents like Kosmos and others will also find themselves targets of the majors' mergers and acquisitions (M&A) activities. XTO is a case in point: XTO made a financial and technological commitment to shale gas early in the technology cycle, had some success, and was acquired by ExxonMobil in 2009.

Industry Insight: Kosmos Energy

Kosmos Energy, an independent E&P firm, was founded in 2003. The company was started with $300 million in capital and is backed by two large private equity firms. Kosmos says that it "thrives on risk." The company's mission is to find "difficult oil" in areas where the majors and other companies have no interest. The strategy includes several elements:*

- Create and nurture a corporate culture that encourages and rewards informed exploration risk taking.
- Build a first-class exploration portfolio by capturing acreage positions in emerging basins and established provinces overlooked by risk-averse oil and gas companies.
- Maximize an asset's value through accelerated, prudent appraisal.

- Implement fast-paced production and development programs, including the design and installation of complex subsea architecture and the utilization of floating production, storage, and offloading vessels (FPSOs).

In 2007, Kosmos discovered the deepwater Jubilee Field off the coast of Ghana. Jubilee was the largest oil discovery in the world in 2007, with recoverable reserves of more than 1 billion barrels. Partners in the field include two independents, Tullow Oil and Anadarko Petroleum, and Ghana National Petroleum Corporation. First oil was expected by the end of 2010.

Kosmos was very clear in stating that it was not likely going to be an oil producer. Kosmos finds the oil, brings it to the point of development, and then sells a stake in the project to companies "designed to be there for the life of the project." In the words of the CFO, "Organic growth through the drill bit—coupled with substantial creativity and capital commitment by the appropriate entities—promotes independence and entrepreneurship within E&P companies and encourages an exploration and development process second to none."

---

* Private Equity Funding Empowers Kosmos Energy, *Oil & Gas Financial Journal*, September 2008.

Ruminations on Valuation

In chapter 1 we noted many of the world's leading oil and gas companies by market capitalization. In this chapter, we focused on financing and financial performance, particularly the return on sales and return on capital generated by global oil and gas companies. But what of that critical market valuation item—share price?

The market capitalization of a company is its total value in the publicly traded market:

$$\textit{Market capitalization} = \textit{Share price} \times \textit{Shares outstanding}$$

Although on the surface this may seem to be an extremely important firm metric, aside from bragging rights about which firm has the greatest market capitalization, it is not that significant for either management or shareholders. Shareholders are most interested in capital gains that come from changes in the share price alone. Management's interest is on its compensation, and its compensation is traditionally based on financial performance like profit or return on sales (although senior management may hold stock options and therefore be interested in share price as well).

If share price is of more significance to shareholders and top management than market capitalization, what is it that drives share price? (We must note that if we really knew the answer to this question, we would not teach and most certainly would not answer the question in a book.) According to the theory of finance, the price of any asset today—including share prices—is the present value of all expected future cash flows (ECF_t) to be generated by that asset.

$$Share\ Price_t = \frac{ECF_{t+1}}{(1+k)^1} + \frac{ECF_{t+2}}{(1+k)^2} + \frac{ECF_{t+3}}{(1+k)^3} + \ldots \text{ through time}$$

Therefore the share price of Chevron or Baker Hughes or any other oil and gas firm is based on what the firm expects to generate in actual net operating cash flow out into the indefinite future. This means that, for example for Chevron, the market has formed an opinion on all asset-producing properties the company currently has in its portfolio and all others that it is expected to acquire, the operational efficiencies and costs of developing and producing those assets, not to mention the market prices that those assets are likely to garner. Obviously, if markets can do all of that with a high degree of accuracy and competence, they are indeed very smart.

In practice, management and markets do place a heavy emphasis on profits—earnings—and whether the firm is able to grow earnings steadily over time. Since earnings do indeed make up a large portion of the operating cash flows of all firms, this is reasonable and consistent with most of financial theory. The challenge, therefore, is to generate greater earnings if you are management and predict future earnings if you are an investor. In principle, any expected event or activity that is likely to alter future operating cash flows in a material way, whether it be the expectation of successful exploration off the coast of Ghana or the potential legal liabilities associated with the *Deepwater Horizon* disaster in the Gulf, will affect share price.

Summary Points

- As a capital-intensive industry, oil and gas firms must continually compete for capital from global markets. They must do so when investing in projects requiring more capital in ever-riskier environments.

- Most of the oil majors use return on capital employed (ROCE) as one of their primary measures of financial performance. Despite its popularity, many academic studies have concluded that although ROCE may be a good measure for aiding management and possibly investors in determining the capital efficiency of an oil and gas firm, ROCE values have shown little ability to predict or even correlate with share price.

- Examined over a number of years, oil and gas companies do not have above-average profits. Return on sales (ROS), the most common measure of profitability across industries, is not particularly high for the oil and gas industry relative to other industries. ExxonMobil, the world's most profitable public company in the 2006 to 2009 period, did not demonstrate much more than a slightly above average ROS during that period.

- Ratings agencies use many financial and operating characteristics of oil and gas companies in determining credit quality. Although they include a variety of financial variables, it is still the underlying drivers of company competitiveness and production capability—oil and gas reserves and the ability of companies to replace production—that dictate a large portion of the rating.

- The oil and gas industry—the thousands of firms operating globally—relies primarily on corporate finance for its funding. This means traditional debt and equity (both public and private) for financing the seemingly insatiable appetite for competing in the global markets.

- In specific cases of large, long-term, stand-alone, high capital projects, project financing structures are used to fund the projects. Although much more complex to arrange, and often requiring more governmental approvals and some government agency lending, project finance has been successful in funding many major petroleum and gas projects. Pipelines have been one area of particularly high visibility, including the Alaska pipeline, the Chad-Cameroon pipelines project, and the BTC pipeline.

- Private equity has been a source of capital critically important to the periodic rejuvenation of the industry. Especially during boom periods of high prices and great promise, private equity capital has allowed many small firms to expand. In recent years, private equity investors have shown significant interest in small international E&P firms.

Notes

1. This assumes that when the firm makes the sale, the buyer does not pay in cash. If a cash payment, the proceeds would go directly to the "cash and marketable securities" line item, not receivables. Global practice is, however, the same as most businesses, the buyer wishes to pay later, and therefore accounts receivable (A/R) are created regularly.
2. ExxonMobil Corporation, *2009 Summary Annual Report*, p. 44.
3. ExxonMobil Corporation, 2009 *Summary Annual Report*, pp. 44–45.
4. One such study is Petter Osmundsen, Frank Asche, Bård Misund, and Klaus Mohn, "Valuation of International Oil Companies—The RoACE Era," CESifo Working Paper No. 1412, February 2005.
5. "Rating Methodology: Global Integrated Oil & Gas Industry," Moody's Investor Services, November 2009.
6. Moody's calculates ROCE on a pretax basis (the numerator), rather than an after-tax basis like that of ExxonMobil.
7. Moody's Investor Services, October 2005, p. 9.
8. Moody's Investor Services, October 2005, p. 4.
9. Salvatore Lazzari and Robert Pirog, "Oil Industry Financial Performance and the Windfall Profits Tax," Congressional Research Service, September 30, 2008.
10. Ibid.
11. Y. Lee, "Cnooc Shuffles Management: Fu to Give Up Role as CEO, Making Way for Younger Leadership; Net Doubles," *Wall Street Journal*, August 20, 2010.
12. http://www.energytrustpartners.com/.
13. http://www.3i.com/sectors/oil-gas-and-power.html.
14. "HitecVision Launches New $420 Million Fund to Invest in Offshore Service Industry," *Oil & Gas Journal*, June 2, 2010.
15. "Private Equity Investments Crucial to Energy Industry," *Oil & Gas Financial Journal*, December 1, 2009.
16. J.J. Dooley, "Trends in US Venture Capital Investments Related to Energy: 1980–2007," Pacific Northwest National Laboratory, Prepared for the US Department of Energy, PNNL-17953, October 2008.

17. Robert W. Shaw, Jr., "Energy Venture Capital: The New Wave," Aretê Corporation, in *Energy Venture Capital Best Practices*, Boston: Aspatore Books, Inside the Minds series, May 2006, p. 8.

18. Rolf Wustenhagen and Tarja Teppo, "What Makes a Good Industry for Venture Capitalists? Risk, Return, and Time as Factors Determining the Emergence of the European Energy VC Market," University of St. Gallen, IWO Discussion Paper No. 114, October 2004.

19. Nicole Weygandt and Taylor-DeJongh, "Accessing and Raising Capital for Jr. Oil and Gas Companies," *Commentary*, December 2008, p. 2.

20. Ibid.

21. There are a number of excellent sources of information for those interested in greater depth in understanding project finance, including E. R. Yescombe, *Principles of Project Finance* Academic Press, London, 2002, and John D. Finnerty, *Project Financing: Asset-Based Financial Engineering*, second edition, Upper Saddle River, NJ: Wiley, 2007.

22. http://www.equator-principles.com/.

23. "Angola's Elusive Oil Riches," *The New York Times*, June 15, 2004.

24. Jad Mouawad, "Nowadays, Angola Is Oil's Topic A," *The New York Times*, March 20, 2007.

25. Jad Mouawad, "Angola, One of the Poorest Places on Earth, Is an Oil Industry Darling," *The New York Times*, March 19, 2007.

Chapter 8

NATURAL GAS

Natural gas is really well-suited to meet that growing power generation demand, both from the standpoint of its lower environmental impact, but also its capital efficiency and its flexibility.

—Rex Tillerson, chairman and chief executive officer, ExxonMobil[1]

What is worse than drilling a dry hole? Finding gas.

—Wildcatter's lament

Although the industry has long been labeled *oil and gas*, and most of the major firms operating in the industry have always been described as *oil and gas companies*, the oil and gas industries are in many ways very different.[2] Oil and gas are indeed often found together, though their relative mix and qualities differ dramatically. The qualities of gas found, particularly the residual content, and its geographic proximity to markets, determine whether the gas is economically recoverable and commercially saleable.

Natural gas is considered the most environmentally friendly of all fossil fuels. Gas has the lowest carbon emissions (CO_2) per unit of energy, and it can be used in a variety of power applications of relatively high efficiency. Gas provides roughly 20% of all global energy, primarily in nontransportation energy demand uses. Natural gas demand is dominated by three sectors: residential and commercial consumers, industrial consumers, and power generation. Unlike oil, the market for natural gas has in the past been highly regional and distinctly not global.

The primary factor differentiating oil from gas is *transportability*: oil is a liquid that can be transported economically by truck, rail, ship, and pipeline. Natural gas, although convertible to liquids, has historically needed to be in close proximity to major industrial and commercial markets to make pipeline infrastructures for transportation economically justified. The economics of natural gas meant, for many years, that gas discovered in the process of searching for oil was seen as a failure and was often either not produced, abandoned, or shut in. That is now changing, as the development of liquefied natural gas (LNG) is quickly making gas a globally transportable and tradable commodity.

This chapter first describes the composition of natural gas—its chemical and physical properties, and then follows with sections on its production, use, and the markets for its many associated products.

Natural Gas: Chemistry and Form

Oil and natural gas are produced by the same geological process—the anaerobic decay of organic matter deep under the earth's surface, under immense pressure and intensive heat for millions of years. As a result, oil and natural gas are often found together. Common terminology distinguishes oil and gas by relative quantities per deposit or field. Deposits rich in oil are known as *oil fields* and deposits rich in natural gas are called *natural gas fields*. As a result, oil and gas are found both onshore and offshore.

Quantities of natural gas are measured in normal cubic meters or in standard cubic feet. The gross heat of combustion of 1 normal cubic meter of commercial quality natural gas is around 39 megajoules (which equals 10.8 kilowatt hours or kwh). In US units, one standard cubic foot of natural gas produces around 1,028 British thermal units (Btus). The average barrel of crude oil is equal to 6,040 cubic feet of average natural gas and termed a *barrel of oil equivalent* (BOE). This conversion metric is frequently used to compare the total reserves or production of countries and companies that have differing amounts of oil and gas.

In general, organic sediments buried in depths of 1,000 m to 6,000 m (at temperatures of 60°C to 150°C) generate oil, while sediments buried deeper and at higher temperatures generate natural gas. The deeper the source, the *drier* the gas—the smaller the proportion of condensates in the gas (to be defined later). Because both oil and natural gas are lighter than water, they tend to rise and migrate upwards from their source rock formations until they either seep to the surface or are trapped by a nonpermeable layer of rock. Once trapped, they can be extracted and produced by drilling.

Natural gas is itself often a complex combination of many hydrocarbons and nonhydrocarbons. The hydrocarbons (organic compounds consisting exclusively of carbon and hydrogen) methane, ethane, propane, and butane are the combustible components of gas and are the primary objective of natural gas production. As detailed in table 8–1, methane is by far the most common hydrocarbon and is burned in homes and industry. Some natural gas fields may yield nearly pure, approximately 98%, methane gas. The other hydrocarbon gases frequently present in natural gas are often referred to as *raw make* and include ethane, propane, and butane. These gases occur in much smaller quantities and have more specialized uses. The composition of raw make is typically 39% ethane, 26% propane, 20% butane (combined butane and isobutane), and 14% field natural gas. If used for heat-producing purposes, all burn hotter than methane (as the C content increases, heat production power increases).

Commensurate with heat content, the value (price) of these assorted hydrocarbon gases generally rise with carbon, C, and heat contents. Methane and ethane are gaseous at ambient temperatures and cannot be readily liquefied by pressure alone. Propane, however, is easily liquefied and is used commercially as a liquid and sold in propane bottles. Butane is easily liquefied and provides a safe but volatile fuel for a variety of applications, including pocket lighters. Pentane, a clear liquid at room temperature, is a commonly used odorless solvent in a variety of industrial applications.

Table 8–1. Natural gas composition

Hydrocarbons		Typical	Attributes and Uses
Methane	CH_4	70% to 98%	commercial gas for residential and industrial use
Ethane	C_2H_6	1% to 10%	colorless, odorless gas used as feedstock for ethylene
Propane	C_3H_8	trace to 5%	burns hotter than methane; common liquid fuel; LPG
Butane	C_4H_{10}	trace to 2%	safe, volatile, used in pocket lighters; LPG
Pentane	C_5H_{12}	trace	commonly used solvent
Nonhydrocarbons			
Water vapor	H_2O	Inert	occasionally used for reinjection
Carbon dioxide	CO_2	Inert	colorless, odorless; used for reinjection
Nitrogen	N	Inert	colorless, odorless; used for reinjection
Helium	He	Inert	colorless, odorless, light gas; specialty uses
Hydrogen sulfide	H_2S	—	poisonous, lethal, foul odor; corrosive; gas with low sulfur levels is sweet, high levels is sour

Natural gas in its raw state will also contain a variety of nonhydrocarbons, which are gaseous impurities that do not react or burn and are categorized chemically as *inert*. Water vapor is commonly present in varying quantities. Carbon dioxide can be present in some gas fields in extremely high amounts and because of its inert properties, may significantly reduce the combustion properties and commercial value of the gas. Both carbon dioxide and nitrogen can be reinjected into wells that are losing pressurization and depleting. Helium, although not always present, may be a valuable associated product of raw natural gas, as it has a number of specialized industrial applications in semiconductor and electronic circuitry manufacturing. In the end, regardless of their associated value, nonhydrocarbon components need to be separated and removed from the natural gas.

The last nonhydrocarbon listed in table 8–1, hydrogen sulfide, is highly problematic when present. Hydrogen sulfide is not inert and, if present in sufficient quantity, is highly poisonous and corrosive. It is easily detectable in even small amounts by the strong sour odor it releases similar to that of rotten eggs. Besides the odor, however, it is potentially lethal and has been the cause of numerous deaths at drilling sites around the world. Because of its highly corrosive qualities, it must be removed before gas can enter pipelines.

Industry Insight: The New London Catastrophe of 1937

New London High School was one of the most modern educational institutions in East Texas in the 1930s. It had been constructed using funds generated by the oil boom in the East Texas oil patch of Rusk County. But sometime after 3 p.m. on March 18, 1937, all that came to a horrible end when the school exploded as a result of a buildup of natural gas underneath its foundation. More than 300 students and faculty died (the exact death toll was never determined) as a result of a leakage in the school's pipes, which were attached to gas tapped from local oil fields. The school had recently changed its gas supplier from a commercial provider to a collection of unofficial tap lines that drew associated gas off nearby oil producing fields ("raw wet gas") which was considered waste and would normally have been flared.

The natural gas buildup, although causing a number of headaches among students and employees in the days prior to the explosion, went undetected as a result of its odorless form. The building was literally lifted into the air as a result of the explosion. The new steel and concrete facility fell into a massive pile of rubble.

As a result of the New London catastrophe, new laws were implemented requiring that mercaptan be added to all gas produced to give natural gas a strong, rotten-egg odor in order to make future gas leakages detectable.

Natural gas, or raw natural gas, is produced from three types of wells: gas wells, oil wells, and condensate wells. Wells designed to produce only or mainly gas are termed *gas wells*. Gas produced from oil wells, *associated gas*, is either *free gas* that separates from the oil within the reservoir or is *dissolved gas* within the crude oil. Free gas is separated from the oil while sub-surface (below ground). Dissolved gas will often separate from the crude oil as it is pumped to the surface, releasing pressure and thus the gas. Condensate wells produce natural gas along with a semi-liquid hydrocarbon called *condensate*.

Condensate, also commonly called *field natural gas*, condenses out of natural gas when pressure and temperature decrease as the gas is raised to the surface in production. The resulting liquid, condensate, can be nearly pure gasoline. It is of considerable value when present and is technically classified as crude oil when produced from natural gas in the field. Natural gas containing condensate is termed *wet gas*; natural gas without condensate is *dry gas*. Condensate and other nonmethane hydrocarbon components of natural gas are collectively referred to as *natural gas liquids* (NGLs).

Raw natural gas production

Depending on the type of gas and location of wells, raw natural gas may be first processed at or near the wellhead, termed *field processing*. At the wellhead itself, scrubbers and heaters are often used to eliminate various particulate components, such as sand, and to maintain the heat levels of the gas after extraction to prevent various types of ice crystals from forming and impeding fluid and gas transfers. If the gas is to be transported via pipeline, a number of the nongas components will need to be removed from the gas prior to entry into the pipeline. The process of removing the various impurities in raw natural gas to create a dry gas quality for pipeline transportation is quite complex, but typically includes four main processes to remove 1) oil and condensate, 2) water, 3) NGLs, and 4) sulfur and carbon dioxides. Figure 8–1 shows a schematic flow diagram as to how this may be accomplished.

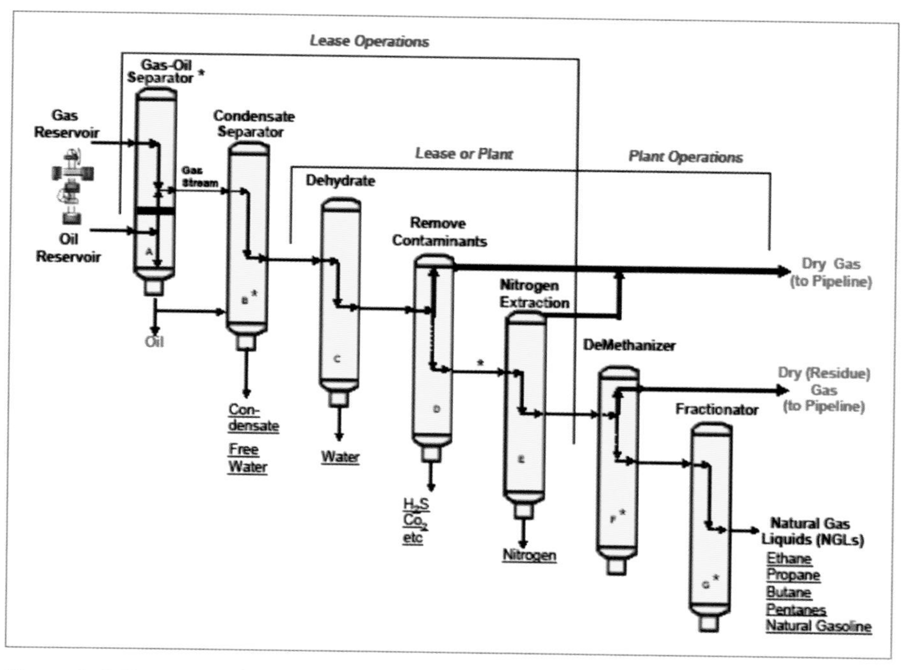

Figure 8–1. Gas separation and derivative products
Source: Unigas © 2009.

Inevitably, the gas will be final processed and separated into its components at a centralized processing facility that may draw upon numerous pipelines from the producing area. A complex gathering system can consist of hundreds or even thousands of miles of pipes, which connect the central processing plant to hundreds of producing wells in a region.

Oil and gas separation. The actual separation of oil from gas, when associated in the well, depends on the specific reservoir. Some oil and gas deposits separate naturally with production, as the depressurization of production allows gravity to separate the two according to density. In some facilities this is aided with the use of a simple holding tank that allows gravity separation to be completed. In more complex cases, specialized equipment is needed. One example is the low-temperature separator (LTX), used when producing a high-pressure gas field that also contains a light crude oil or condensate. LTX separators use pressure to cool the wet natural gas and separate the oil and the wet condensate.

Water separation. The associated water from the natural gas must be removed. Most of the liquid-free gas associated is separated using simple methods at the wellhead. Water vapor removal requires a more complex treatment process of dehydrating the natural gas. There are two major methods of dehydrating, absorption and adsorption. *Absorption* uses an active agent to separate the water and *adsorption* occurs when the water vapor is condensed and collected on the surface.

NGL extraction. There are two principle techniques used for removing NGLs from the natural gas stream: *t*he absorption method and the cryogenic expander process. The absorption method of NGL extraction is very similar to using absorption for dehydration. In NGL absorption, an absorbing oil is used as opposed to a chemical agent like glycol. This process allows for the recovery of around 75% of butanes and 85% to 90% of pentanes and heavier molecules from the natural gas. Cryogenic processes are also used to extract NGLs such as ethane from natural gas. Although absorption methods can extract almost all of the heavier NGLs, lighter hydrocarbons like ethane are more complex. Under certain conditions it may simply be more economical to leave the lighter NGLs in the natural gas stream.

Sulfur removal. Natural gas from some deposits contains significant amounts of sulfur. This natural gas, called *sour gas* (if the sulfur exceeds 5.7 milligrams of H_2S per cubic meter of gas) as a result of its sulfur smell, is both corrosive and dangerous. The sulfur is sometimes sold as a derivative product after separation. In the United States, for example, approximately 15% of all commercial sulfur is derived from natural gas processing. Sulfur extraction processes are similar to those used for both dehydration and NGL absorption.

Liquefied natural gas (LNG)

Natural gas can be liquefied, and as a result, can resolve the major development challenge it has faced historically, transportability. *Liquefied natural gas*, commonly referred to as *LNG*, is natural gas that has been processed to near pure methane (CH_4), often 90% or more, and then cooled. In its liquid form, LNG is colorless, odorless, nontoxic, and noncorrosive.

The initial process of moving LNG to market, summarized in figure 8–2, is similar to normal natural gas production. Gas is piped from fields and wells to a central processing center. Here the process becomes complex and expensive. The gas must have all nonmethane components removed, which is a more rigorous separation process than that for natural gas entering a pipeline (all oil and gas pipelines have specific content specifications and requirements in order to avoid corrosion, blockages, or spillages).

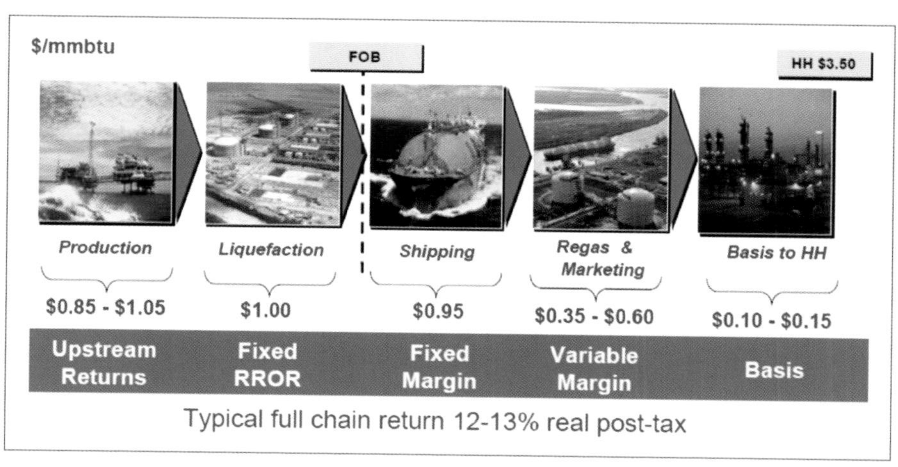

Figure 8–2. LNG supply chain

Source: "Oil & Gas for Beginners," Deutsche Bank, January 7, 2008, p. 187, Fig. 240. Copyright © 2008 Deutsche Bank AG.

The next component, the LNG plant, is the critical technology and capital-intensive step in the entire LNG value chain. The LNG plant consists of one or more *LNG trains*, each train being a stand-alone gas liquefaction unit. These units cool the gas to –163°C (–260°F), converting the gas into a liquid that can be stored and transported. (The volume of the resulting liquid is 625 times smaller than the original gaseous form.) The LNG is loaded onto specially designed LNG ships—the LNG must be kept in liquid form throughout its transportation cycle—for transportation and delivery to a regasification terminal. There the LNG is reheated and returned to its natural gas state for distribution via a pipeline distribution network to local industrial, commercial, and residential users. We explore the LNG value chain in detail in chapter 9.

Unconventional gas

> *Right at the time oil prices are skyrocketing, we're struggling with the economy, we're concerned with global warming, and national security threats remain intense, we wake up and we've got this abundance of natural gas around us.*[3]
>
> —Aubrey K. McClendon,
> Chairman and CEO Chesapeake Energy

As energy demand continues to grow and traditional or "conventional" sources of oil and gas become increasingly scarce or costly, unconventional sources of fuel grow in significance. *Unconventional gas*, although a rather imprecise term, is used to describe a number of different types or deposits of gas that have traditionally been considered too costly or too difficult technologically to produce.[4]

Deep gas. This gas is typically 15,000 feet or deeper underground and much deeper than conventional gas deposits. As exploration and drilling technology advances, finding and developing these gas deposits is increasingly economic. Deep gas is, however, still more expensive to produce than conventional natural gas, and market conditions will determine commercial viability to justify large-scale development of deep gas deposits.

Tight gas. Tight gas is gas that is stuck in a very tight formation underground, trapped in unusually impermeable hard rock, or in a sandstone or limestone formation that is impermeable and nonporous (tight sand). In a conventional natural gas deposit, once drilled, the gas is usually easily extracted. A great deal more effort has to be put into extracting gas from a tight formation. Several significantly more costly techniques exist that allow natural gas to be extracted, including fracturing and acidizing. Like all unconventional natural gas, market conditions and prices will ultimately determine the commercial competitiveness of tight gas. It is estimated that more than 20% of all US recoverable natural gas is tight gas.

The Barnett Shale field in northeastern Texas in the United States may be one of the largest reserves of tight gas in the world. This field already has proven reserves of 2.5 trillion cubic feet of natural gas. Although difficult to produce given the hardness of the shale, recent technological changes and commercialization of those technologies will allow hydraulic fracturing ("fracking") and horizontal drilling to produce the Barnett Shale field.

Devonian shale gas. Natural gas can also exist in Devonian shale deposits. Devonian shales are formed from the mud of shallow seas that existed about 350 million years ago during the Devonian period of the Paleozoic era. Shale is a very fine-grained sedimentary rock, which is easily breakable into thin, parallel layers. It is a very soft rock but does not disintegrate when it becomes wet. These shales can contain natural gas, usually when two thick, black shale deposits "sandwich" a thinner area of shale. Because of some of the properties of these shales, the extraction of natural gas from shale formations is more difficult than extraction of conventional natural gas. Although estimates of the amount of natural gas contained in these shales are high, it is expected that only about 10% of the gas is recoverable.

The Marcellus Formation in the Appalachian Basin in the eastern United States (primarily Pennsylvania and Ohio) is one of the largest Devonian shale fields already under production. The Devonian or black shale layer varies anywhere from 30 to 300 feet in depth and is being produced primarily with high-pressure hydraulic fracturing. Companies like Talisman (Canada) have redirected much of their corporate strategy towards the development of the Marcellus field. The development of the Marcellus Formation has also reignited environmental concerns over the potential contamination of groundwater and aquifers as possible side effects of the hydraulic fracturing production techniques. The problem has been magnified as the formation's development is in a number of places close to population centers.

Coalbed methane. Many coal seams also contain natural gas, either within the seam itself or the surrounding rock. Coalbed methane is trapped underground and is generally not released into the atmosphere until coal mining activities unleash it. Historically, coalbed methane has been considered a nuisance in the coal mining industry. Once a mine is built and coal is extracted, the methane contained in the seam usually leaks out into the coal mine itself. This poses a safety threat, as too high a concentration of methane in the mine creates dangerous conditions for coal miners. In the past, the methane that accumulated in a coal mine was intentionally vented into the atmosphere. Today, however, coalbed methane has become a popular unconventional form of natural gas. The methane can be extracted and injected into natural gas pipelines for resale, used as an industrial feedstock, or used for heating and electricity generation.

Geopressurzied zones. Geopressurized zones are natural underground formations that are under unusually high pressure for their depth. These areas are formed by layers of clay that are deposited and compacted very quickly on top of more porous, absorbent material such as sand or silt. Water and natural gas that is present in this clay is squeezed out by the rapid compression of the clay and enters the more porous sand or silt deposits. Due to the compression of the clay, the natural gas is deposited in sand or silt under very high pressure (hence the term *geopressure*). In addition, geopressurized zones are typically located at great depths, usually 10,000 to 25,000 feet. The combination of these factors makes the extraction of natural gas in geopressurized zones quite complicated. However, of all of the unconventional sources of natural gas, geopressurized zones are estimated to hold the greatest amount of gas.

Methane hydrates. Methane hydrates are the most recent form of unconventional natural gas to be discovered and researched. These interesting formations are made up of a lattice of frozen water, which forms a sort of "cage" around molecules of methane. These hydrates look like melting snow and were first discovered in permafrost regions of the Arctic. Research into methane hydrates has revealed that they may be much more plentiful than first expected. Some experts, including the United States Geologic Services, estimate that methane hydrates may contain more organic carbon than the world's coal, oil, and conventional natural gas combined. It is not yet known what kind of effects the extraction of methane hydrates may have on the natural carbon cycle.

Unconventional natural gas constitutes a large proportion of the natural gas that is left to be extracted globally, but it is expected to represent a growing source of fuel in the coming decade. In the United States alone, as illustrated in figure 8–3, unconventional gas will be a significantly larger player in the near future. As technology advances and new methods of extraction and use are developed, the resource potential of unconventional natural gas is believed so significant that projections of both reserves and production are considered very unreliable.

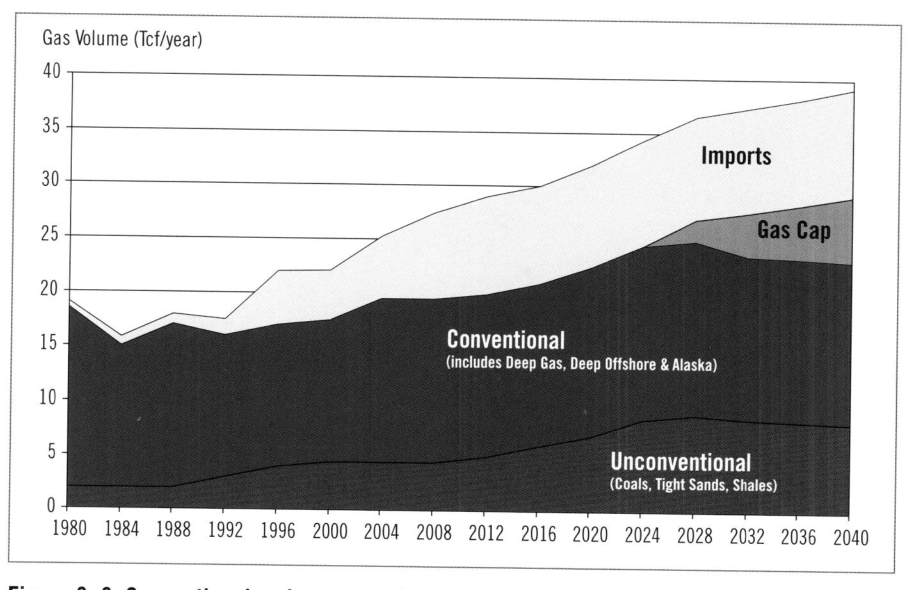

Figure 8–3. Conventional and unconventional US gas reserves
Source: Historical data, EIA and Annual Energy Outlook 2003; Projections, NETL Extrapolation.

Natural Gas Reserves, Fields, and Production

Unlike the global oil industry, the global natural gas industry is in its infancy. Gas, although "accidentally" produced for over a century as a result of its association with oil deposits, has only been produced for its own sake for a little more than four decades. And, during that short time, its development has largely been dictated by proximity to major customers. Although LNG is rapidly changing the industry, much of the world's gas reserves and production to date have been driven by the lack of gas transportability.

Natural gas reserves

The world's major natural gas can be measured and characterized a variety of ways, with *proven reserves* and *current production* being the most frequently used. Global gas reserves more than doubled between 1980 and 2007, increasing from 82 trillion cubic meters (tcm) in 1980 to 177 tcm in 2007. Table 8–2 provides a breakdown of global proven gas reserves by major geographic region at end-of-year 2007. It is interesting to note that, as opposed to global oil reserves, gas reserves seem to be more highly concentrated in a few regions. Although there are major oil resources and major new oil producing fields in Africa and South

America, there are no proven gas reserves of equal significance in these regions. Clearly, at least for now, the Western Hemisphere is not thought to be the home for much of the conventional natural gas resources of the future, while the Eurasian and Middle Eastern regions make up roughly 75% of world gas reserves.

Table 8–2. Proven gas reserves by region

Region	Tcm of Gas	Percent
North America	7.98	4.50%
South and Central America	7.73	4.40%
Europe/Eurasia	59.41	33.50%
Middle East	73.21	41.30%
Africa	15.58	8.20%
Asia Pacific	14.46	8.20%
Global total	178.37	100%

Source: *BP Global Energy Outlook, 2008.*

Gas is even more specific in its country of residence. As illustrated by figure 8–4, three countries dominate the world's proven reserves: Russia, Iran, and Qatar. As of 2007, Russia was home to 44.7 tcm of gas, Iran held 27.8 tcm, and Qatar, the small Persian Gulf Arab nation, 25.6 tcm. These three countries constitute 55% of the world's natural gas reserves. Longtime producers and consumers of gas like the US and Venezuela make up a very small percentage of reserves. It is also clear from figure 8–4 that although the reserves of most countries have grown marginally over the 1970 to 2007 period, the largest addition to proven reserves has been that of Qatar. Although both Iran and the Russian Federation continue to add to their proven reserve bases, there have been no individual remarkable gas discoveries in recent years to change the basic natural gas footprint globally.

Like oil, natural gas is defined by its fields or reservoirs, which are quite distinct from simply the country of the reserve's governance or sovereign ownership. Natural gas fields are, not surprisingly, largely found in or around the world's major oil fields—and the world's largest fields are largely found in Russia and the Middle East. With gas, size does indeed matter. Gas development is still heavily dependent on very large initial capital investments that can only be justified when the gas field itself is of such a magnitude that many years of production are assured for the repayment of the investment.

The major gas producing regions of the world are quite diverse, and yet, consolidated. Although figure 8–4 lists a multitude of countries possessing gas reserves, it also makes it very clear that three countries—Russia, Iran, and Qatar, hold the lion's share of reserves.

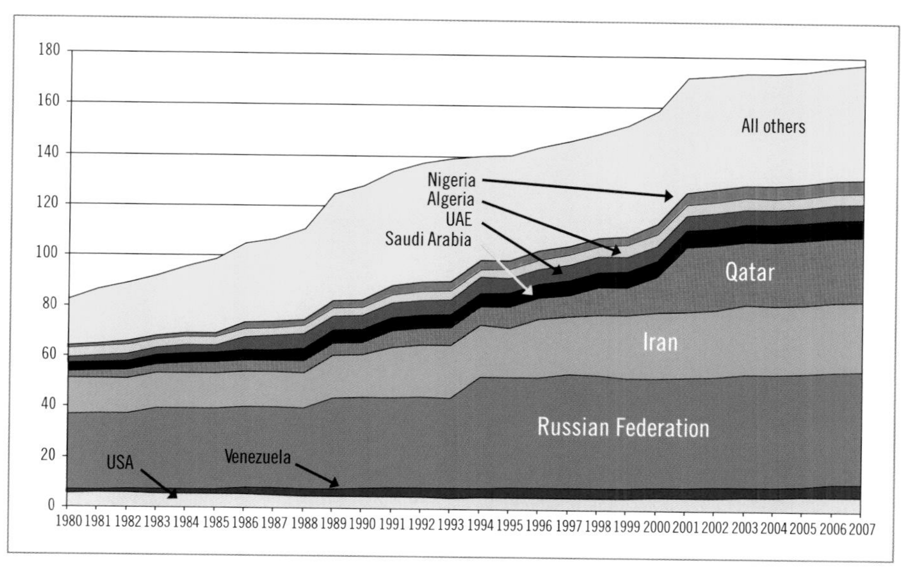

Figure 8–4. Global proven gas preserves (trillion cubic meters)
Source: *BP Global Energy Outlook*, 2008.

Gas production in most countries and regions of the world follows traditional field-pipeline-distribution systems. Notable developments are the gas fields in Mexico and the Gulf Coast region of the United States, the gas producing areas of western Russia, the North Sea production by Norway and the United Kingdom, and Libya and Algeria's production piped into the European market. The production in Indonesia, Australia, and much of the Persian Gulf (Qatar and the UAE) is following the new path—LNG production and distribution. Although many of the LNG facilities are not yet online, liquefaction facilities are under development in many countries. As those facilities and LNG shipping and gasification receiving terminals come online, the separation of the gas from the customer will likely grow greater and greater.

The world's largest proven gas field is North Dome, which lies offshore in the waters between Qatar and Iran, as shown in table 8–3. At more than 1.2 tcf of reserves, the field dwarfs all other known fields. (There is some debate as to the size of North Dome versus the size of the South Pars field, number 12 on the list in table 8–3. The two fields are either overlapping or immediately adjacent.) It is estimated that if produced at an optimum rate, the field has a minimum life of 200 years. All of the other major top gas fields by size are Russian, with the exception of the Umm Shaif/Abu el-Bukush field of Abu Dhabi.

Table 8–3. World's largest gas fields (tcm)

Rank	Field	Reserves	Location
1	North Dome	1,200	Qatar/Iran
2	Urengoy	275	Russia
3	Yamburg	200	Russia
4	Orenburg	200	Russia
5	Shtokman	200	Russia
6	Umm Shaif/Abu el-Bukush	175	Abu Dhabi
7	Zapolyarnoye	150	Russia
8	Kharasevey	150	Russia
9	Bovanenko	125	Russia
10	Medvezh'ye	100	Russia
11	Hassi R'Mel	100	Algeria
12	South Pars	100	Iran
13	Panhandle-Hugoton	80	USA

tcm = trillion cubic meters of gas

Although most gas production and development in the past has focused on fields either close to the consumers and consumer markets, or of a scale so significant that they justify LNG investment, the story is increasingly complex as more and more mid-size fields are now potentially being developed for the growing global LNG market.

Gas development

Although the basic methods of producing oil and gas have stayed the same for many years, technology continues to advance. For example, a single well in 2005 could produce twice the volume of gas compared with a similar well in 1985. Every aspect of gas production—exploration techniques, geologic data and understanding, reservoir engineering and diagnostics, drilling pad size and exploration footprints, drilling waste reduction and recycling, collection, and transportation—all have improved dramatically in recent years.

One example of technical change is the number of new fracturing techniques used since the 1970s to increase the amount of gas extracted from underground formations. CO_2-sand fracturing involves using a mixture of sand and liquid CO_2 to fracture formations, allowing any oil and gas in the formation to flow more freely for production purposes. Because the process uses only sand and CO_2, there are no other residual or waste products, making the technique both production and environmentally sound. This type of fracturing effectively opens the formation and allows for increased recovery while not damaging the deposit, generating no ground wastes, and protecting groundwater.

A number of new drilling techniques, such as coiled pipe drilling and slimhole drilling, allow companies to drill into new and existing reservoirs with smaller footprints, smaller drilling crews, faster, and with less environmental impact. *Coiled pipe drilling* uses a longer and more flexible drill pipe, requiring much less drill mud and other material resources in addition to smaller drilling rigs and rig footprints. *Slimhole drilling* uses a smaller diameter drill and drill pipe (typically less than 6 inches in diameter as opposed to the usual 12¼-inch diameter) for more than 90% of the drill. It is particularly useful in increasing the recovery of existing reservoirs and reaping the benefits of other secondary and tertiary recovery techniques like fracturing.

The technology utilized to extract and transport offshore natural gas is different from land-based fields in that a few very large rigs are usually used, due to the cost and logistical difficulties in working over water. Rising gas prices have encouraged drillers to revisit fields that previously were not considered economically viable. For example, McMoran Exploration has passed a drilling depth of over 32,000 feet (the deepest test well in the history of gas production) at the Blackbeard site in the Gulf of Mexico. ExxonMobil's drill rig had reached 30,000 feet by 2006 without finding gas, after which the site was abandoned.

A reserve of gas that is discovered, but then determined not economically producible, is said to be *stranded*. In some cases the gas field may be too remote from gas markets, usually large industrial or commercial centers, which makes the construction of pipelines prohibitively expensive. A field may be found in or near an established market, but one where the demand for natural gas is already saturated. Although exporting gas is possible, the costs of transportation and export often prove prohibitive. The gas field may then be closed up—*shut in*—preserving the possibility of its extraction and production at some future date. There is also the possibility that a gas deposit or reservoir is physically trapped, because its depth or access is prohibitive. Technological developments in all aspects of gas production and liquefaction constantly change many of these barriers and dynamics. Many industry analysts estimate that more than half of the discovered natural gas on earth is stranded.

Because the market for oil is much older, larger, and deeper (pardon the use of the term) than that of natural gas, oil discoveries are often developed *faster* than gas—faster in initial production and faster in rate of production through time. As illustrated in figure 8–5, the result is that after discovery, gas fields start production much later than oil fields. The primary metric illustrated in figure 8–5 is the production rate, which is annual production as a percentage of the hydrocarbon reserve. Whereas a typical oil field development may rise rapidly to a peak of 13% or 14% of total reserves with a few years, a gas field will normally be produced at only 2% or 3% of total reserves at a constant rate over an obviously much longer lifespan.

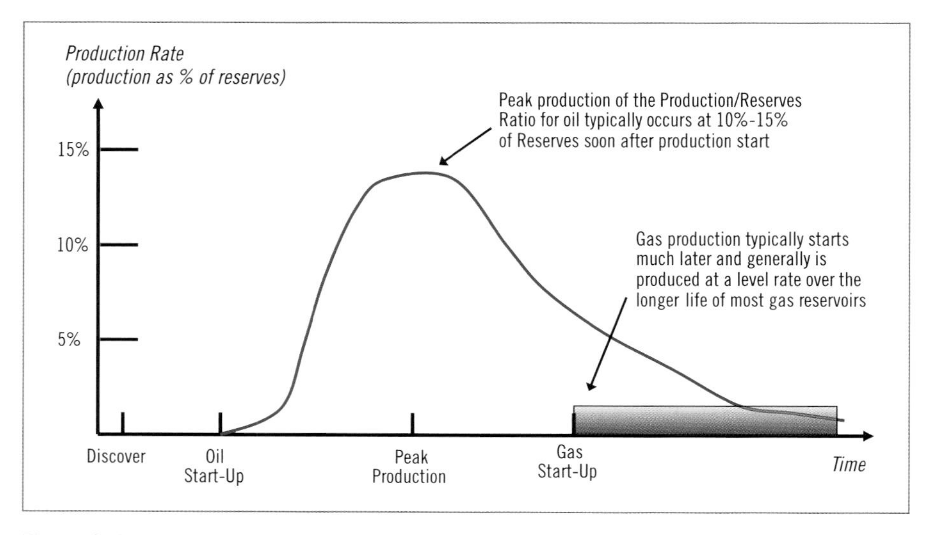

Figure 8–5. Typical oil and gas production profiles

Source: *International Exploration Economics, Risk, and Contract Analysis*, Tulsa: PennWell, p. 117.

One of the most distinct differences between oil production and gas production is recovery. Oil reservoir recovery has historically been quite low, with only 25% to 30% of the oil in a reservoir actually recovered using primary production techniques. Secondary or enhanced recovery techniques, such as the use of water-, gas-, or steam-flooding, may increase oil recovery an additional 5% to 15% but at an appreciable cost (see the discussion of enhanced recovery at Chevron's Kern River Field in chapter 5). A gas reservoir may see recovery rates of 70% to 80% without the use of complex and expensive secondary and tertiary recovery additives, and possibly 90% with their use.

Unfortunately, for many years, gas recovered in the process of producing oil could not be profitably sold and was simply burned or *flared*. It is now illegal in many countries to flare gas, and there are a number of ways in which stranded gas can now be liquefied for transportation and sale or reinjected into wells or reservoirs to either increase total recoverable oil or preserve the gas for future production. That said, flaring is still extremely widespread. Globally, it is estimated that nearly 10 bcf of gas is flared every day, adding more than 400 million tons of carbon dioxide into the atmosphere.[5] Flaring of gas is so large in a number of producing regions around the world that it is visible at night via satellite photographs. Flaring sites include the Nigerian and Angolan coastlines, central Libya, Baku, Azerbaijan, and large parts of the Ukraine, Yucatan peninsula, and the Peruvian coast. A number of organizations such as the Global Gas Flaring Reduction (GGFR), a self-described public/private partnership, are working to reduce and eliminate gas flaring.

Natural Gas Use and Markets

Natural gas demand is dominated by three sectors: residential and commercial consumers, industrial consumers, and power generation.[6]

- Residential gas use is for heating, cooking, and cooling, although heating is by far the largest segment. As a result, weather conditions and cycles have significant impacts on the demand for gas in the residential sector. Residential users have limited access to substitutes.

- Gas demand by the industrial sector is much more stable over time, as use is primarily heating, melting, feedstock, and power generation. The industrial sector may have more substitutes, optimizes its use constantly, and may change production or production location in response to gas prices.

- Power generation uses natural gas, oil, hydro, nuclear, and coal for its energy sources. The development of the combined cycle gas turbine (CGGT) has been a technological breakthrough for natural gas, with gas power efficiency now more than 60%. Although gas demand for power generation has surged, its price relative to electricity—the *spark spread*—is critical for continued growth and development.

Within these categories lies a multitude of active uses and demands for natural gas with many different evolutionary tracks of development.[7] The Industry Insight on gas lighting details one of the first uses for natural gas, a use that has largely disappeared.

Industry Insight: Gas Lighting

The first uses of natural gas were in lighting. The *thermolampe*, a lamp that used distilled gas from wood, was first patented in 1799. In the 1790s, Mathew Boulton and James Watt of the Soho Foundry in Birmingham, England began experimenting with gas lighting. In 1798, they lighted the foundry with coal gas and in 1802, lit the outside grounds surrounding the building, creating a public stir.

Officially, the first public street lighting was at Pall Mall, London, in 1807. At the same time, Phillipe Lebon of Paris used gas lights to illuminate his house and gardens. By 1820, the city of Paris adopted gas street lighting, followed soon after by Baltimore, Maryland. By the turn of 20th century, most cities in the United States and Europe had gas lighting. Slowly over the following two decades, gas lighting was replaced with electric lighting with the mass introduction of small incandescent electric lamps.

Power generation. Natural gas is a major source of electricity generation through the use of gas turbines and steam turbines. Particularly high efficiencies can be achieved through combining gas turbines with a steam turbine in combined cycle mode. Natural gas burns cleaner than other fossil fuels, such as oil and coal, and produces less carbon dioxide per unit energy released. For an equivalent amount of heat, burning natural gas produces about 30% less carbon dioxide than burning petroleum and about 45% less than burning coal. Combined cycle power generation using natural gas is the cleanest source of power available using fossil fuels, and this technology is widely used wherever gas can be obtained at a reasonable cost. Fuel cell technology may eventually provide cleaner options for converting natural gas into electricity, but as yet it is not price competitive.

Residential domestic use. Natural gas is supplied to homes, where it is used for such purposes as cooking in natural gas–powered ranges and/or ovens, natural gas–heated clothes dryers, heating/cooling, and central heating. Home or other building heating may include boilers, furnaces, and water heaters. Compressed natural gas (CNG) is used in rural homes without connections to piped-in public utility services, or with portable grills. Because CNG is more expensive than LPG, LPG (propane) is the dominant source of rural gas.

Natural gas vehicles. Compressed natural gas (methane, CNG) is a cleaner alternative to other automobile fuels such as gasoline and diesel. As of 2005, the countries with the largest number of natural gas vehicles were Argentina, Brazil, Pakistan, Italy, Iran, and the United States. The energy efficiency is generally equal to that of gasoline engines, but lower compared with modern diesel engines. Gasoline/petrol vehicles converted to run on natural gas suffer because of the low compression ratio of their engines, resulting in a cropping of delivered power while running on natural gas (10%–15%). CNG-specific engines, however, use a higher compression ratio due to this fuel's higher octane number of 120–130.

Hydrogen. Gas can be used to produce hydrogen, with one common method being the hydrogen reformer. Hydrogen has various applications: it is a primary feedstock for the chemical industry, a hydrogenating agent, an important commodity for oil refineries, and a fuel for hydrogen vehicles.

There are many additional and emerging uses for gas. Natural gas is a major feedstock for the production of ammonia for use in fertilizer production, and has been used in recent years in the manufacture of fabrics, glass, steel, plastics, paint, and other products.

Gas-consuming countries and markets

The major natural gas–consuming countries are in the three major global gas markets: North America, Western Europe, and Japan/East Asia. Figure 8–6 shows the top gas-consuming nations. Combined with the earlier exhibits, the differences can be seen between nations consuming gas, nations with gas reserves, and nations producing gas. Clearly, the United States and a handful of other major industrialized countries are the significant consumers of natural gas today. The US makes up more than 22% of global gas consumption.

The gas industry has been growing rapidly for many years. As detailed in figure 8–7, global gas consumption has increased from 63.4 bcf per year in 1965 to more than 282 bcf per year in 2007 (sum of all listed categories), a fourfold increase in 40 years. The majority of the increase in consumption has not been in the two largest consuming markets, the United States and the Russian Federation. Rather, smaller countries (only Canada and Iran are shown here) have caused a steady and rapidly increasing consumption of gas.

Natural gas is increasingly a globally traded commodity. Historically, it has been regional in nature, as gas had to be produced and then transported by pipeline or truck to a reachable market. But the growth of LNG has started to decouple the production from the consumer in the regional marketplace. The price of gas in its various forms has shown an increasing correlation across markets, reflecting the growing integration and substitutability across regional boundaries. In short, natural gas globalization is occurring.

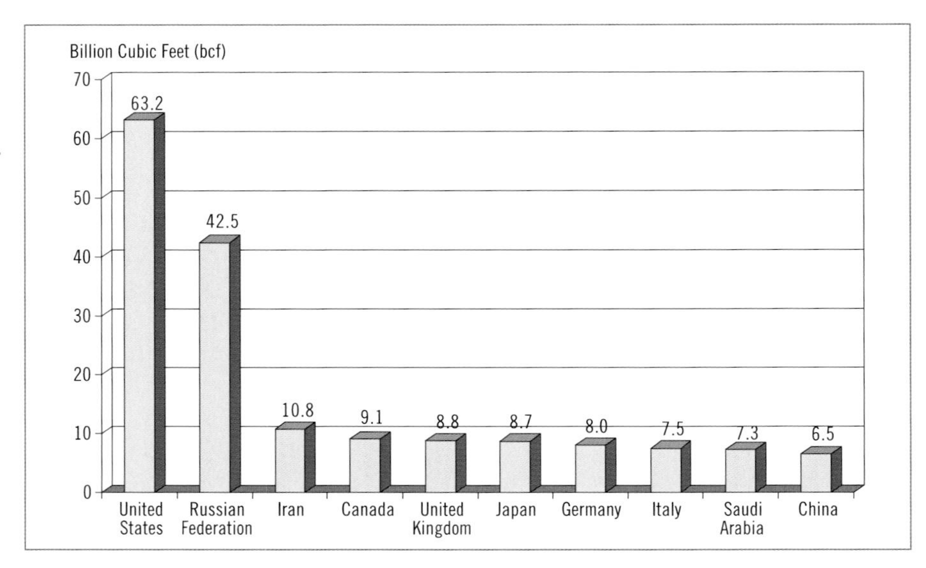

Figure 8–6. Top gas-consuming countries, 2007

Source: *BP Global Energy Outlook, 2008.*

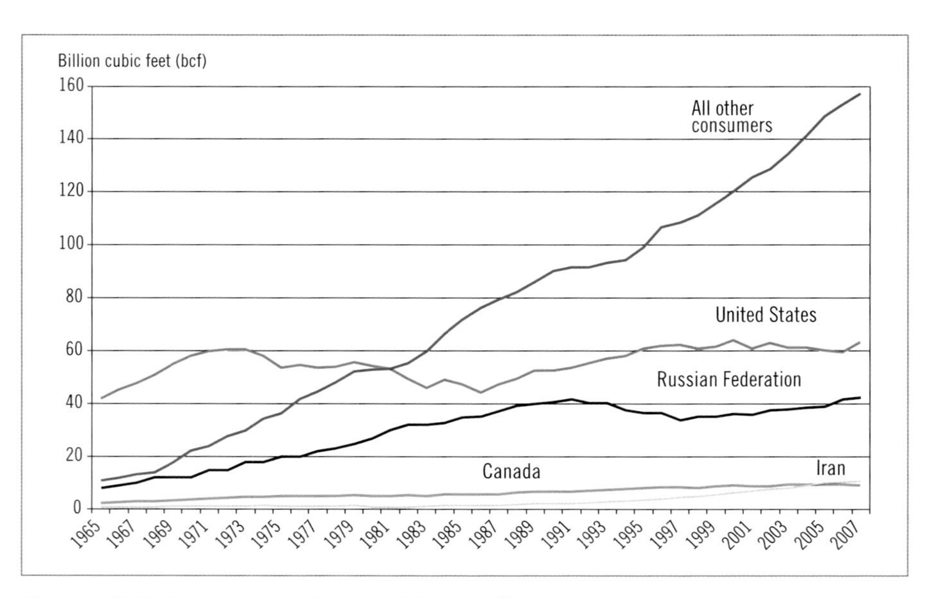

Figure 8–7. Major gas consuming countries over time

Source: *BP Global Energy Outlook, 2008.*

The regional markets have always been characterized by specific market structures. One way to portray this development is to assume that the global market for natural gas can be simplified to three markets: North America, the Pacific, and Europe.

- The North American region has traditionally grown around pipelines, reaching out in a web to commercial and industrial consumers. The rapid growth of the oil industry in the United States led to increasing efforts to use the large quantities of associated gas produced. Pricing in the US is based on Henry Hub, the subject of the following Industry Insight.

- Simultaneously, the Pacific region, led by Japan, has focused on LNG consumption and utilization. As a result of near-100% imported energy dependency in Japan, LNG was logical for market development. For example, in 1990, Japan made up 66% of all LNG imports, which then dropped to 48% by 2002 as other countries joined the aggressive development of LNG markets.

- The European market for gas has evolved rapidly. In the 1970s and 1980s the European appetite for natural gas for both industrial and consumer use grew rapidly. The majority of gas was supplied by the North Sea and Russian market. In the later 1980s, as the North Sea showed signs of maturity and depletion, European consumers sourced increasing quantities of gas from Russia via pipeline in a highly profitable relationship for both parties. In the post-2000 energy market, the growing consumption needs of Europe led to a growing importation of LNG, to replace faltering supplies from the North Sea and to diversify sourcing from dependence on the Russian market alone.

Industry Insight: Henry Hub

Henry Hub is the pricing point for natural gas futures contracts traded on the New York Mercantile Exchange (NYMEX). It is a major pipeline interchange—hub—on the natural gas pipeline system in Erath, Louisiana. Since futures contracts on physical products must assume physical delivery of the product on maturity, pricing must be based on the delivered cost to some geographic point. As a result, Henry Hub, a long-time central transshipment point for natural gas in the United States, became the benchmark delivery point for pricing quotations as traded on the NYMEX.

The electrical power sector is the most dominant area of natural gas demand growth. The demand for electricity, driven by population and economic growth, has boomed in the past two decades. Total power generation capacity grew a remarkable 35% among the OECD countries between 1990 and 2004. Of that growth, 64% was high-efficiency combined cycle gas turbines (CCGTs). As illustrated in figure 8–8, this far outstripped new generating capacity in all other major fuel sources and power generation types.

Natural gas is now considered the "fuel of choice" for most new power generation capacity. The technological improvements that resulted in much more efficient turbine development, combined with relatively large and cheap sources of natural gas near major demand markets and gas's environmentally clean-burning characteristics, made it the primary fuel driver for both base-load and peak-load power generation additions in recent years. New gas-powered generation also requires substantial less capital upfront for power generation, making its use cheaper in capital requirements for initial development and therefore more flexible and more affordable for developing markets. Although there is renewed interest in both coal and nuclear for electrical power in a number of the major markets, the lead times required for their development will assure gas's primary position for new generation capacity through the period 2012 to 2015.

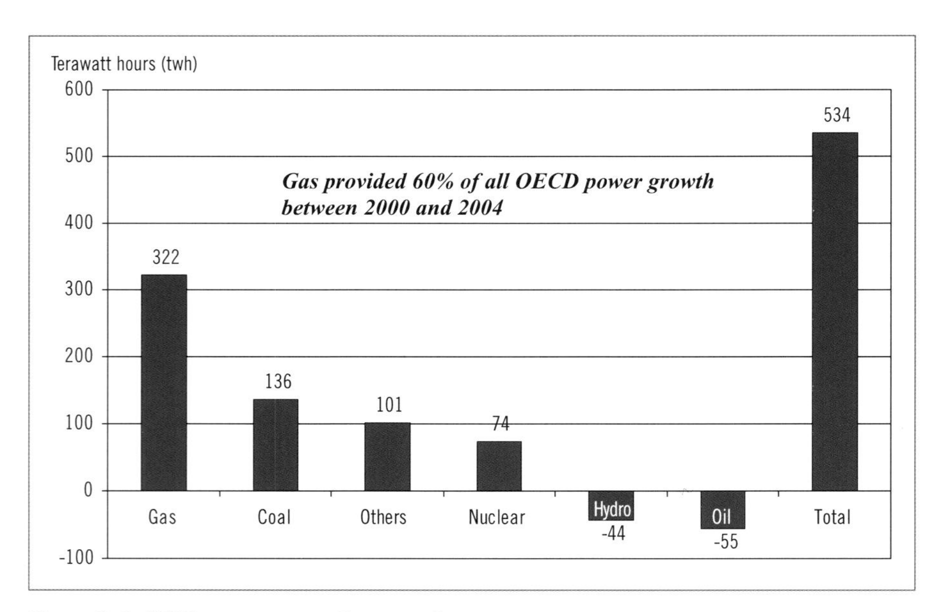

Figure 8–8. OECD power generation growth

Source: *Natural Gas Market Review*, 2007, EIA, p. 121.

The North American market in recent years has played a pivotal role in global LNG markets, as it first set the floor price for sales of all kinds in most markets as a result of its own depth and liquidity, then swung 180 degrees and dumped LNG on the world spot market. In a number of LNG pricing cases in recent years, from Japan to the United Kingdom, prices were clearly pegged to Henry Hub price (the benchmark US natural gas price). In 2009 and 2010, the US market's rapid development of shale gas resulted in a large drop in LNG imports, causing much of the new LNG product originally intended for the North American market to float towards European opportunities. This dampened LNG prices dramatically.

Although the market potential and opportunity for natural gas, particularly LNG, is strong, investment commitment in LNG trains for liquefaction and regasification has fallen with gas prices. Given the lead time needed for regulatory approval and construction to start up, any renewed interest in LNG capacity and delivery growth will suffer from a significant delivery lag. Starting in 2006, the industry saw a number of LNG project delays averaging nearly a year, as well as cost overruns of more than $2 billion per project. A second level of concern is the rising cost of the basic investments. The rapidly rising cost of raw materials and construction talent is increasingly pushing up the cost of new projects. In short, the industry is getting much less bang for the buck.

Prices, Trading, and Markets

With once-regional natural gas markets now turning global, consumers are increasingly exposed to gas exporters who link their prices to oil, overriding the play of supply and demand in the local market and frustrating regulators who fret that they can only counter the link indirectly.

"Gas producing countries have benefitted from this, because prices have been going up for reasons that had nothing to do with the gas," a European Commission spokesman said, referring to the surge in oil prices to record levels this year before they receded amid the current global financial woes.

—"Natural Gas Price Links to Oil Markets Frustrate Regulators' Efforts to Develop Competition," *International Herald Tribune*, October 29, 2008

The price of natural gas varies significantly depending on location, type of consumer, LNG versus non-LNG origin, competing fuels, and available substitutes. Pricing complexity for natural gas is primarily a result of the fact that it is not yet a widely transportable global commodity.

Gas pricing fundamentals

Whether it's called "supply and demand" or "buying and selling," a market for any product requires access and activity by a large number of participants. Market openness, combined with a critical mass of participants and degree of activity, give rise to a market price. Without openness or activity, a market price is not achievable. The existence of some form of a single market price has been the primary characteristic differentiating natural gas prices in regional markets around the world. As a result, many gas prices and contracts have long been linked to the price of oil—gas-to-oil—rather than gas-to-gas. Figure 8–9 provides a brief overview to the regional pricing differences that dominate natural gas markets.

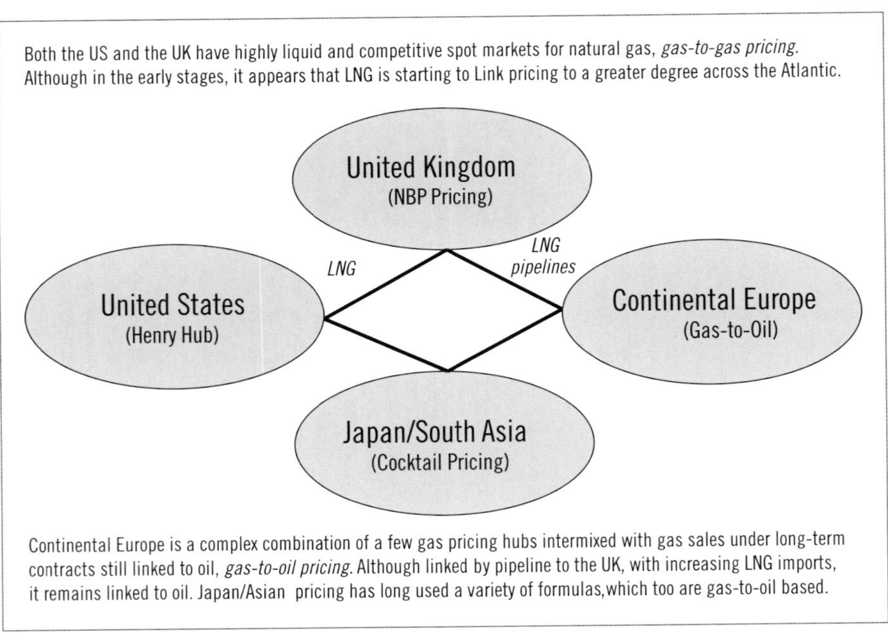

Figure 8–9. Gas-to-gas or oil-to-gas pricing?

Markets are historically regional in character and subject to localized supply and demand factors. For example, in the United States, there is a complex system of intrastate and interstate gas pipelines, allowing a relatively high degree of competitive pricing to occur. As a result, wellhead prices reflect base gas prices on a supply and demand basis. The US markets also find added depth and breadth through natural gas futures contracts traded on the New York Mercantile Exchange.[8]

The United Kingdom has its own version of benchmark pricing in the form of the national balancing price (NBP). The NBP, unlike Henry Hub, is a virtual or digital market for the sale and purchase of UK natural gas. NBP prices are quoted in pence per therm. The NBP is also the official pricing and "delivery point" for the natural gas futures contracts traded on the Intercontinental Exchange (ICE). The NBP is the most liquid gas trading point in Europe, but affects prices only in the UK itself, as Continental European gas prices have a distinctly separate pricing mechanism, although actual sales volumes are indeed connected to both the UK market and the growing global market for LNG.

In addition to NBP in the UK, two hubs are gaining more and more market attention in Continental Europe: the Dutch Title Transfer Facility (TTF) and the Zeebrugge Hub. The Dutch TTF is a virtual trading point for natural gas in the Netherlands. Organized by Gasunie in 2003, it is very similar to the NBP in the UK, providing a real-time platform for gas trading in the Dutch network and gas futures contracts traded through Endex. Gas at TTF usually trades in euros per megawatt hour. The Zeebrugge Hub is the natural gas trading point in Belgium. Zeebrugge is linked to the NBR via the Interconnector and has gained increasing activity as a result of increasing purchases of gas by the UK through the link.

Continental Europe: The gas-to-oil linkage in natural gas pricing

The pricing and contracting practice used in natural gas in Continental Europe today developed in the 1960s in a number of long-term Dutch contracts. The market gained momentum immediately following the Middle East–based price increases for oil in the years 1973 to 1974. The *market value principle*, also termed the *netback market principle*, links the price of natural gas to the delivered price of the next-cheapest alternative. This has been reduced to essentially the price of crude oil for more than three decades. The formulation as defined by the IEA is presented in the following example, called the netback market principle example:

The netback market value of gas to a specific customer at the beach or border is defined as follows:

Netback = Delivered price of cheapest alternative fuel to the customer (including any taxes) adjusted for any difference in efficiency or in the cost of meeting environmental standards/limits;

minus Cost of transporting gas from the beach or border to the customer;

minus Cost of storing gas to meeting the customer's seasonal or daily demand fluctuations;

minus Any gas taxes.

The weighted average netback value of all customer categories is used as the basis for the negotiation of bulk prices at the beach or border.

Source: International Energy Agency, 1998, p. 32.

The Continental European market for natural gas experienced a number of major liberalization steps in the late 1990s, prompting many market analysts to predict that long-term natural gas contracts would move away from their historical linkage to oil products and begin what is termed as "gas-to-gas competition."

The pricing of natural gas contracts in North America and the United Kingdom is fundamentally different from that practiced in Japan/Asia and Western Europe.[9]

- North America has a very well-developed and liquid spot market for both crude oil and natural gas. In the case of gas, Henry Hub has long served as a single point of spot market trading and pricing. Its position is preserved as an indicator of market prices for natural gas by wide-scale third-party access to the hub. As a result, gas prices and crude oil supply prices are inherently separated markets.

- Western Europe has no single equivalent to Henry Hub. Without a single spot market trading post, the price of gas used in long-term gas supply contracts has looked to the price of oil as its indicator or link to market pricing. The prices of gas as indexed to crude oil are set independently of actual supply-and-demand conditions for natural gas itself. This is effectively also the case for Japan and South Asia.

A number of significant events in both energy markets and the general global economy in the years 2007 to 2009 raised serious questions over the sustainability of current gas pricing practices in Europe:

- The global financial crisis of 2008 evolved into a global recession in 2009. The demand for natural gas by the residential and industrial sectors fell dramatically in all three major regional markets of North America, Western Europe, and Japan/Asia.

- Discovery and development of unconventional natural gas reserves in the United States, the "shales," has reduced the US demand for imported gas, specifically LNG.

- A number of major new LNG projects came online in 2008 and 2009 with more to come in 2010 and future years. These projects involve liquefaction capability, regasification, and receiving capability in the three regional markets.

The result of these three factors is that LNG afloat on the international market is looking for buyers. The European market, with most long-term gas contract prices tied to the price of crude oil, is now seeing these floating supplies of LNG available at much lower prices than what their take-or-pay contracts require of them. At the same time, with the severe drop in gas and power demand, many of the long-term take-or-pay contracts are in effect "paying," contributing to a growing unhappiness with buyers. A decoupling of gas prices from crude prices may be happening.[10]

Changing European question

For many years the accepted principle of natural gas pricing in Europe was based on one fundamental assumption: end-users had a real choice between burning gas and oil products and would switch to the latter if given a price incentive to do so.

Many believe that this assumption is no longer true. First, oil products have already been virtually eliminated from many stationery energy sectors in these markets. In fact, oil products are no longer present. Secondly, the cost and inconvenience of maintaining oil-burning equipment and substantial stocks of oil products have become prohibitive. This means that *switching*—converting from gas- to oil-burning as a result of price or other market conditions, is not really feasible now (if it ever truly was). Third, the emergence of modern gas-burning equipment that is much more efficient in its energy production has made moving to oil-burning even less attractive. Fourth, the heightened concern, awareness, and regulation of environmental impacts in many markets of oil-burning emissions mean oil is not a realistic substitute for gas.

The January 2006 Russia-Ukraine crisis, combined with the bitter-cold February 2006 weather and the fire at the UK's Rough gas storage facility, resulted in rising NBP and Continental European spot gas prices. Anyone in Europe who could have switched to oil would have done so. With a few exceptions, nobody switched. Gas and oil prices are in effect no longer really substitutes.

Gas OPEC, Gas troika (Russia, Iran, Qatar)

The growing discussion over breaking the oil-price link has not gone unnoticed by the major gas producers who feed the European market. In 2008 and 2009 there was a growing movement towards a potential "OPEC for gas," as the three major producers, Russia, Iran, and Qatar, attempted to hold the line on gas pricing.

Even if the pricing convention changed and a new effective hub price developed in Europe, two things should be kept in mind. First, this would only mean a new price-setting mechanism for long-term gas contracts in Europe, not the end of long-term gas contracts. Second, the contractual abolishment of a formal linkage between oil prices and gas prices would not necessarily eliminate any true or fundamental linkage that may still exist, or be reestablished.

United States

In the United States, retail sales are often in units of therms (th); 1 therm = 100,000 Btu. Gas meters measure the volume of gas used, and this is converted to therms by multiplying the volume by the energy content of the gas used during that period. Wholesale transactions are generally done in decatherms (Dth), or in thousand decatherms (MDth), or in million decatherms (MMDth). A million decatherms is roughly 1 billion cubic feet of natural gas. The caloric value of natural gas is roughly 1,000 Btu per cubic foot.

As illustrated by figure 8–10, it has become more and more difficult to identify a "typical" price of natural gas in the US market. Volatility is increasingly a state of mind. In June 2008, US wellhead prices peaked (as measured on a monthly basis) at $10.82 per 1,000 cubic feet (28 m^3) ($10.82/mmbtu), before starting a precipitous fall. By September 2009, in only 15 months, prices had plummeted to $2.92/mmbtu, a level not seen since 2003. Although prices have recovered marginally since then, the volatility has had enormous repercussions on the prospects for LNG imports into the United States, as well as the competitive outlook for unconventional gas development like gas to liquids, topics we address in chapter 9.

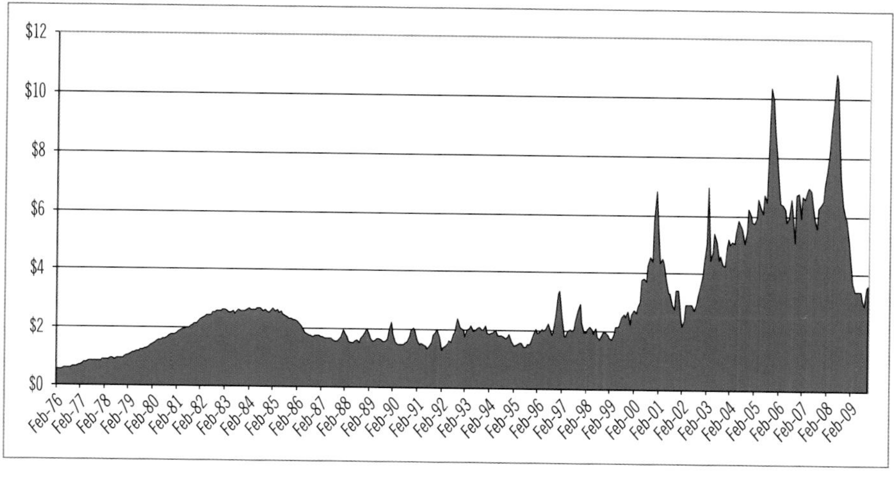

Figure 8–10. US wellhead prices, 1976–2009 (US$/mcf)

Bottom-line competitiveness

Natural gas is still largely positioned as the next-best alternative to oil in many electrical power uses and markets. As a result, there is a competitive positioning for it that is rather unlike any other major hydrocarbon.

Summary Points

- Natural gas was long considered an afterthought of crude oil production. Associated gas, or gas inadvertently discovered in the process of exploring and developing oil, was often stranded or flared.

- Historically, only natural gas deposits of significant size and located within reach of major industrial or consumer centers for use via pipelines were actually developed.

- Increasing concern over the cost and future availability of oil has now renewed interest in the global potential of natural gas. Technological changes such as the development of the combined-cycle generation turbine have motivated many power developers to see natural gas as a more affordable and more environmentally sustainable power solution.

- The price of natural gas varies significantly depending on location, type of consumer, LNG versus non-LNG origin, competing fuels, and available substitutes. Pricing complexity for natural gas is primarily a result of the fact that it is not yet a widely transportable global commodity.

- The massive and rapid development of liquefied natural gas, LNG, is now making natural gas much more of a globally traded commodity.

- A clear result of LNG development is the ability of previously stranded or abandoned natural gas reservoirs all over the world to now be available for potential commercial development.

- *Unconventional gas*, although a rather imprecise term, is used to describe a number of different types or deposits of gas that have traditionally been considered too costly or too difficult technologically to produce. This includes so-called tight gas, deep gas, and the highly publicized Devonian gas discoveries.

Notes

1. Ann Davis, Ben Casselman, and Rebecca Smith, "Has Natural Gas's Moment Arrived?" *The Wall Street Journal*, December 16, 2009.
2. This chapter draws upon a variety of sources including *Natural Gas Market Review, 2007*, International Energy Agency, OECD, 2007; Norman J. Hyne, *Petroleum Geology, Exploration, Drilling, and Production*, second edition, Tulsa: PennWell, 2007; Daniel Johnston, *International Exploration Economics, Risk, and Contract Analysis*, Tulsa: PennWell, 2007; *The Global Liquefied Natural Gas Market: Status & Outlook*, Energy Information Administration, US Department of Energy, Washington D.C., 2003.
3. Clifford Krauss, "Drilling Boom Revives Hope for Natural Gas," *The New York Times*, August 25, 2008.
4. *Oil and Natural Gas*, energy4me.org, Society of Petroleum Engineers, DK Public, Inc., 2007.
5. "Financing Oil & Gas Sector Gas Recovery and Flare Reduction Projects," GGFR, Monterrey, Mexico, January 28, 2009.

6. *Natural Gas Market Review, 2007*, International Energy Agency, OECD. This is annually one of the most comprehensive and detailed analysis publications of the natural gas market, including LNG.
7. Charles Augustine, Bob Broxson, and Steven Peterson, "Understanding Natural Gas Markets, A Policy Analysis Study by Lexecon Undertaken for API," 2006.
8. Each contract on the NYMEX is for 10,000 gigajoules, or 10 billion BTUs. Thus, if the price of gas is $10 per million Btus on the NYMEX, the contract is worth $100,000.
9. Jonathan Stern, "Is There a Rationale for the Continuing Link to Oil Product Prices in Continental European Long-Term Contracts?" Oxford Institute for Energy Studies, NG19, April 2007.
10. Jonathan Stern, "Continental European Long-Term Gas Contracts: Is a Transition Away from Oil Product-Linked Pricing Inevitable and Imminent?" Oxford Institute for Energy Studies, NG 34, September 2009.

Chapter 9

LIQUEFIED NATURAL GAS (LNG)

Many NOCs are not only major international companies in their own right, but also have decades of experience in the development of LNG projects. This has raised questions as to whether the "old bargain" of allowing IOCs access to gas reserves in return for finance, technology, and LNG project management and development skills, will continue to be attractive to NOCs.

— David Ledesma, "The Changing Relationship between NOCs and IOCs in the LNG Chain," Oxford Institute for Energy Studies, NG32, July 2009

The growth and commercialization of *liquefied natural gas*, LNG, has rapidly created a new sector in the global oil and gas industry. Natural gas has long possessed a collection of qualities seen as highly desirable in major applications such as commercial power production and residential heating. Lack of transportability, a primary technical drawback for many years, was resolved more than 50 years ago with liquefaction and regasification development. The commercialization of that technology, however, has only recently reached the cost efficiencies and industry scale that allow true global competitiveness. This chapter provides an overview of LNG and highlights its complex value chain and continuing competitive pressures.

The Integrated LNG Project

Capturing value from an LNG project requires the integration of a chain of activities extending from the gas well to the ultimate consumer. The various links in the chain, shown in figure 9–1, vary in complexity, cost (capital and operational), and ownership.

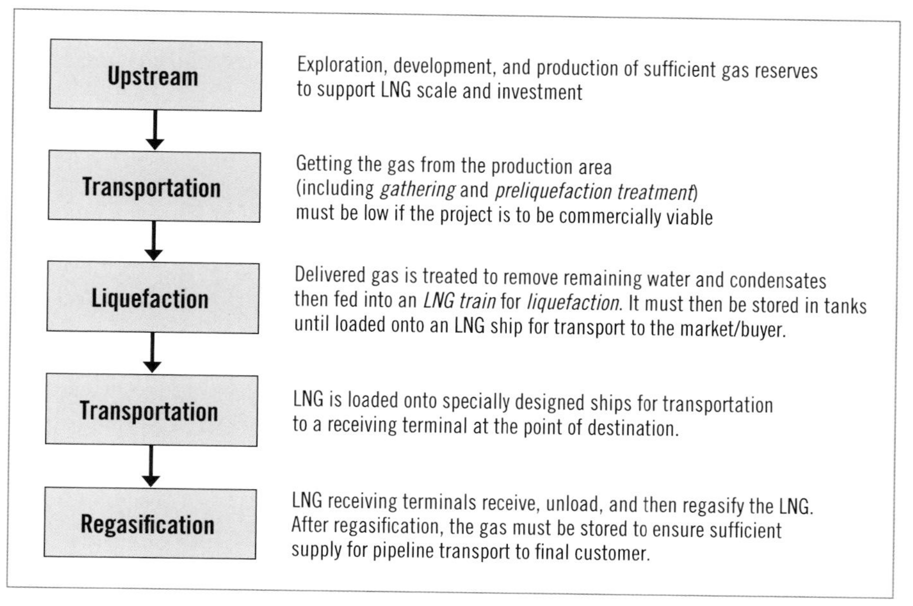

Figure 9–1. The integrated LNG project

The capital intensity and cost of LNG have resulted in integrated contractual arrangements that often commit gas, the LNG plant, the LNG vessels, and the LNG terminus facilities to a specific customer for a 20- to 30-year period.

The contracts are frequently strict in form and inflexible on price and volume and include automatic escalator costs for fuel cost changes, input, and labor cost charges over time. The contracts also use a *take or pay* framework: if the supplier meets all contractual commitments, the customer receiving LNG is required to pay for the gas regardless of whether they need it or take it. The intermediate steps are covered by a sale and purchase agreement (SPA) between the supplier and receiving terminal, and a gas sale agreement between the receiving terminal and end-users.

If an LNG project is a new development throughout the entire chain, integrating the elements successfully is extremely difficult in terms of contracting, timing, and funding. If, however, any of the various links exists prior to the individual project's execution, such as a receiving and regasification terminal ready to receive LNG from a variety of sources, project development may be simplified.[1]

Upstream

The initial stage of LNG development involves the upstream activities of exploration, development, and production. The gas reserves identified for LNG development must meet three fundamental criteria: composition, size, and sustainability.

The gas must be of a certain composition that is economic to produce. Natural gas is often found with a combination of associated components. Some of these components, like associated liquids (liquefied petroleum gases [LPGs] and condensates), are of considerable commercial value and add significantly to the revenues earned from gas development. Impurities in the gas such as hydrogen sulfide, carbon dioxide, or mercury must be removed before liquefaction and add cost to gas processing.

The size of the gas field for development (i.e., the proven gas reserve base) must be large enough to support at least 1 million tons of LNG production per annum (mtpa) for 20 years (1 Tcf or 28 Bcm). Most LNG liquefaction facilities are devoted to specific gas reserves and are dependent on the reserves being sufficient to supply production (and economic) needs over time. Interruptions in gas supplies in an LNG chain are extremely costly. The size of the reserve base must also consider the 10% to 15% of the gas lost/used in the LNG chain. Finally, the field must be sustainable over time. The field must possess a sufficiently large reserve base left in the field at the end of the project life span in order to maintain the field's production over time (plateau level). This may mean a capacity of at least 8 million tons per annum, which could support two large LNG trains—a minimum proven gas reserve of 10 Tcf (280 Bcm).

Transportation to liquefaction

Getting gas from the field to processing and liquefaction facilities is a critical activity. Natural gas has often been discovered in association with crude oil, often in insufficient quantity or quality to justify the infrastructure and cost of moving the gas to a liquefaction facility.

If quantity and quality are not a limitation, the development and transportation of gas to liquefaction passes on to the development and production agreements associated with reserve development. The type of fiscal regime in use (royalty/tax regime or production sharing agreement), as well as the business structure, dictate various concerns and potential outcomes.

There are three business structures typically used in LNG production and transportation: integrated projects, transfer pricing agreements, and throughput agreements.

- **Integrated projects.** In an *integrated project*, the ownership structure is the same for both upstream development and liquefaction. This structure allows a high degree of alignment and the business agreements may be simpler than for the other structures. For example, unless there is a government requirement, there is often no need to establish a gas transfer price between the gas producer and the liquefaction facility. If the liquefaction facility resides in a different legal domicile, a transfer price may need to be set for royalty, tax, or other fee-based charges.

- **Transfer pricing agreements.** A *transfer pricing agreement* is often used when ownership of gas reserves is separate from the LNG project sponsors. This is frequently the case in countries using production sharing agreements (PSAs) or production sharing contracts (PSCs) and the government retains ownership of the gas. The different ownership interests will require a transfer pricing agreement to set the price of the natural gas feedstock for transportation and liquefaction. Obviously, the actual transfer price will have a significant impact on the returns to the two parties. An inability to negotiate acceptable transfer prices could prevent project development.

- **Throughput agreements.** A *throughput agreement* is a contractual structure in which the owner of the upstream reserves pays a contractual toll for transportation and liquefaction, retaining ownership of the gas for postliquefaction marketing and sale. Although rarely used to date, this structure is of growing interest as upstream owners—sovereign governments—attempt to retain more ownership and control over the sale and profitability of their gas. Alternatively, in recent years a number of

IOCs have purchased or leased their own ships and are now pursuing throughput agreements in order to do their own marketing and sales of LNG in the global market.

Figure 9–2 details the commercial structure of the Angola LNG project.[2] Angola LNG is an integrated project with a complicated twist. The project was created to utilize the massive amounts of gas being flared from Angola's offshore oil platforms. Flared gas will be captured via a network of pipelines and then liquefied. Both the upstream gas and the liquefaction plant are owned by the same groups. Angola, or more specifically Sonagol, the country's NOC, owns all the oil and gas produced in Angola. As a result the gas previously flared is "purchased" for free from the operators of Angolan blocks in the Lower Congo Basin. The LNG plant itself will be located south of the Congo River near Soyo in northern Angola.

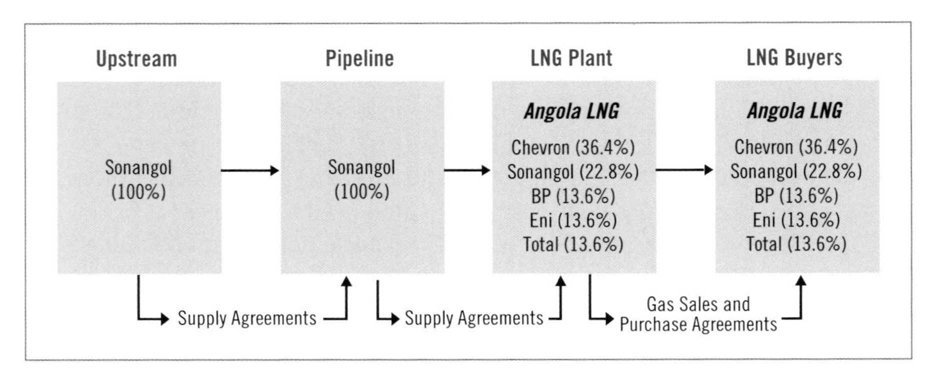

Figure 9–2. Angola LNG commercial structure. The Angola LNG project has one of the more complex commercial structures due to the project's wider scope. The project's motivation from the beginning was to recapture for liquefaction and sale gas that was being flared from offshore platform production. The project's scope has since grown to be a full-scale LNG development.

Source: Wood Mackenzie, December 2010.

The cost associated with transporting gas to liquefaction facilities is a key element in determining the economic competitiveness of LNG projects. High transportation costs to liquefaction can easily doom an LNG project's prospects.

Liquefaction

Once gas feedstock reaches a liquefaction facility, it enters a series of processing and storage steps called the *LNG train*. Past LNG developments often consisted of at least two trains to gain economies of scale and still retain the operational efficiencies, flexibility, and reliability of individual trains. Recent technological developments have resulted in much larger single trains, making single-train LNG facility developments economic.[3] Early developments also retained liquefaction reliability by having two different trains that could operate independently. Different trains helped fulfill long-term supplier agreements.

The first step in liquefaction is the removal of any remaining condensates and impurities from the natural gas feedstock. The gas is then liquefied when it passes through a heat exchange where it is cooled to –161°C. This cooling is achieved in the same way most refrigerators work. A refrigerant gas, typically propane, methane, or ethylene, is cooled by compression and then released through a valve lowering its temperature (the *Joule-Thompson effect*).

There are two primary liquefaction processes used globally: the multi-component refrigerant (MCR) process and the Phillips Cascade process. The MCR process, in use since 1970, uses propane to precool the gas to –35°C before passing it to a second stage for final liquefaction. The Phillips Cascade process is a three-stage process using different refrigerant gases for sequential cooling. Once in liquid form, LNG is stored in insulated tanks where it rests at atmospheric pressure until loaded on LNG ships for transport.

Liquefaction is expensive. Capital costs of LNG trains are by far the most significant cost, making up on average more than 80% of liquefaction cost. The industry has seen real capital cost reduction over time, falling from more than $500 per ton per year in the 1970s to under $240 per ton per year in 2004. Much of this capital cost reduction has been achieved by expanding the scale of trains from under 2.0 mtpa in 1970 to nearly 8.0 mtpa in 2008.

The construction cost for an LNG plant varies dramatically depending on where it is located, government requirements and restrictions, and size. A 390 bcf per year facility (8.2 million tons per year) may cost between $1.5 and $2.0 billion. This cost is roughly 50% construction, 30% equipment, and 20% bulk material costs. Once under operation, ongoing liquefaction operating costs are roughly 50% operation of the train, 24% storage and loading costs, 16% utilities, and 11% other.

Shipping

Once liquefied, LNG must be transported to market. The LNG may be sold under any one of three major shipping agreements: 1) on a *free-onboard basis* (FOB), where the shipper purchases the LNG at origin and accepts all responsibility for reaching the buyer; 2) *cargo, insurance, and freight* (CIF), where shipping costs are the responsibility of the seller; or 3) *delivered ex-ship* (DES), where an integrated project agreement may be used in which the owner of the gas, the liquefaction process, and transportation to market is controlled by the same entity. The ownership of LNG ships is usually separate from LNG ownership. Gas shippers may own, lease, or charter the ships used for LNG transport.

The design and development of innovative LNG ships was necessary to support growth in the global LNG industry.[4] LNG remains a liquid during transportation using insulation in the ships, not by cooling or compression. The LNG itself contributes to its own cooling, as roughly 0.10% of the LNG boils off daily, adding to the overall cooling. Additional boil-off gas is usually used to power the ship's engines. An additional small amount of LNG is left in the ship's storage tanks after unloading at the receiving terminal to keep the storage tanks cooled on their return trip to the liquefaction facility. This saves time and cost in recooling the ship's storage tanks.

There are two basic ship designs used for LNG shipping. The Kvaerner-Moss design (shown in figure 9–3) stores the LNG in large spherical tanks welded into the ship's hull. The tanks extend vertically far above the ship's deck, giving the ships their unmistakable look of a series of balls. Designs vary between four and six spherical tanks. The second structure, the *membrane design*, has LNG tanks built into the insulated hull of the ship. This design also has the ability to reliquefy the boil-off gas to reduce volume loss. This has proved important in cost reduction, particularly in longer delivery, making distant markets competitively viable. Both designs use steam turbine engines that burn the boil-off gas of the cargo along with heavy fuel oil, as opposed to most ocean-going shipping using gas turbine engines. As of 2008, the global LNG ship fleet was evenly split between the two designs.

LNG ships, like LNG trains, continue to grow in size, as shown in figure 9–4. The first LNG ships were only 27,500 m^3 in capacity, but in recent years a number of the newer ships delivered are more than 250,000 m^3. The ships have proven to be long lasting, with many still operating after nearly 40 years of service. Newer LNG ships can be fully loaded and unloaded in 12 hours. Continuing improvements in transportation have resulted in many LNG ships spending less than 24 hours for turnaround at both ends of the cycle.

LNG ship construction costs have varied over time depending on the capacity utilization rates of the shipbuilding industry. Costs fell steadily over the first half of the 1990s, and a 135,000 m^3 capacity ship fell from $250 million to as low as $150 million by the late 1990s. In recent years costs have risen again, with a 135,000 m^3 ship being delivered at $175 million.

Figure 9–3. Kvaerner-Moss LNG Tanker

Source: http://matex-international.com/page5.html.

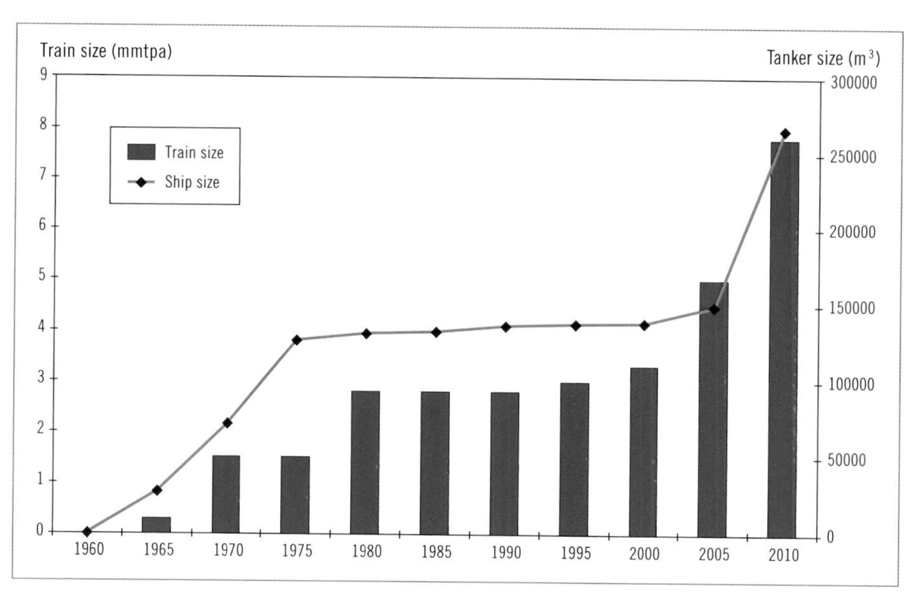

Figure 9–4. Growing train & LNG tanker size

Source: Wood Mackenzie, December 2010.

Transportation costs of LNG to market can prove to be the critical element in project competitiveness. Many LNG developments result in very similar FOB costs of LNG, and as result, whichever can reach a specific market, a receiving and regasification terminal, at lowest delivered cost can win the long-term supply agreement. With a growing fleet and occasional oversupply of LNG ships on the global market, distance from liquefaction facility to receiving terminal can easily determine the competitive winner.

The LNG global fleet has been growing at an extremely rapid rate in recent years, as illustrated by figure 9–5. As recently as 1980, there were still only 50 operating LNG tankers in the world. As of early 2009, that number had grown to over 300. Including those tankers that were scheduled for delivery in the 2009 to 2013 period, the global fleet should near 400 ships by 2013.

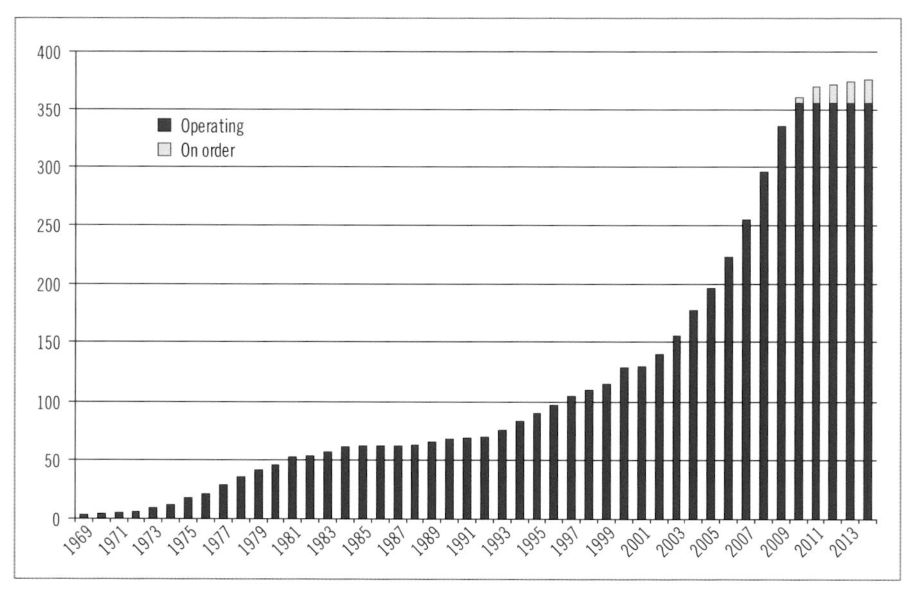

Figure 9–5. Global LNG tanker fleet

Source: Wood Mackenzie, December 2010.

Regasification

The cost of receiving terminals and regasification facilities, like liquefaction facilities, are very site-specific. A small facility may cost as little as $100 million, whereas a very large state-of-the-art facility in Japan (one of the most stringent safety and security regulatory environments) may cost as much as $2 billion. In the United States, LNG receiving terminals range from $200 million (185 bcf) to $600 million (700 bcf). An industry rule-of-thumb is that a complete receiving terminal and regasification plant cost will run $1 billion per billion cubic feet (bcf) of gas per day capacity. If the receiving terminal requires a variety of marine upgrades, like dredging to deepen water for tanker access, costs can easily rise $100 million.

Regasification of LNG is a much simpler and less costly process than liquefaction. A receiving terminal consists of a jetty and berth for the LNG tanker, in addition to the storage tanks and vaporizers needed for moving the regasified LNG into a gas distribution network or power plant. Storage tanks make up the single largest component of capital costs (about one-third of the costs), as significant storage capacity is needed to assure a dependable supply of gas to retail customers. The roughly 50 receiving and regasification terminals around the world are all slightly different, ranging in size from just 0.5 mt to over 8.0 mt. Storage capacity is critical, with the Sodegaura terminal in Tokyo possessing the largest storage capacity in the world at 2.66 million m^3 of LNG, or about 20 standard shiploads of LNG.

There have been several significant innovations in recent years. One is the development of the floating *regas* (regasification) vessel, which is an LNG carrier with onboard LNG vaporizers. Already in use in a variety of countries (e.g., Argentina and Kuwait), this reduces the entry costs for LNG into many markets and countries (there are also stationary regas vessels that operate much the same as a terminal). This allows faster and more cost-effective switching from traditional pipeline gas or other fuel. A second innovation is offshore regasification terminals that take advantage of limited onshore space and "not in my backyard" concerns about terminals on land. ExxonMobil has built the first offshore terminal in the northern Adriatic Sea. The Adriatic LNG terminal is a joint venture between Qatar Petroleum (45%), ExxonMobil (45%), and Edison (10%). A third innovation is offshore LNG plants. Processing gas in LNG at sea could open up new opportunities for gas developments far from shore. In 2009, Shell began early-stage design work on what could be the world's first operational floating LNG platform.

The cost of transportation to market remains significant for LNG. Figure 9–6 provides one analyst's competitive cost analysis across fuels when considering the actual Btu value per kilometer of movement. LNG is clearly the most expensive, and as the analysis in the following section details, transportation costs have largely dictated the relative competitiveness of LNG across fuels and across LNG sources themselves.

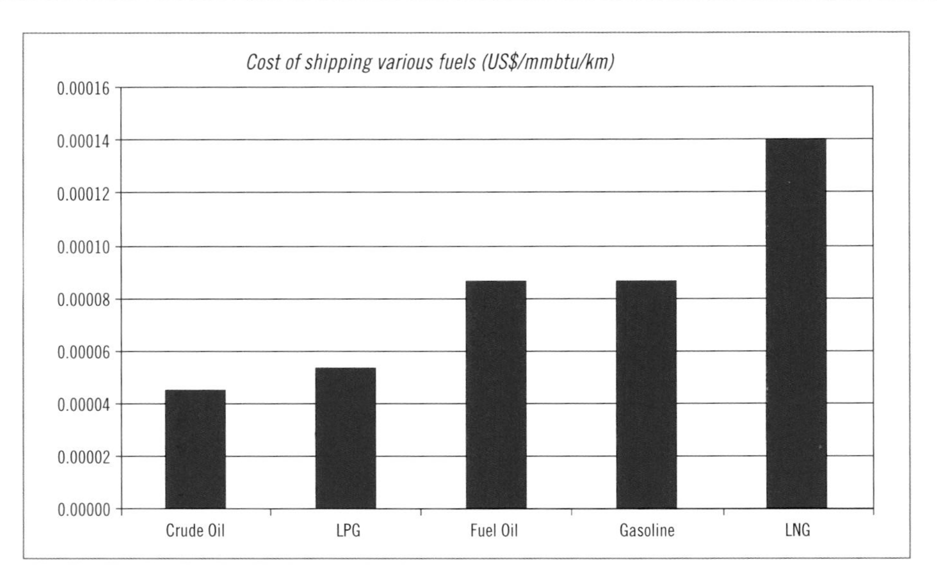

Figure 9–6. High cost of LNG transportation

Source: G.M. Schuppert, Jr., ConocoPhillips, January 2011. Based charter rates for long-term contracts and for average-size ships and estimated average fuel consumption and port costs and for return trips. Used with the permission of ConocoPhillips Company.

Competitive cost analysis

To date, LNG projects have been demand/buyer driven. As a result, once a true buyer is identified, particularly for long-term purchase agreements, the lowest delivered cost of a sufficiently large and reliable supplier will prevail. Therefore, many LNG projects are driven by the buyer's willingness to sign a long-term purchase agreement with the most competitive long-term supplier.

The nature of demand-driven projects is illustrated by the competitive analysis in figure 9–7. The figure shows the estimated cost of LNG ($/mmbtu) delivered from 12 different origination points around the world to the receiving terminal at Lake Charles, Louisiana (United States).[5] Delivered cost has three components: FOB costs, transportation costs, and regasification or regas cost. Regas costs are the same across all 12 given the same assumed regas and receiving facility at Lake Charles ($0.35/mmbtu).

Two important conclusions can be drawn from figure 9–7. First, FOB cost differs dramatically across sources. The lowest cost source, Algeria LNG at $0.45/mmbtu, is one-sixth of the FOB cost originating from the Norway Snohvit LNG liquefaction facility, at $2.90/mmbtu. Second, the shipping tariff varies, with the highest from QatarGas II (Qatar in the Persian Gulf) at $1.90/mmbtu and the lowest of the ALNG facilities (Atlantic LNG facilities, Point Fortin, Trinidad and Tobago) at $0.50/mmbtu. The dominance of the ALNG cost is the result of a low FOB cost and a highly competitive shipping tariff. The low shipping tariff reflects the short distance to the Lake Charles destination.

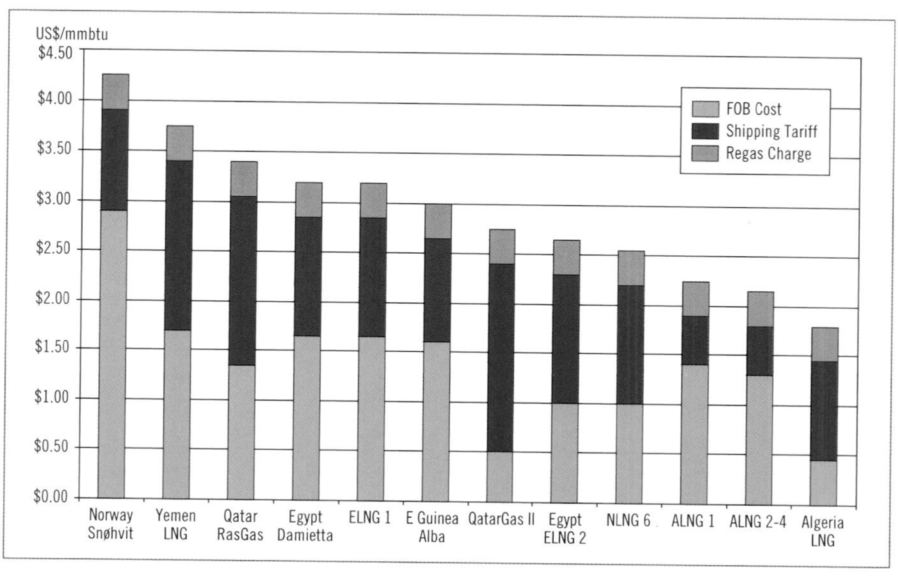

Figure 9–7. Estimated cost of LNG delivered to Lake Charles, LA

Source: "Oil & Gas for Beginners," Deutsche Bank, January 7, 2008, fig. 250, p. 188. Copyright © 2008 Deutsche Bank AG.

LNG contracting

The LNG market was created from the older natural gas market. Natural gas has long been sold under long-term agreement between producer and buyer, shipped via pipeline. But, as the LNG market grew, new forms of contractual arrangements were needed because of the following:[6]

- After securing long-term concessions or PSAs for the development of upstream gas, LNG developers put massive amounts of capital at risk by building pipelines and liquefaction facilities. The developers need to secure long-term sale agreements that assure them of a return on their sizeable capital investment.

- LNG buyers without access to pipeline gas must secure long-term supplier assurances of regular shipments in order to supply electrical power plants and residential and commercial gas needs. In the event of gas interruptions or supplier failure, alternative energy supplies are not easily available.

The growth and development of the LNG industry has led to innovative thinking and less-restrictive development relative to the pipeline gas industry. Buyers and sellers use a series of sale and purchase agreements that tend to escalate towards a full commitment of resources—specifically the capital associated with the construction of pipelines, liquefaction facilities, shipping arrangement, and regasification terminals—for completion of the LNG chain.

The marketability and competitiveness of LNG is based on its delivered cost to the customer. If LNG can be delivered to a customer at a lower or competitive price relative to other energy sources and if sustainability and security of delivery over time is assured, LNG will be competitive. There are three types of contracts commonly in use in the LNG markets today: free on board (FOB); cargo, insurance, and freight (CIF); and delivered ex-ship (DES). Figure 9–8 provides an overview of the distinction between the three.

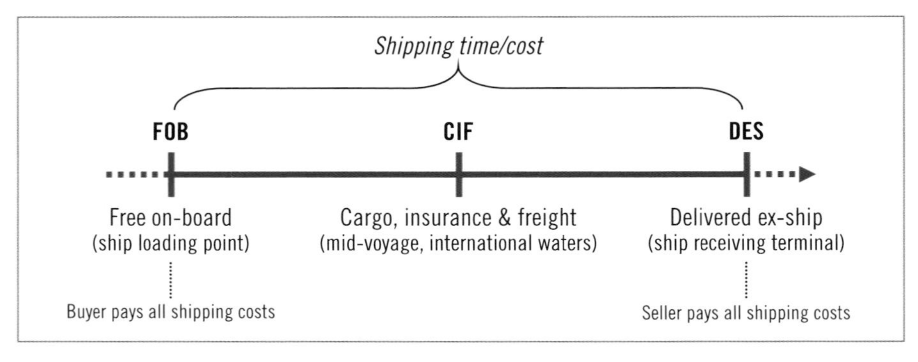

Figure 9–8. Shipping contract terms. Since 2000 LNG shipping contracts have been roughly evenly split between FOB and DES. Although there are a variety of strategies, buyers who do not wish to invest in tankers generally wish suppliers to compete on DES price. Other buyers, wishing to reduce total delivered cost through efficiencies and control, prefer FOB arrangements.

Free on board (FOB). A *free on board* contract obligates the buyer to be responsible for shipping the LNG from the liquefaction facility to receiving terminal. Ownership of the LNG passes from the producer to the buyer at the loading point at the liquefaction plant. All costs of shipping and insurance are born by the buyer.

Cargo, insurance, and freight (CIF). A *CIF* sale agreement makes shipping costs the responsibility of the seller. LNG ownership is transferred in the midst of the voyage, typically in international waters (and not at the point of delivery). This transfer of ownership is sometimes of value to buyers who for various reasons prefer not to take possession in the home port of the LNG liquefaction facilities.

Delivered ex-ship (DES). Under a *DES* sale, the seller is responsible for all shipping and delivery costs, the transfer of LNG ownership not occurring until receipt by the buyer at the receiving terminal. For pure buyers wishing to stimulate competitive bidding for delivery, DES has been a frequent shipping choice.

In recent years contracts have tended to be predominantly FOB and DES, as CIF sales have declined with increasing political stability and greater LNG traffic globally. As shipping costs and options have improved, buyers have been willing to push competitive sourcing costs back down to the FOB point in the LNG chain.

LNG Markets and Pricing

The LNG market is only about 40 years old, so the market's structure, players, and pricing continue to evolve. This section discusses the Asia Pacific and Atlantic Basin markets and some recent shifts in the competitive environment.

LNG's regional markets

The LNG industry evolved around two separate regional markets, the Atlantic Basin and the Asia Pacific.[7] These markets, which were very "thin" in structure originally with few sellers and few buyers, developed geographically along the routes and linkages illustrated in figure 9–9.

Asia Pacific market. The Asia Pacific market is the larger of the two regional markets and is driven by Japan (the world's largest LNG buyer), Korea, and Taiwan. As illustrated in figure 9–9, LNG markets in 1990 were serviced primarily by producers in Indonesia and Malaysia. More distant producers in Alaska, the Persian Gulf, and North Africa (Algeria and Libya) also played a role.

Pricing in this market was based primarily on the *Japan crude cocktail (JCC)* in Japan and South Asia, as imported crude oil was deemed the benchmark competitive alternative.[8] The Japan customs-cleared crude is the average price of customs-cleared crude oil imports into Japan (formerly the average of the top 20 crude oils by volume) as reported in customs statistics—nicknamed the *JCC*.[9] The JCC is a commonly used index in long-term LNG contracts in Japan, Korea, and Taiwan. It is published by the Japanese government every month as an average crude oil import price in kiloliters per yen. LNG prices also reflect the price of major competing fuels in most markets. LNG prices have generally been the highest in the Pacific (e.g., US$4/mmbtu as opposed to $3/mmbtu in the Atlantic) as a result of limited competing fuels and the structure of long-term contracts. With few exceptions, all volumes in the Asia Pacific were under long-term commitments to justify the substantial capital costs on both ends of the LNG chain (liquefaction, transportation, and regasification). Shipping costs were high and made up a large portion of the delivered cost of the gas.

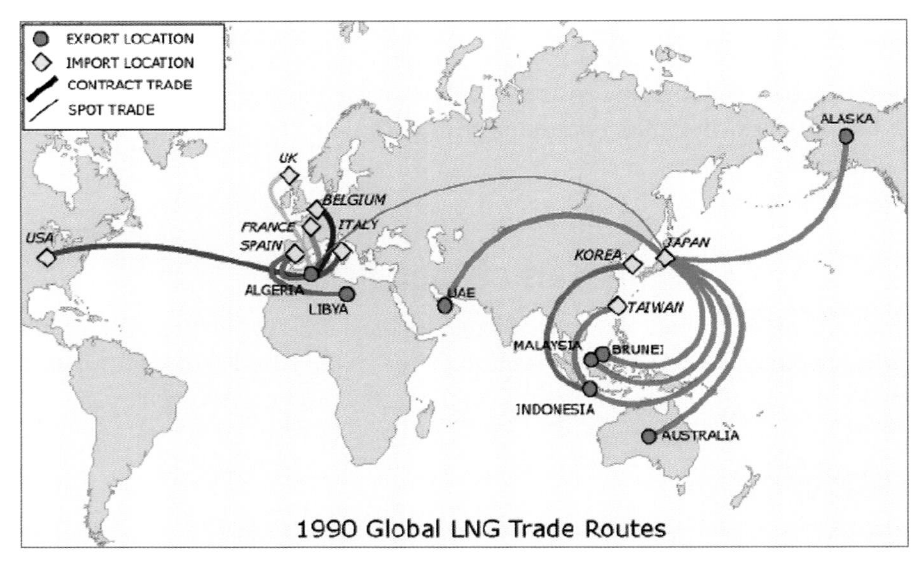

Figure 9–9. Global LNG trade, 1990

Source: From Stephen Thompson, *The New LNG Trading Model: Short-Term Market Developments and Prospects*, Copyright ©2009 Poten & Partners, Inc. Reprinted with permission.

In the 1990 period, the market had few short-term trades given the predominance of long-term contracting. Short-term divertible cargoes (i.e., cargoes that were sold into a spot market) were limited to early volumes of new liquefaction facilities (new trains often produce at full capacity before some buyers are capable of taking full contract volumes) and added volumes from existing trains that produced greater volumes as a result of debottlenecking completions. All in all, the market was long-term-oriented and inflexible.

Atlantic Basin. The Atlantic Basin LNG market developed later than the Asia Pacific market. Buyers were primarily European-based and attempting to supplement pipeline gas from the former Soviet Union. The United States also became part of this market but did not purchase significant volumes. The market was supplied nearly exclusively out of North Africa.

Pricing in the Atlantic Basin was significantly more complex than in the Asia Pacific. Since LNG agreements generally follow gas pricing structures used in natural gas markets, this meant that prices reflected traditional gas pricing by region: 1) the mix of NBP pricing in the UK and long-term pipeline pricing agreements in Continental Europe, and 2) Henry Hub in the United States, where pipeline gas was the widely available competitive alternative.

The LNG market: The shift toward a global commodity

Over the last 10 to 15 years, the LNG market has shifted from its traditional regionalization to a market that is increasingly global. Figure 9–10 shows how LNG has expanded—in where it is produced and used.

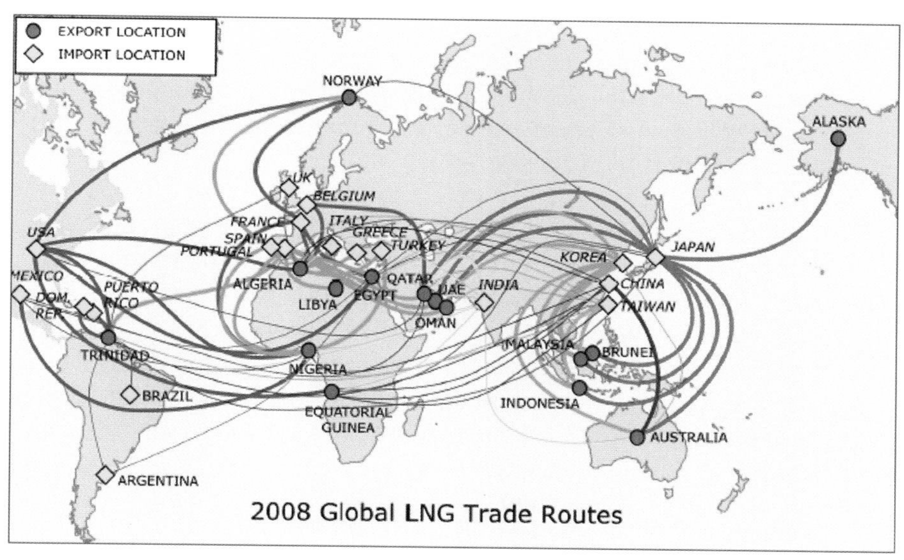

Figure 9–10. Global LNG trade, 2008

Source: From Stephen Thompson, *The New LNG Trading Model: Short-Term Market Developments and Prospects*, Copyright ©2009 Poten & Partners, Inc. Reprinted with permission.

LNG production has expanded far beyond the original North African, small Persian Gulf, and large Southeast Asian liquefaction centers. Significant expansion in all three of the original production areas has been supplemented by new LNG liquefaction facilities in West Africa and the Caribbean. At the same time, LNG regasification and receiving facilities have expanded rapidly in Europe (particularly Mediterranean Europe), the United States, Mexico and the Caribbean, China, Taiwan, and India. Volumes move across the traditional regions on both a long-term contract basis and increasingly a short-term spot market basis.

The growth in the short-term spot market for LNG is an added sign that the LNG market is moving towards a more global structure. The drivers for change have arisen from both the sellers and the buyers.[10]

- **Buyers.** Unexpected supply disruptions like that of Hurricanes Katrina and Rita on the US Gulf Coast in 2005 disrupted production significantly. Supply disruptions have led to serious price spikes, pushing buyers to seek more short-term supply options.

- **Suppliers.** Many long-term production areas have experienced declining production (such as Aceh province in Indonesia). As a result, buyers have sought short-term alternatives to fill growing supply gaps. A new set of producers is increasingly filling the gaps.

- **Buyers.** LNG buyers have experienced large fluctuations in demand for gas. Extremely cold weather in January 2008 forced Russia to reduce offered volumes to Eastern and Western European markets, forcing buyers to look for short-term alternatives.

- **Suppliers.** As more production capability comes online, there is an increasing number of delivered cost-competitive alternatives for buyers. For example, the recent development of shale gas in North America has freed up large volumes of LNG originally destined for North America to search out short-term sale alternatives in Europe or the Asian-Pacific. Major new gas projects such as those in Papua New Guinea and Western Australia (see the Gorgon case study later in the chapter) will bring large new LNG volumes to the market over the next decade.

A complex combination of events in Asia in the 2007 to 2009 period was particularly instrumental in expanding the short-term flexibility in the LNG market. Market demand in Asia unexpectedly soared because of various factors, including a large nuclear power outage in Japan. Gas supplies in the Asia Pacific basin suddenly and unexpectedly declined as several producers encountered dwindling production and others suffered new construction delays. The market shortfall was rapidly filled by volume from the Atlantic Basin, roughly 15% of its continuing volume. The ability of Atlantic Basin producers to respond quickly to a short-term supply problem is evidence of new market flexibility and, in particular, of producers' willingness to create divertible volumes.

This influx of short-term volumes into a market traditionally dominated by long-term supplier agreements and long-term contract prices upset much of the regional pricing structures. Most of the Atlantic Basin volumes diverted to the Asian market have been at spot market prices based purely on supply and demand on the open market, something nearly unknown in the prior 40 years of LNG trade. The spot prices, which were substantially higher than most long-term contract prices in either the Atlantic Basin or the Asian Pacific, also represent the increasing use of contractual structures that allow producers to selectively respond to short-term market price spikes and diversion clauses.

LNG breakeven: An aside

The cost and complexity of many LNG projects have led to a significant spread of break-even prices on the actual gas produced by many LNG field projects. The variance in break-even costs and prices may contribute an additional impetus to the growth and development of global LNG trade in the near future.

LNG Case Study: The Gorgon Project

"The decision of the Gorgon Joint Venture Partners to invest in the Gorgon project will deliver a major economic boost for Australia, create thousands of jobs and can significantly reduce global greenhouse gas emissions."

—Belinda Robinson, chief executive, Australian Petroleum Production & Exploration Association, September 14, 2009 [11]

"Outside of Qatar, nobody has ever developed an LNG project of this scale in one go."

—Frank Harris, director of LNG Consulting, Wood Mackenzie[12]

There is probably no better example of the many challenges faced by LNG producers than the Gorgon project deep under the Indian Ocean off the northwest coast of Australia. What once was considered stranded gas will now be part of one of the most complex and expensive LNG projects in the world. Gorgon is representative of the technical, environmental, regulatory, contractual, financial, and market dimensions that the gas industry must confront in the twenty-first century.

Project overview

Gorgon is the first major field development in the Carnarvon Basin off the northwest coast of western Australia. The project is a joint venture between Chevron (50% and the project operator), ExxonMobil (25%), and Shell (25%). The project cost is estimated at more than AUD 43 billion (USD 37 billion, JPY 3.4 trillion).[13] Gorgon is thought to hold more than 13.8 trillion cubic feet (tcf) of hydrocarbon reserves (2.25 billion barrels of oil equivalent) and is expected to have a production life close to 40 years. Figure 9–11 provides an overview of the Gorgon project.[14]

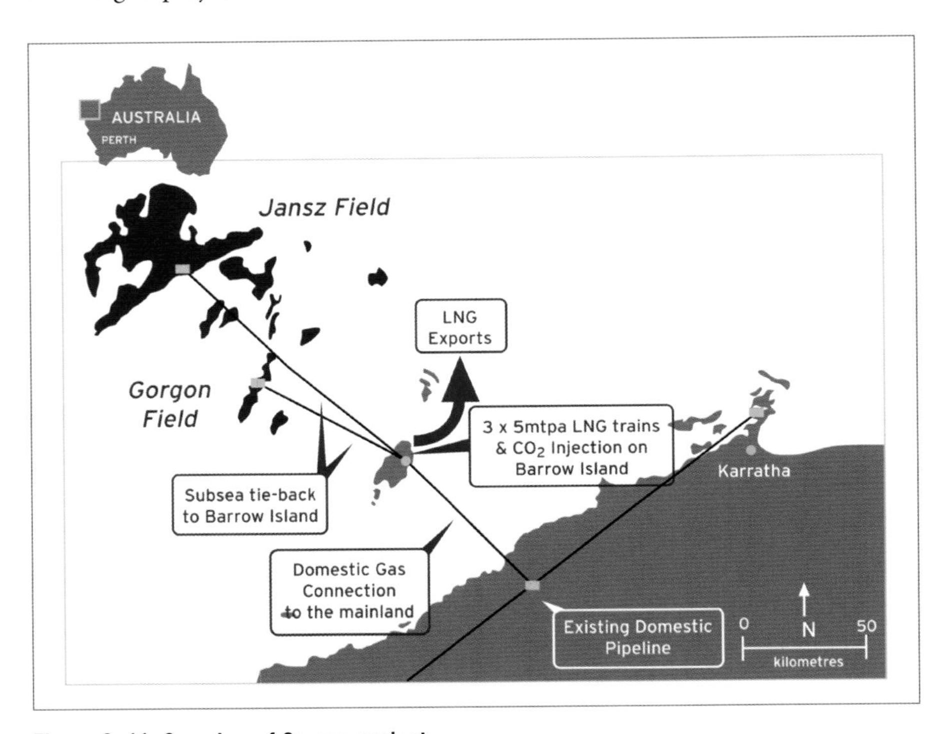

Figure 9–11. Overview of Gorgon project

Courtesy of Chevron.

The Gorgon field is located between 130 km and 200 km off the northwest coast of Australia. The gas will be produced from sub-sea wellheads at roughly 250 meters depth, and the gas will then flow via a trunk line ("tie-back") 180 km to the gas processing facilities on Barrow Island, as shown in figure 9–12.

On Barrow Island, the raw gas will be processed, removing carbon dioxide, water, mercury, and other nonmethane components. The methane gas will then be liquefied for export using three LNG trains of 5 million tons per annum (mtpa), a total of 15 mtpa production capacity. In a later stage of development, up to 300 terrajoules of natural gas will also be piped to the Australian mainland, a distance of 70 km, where it will enter the existing Australian natural gas pipeline and distribution system. LNG will be stored in tanks on Barrow and then piped through a 2.2-km jetty to awaiting LNG tankers for loading and export. First gas is expected in 2014.

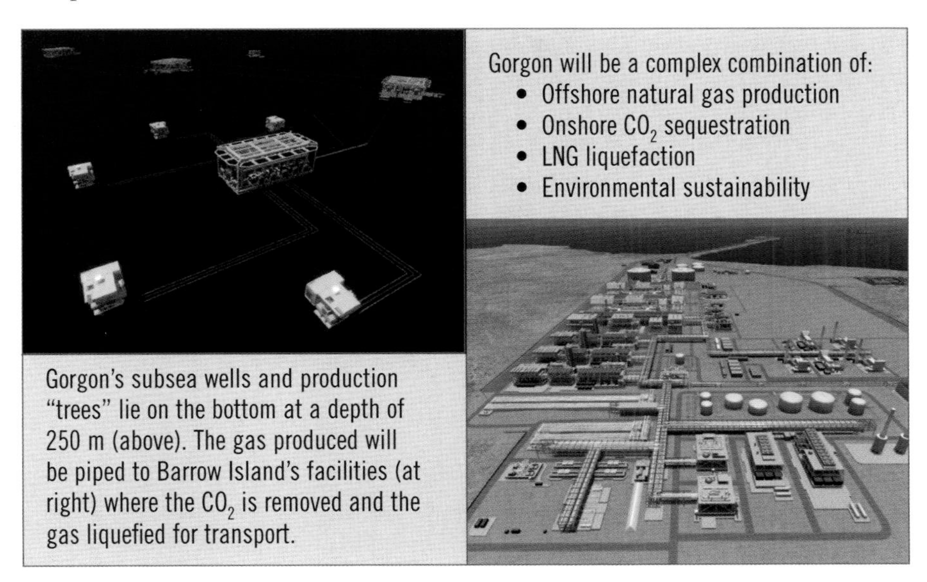

Figure 9–12. Gorgon Subsea/Barrow Island Structure
Courtesy of Chevron.

One of the Gorgon's major challenges is the design for carbon dioxide capture and storage. Gorgon has a high carbon dioxide content at 14% or 15%. (Interestingly, the adjacent Jansz-Ilo Field has less than 1% carbon dioxide content.) The project development plan is to separate the CO_2 from the hydrocarbons and then inject it into a subterranean cavern, the Dupuy Formation, more than 2,000 meters below Barrow Island, for permanent storage.[15] This CO_2 re-injection for permanent storage, termed *CO_2 geosequestration*, will make Gorgon the largest carbon capture project in the world.

Project approval

Oil and natural gas were first discovered in the Gorgon area in 1981 by Western Australian Petroleum (WAPET). WAPET was originally an operator on behalf of Chevron, Texaco (Chevron merged with Texaco), Shell, and Ampol (later acquired by Mobil, who merged with Exxon). Chevron became the operator of

WAPET's assets. Gorgon is the first of the five fields discovered to date to be developed. The five are Gorgon, Chrysaor, Dionysus, West Tryal Rocks, and Spar (collectively referred to as *Greater Gorgon*). The Gorgon field was discovered in 1981.

One of the largest regulatory hurdles for the project was gaining the permits to construct the gas processing and liquefaction facilities on Barrow Island, a Class A nature reserve. The environmental management plan had to contend with a variety of challenges. The island is home to more than 20 endangered species, including the flatback turtle, the spectacled hare-wallaby, and the golden bandicoot.[16] The final project plan is for the total surface footprint on Barrow to make up less than 1.3% of the island's total surface, with most construction and operational impacts minimized by construction of new port facilities adjacent to the complex. The project sponsors themselves only gave final project approval—the financial investment decision (FID)—after securing both the Australian government's various approvals and long-term sales contracts for at least 3 million tons per year of LNG exports.[17]

The Asian LNG market

> *Some analysts have warned about the raft of other new LNG projects planned around the region and said the market was unlikely to be able to accommodate all the new capacity, most of which was targeted to come on-stream on 2015–2016. "The previously expected rise in US LNG demand won't be around anymore because of the surge in domestic gas supplies. So a lot of the gas in the Middle East will have to make its way to Asia, where demand growth for LNG too has its limits," said an analyst.*
>
> – "Chevron Approves $37 BN Gorgon LNG Project," Reuters, September 14, 2009

Demand for LNG in the Asia Pacific region is expected to more than double between 2005 and 2015. The major Asia Pacific markets, and the only countries importing LNG in 2000, Japan, South Korea, and Taiwan, will continue to make up the majority of Asia Pacific LNG purchases. Substantial new demand is expected from China, India, Mexico, and Singapore.[18] Demand from China, which imported no LNG as recently as 2005, is expected to rise rapidly by 2015.

The South Pacific region is currently experiencing massive gas exploration and development. The proximity of these production areas to the major Chinese, Korean, and Japanese customers has provided a competitive advantage for securing long-term sales agreements over the other hotbed of LNG development in the Persian Gulf.

Gorgon joint venture structure

The joint venture project sponsors are charged with the marketing of their respective shares of Gorgon's LNG production. By September 2009 the project had secured sufficient sales and purchase agreements (SPAs) and *heads of agreements (HOAs)* to assure a minimal critical level of sales for the project. (An *HOA* is a nonbinding document outlining the main issues relevant to a tentative partnership or other business agreement.) As detailed in figure 9–13, ExxonMobil had fulfilled its full equity share of the project, while Chevron and Shell still had uncommitted volumes of their equity participations of the 15 mtpa production.

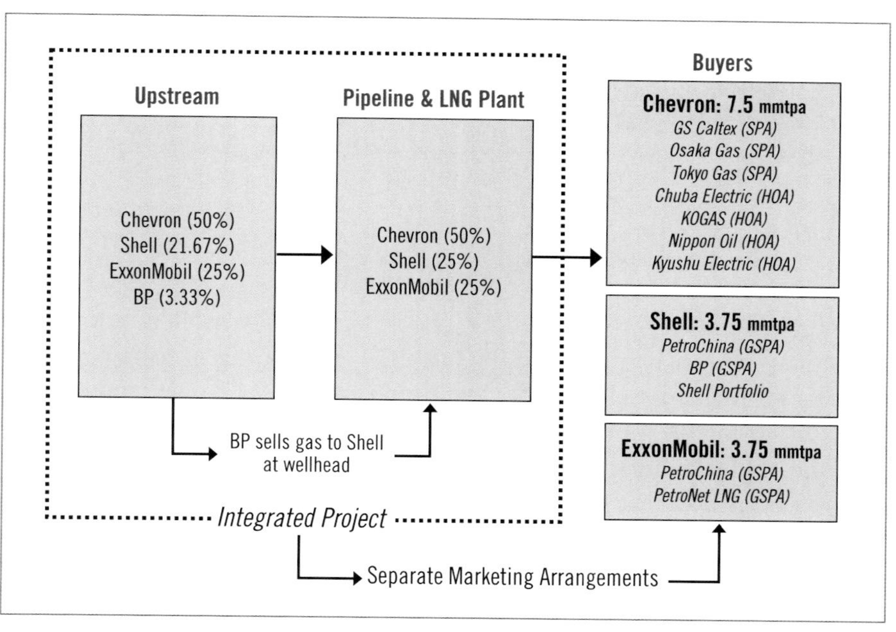

Figure 9–13. Gorgon's commercial structure
Source: Wood Mackenzie, December 2010.

Gorgon involves many international contractors and subcontractors. For contracting purposes, the project has been divided into upstream and downstream components. The project's downstream contractor is Kellogg Joint

Venture (KJV), an unincorporated joint venture between KBR, JGC Corporation, Hatch, and Clough. KJV was awarded an AUD2.7 billion contract to engineer, procure, and construction manage (EPCm) the LNG downstream and logistics portion of Gorgon. The upstream EPC contractor is Gorgon Upstream Joint Venture, a 50/50 joint venture between Technip and JP Kenny.

Other contractors include:

- Boskalis Australia, a subsidiary of Royal Boskalis Westminster N.V., was awarded a €500 million contract for the construction of a port on Barrow Island. The contract includes the construction of a material offloading facility with associated dredging and marine preparation activities, as well as logistical and program management responsibility. Work began in 2010 and will be completed by end-of-year 2011.[19]

- Compass Australia won an AUD150 million three-year contract to feed workers on the gas project and to maintain the accommodation that is to be built.

- Neptune Marine Services was awarded a $5 million initial works contract to provide diving, survey, and vessel support services for Gorgon. Work will commence in the first quarter of 2010.[20]

Gas to Liquids (GTL)

Another liquefaction process gaining interest and investment in recent years is the process known as *gas to liquids* (GTL). GTL turns natural gas into a clean-burning synthetic diesel fuel. GTL fuels ignite more easily than conventional fuels, improving the performance of car engines. An Audi race car powered by GTL diesel blended with Shell GTL has won the Le Mans 24-hour endurance race in France in recent years. Although GTL fuel is clean burning, the process generates significant carbon dioxide emissions that, if regulated, could prove extremely costly. The International Energy Agency says the cost per barrel of producing GTL is in the range of $40 to $90. Depending on the current market price of crude, GTL's economics are questionable.

Shell's Pearl GTL in the Middle East is the largest GTL project to date. Due to come onstream in 2010, it is Shell's single largest investment project globally. The Pearl GTL plant is under construction at Ras Laffan, the industrial city 50 miles northeast of Doha, Qatar. Some 35,000 workers are employed at what is one of the world's largest construction sites. When finished, the complex will boast four cricket pitches for its workers, three soccer fields, an outdoor movie theater, and its own mayor.

Shell started experimenting with GTL technology in 1993 when it opened its first GTL demonstration plant in Bintulu, Malaysia. The project was derailed in 1997 by a massive explosion caused by a profusion of carbon molecules in the air as a result of extensive forest fires in Indonesia. It took three years to repair the damage. Since then, fuel from the Bintulu plant has built up a small but growing presence on the market. Shell's V-power diesel, which is blended with GTL fuel, has enjoyed a growing following in Europe, although it does sell at a premium to regular motor fuel.

Other companies have looked at GTL but backed away because of technological and cost concerns. ConocoPhillips, Marathon Oil Corp., and ExxonMobil have all jettisoned planned GTL projects in Qatar. In 2009, Algeria canceled tenders for a GTL project in Tinrhert. Chevron and Nigerian National Petroleum Corporation are moving ahead with their Escravos plant in Nigeria, but costs have increased substantially and the start-up date has been pushed out a year, to 2011. Figure 9–14 provides a brief overview of major GTL projects currently underway.

Escravos Gas-to-Liquids Project
Niger Delta, Nigeria

Ownership: Chevron Nigeria (75%), Nigerian National Petroleum (25%)
Objective: Natural gas into clean liquid fuels, primarily GTL diesel and naptha
Production: 34,000 bpd (120,000 in future)
Operational startup: 2011
Estimated cost: $1.70 billion

Pearl Gas-to-Liquids Project
Ras Laffan, Qatar

Ownership: Qatar Petroleum (50%), Royal Dutch Shell (50%)
Objective: Natural gas into naptha, GTL fuel, paraffins, and kerosene
Production: 140,000 bpd
Operational startup: 2010/2011
Estimated cost: $6.0 billion

MossGas Gas-to-Liquids Project
Mossel Bay, Bredasdrop Basin, South Africa

Ownership: South African Coal and Oil, Ltd (Sasol)
Objective: Natural gas into liquid fuels, primarily premium-quality diesel
Production: 25,000 bpd
Operational startup: 1993
Estimated cost: $388 million

Bintulu Gas-to-Liquids Project
Bintulu, Malaysia

Ownership: Shell (72%), Mitsubishi (14%), Petronas (7%), Sarawak State (7%)
Objective: Natural gas into liquid fuels, specialty chemicals, and waxes
Production: 14,700 bpd
Operational startup: 1993
Estimated cost: $1 billion

Figure 9–14. Selected gas-to-liquids projects (GTL). ***Gas-to-liquids*** **is the term used to describe a number of developing technologies that seek to use natural gas to create liquid fuels. It may prove particularly valuable in providing commercial uses for stranded gas deposits or gas currently flared with petroleum production.**

The Future

LNG is slowly but surely revolutionizing the global natural gas industry. In the recent past, gas fields were developed only after identifying a relatively nearby market and customer for the gas. That customer had to be willing to sign a long-term purchase agreement that would justify the construction of a gas pipeline to move the gas from field to market. LNG has changed much of that. As LNG trains are constructed at the largest fields and more markets develop that support receiving terminals and regasification facilities, LNG is becoming a truly transportable global commodity similar to oil. Tradable LNG benchmarks may join crude oil benchmarks as a daily staple of business news.

Several other factors could have a major impact on LNG markets. In response to the increasing openness in LNG markets, the major gas producers have had early discussions about forming an OPEC-like gas cartel that would seek some control over price and supply. At this point a gas cartel looks unlikely, but OPEC also looked unlikely 50 years ago. The growing importance of shale gas will also have an impact on LNG markets. As noted earlier, shale gas volumes in the United States have resulted in a reduction in US demand for LNG. Several US LNG projects involving receiving terminals and regasification plants have been cancelled because the LNG is no longer needed. Should shale gas discoveries be made in significant volumes in Europe and Asia, LNG markets could be significantly disrupted, and projects like Gorgon may prove noneconomic. Shale gas developments require very little initial capital and can be shut down or suspended when demand drops. When demand rises, drilling can restart. In contrast, LNG megaprojects require billions in capital up front for a multi-decade expectation of production. Should demand drop, the capital has been spent, and much of the production will be locked into long-term sales contracts.

Summary Points

- Natural gas was long considered an afterthought of crude oil production. Associated gas, or gas inadvertently discovered in the process of exploring and developing oil, was often stranded or flared.

- Historically, only natural gas deposits of significant size, and located within reach of major industrial or consumer centers for use via pipelines, were actually developed.

- Increasing concern over the cost and future availability of oil has renewed interest in the global potential of natural gas. Technological changes such as the development of the combined-cycle generation turbines have motivated many power developers to view natural gas—specifically LNG—as a more affordable and environmentally sustainable power solution.

- The massive and rapid development of LNG is making gas much more of a globally traded product. The development of LNG markets means that previously stranded or abandoned natural gas reservoirs can be evaluated for potential commercial development.

- LNG markets traditionally involved regional markets characterized by long-term contracts and long-term contract pricing.

- LNG markets have recently shown significant growth and development of short-term sales and true short-term supply and demand pricing, primarily a result of diversions of long-term LNG to buyers driven by short-term demand. This is considered by many to be the first major step for LNG to become a globally traded commodity.

Notes

1. Andy Flower, LNG Today, 2004 edition, The Energy Publishing Network, Gas Strategies, London: August 2004.
2. "Angola LNG Commercial Overview," Wood Mackenzie, January 2009.
3. Anthony Edwards, Rick Hernandez, and Allyn Risley, ConocoPhillips Company, Phil Hunter, Amos Avidan, and John Duty, Bechtel Corporation, "Lowering LNG Unit Costs Through Large and Efficient LNG Liquefaction Trains—What Is the Optimal Train Size?" Spring AIChE Meetings, New Orleans, LA, April 25–29, 2004.
4. *Transportation of Liquefied Natural Gas*, Office of Technology Assessment, NTIS, September 1977.
5. Deutsche Bank, 2008.
6. *The Global Liquefied Natural Gas Market: Status and Outlook*, Energy Information Administration, US Department of Energy, Washington, D.C., December 2003.
7. Stephen Thompson, "The New LNG Trading Model: Short-Term Market Developments and Prospects," Poten & Partners, 2010.
8. Akbar Nazemi, "New Mechanisms of LNG Pricing in Asia," National Iranian Gas Company, undated.

9. Akira Miyamoto and Chikako Ishiguro, "A New Paradigm for Natural Gas Pricing in Asia: A Perspective on Market Value," Oxford Institute of Energy Studies, NG28, February 2009.
10. This section draws upon a multitude of sources including Stephen Thompson, "The New LNG Trading Model: Short-Term Market Developments and Prospects," Poten & Partners, 2010; Rob Fenton and James Ball, "Can Price Terms in Yesterday's LNG Contracts Survive the Upheaval of Today's Markets?" Gas Strategies; Akbar Nazemi, "New Mechanisms of LNG Pricing in Asia," National Iranian Gas Company, undated; Polina Zhuravleva, "The Nature of LNG Arbitrage: An Analysis of the Main Barriers to the Growth of the Global LNG Arbitrage Market," Oxford Institute for Energy Studies, NG 31, June 2009.
11. "Gorgon Go Ahead Is Good News for the Economy and the Environment," *Media Release*, Australian Petroleum Production & Exploration Association, Limited, September 14, 2009.
12. Jim Redden, "Australia Outback Could Be Next World-Class Unconventional Play," *JPT*, December 2009, p. 29.
13. "Tokyo Gas Inks Agreements in Connection with Gorgon LNG Project," *TG Minutes*, Tokyo Gas, November 2009.
14. "Gorgon, Northern Carnarvon Basin, Australia, offshoretechnology.com. Accessed February 17, 2010.
15. "Media Fact Sheet Gorgon Project," Chevron, September 2009.
16. "Chevron Approves $37 BN Gorgon LNG Project," *Reuters*, September 14, 2009.
17. "Final Investment Decision for the Gorgon LNG Project," Tokyo Gas Co., Ltd, September 14, 2009.
18. Chevron Australia, "The Gas Market," www.chevronaustralia.com.
19. "Boskalis Wins €500 Million LNG Project in Australia," Press release, Royal Boskalis Westminster NV, October 16, 2009.
20. "Neptune Nabs Support Services Gig for Gorgon LNG Project," *Rigzone News*, January 12, 2010.

Chapter 10
THE MARKET FOR CRUDE OIL

It should be remembered that oil is not an ordinary commodity like tea or coffee. Oil is a strategic commodity. Oil is too important a commodity to be left to the vagaries of the spot or the futures markets, or any other type of speculative endeavor.

—Saudi Arabian Finance Minister, Sheik Ahmed Azaki Yamani, 1983

Bubbles are episodes of collective human madness—euphoria over investments whose skyrocketing values are unsustainable. They tend to arise from perceptions of pending shortages (as happened last year, with the oil bubble); from glamorized new technologies or investment frontiers (like the dot-com bubble of the 1990s, the radio bubble of the 1920s or the multiple railroad bubbles of the 19th century); or from faddish cultural obsessions (like the Dutch tulip bubble of the 17th century, or the more recent Beanie Babies bubble).

—"This Bubble is Different," *The New York Times*, September 13, 2009

The markets and pricing for crude oil have long been considered something of an enigma. The question of whether crude is a tradable commodity, subject to the whims and tactics of the trading world, continues to confound the market's analysts. As Sheik Yamani noted more than 25 years ago, a second level of debate is whether the price of crude should be left purely to the forces of the markets and its various participants. Those markets also have been characterized by sudden and large price swings throughout the history of crude oil. These swings have led most of the world's crude oil experts, including the major oil companies, to conclude that they cannot forecast the price of crude oil. Even Wall Street has not been able to find much clarity in its crystal ball over the future price of crude.

After crude oil is produced, it must be sold and transported to a refinery before it is of real value. Although some crude is both produced and refined by the same company, it is more likely that the crude is sold into the global oil market. The sale of crude oil requires an understanding of oil prices, one of the more complex and perhaps most misunderstood areas of the petroleum value chain. When we read that the oil price is $78 per barrel or $147 per barrel, how was the price determined, who is actually paying the price, what is the role of futures markets, and how does the price impact the eventual price of products like gasoline and jet fuel? In this chapter, we examine crude oil markets and the prices they produce.

Crude Oil Fundamentals

According to Merriam-Webster's, petroleum is "an oily flammable bituminous liquid that may vary from almost colorless to black, occurs in many places in the upper strata of the earth, is a complex mixture of hydrocarbons with small amounts of other substances, and is prepared for use as gasoline, naphtha, or other products by various refining processes."[1] The name itself is derived from Latin—*petr* meaning "rock," *oleum* meaning "oil." "Crude oil" is self-explanatory and is generally defined as unrefined petroleum. According to the Energy Information Administration (EIA), crude oil is "a mixture of hydrocarbons that exists in liquid phase in natural underground reservoirs and remains liquid at atmospheric pressure after passing through surface separating facilities."

The demand for oil

The demand for oil is the demand for its derivative products. The demand for gasoline, diesel, jet fuel, and lubricants, to name but a few, has been rising steadily in all economies, industrialized and developing, for more than a century. The introduction and commercial acceptance of the diesel locomotive, the automobile, plastics, and petrochemical fertilizers are all examples of technological and product innovations that have fueled demand. Once these technologies and uses have been widely adopted in industrial economies, the cost, time, and ability of users to switch away from them, and their oil needs, is difficult.

The field of economics has long described the responsiveness of quantity demanded to price changes as the *price elasticity* of demand. For example, the price elasticity of demand for oil has been long been considered to be relatively low, a value less than 1, where a 1% increase in the price of oil results in a less than 1% decrease in the quantity of oil demanded. This reflects, at least in the short run, the lack of alternative fuels or technologies to act as substitutes for crude oil derivative products. As time passes, giving the consumers of oil-based products more time to explore and develop substitutes, the price elasticity of demand for oil falls.

The supply of oil

The supply of oil to the marketplace reflects the ability of oil exploration and development organizations to not only find the oil, but also to recover it and move it to the refining and processing sectors that will turn the true crude into the multitude of derivative products that drive its demand.

Although the resources and efforts devoted to the exploration and production of oil have remained very high for many years, the supply of oil has always been subject to a variety of supply-side shocks that have frequently resulted in rapid price changes. These rapid price changes—the *price volatility* of oil—have historically resulted from four categories of supply-side events:

- Massive new field discoveries like the East Texas discoveries of the 1930s flood crude oil markets, causing prices to plummet, in some cases below the cost of production.

- Producer collectives restrain quantities, as illustrated by OPEC's actions of the 1970s and 1980s. OPEC restricted the availability of crude oil in times of growing market demands and global political pressures, driving prices upwards.

- Sudden disruptive man-made forces such as the Arab-Israeli conflicts and the Persian Gulf War of 1991 pull large quantities of crude supplies from the market in a short time, driving prices up.

- Disruptive forces of nature, as illustrated by the hurricanes of 2006 that ravaged the US Gulf Coast, damaging both production facilities and platforms in the Gulf of Mexico, as well as refining and processing facilities all along the Gulf Coast. Again, the result was upward pressure on prices.

Any and all of these forces can have sudden and significant impacts on the price of oil. Thus, as global oil demand edges continually upwards, supply flows through a never-ending cycle of plenty and shortage.

Transportation costs

Getting the oil from well to refinery is an important part of the global value chain. Many of the world's largest crude-producing fields are distant from the refining infrastructure so critical to the crude's ultimate value. Often the land-locked isolated fields create the biggest challenges. An oil field producer in south central Russia, or Chad in central sub-Saharan Africa, or land-locked Bolivia in South America is confronted with a significant cost challenge in getting the crude to market. Again, the true value of crude can only be determined once the oil gets to the refining market.

The predominant transport method for decades, the construction of an oil pipeline, has proven cost-effective once two major hurdles are overcome: first, acquiring the political rights and alignments needed for pipeline transit; and second, raising the massive amount of capital needed for construction. The development of offshore production has actually been somewhat easier politically, although the capital costs are frighteningly large in their own way. The growth of offshore production, though, has actually allowed quick access of crude oil tankers to both the producing platforms and storage vessels, from which they can obtain their loads and move to refining facilities located on coastal geographies around the world.

Transportation costs have an enormous impact on the price paid to the producer at the well head. If market prices are relatively competitive for similar crudes delivered to any specific refining facility, the actual price paid to the producer is essentially "backed out" of the refining purchase price—for that specific grade of crude, including transportation cost.

$$Price_{Producer} = Price_{Crude\ refinery} - Transportation\ cost$$

For example, if the refinery price is \$40/bbl, but it takes a \$10/bbl transportation cost to get it to the refinery, the producer price—the return to the oil owner—is $30/bbl. Of course this is all within the framework of a dynamic marketplace in which price is changing daily, as well as the discount or premium associated with the supply and demand for any specific grade of crude.

The Price of Crude

The price of crude oil has exhibited major swings over the past 150 years, the age of its industrial and commercial development. As illustrated in figure 10–1, the general rise in price, and its commensurate volatility in the post-1970 period, represent a fundamentally new era in oil prices.

A short history of price

Graphics like figure 10–1 can be somewhat misleading when it comes to understanding the development of the markets for crude oil. Prior to World War I, there was minimal trading of crude oil outside of regional markets. The first recorded shipping of oil between countries, the shipment of crude from Pennsylvania to London on the sailing ship *Elizabeth Watts* in 1861, did indeed signal a new era in market development. But global trade in crude oil and the determination of a single price of crude was still very far off. In fact, as the following section will detail, there is still little evidence of a single global market price for crude oil.

Early prices, such as those shown in figure 10–1 in the 1860s and 1870s, were largely driven by the cost of transporting the crude to refining markets, not by the expansion of commercial demand for crude oil products or even the cost of crude production. During this era the cost of a wooden barrel often exceeded the value of the crude it contained. As a result, the development and expansion of oil pipelines was instrumental in changing the market drivers and dynamics for many years to come.

It should also be noted when reviewing crude price history that price volatility has always existed and is not just a post-1970 phenomena. Although prices were low from 1860 to 1970 by modern standards (e.g., $147/bbl in July 2008), the price changes experienced in the industry were large and often resulted in massive infrastructure and industrial changes on both the production and consumption sides of the market. For example, a simple average percentage change of annual average prices from 1860 to 2008 (data depicted in figure 10–1) is 8.4%; the same average for the 1860 to 1970 period is 5.7%. This, however, includes price changes such as +200% in 1863, –51% in 1878, and –45% in 1931.

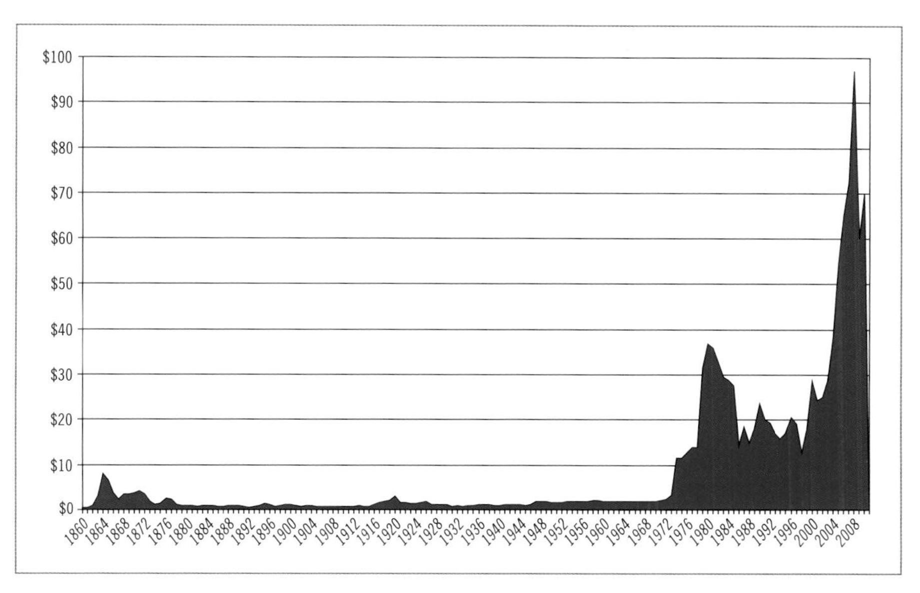

Figure 10–1. The price of oil, 1860–2010 (US$ per barrel)

Source: Annual average prices in US$ per barrel. Based on "BP Statistical Review of World Energy," June 2009. Value assumed for 2009 is $60/bbl.

The two crude price eras

The history of crude oil prices can be divided into two major eras: the pre-1970 and post-1970 periods. The pre-1970 period was dominated by major oil discoveries and developments (like the development of Pennsylvania and the Baku fields in the 1860s), and the major changes in market with the First and Second World Wars. The US Supreme Court's order to break up Standard Oil in 1911, the result of years of political turmoil over the control of access and pricing in oil markets in the US, was to have lasting impacts on the structure and identity of the major players in global markets for years to come.

Prior to World War II, crude markets were largely regional, with major new oil discoveries like that of East Texas in the 1930s affecting continental prices. The East Texas development itself was instrumental in the development of two different dimensions of crude oil production and marketing: first, the development of unitary production principles in order to increase the ultimate recoverable crude from large reservoirs, and second, the use of regulatory bodies (in this case the Texas Railroad Commission) to restrict production and preserve price. The second principle would prove instrumental in the development of OPEC in the 1960s.

In the post–World War II period, oil prices stabilized in the United States and globally. From 1950 to 1970 the price of oil remained within a very tight band between $1.70/bbl and $2.10/bbl. At these prices, profits were low, particularly to the major producing states—Mexico, Venezuela, Saudi Arabia, Iraq, Iran (Persia), and Kuwait. The producing states were largely restricted in their returns to the long-term contracts that the major oil companies were willing to pay. The producing states had not yet realized their own market power, which would eventually come through coordinated action. The majors, in turn, found much of their profit potential limited by heavy price regulation in major consumer markets like the United States and Great Britain. The major oil companies—the so-called Seven Sisters—balanced price with world production and the demand for refined products. The *Seven Sisters* was a reference first popularized by Italian oil tycoon Enrico Mattei to refer to the dominant global oil companies of the 1950s and 1960s: Exxon (Esso), Shell, BP, Gulf, Texaco, Mobil, Socal (Chevron), and sometimes an eighth, the Compagnie Francaise Des Pétroles (CFP-Total).

But change was in the wind. As illustrated in figure 10–2, pricing power began shifting from the Texas producers to OPEC in the 1970s. The 1983 statement by Saudi Arabia's oil minister Ahmed Azki Yamani that opened this chapter ("Oil is not an ordinary commodity like tea or coffee. Oil is a strategic commodity. Oil is too important a commodity to be left to the vagaries of the spot or the futures markets . . .") is indicative of the growing complexity of the role of crude oil markets in global economics and politics. At that time, those who owned the oil production or controlled access to the oil controlled the price. Oil was still not traded like other commodities such as copper or wheat.

OPEC exercised enormous pricing power in the 1970s and early 1980s and essentially used a fixed price system. In the fall of 1973, OPEC increased prices from $2.90 a barrel to $11.50 within a few months, precipitating the first global oil crisis. In 1980, OPEC capitalized on turmoil created by the Islamic revolution in Iran and the Iran-Iraq war and increased prices to $32/bbl. Sheik Yamani's warnings that high prices would lead to reduced demand fell on deaf ears. As it turned out, 1980 was probably the peak of OPEC's pricing power. After the very high prices of the early 1980s, demand declined and supply increased, leading to significant price declines. In addition, new oil producing players (both countries and companies) entered the market. After 1980, oil prices began a six-year decline to as low as $7/bbl, and of relevance for this chapter, oil began trading on futures markets.

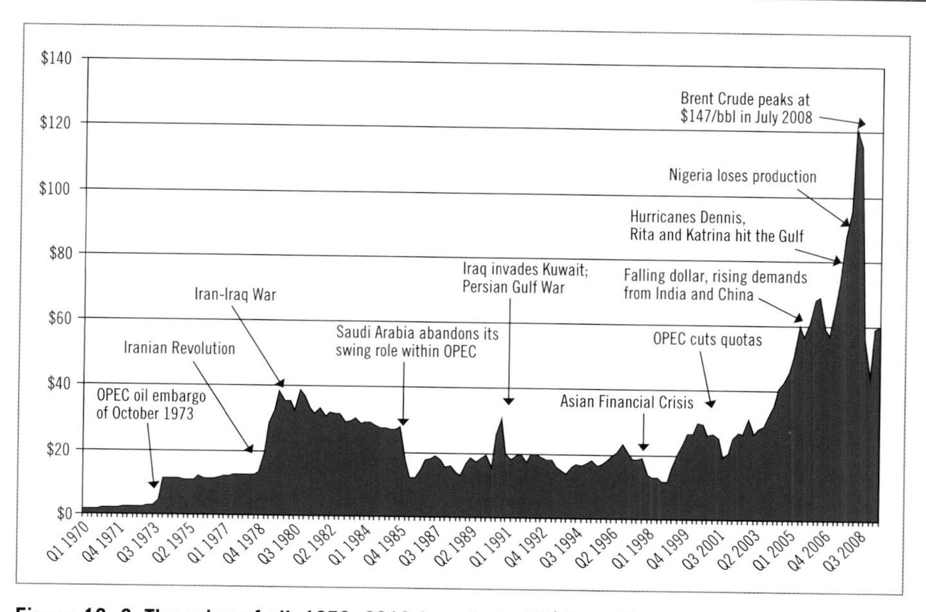

Figure 10–2. The price of oil, 1970–2010 (quarterly, US$/barrel)

Source: Price data, quarterly, from the International Monetary Fund, *International Financial Statistics*, imf.org.

In 1981, President Reagan abolished the remaining price and allocation controls on refined oil products and crude oil. The New York Mercantile Exchange (NYMEX) launched heating oil futures contracts in 1978, gasoline contracts in October 1981, and the crude oil futures contract in 1983. According to Michael Marks, a NYMEX director, when crude oil futures began, "There was a feeling that there was enough volatility and risk in the marketplace to create the support for a futures market."[2] In other words, what Sheikh Yamani said would never happen—did. In March 1983, OPEC announced the first official price cut in its 23-year history: a drop of $5 a barrel to a new posted benchmark price of $29 a barrel. This was just two weeks before the first futures contracts started trading.

According to Daniel Yergin, author of *The Prize: The Epic Quest for Oil, Money, and Power*, the futures contract undermined OPEC's price-setting power.[3] Pricing power shifted to the oil consumers. Evidence since 1983 suggests that OPEC has never regained its former power, and oil has become the world's most actively traded physical commodity (besides NYMEX, the other major energy futures exchange is London-based ICE Futures Europe). Without the ability to set prices arbitrarily, OPEC began using production controls to try to influence supply. However, even when OPEC seeks to control production, the challenge is to get the various members to agree to production cuts—and not cheat. Often, Saudi Arabia, the largest OPEC producer, will take the lead, a position that several decades ago Saudi Arabia said it would never take.

Is crude oil a commodity?

In contrast to the popular notion that there is one standard crude oil, and that "crude is crude" regardless of where it is produced, there are many different types of crude, with more emerging as new oil-producing areas come on stream. There are in fact roughly 160 different grades of crude oil traded internationally.[4] Table 10–1 provides an overview of the predominant crudes and their characteristics.

Table 10–1. Selected crude oils and key characteristics

Country	Crude Oil	API Gravity	Sulfur	Primary Loading Port
Canada	Cold Lake	21.2	3.70	Westridge Terminal
Chad	Doba	21.1	0.10	Kome Kribi 1 FSO
Dubai	Dubai	30.4	2.13	Fateh Marine Terminal, Dubai
Iran	Iran Heavy	31.0	1.70	Kharg Island
	Iran Light	33.8	1.35	Kharg Island
Malaysia	Tapis	44.0	0.08	Terengganu
Mexico	Isthmus	32.9	1.60	Pajaritos, Dos Bocas, Cayo Arcas, Salina Cruz
	Maya	22.1	3.50	Pajaritos, Dos Bocas, Cayo Arcas, Salina Cruz
Nigeria	Bonny Light	32.6	0.16	Bonny Terminal
Oman	Oman	33.0	1.14	Mina Al Fahal
Saudi Arabia	Arab Extra Light	39.4	1.09	Ras Tanura
	Arab Heavy	27.5	2.92	Ju'aymah, Ras Tanura
	Arab Light	32.8	1.97	Ju'aymah, King Fahd, Ras Tanura
	Arab Medium	30.2	2.59	Ju'aymah, Ras Tanura
	Arab Super Light	50.6	0.04	King Fahd
United Kingdom	Brent Blend	38.3	0.37	Sullom Voe
United States	West Texas Intermediate	39.6	0.24	Cushing, Midland
Venezuela	Petrozuata	19–25	2.65	TAEJ-Jose Platform

Source: Various company documents, EIA, *Oil & Gas Journal Data Book 2008*, Energy Intelligence.

Crude oil varies in characteristics, quality, production level, and market penetration. The two most important qualities of crude oil are *viscosity* (thickness or density) and sulfur content. Crude oils with lower density are called *light crude* and usually yield a higher proportion of more valuable final refined products, such as gasoline and other light petroleum products. The density or weight compared to water is measured by API gravity (American Petroleum Institute). As illustrated in figure 10–3, an API gravity below 10 is heavier than water and above 10 is lighter than water (as seen by oil slicks on an ocean surface). *Heavy crude* oils have a lower share of light hydrocarbons and require more complex refining processes to produce similar proportions of the more valuable products.[5] Although sulfur is a naturally occurring element in crude oil, it is an undesirable property. The greater the sulfur content, the more costly the refining process to remove it. Crude oils with high sulfur content are referred to as *sour crudes,* while those with low sulfur content are known as *sweet crudes.*

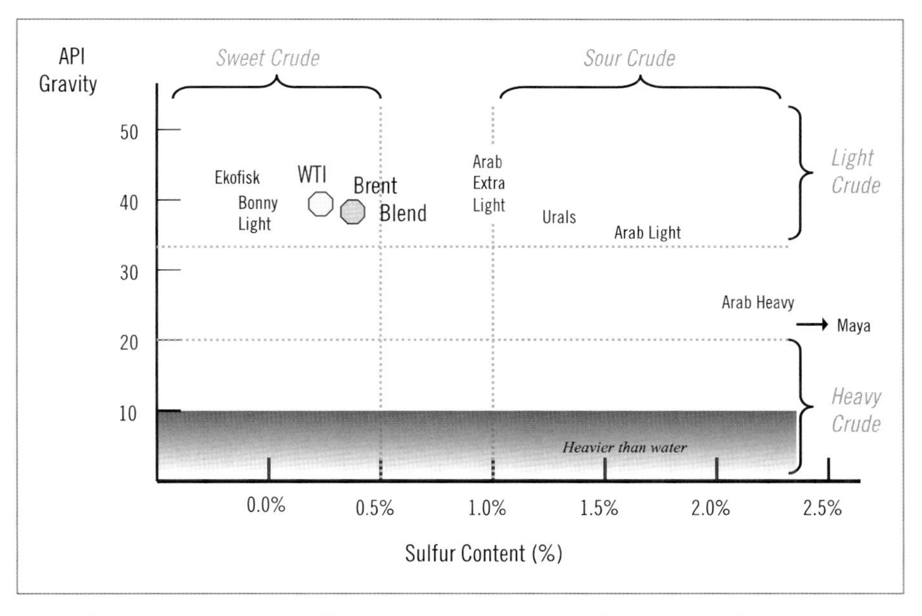

Figure 10–3. API gravity and sulfur content. New crude oil discoveries in the past two decades are increasingly sour and heavy.

Benchmark crude oils

A variety of crude oils serve as benchmarks, or *oil markers*, for trading and pricing: West Texas Intermediate or WTI, Brent Blend (North Sea), Dubai (also known as Fateh), Tapis (Singapore), Bonnie Light (Nigeria), Maya (Mexico), and Isthmus (Mexico). The two most important oil markers are WTI and Brent Blend. WTI crude oil is a high quality oil that is excellent for refining gasoline. WTI has an API gravity of 39.6°, making it a light crude, and contains about 0.24% sulfur, making it a sweet crude. This combination of characteristics, combined with its location in the United States, make WTI an ideal crude oil for US refining.

Most WTI crude oil gets refined in the Midwest region of the country, with some refined in the Gulf Coast region. Although the production of WTI crude is declining, it remains the major benchmark for crude oil in the Americas. Figures 10–4 and 10–5 illustrate how crude prices differ across grades.

Brent Blend crude futures began trading in 1988, and today about two-thirds of the world's crude sells at prices tied to Brent. The name *Brent* comes from Shell, which named all of its fields after birds (in this case the Brent goose). Brent crude is a combination of crudes from 15 different oil fields in the Brent and Ninian systems located in the North Sea. Brent's API gravity is 38.3° (making it a light crude, but not quite as light as WTI), and its sulfur content is about 0.37% (a sweet crude, but slightly less sweet than WTI). Brent Blend is ideal for making gasoline and middle distillates, both of which are consumed in large quantities in Northwest Europe, where Brent Blend crude oil is typically refined. However, depending on the market pricing, Brent and WTI are also exported. Although the production of Brent, like WTI, is on the decline, it remains the major benchmark for other crude oils in Europe or Africa. ICE Futures has also created the ICE Middle East Sour Crude futures contract. In 1999, the Brent Weighted Average (Bwave), a daily price averaging the prices of all trades on a given day, was introduced by ICE Futures Europe. The Bwave became a reference price in its own right, with Saudi Arabia, Kuwait, and Iran pricing against Bwave.

OPEC also has its own benchmark—the OPEC reference basket—constructed as a weighted average from its own producing members. The basket is currently composed of the following oils: Saharan Blend (Algeria), Girassol (Angola), Oriente (Ecuador), Minas (Indonesia), Iran Heavy (Islamic Republic of Iran), Basra Light (Iraq), Kuwait Export (Kuwait), Es Sider (Libya), Bonny Light (Nigeria), Qatar Marine (Qatar), Arab Light (Saudi Arabia), Murban (United Arab Emirates), and BCF 17 (Venezuela). OPEC ties its production management activity to the goal of maintaining the OPEC reference basket price within a predetermined range.

In Asia there is not a specific futures exchange or benchmark for crude pricing. The pricing mechanism for a crude like Tapis from Malaysia, a marker for light sweet crudes in the region, is based on an independent panel approach where producers, refiners, and traders are asked for information on Tapis crude trades.

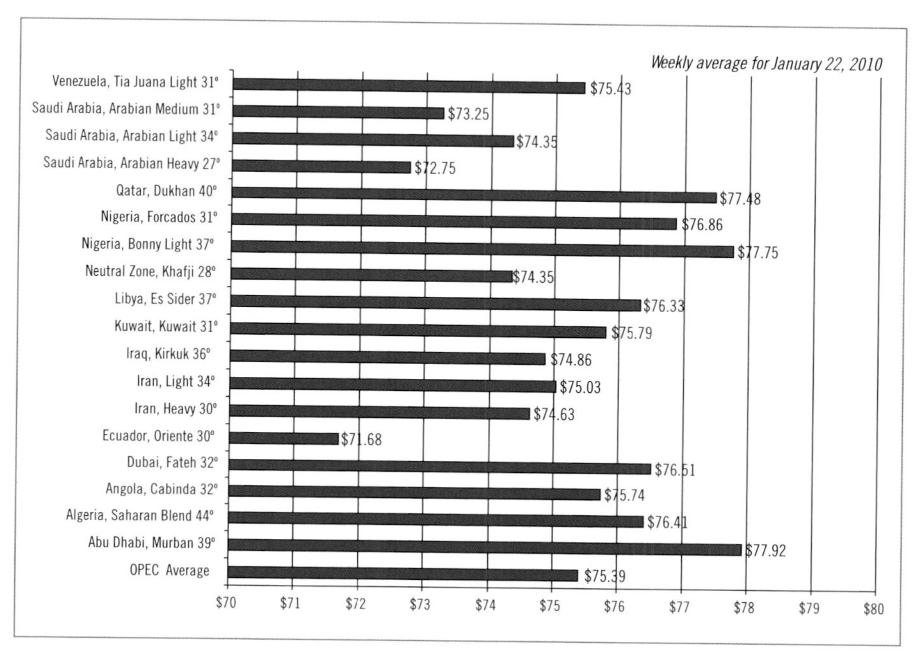

Figure 10–4. Spot prices for OPEC crude (US$/per barrel)

Source: Energy Information Administration.

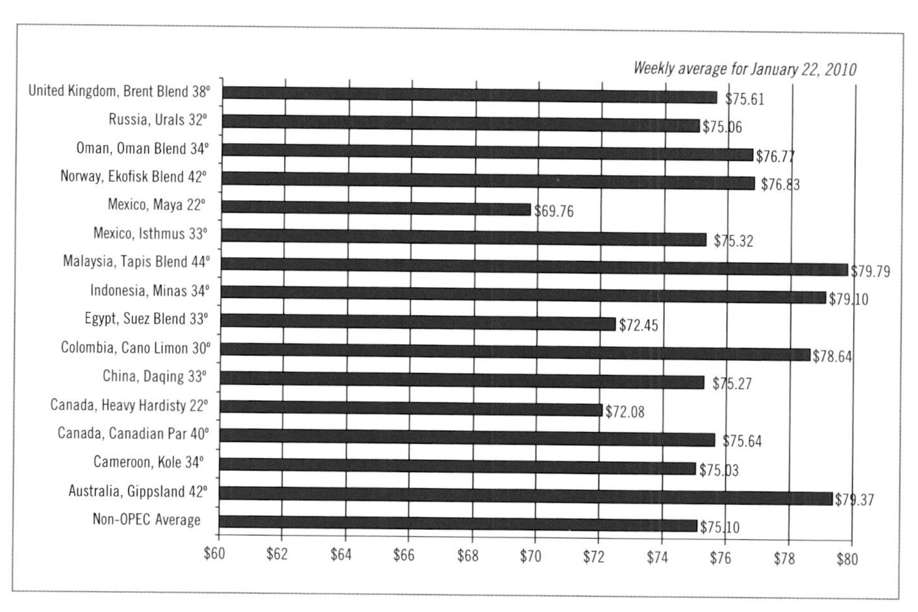

Figure 10–5. Spot prices for non-OPEC crude (US$/per barrel)

Source: Energy Information Administration.

Crude benchmarks and pricing

WTI and Brent crudes play a vital role in pricing. Prices for other crude oils are usually priced as a differential to WTI or Brent. WTI and Brent are sweet/light crudes, and generally trade at a premium to sour/heavy crude varieties (Dubai). WTI often enjoys a \$5 to \$6 per barrel premium to the OPEC Basket price and about \$1 to \$2 per barrel premium to Brent (figure 10–6), although on a daily basis the pricing relationships between these can vary greatly.

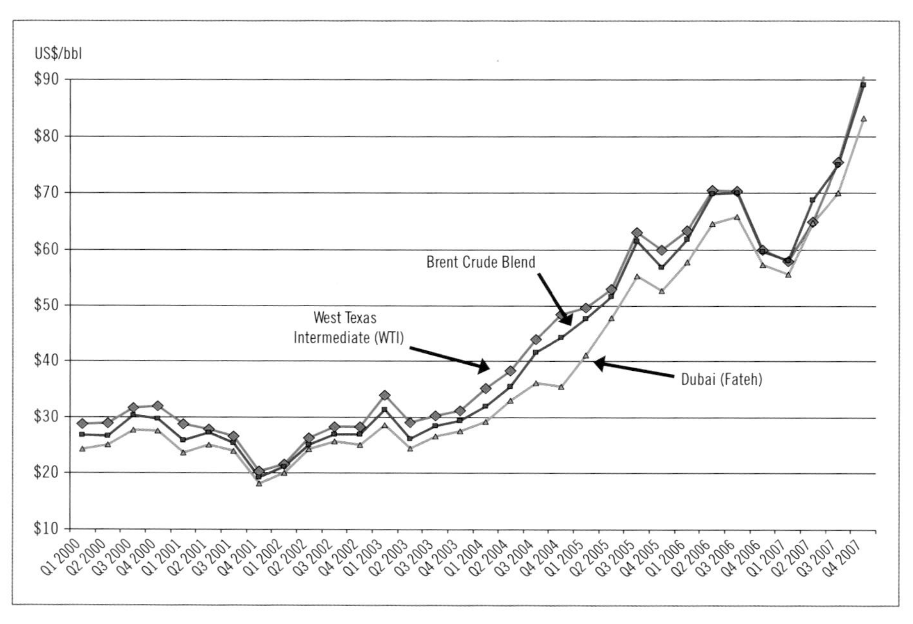

Figure 10–6. Crude benchmark prices (quarterly, 2000–2007)
Source: *International Financial Statistics*, monthly, International Monetary Fund, www.imf.org.

The differential in crude prices is also impacted by the availability of refining capacity. Because refining margins over the past decade have been quite low, refiners have been reluctant to invest in upgrades that would allow them to process heavier crudes such as those from Venezuela. This considerably widened the differential between light/sweet and heavy/sour crude oils, commonly called the *clean-dirty spread*. Most of the major new crude discoveries and developments in recent years are heavier and dirtier, and as a result, the clean-dirty spread will likely be more and more noticeable in crude markets.

In recent years, many have questioned the continued use of WTI as a benchmark. Because of unique supply and demand situations, WTI can occasionally be priced lower than Brent. In April 2007, the discount of WTI to Brent reached more than $6.50.[6] According to one report, "The standard barrel of oil no longer has a straightforward price, and some oil-market experts say price disparities are set to persist or possibly worsen in the near term."[7] Whereas WTI makes up only a mere fraction of global oil output, it is now estimated that Brent crude is used as the price benchmark for more than two-thirds of international sales. However, changing a benchmark is not easy. Previous attempts to introduce Middle Eastern crude oil benchmarks have failed because they allowed traders to speculate but did not involve physical delivery of crude.

Crude Oil Prices and Transactions

The price of crude oil is determined by many different factors and is the outcome of thousands of transactions that occur every day around the world involving crude producers, oil traders, refiners, speculators, OPEC, and many others players. With the end of the old fixed-price system for crude, oil prices became both more volatile and more closely tied to supply and demand for finished products. Fear of the future can also have a powerful impact on price. When supplies are constrained, short-term price spikes can be huge, especially in reaction to worst-case political and economic scenarios created by wars, terrorists, disasters, or other unpredictable events. When supplies are plentiful and investors are unloading financial positions, prices can fall rapidly, as they did in mid- to late 2008.

Crude oil transactions

The sale of crude oil occurs in three different types of transactions: spot transactions, futures markets, and contract arrangements.[8]

Spot transactions. A *spot market* is a market where goods are sold for cash and are immediately deliverable. If the goods cannot be immediately deliverable, such as natural gas produced in Nigeria, a spot market cannot develop (in the US, pipeline gas is actively traded and a spot market exists). A spot transaction is an agreement to sell or buy a shipment of oil under a price agreed upon at the time of the arrangement. In the crude oil market, spot contracts typically involve delivery of crude over the coming month, e.g., a contract signed in June for delivery in July. For a producer or refiner of crude oil, the spot market is used to balance supply and demand. When a company temporarily has too much supply for its own needs, it may offer excess crude for sale in the spot market. Likewise, if a refiner needs additional volumes to meet a demand spike or because supply is unexpectedly curtailed, it will purchase oil on a cargo-by-cargo, shipment-by-shipment basis. In recent years, merchant refiners not tied to crude oil production networks, such as Petroplus in Europe, have depended on viable spot markets to access regular crude supplies. These independent refiners manufacture products not for their own marketing networks but to sell via third-party transactions to the highest bidder.

Spot market prices send a clear signal about the current balance between supply and demand. Rising prices indicate that more supply is needed, and falling prices indicate that there is too much supply for the prevailing demand level. There are a number of regional spot markets, such as Rotterdam/Northwest Europe, New York Harbor/US Northeast, Chicago/US Midwest, Singapore/South East Asia, and the US Gulf Coast. The evolution of a regional market into a pricing center is largely based on logistics. For a regional market to be a pricing center, there must be a ready supply, transportation choices, storage facilities, and many buyers and sellers. Spot prices are reported for transactions in the different markets, and prices in spot markets are quite transparent in that they are reported by a number of sources and are available in a variety of media. While some of the most active spot markets offer deals on supplies that will be available in the future (a forward physical market), most focus on immediate delivery of readily available volumes.

Futures contracts. A *futures contract* is a promise to deliver a quantity of a standardized commodity at a specified place, price, and time in the future. In practice, oil is seldom actually delivered under a futures contract. Futures exchanges such as the New York Mercantile Exchange (NYMEX) record the pairings of buyers and sellers and report the transaction prices. Electronic services then report these prices with minimal lag. Prices are available throughout the day from the exchanges via the Internet, are published in specialty trade publications and daily newspapers throughout the country, and are reported on a weekly basis by the US Energy Information Administration. The ready availability of the reported prices enhances price transparency, which is necessary for markets to function efficiently.

NYMEX describes the futures business as follows:[9]

> *The futures markets help businesses manage their price risk by providing a means of hedging; matching buyers and sellers of a commodity with parties who are either more able and willing to bear market risk, or who have inverse risk profiles. A crude oil producer, for example, might sell a crude oil futures contract to protect his sales price while a petroleum refiner might buy crude oil futures to lock in his raw materials cost.*

Because futures are traded on exchanges that are anonymous public auctions with prices displayed for all to see, the markets perform the important function of price discovery. The prices displayed on the floor of the Exchange and on its electronic platforms are disseminated to information vendors and news services worldwide. They reflect the marketplace's collective valuation of how much buyers are willing to pay and how much sellers are willing to accept. The diverse views of many market participants are distilled to a single price.

According to ICE Futures Europe, futures trading in "a contract parallel to the physical market" gives the oil industry "the opportunity to separate its pricing from supply arrangements."[10] Futures markets provide a projection of expected prices in the future and are a mechanism designed to distribute risk among participants on different sides.

Futures prices are not price predictions, but are the collective current opinion of the marketplace of where prices appear to be heading. That opinion, and the direction of prices, can change in an instant, which makes trading these markets so challenging and potentially rewarding.[11]

Traditionally, futures markets serve three purposes. First, they provide information about future expectations of oil supply and demand conditions. Second, future expectations are transparent and therefore provide a very efficient vehicle for price discovery. Third, crude oil producers, marketers, refiners, and others are able to use futures contracts to manage the risks resulting from the increasing participation of investors and speculators without a commercial interest in the petroleum industry (i.e., no capacity to produce, refine, store, or sell physical volumes of crude or petroleum). It is this third function that has come under heated debate in recent years, as more and more of the futures market activity is dominated by those market participants without a commercial interest in oil—other than speculation.

The prices paid on futures markets enhance the availability of price information for all parts of the oil industry value chain. In times of tight supply, premiums for high-quality crude will rise and gradually pull up the marker crude price. In times of surplus supply, a reduced premium or even a discount will drag down the marker crude price. Big changes, announcements, or events like wars, hurricanes, and refinery shutdowns can significantly influence crude supply levels and may result in steep changes in the prices of crude oil.

How futures work

Futures markets allow participants to lock in the prevailing price for future deliveries, such as heating oil prices for the winter heating season. Such a strategy, called a *hedge*, involves a series of transactions, offsetting profits or losses on a futures transaction against losses or profits on the physical purchase or sale of oil. By limiting the uncertainty over future costs, the hedge allows companies or consumers to make other choices. For example, Southwest Airlines has reported in SEC filings that "the Company believes there is significant risk in not hedging against the possibility of such fuel price increases." In 2007, Southwest's hedging program generated a reduction to fuel expense of $686 million. Unfortunately, hedging can be very costly in a falling price environment. Falling oil prices in 2008 required Southwest to record a third quarter fuel hedging charge of $247 million for an unrealized loss on hedging contracts.[12]

The price discovery and risk management capabilities of futures contracts are obvious when working through some basic numerical examples. Table 10–2 provides NYMEX quotes on the crude oil futures on December 31, 2008. The table illustrates a variety of near-term contract maturities—beginning with the February 2009 contract. (The actual maturity of a futures contract on the NYMEX occurs at the conclusion of the third business day preceding the 25th day of the month preceding the maturity month listed. For example the February 2009 contract would actually mature at the close of trading on Wednesday, January 21, 2009.) "Last" refers to the last price at which a trade was executed during the trading period, with "Open High" and "Open Low" referring to the high and low extremes seen since trading began on that contract this day, with the "High" and "Low" values being the lifetime high and low extreme values for the specific maturities. "Most Recent Settle" is the closing price from the previous trading day.

Table 10–2. NYMEX oil futures (12/31/08)

	Last	Open High	Open Low	High	Low	Most Recent Settle	Change
Feb. 2009	40.86	39.17	39.17	40.98	36.94	39.03	1.83
Mar. 2009	44.70	43.09	43.09	44.84	40.89	42.76	1.94
April 2009	46.75	45.04	45.04	46.92	43.19	44.80	1.95
May 2009	47.94	46.23	46.23	48.33	45.12	46.33	1.61
June 2009	49.50	47.67	47.67	49.70	46.36	47.67	1.83
July 2009	50.36	48.82	48.82	50.50	47.61	48.91	1.45

Source: NYMEX.

The price discovery use of futures is apparent from the "Most Recent Settle" prices posted for the various future maturities listed. Simply put, the market is now expecting the spot price of WTI to be at \$39.03/bbl in February 2009, \$42.76/bbl in March 2009, etc. The futures market provides an insight into what many market participants believed the future spot price to be on the basis of all their various constituent opinions and information. And, for what it is worth, the market provides a "price curve" for the coming future.

The risk management use of the futures market (or speculative use, whichever applies) can be seen by following an individual agent's actions using these prices. Assume Odessa Oil believes that the price of oil will rise in the coming months, and believes that the WTI spot price will be higher than the "Last" price of \$40.86/bbl listed for February 2009. Odessa could buy a February 2009 contract. This contract, termed a *long position*, would guarantee Odessa the ability to purchase 1,000 barrels of WTI at a price of \$40.86/bbl (assuming it can contract at the same price as the most recent trade). If Odessa's expectations prove correct, and the actual spot price at contract maturity is higher, for example \$42.50/bbl, then the final contract value at maturity would be calculated as follows:

$$\begin{aligned} \textit{Value} &= \textit{Contract principal} \times (\textit{Spot price} - \textit{Futures price}) \\ &= 1{,}000 \times (\$42.50 - \$40.86) = \$1{,}640 \end{aligned}$$

This value can be seen as either assuring Odessa of purchasing crude oil at the specific contract price (in this case \$40.86/bbl) or of allowing it to profit from an expectation over the future price of oil.

Alternatively, if Odessa believed that oil prices would fall over the coming months, it could sell a February 2009 contract, creating a *short position*. If Odessa's expectations proved correct, and the spot WTI price in February 2009 was \$36.00/bbl, the final contract value at maturity would be:

$$\begin{aligned} \textit{Value} &= \textit{Contract principal} \times (\textit{Spot price} - \textit{Futures price}) \\ &= -1{,}000 \times (\$36.00 - \$40.86) = \$4{,}860 \end{aligned}$$

This value can be seen as either assuring Odessa of physically selling crude oil at the specific contract price (in this case \$40.86/bbl) or of allowing it to profit from an expectation over the future price of oil.

NYMEX futures contracts are traded for each month for 18 months in the future (trading volumes are low for the most distant months). The core energy futures traded on the NYMEX stipulate physical delivery (less than 1% of trading volumes involve delivery). Prices of futures contracts are connected to prices in the physical market because futures positions that are not closed out will lead to either delivery or receipt. Thus, the closing futures price for any given month must equal the physical price at the time trading in the futures contract ends. With delivery, the futures price effectively becomes a physical price at the time the futures contract matures. For example, the closing futures price for delivery in June must equal the spot price for oil in June. If the prices differed, a trader would buy in the market in which the price is lower and immediately sell it into the market where the price is higher and earn a profit.[13]

The futures trading process provides transparent price information to firms that can use this information. When the prices of futures contracts with early delivery dates exceed those with later delivery dates, the market consensus is for future prices to fall. If prices are expected to fall, there is a financial incentive to reduce inventory today. When prices of futures contracts with early delivery dates are lower than those with later delivery dates, the market consensus is for prices to rise in the future. With rising prices there is an incentive to build inventories as long as higher futures prices can cover the cost of storage. Thus, futures prices provide information about expected future supply and demand conditions that producers and consumers can act on today. These actions shift the supply of crude oil from periods of lower prices to periods where crude oil prices are expected to be higher.

Futures markets also create an opportunity for speculation. In 2008, *speculators*, traders who seek to profit through the assumption of risk, were widely blamed for the huge run-up in oil prices (see *Industry Insight: Was There an Oil Bubble in 2007 and 2008?*). While speculators play a role in the pricing equation, speculation is only one of the myriad factors that drive crude prices. According to one analyst, without speculators, it is likely that energy prices would be even higher due to higher costs associated with higher risk mitigation.[14]

Over the past few years, the number of oil futures contracts outstanding tripled on the NYMEX as hedge funds and other institutional investors got involved in trading. Daily trading volumes of WTI contracts are about 700,000 contracts and have been as high as 1.2 million contracts a day. (One contract represents 1,000 barrels of oil.) The widely reported NYMEX futures price for crude oil provides a market-determined value of a futures contract to either buy or sell 1,000 barrels of WTI or some other light, sweet crude oil at a specified time. The NYMEX futures prices usually track within cents of the WTI spot price. On the settlement date or the expiry of futures contracts, settlement can be carried out in two ways: through the actual delivery of oil into a predefined location or through a cash settlement. WTI futures contracts involve the physical delivery of oil in Cushing, Oklahoma, thrusting this small town into the middle of the global oil business (see *Industry Insight: Cushing, Oklahoma*). ICE Brent futures contracts are somewhat different from WTI futures. Brent futures are deliverable contracts with an option to settle with cash. This enables companies to take delivery of physical crude supplies if they wish, or more commonly, open positions can be cash settled after expiry against a physical price index.

Industry Insight: Was There an Oil Bubble in 2007 and 2008?

In the last decade there have been many new entrants in oil trading, with the vast majority engaged in financial trading and pure speculative behavior. As oil prices soared to almost $150, many people insisted that speculators, and especially new market entrants, had created a bubble. In a May 12, 2008 *New York Times* article entitled "The Oil Nonbubble," Nobel Laureate Paul Krugman argued there is no evidence of a speculative oil bubble because if speculators are driving up the price of oil above the price "justified by fundamentals," then a market adjustment would occur where "drivers would cut back on their driving; homeowners would turn down their thermostats; owners of marginal oil wells would put them back into production . . . [and the resulting] excess supply would . . . drive prices back down. . . . This tells us that the rise in oil prices isn't the result of runaway speculation; it's the result of fundamental factors, mainly the growing difficulty of finding oil and the rapid growth of emerging economies like China." He concluded by saying, "A realistic view of what's happened over the past few years suggests that we're heading into an era of increasingly scarce, costly oil."*

In contrast to Krugman's arguments, on June 10, 2008, OPEC President Chakib Khelil maintained that the price of oil would be $70/bbl if the element of speculation was taken out of the market. He said, "In terms of market fundamentals, there is no problem with supply and demand. Prices have to do more with this speculative bubble, which is due to the depreciation of the value of the dollar and geopolitical tensions, which are expected to increase in the future."**

So, was there a bubble? By November 2008, oil prices had fallen sharply and rather than talking about bubbles, OPEC was cutting production to stop prices from falling.

---

* Paul Krugman, "The Oil Nonbubble," *New York Times*, May 12, 2008. ** Lies Sahar, "OPEC President Khelil Puts Oil Price at $70/b Without Speculation; Disagrees With US' Paulson on Impact of Weak US Dollar," *Platts Oilgram News*, June 11, 2008, Vol. 86 No. 114, p. 13.

Industry Insight: Cushing, Oklahoma

Cushing, Oklahoma, population 8,500, including 1,000 prison inmates, occupies an interesting position in the global oil industry. When NYMEX created the first oil futures contracts, Cushing was designated the official delivery point for WTI. A century ago, when oil was discovered, Cushing was one of the most famous frontier towns in the United States. Production peaked in 1915 at 305,000 b/d. By the 1940s, the wells had almost run dry, and oil began to flow into Cushing rather than out. Cushing became a major storage and transit hub for oil from around the world. Cushing is connected to a network of pipelines, including the Seaway Crude Pipeline from the Texas Gulf Coast, and has enormous storage tanks, the largest holding 575,000 barrels of oil. At any point in time, Cushing holds about 5%–10% of US oil inventory. Signs in Cushing, a town described as home to a shabby collection of junk stores, a medium-security prison, and a handful of barbecue restaurants, proclaim the town as the "Pipeline Crossroads of the World."* Cushing has laboratories to check the purity and metal content of crude to ensure it meets various grades such as WTI or Western Canadian Select. Some customers ask for different crudes to be blended to a particular formula.

---

* Andrew Clark, "Oil: Business Takes a Turn for the Better at the World's Pipeline Crossroads: The Lifeblood of America Is Revitalising a Sleepy Oklahoma Town," *Guardian*, August 28, 2008, p. 29.

There are some obvious differences between physical traders who expect to take delivery of oil and paper traders. A paper trader in financial futures does not incur large risks because the deal's downside can be hedged and taking physical delivery is never a concern. A physical trader takes on the risk that crude cannot be sold at the expected price. For example, assume a trader with a forward position is considering the purchase of a consignment of crude from the Gulf. If the trader cannot sell the cargo in the course of the 30-day journey to a refinery in Europe, the tanker may have to be diverted elsewhere or the cargo unloaded into storage. These factors place a ceiling on the price a trader will bid for the cargo.

When the futures price is above the physical price plus the cost of storage and transportation, the market is said to be in full-carry *contango*. When this happens, physical traders can build stocks, avoid price risk, and hold oil now to sell later. A contango situation usually indicates that oil-market players expect crude supplies to be less scarce in the short term than they will be in the future. When the market is in contango, "You tend to operate at the top of your tanks."[15] In other words, a refiner might be better off investing in oil storage capacity rather than investing in facility upgrades. *Backwardation*, the opposite of contango, occurs when near-term prices are higher than long-term contracts.

Until fairly recently, most of the Cushing storage capacity was used by oil producers and refiners. More recently, capacity has been acquired by oil pipeline companies and oil traders. Most of the storage tanks are owned by four firms: BP, Enbridge Energy Partners (an affiliate of Canada's Enbridge Inc.), Plains All American Pipeline LP, and SemGroup Energy Partners LP, which declared bankruptcy in July 2008. Much of this capacity is leased to other firms, including Morgan Stanley and other financial firms.

In April 2007, a Lehman Brothers report suggested that WTI could no longer be a gauge for the global oil industry because of various issues, including a shortage of storage capacity in Cushing. By October 2007, millions of barrels of crude had been sold off and Cushing's inventory had fallen by 25%, leading the *Wall Street Journal* to say:

> *The reasons Cushing's crude has been disappearing are surprisingly complex, and shed light on the growing involvement of speculators in the global oil market. Tanks are emptying partly because producers have been straining to keep up with demand. But investment banks and other financial firms also played a part by abruptly shifting their oil-trading strategies this summer. Even the credit crunch sparked by the sub-prime mortgage fiasco had an effect. Until mid-July, unprecedented conditions in the oil market had given oil companies and speculators alike a financial incentive to sock away oil in storage tanks for sale later. Then, almost overnight, it became more lucrative to sell oil immediately, and in short order, the cushion of stored oil shrank.*[16]

Contract transactions. Oil exploration and production (E&P) companies need to find customers for the oil they produce. Although oil is traded like a commodity, and oil producers are price takers, there are many different types of crude and oil must be sold and transported to customers before its value can be unlocked. The purchasers of physical crude are mainly refineries who then process the oil into saleable products. Crude oil marketing and trading is done by a variety of companies, including both the oil companies themselves and specialist firms (see *Industry Insight: The Gunvor Group*). Some of the oil companies, such as BP, Chevron, and StatoilHydro, are major oil traders and sell their services to many smaller E&P firms. For example, StatoilHydro trades above 2 million barrels of crude oil and condensate (light oil) per day, making it the world's third largest net seller of crude oil. The company conducts its trading operations around the clock from offices in Stavanger (Norway), London, Stamford (Connecticut), and Singapore.

Industry Insight: The Gunvor Group

One of the largest oil traders is a secretive firm called Gunvor Group. *The Economist* decribed Gunvor's ownership structure as a "Chinese puzzle."* Gunvor's main activities involve trading, transporting, and storing oil. The company has a diverse range of other activities, including blending fuel oil and gasoline in Amsterdam, Singapore, and other locations, maintaining oil terminals and port facilities, and assessing exploration opportunities. Gunvor's wholly owned shipping company, Clearlake Shipping, is one of the biggest spot charterers in the world. The company has trading activities in Europe, Middle East, Asia, Africa, and South America. Gunvor is headquartered in Amsterdam and was founded by Gennady Timchenko, rumored to be a close friend of Vladimir Putin. In 2000, Gunvor was largely unknown; in 2008, the firm was expected to move about $70 billion of oil, making it the world's fourth-largest independent trader behind Glencore International AG, Vitol SA, and Trafigura Beheer BV. The growth of Gunvor coincided with Russia's increased state control over the oil industry, which began in 2003. The firm opened offices in Singapore, Nigeria, and Amsterdam and bought a Finnish shipping company. Mr. Timchenko bought a mansion in Switzerland overlooking Lake Genva. Both Mr. Timchenko and Gunvor co-founder Torbjorn Tornqvist vigorously denied that their firm's success had anything to do with connections to the Kremlin.

---

*"A Special Report on Russia," *The Economist*, November 29, 2008, p. 10. **A. Higgins, G. Chazan, and A. Cullison, "The Middleman: Secretive Associate of Putin Emerges as Czar of Russian Oil Trading, in First Interview, Timchenko Denies Ties; Rivals Face Hurdles," *Wall Street Journal*, June 11, 2008, p. A1.

The transactions for most of the oil that is sold and delivered are contract arrangements. Both spot markets and futures markets provide critical price information for contract markets. Prior to the emergence of futures markets, contracts covered almost all oil transactions, with terms that were seldom changed. Even the pricing term of the contract was seldom reexamined because prices at all levels of the oil market were relatively stable. But as noted earlier, in the 1980s the game changed. The constant prices in most contracts were higher than purchasers were willing to pay in the abundantly supplied open market. As purchasers rebelled and abandoned contracts, oil producers were forced to create pricing terms tied to market indicators such as the spot or futures markets. Thus, while most oil is sold by contract, its price varies with market conditions.

Most of the oil traded internationally is priced by formula—a base price derived from a market indicator, plus or minus a quality adjustment (i.e., the differential). According to one analyst:

> *Crude oil producers can lose substantial revenues because they do not achieve maximum value from their crude oil sales. Producers, traders, and refiners must be fully aware of key issues that can influence the market and price differentials for crude oil. . . . It is increasingly difficult for producers to understand the key factors affecting price differentials. It is important to understand how much crude oil price relationships are affected by crude supply changes vs. refining value changes due to product price relationships. . . . A crude's refining value is the value a refiner expects to realize for the products, less operating costs, for processing that crude. Subtracting the crude oil cost from the refining value gives a measure of profit or refining margin associated with processing that crude.*[17]

A common pricing method is a base price derived from a spot price published by a particular source or publication. For crude oil sold into the US Gulf Coast, for instance, the base would commonly be the WTI price. Credit terms also affect the realized price. In the United States, some domestic crude is sold at a posted price. *Posted prices* are named for the sheet that was literally posted in a producing field. Posted prices are established by the buyers, usually refiners, but sometimes firms that aggregate supply called *gatherers*. Posted prices generally apply to a crude oil "stream" of standardized quality with quality adjustments where the oil varies from the standard. In decades past, posted prices remained relatively stable even while spot prices fluctuated. Today, posted prices more commonly reflect market conditions quickly. Companies may also add a temporary premium to a posted price to account for transient market conditions.

As an example of how prices are communicated to global customers, consider the case of Saudi Arabia, the world's largest oil producer. Saudi Arabia sells most of its crude production to a set of customers based on monthly allocations of volume. *Business Week* described the pricing process for crude coming out of Saudi Arabia as follows:

Every month executives from Saudi Aramco, Saudi Arabia's national oil company, ring up the likes of ExxonMobil and Royal Dutch Shell, sounding them out about the oil they need and the price they would be willing to pay. The Saudis crunch the numbers, set a price, then call the global customers back to see how much they'd be willing to pay. By the 10th of the following month, customers—there are about 80 in total—are told how much crude they actually get.[18]

Some so-called challenged crudes, heavy crudes or those with high sulfur content or high acidity, may be difficult to sell (see *Industry Insight: Chad Oil Sales*). In times of high inventories and declining demand, price differentials between challenged crudes and the benchmark crudes will increase. Occasionally, tankers are loaded and the crude producers have no buyers. This situation, called a *distressed cargo*, could occur for various reasons, including: the oil is low quality and few refiners exist to process it, a refinery has shut down, or there is damage to the receiving/storage infrastructure. Distressed cargo may result in significantly higher costs and lower revenues, especially if the crude must be heavily discounted in order to sell it.

Given the demand for crude supplies of all qualities, there are a number of major refinery investments underway that are designed to allow the processing of heavy crudes. One of these, Reliance Industries' Jamnagar, India facility, will be the largest refinery complex in the world when completed. Chapter 12 covers petroleum refining in detail.

Industry Insight: Chad Oil Sales

Oil is produced in Chad by the ExxonMobil/Chevron/Petronas consortium. Each of the consortium members markets its share of the crude oil to refineries and oil blenders. The market price for a shipment of Chadian crude is negotiated between the seller—one of the consortium members—and potential buyers approximately two months before oil is shipped. Because Chadian oil, known as Doba Blend, is a heavy crude with high acidity (measured by the total acid number or TAN), it trades at a steep discount to the benchmark crudes. The acidic nature of the crude can corrode refinery pipes, and without sophisticated equipment, a refinery can only turn about one-fourth of a barrel of Chad's Doba Blend oil into high value light products such as gasoline, jet fuel, and diesel. In contrast, even without the sophisticated equipment, at least two-thirds of a Brent barrel of crude oil can be refined into high-demand products.

The Doba crude discount in 2004 and 2005 was about $8–$10/bbl to Brent crude and has been almost as high as $20/bbl. In addition, the producers incurred a transportation cost of about $10/bbl. The poor quality of the Doba crude limits the number of potential customers because only a small number of refineries in the world have the ability to process it. As a result, the consortium members purchased much of the crude for themselves. Chevron, for example, retrofitted a refinery in Pembroke, Wales in 2004 to handle Doba crude at a cost of $12 million.** Safeguards were built into the Chad agreements to ensure that consortium members do not sell to their own refineries at below market prices. To guard against that potential, sales to nonconsortium refineries are used to calculate the average price per quarter. This quarterly average market price is then submitted to Chad's Ministry of Petroleum for approval and used in the calculation of Chad's royalties (figure 10–7).

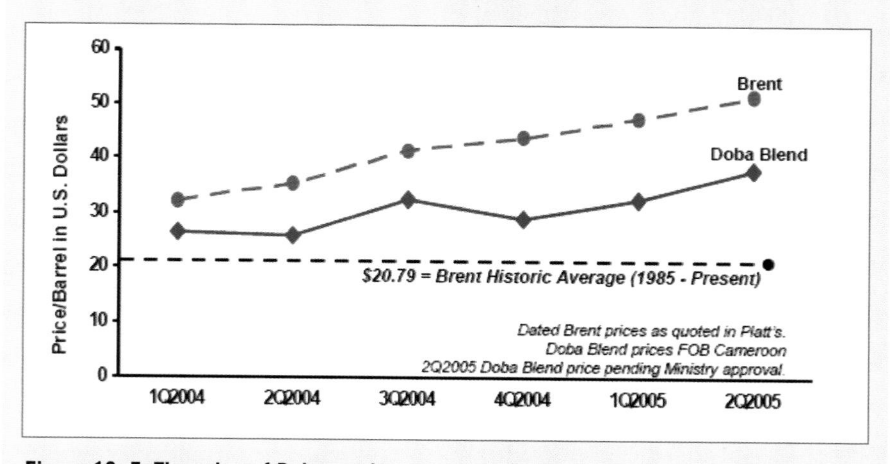

Figure 10–7. The price of Doba crude

Source: Chad/Cameroon Development Project, Project Update No. 18, Mid-Year Report 2006, p. 71.

---

* Price graphic from Chad/Cameroon Development Project, *Project Update No. 18, Mid-Year Report 2006*, p. 71. ** Deborah Kelly, "Doba Finds a Home," *International Oil Daily*, June 25, 2004.

Oil shipments and inventories

Another important factor associated with crude supply and demand is the state of oil inventories and shipments. Specifically, how much crude oil inventory exists in the world, and how much is in shipment at any point in time? In theory, greater inventories of crude oil should exert downward pressure on oil prices (see *Industry Insight: A Bull Market for Oil?*). This inventory could be in a ship, a storage tank, a pipeline, on a train, or in a truck. The problem with the theory is that there is no official measurement of worldwide oil production and supply. Many major oil-producing countries closely guard information about their output as state secrets and regard leaks as breaches of national security. Every month, OPEC publishes an estimate of how much oil its members produced in the previous month. However, because this information is an estimate and does not come from its members, OPEC struggles to ensure that cartel members produce at a level consistent with their assigned quotas.

Industry Insight: A Bull Market for Oil?

In late 2008 oil prices were falling rapidly and crude oil inventories were rising. Nevertheless, futures prices were significantly higher than spot prices, creating the situation known as *contango*. The May 2009 contract for WTI was $50.94, more than $10 higher than the January contract. Futures much farther out were even higher. Higher futures prices provide one indicator for an expectation of rising prices. However, predicting oil prices is far more complex than just looking at futures markets. Futures prices also reflect costs. Storing oil in tankers is more expensive than fixed storage. Financing costs had also risen because of the credit crisis. The forward curve also must account for inflation expectations. The contract farthest out, December 2017, was for $77. If inflation is assumed to run at 1% annually, the 2017 contact is worth $71 in real terms. Assume inflation of 3% and the contract is worth $59.* Thus, concluding that bullish investors were pushing futures prices higher ignores the cost of storing oil, inflation, the effect of OPEC production cuts, overall economic indicators, and a host of other factors.

* L. Denning, "OPEC's Mountain to Climb," *Wall Street Journal*, December 18, 2008, p. C12.

In a falling oil price environment, countries with a high reliance on oil revenues may need more output and, hence, the proclivity of OPEC members to cheat on their quotas. During the falling price environment beginning in mid-2008, members such as Angola, which were attracting new oil and gas investors, were resistant to production cuts. Members with falling production and limited new investment, such as Iran and Venezuela, were pushing for cartel production cuts but were viewed as individually unwilling to cut production because of the resultant damage to their economies.

Oil stocks or inventories are necessary to keep the global supply system operating and are critical to getting the right product to the right place at the right time. The challenge for market participants is getting accurate and reliable information on oil shipments and stocks. Petro-logistics, a one-person company based in Geneva, plays a role in helping oil markets develop knowledge about oil production. Petro-logistics tracks oil production and shipments and calculates OPEC oil output by tracking tanker shipments. The data on OPEC exports come from a network of freelancers who examine bills of lading and other hard-to-obtain shipping documents.

As demand for crude falls, oil industry participants can find themselves holding too much inventory, which may have to be sold at a discount. During the economic downturn in 2008 and 2009, some firms leased VLCCs for oil storage because "sliding demand and poor refinery profit margins have left sellers facing the choice of offering deep discounts to move barrels or risk paying for storage to sell later."[20] At the end of 2008, there was an estimated 40 million barrels being stored in tankers. Producers may also be forced to sell oil at ever lower prices because they want to reduce inventories and raise cash. Earlier in 2008, Iran reportedly sold off vast amounts of heavy crude—at a significant discount—that it had in floating storage offshore at its main Kharg Island terminal. One of Iran's major customers, Reliance Industries (India), cut back on purchases of Iranian crude because of concerns about possible US sanctions tied to the crude. Given that Reliance expected to export refined products to the US, Reliance did not want to risk problems that might occur because of refining Iranian crude.

Summary Points

- Oil markets deal in commodities and are essentially a global auction, which means the highest bidder wins the supply. Sometimes the winning bidder will take delivery of the oil and transform it into saleable products. In other cases, market players get involved in trading activities with no intent of taking delivery. As a result, the global oil market is populated by producers, traders, speculators, governments, regulators, and, ultimately, end customers.

- The price of crude oil is the result of thousands of transactions taking place simultaneously around the world, at all levels of the value chain from crude oil producer to trader to refiner to individual consumer.

- Crude prices behave the same as other commodities; there are wide price swings in times of shortage or oversupply.

- Benchmarks, such as WTI and Brent Blend, play a vital role in the global pricing of crude oil.

- The sale of crude oil occurs in three different types of transactions: spot transactions, futures markets, and contract arrangements. Spot market prices provide signals about the current balance between supply and demand. Futures markets provide a projection of expected prices in the future and help crude oil producers, marketers, refiners, and others to manage their risk. The transactions for most of the oil that is actually sold by producers and delivered to refiners are contract arrangements.

Notes

1. http://www.merriam-webster.com/dictionary/petroleum.
2. Michel Marks, "Crude Oil Futures: A Form of Pricing That Went to the Soul of the Industry," www.nymex.com/energy_in_news.aspx?id=eincrudeprice.
3. Daniel Yergin, *The Prize: The Epic Quest for Oil, Money, and Power*, NY: Simon & Schuster, 1991.
4. *The International Crude Oil Market Handbook*, Energy Intelligence Group, 2007.
5. Bassam Fattouh, *The Dynamics of Crude Oil Price Differentials*, Centre for Financial and Management Studies, SOAS and Oxford Institute for Energy Studies January 2008.
6. B. Fattouh, "WTI Benchmark Temporarily Breaks Down: Is It Really a Big Deal?' *Middle East Economic Survey 49*, No. 20, May 14, 2007.
7. Ann Davis, "The pitfalls of a broken benchmark," *The Globe and Mail*, April 23, 2007, p. B7.
8. www.eia.doe.gov/pub/oil_gas/petroleum/analysis_publications/oil_market_basics/price_transactions.htm.
9. NYMEX, http://www.nymex.com/index.aspx.
10. ICE Futures Europe, https://www.theice.com/homepage.jhtml.
11. NYMEX, http://www.nymex.com/intro.aspx.
12. Ann Davis, "Where Has All the Oil Gone? After Sitting on Crude, Speculators Unload It; The World's Eyes Fall on Cushing, Oklahoma," *Wall Street Journal*, October 6, 1997, p. A1.
13. *Understanding Today's Crude Oil and Product Markets*, API, 2006.

14. R. D. Ripple, "Futures Trading: What is Excessive?" *Oil & Gas Journal*, February 18, 2008.

15. Bhushan Bahree, "Oil Settles Above $70 a Barrel, Despite Inventories at 8-Year High," *Wall Street Journal*, April 18, 2006.

16. Ann Davis, "Where Has All the Oil Gone? After Sitting on Crude, Speculators Unload it; The World's Eyes Fall on Cushing, Oklahoma," *Wall Street Journal*, October 6, 2007, p. A1.

17. Colin Birch, "Achieving Maximum Crude Oil Value Depends on Accurate Evaluation," *Oil & Gas Journal*, January 14, 2002, pp. 54–59.

18. Stanley Reed, "The Saudis and OPEC: Behind the Flare-Up," *Business Week*, October 6, 2008, pp. 45–46.

19. *The Independent Producers Dilemma*, WTRG Economics, http://www.wtrg.com/opec.html.

20. "VLCCs Hired for Storage as Oil Demand Wanes," *Lloyd's List*, November 21, 2008.

Chapter 11
TRANSPORTATION

Life is a journey, not a destination.

—Ralph Waldo Emerson

The early development of the oil industry was characterized by a fragmented, entrepreneurial, under-financed exploration and development part of the business. The real money and power in the oil industry was initially in the refining and marketing, including the transportation components between field and refining. The rapid expansion and growth in power of Standard Oil was mostly in its consolidation and power over refining and marketing, and later over transportation and pipelines, as the industry continued to grow.

— Daniel Yergin, *The Prize: The Epic Quest for Oil, Money, and Power*, 1991

The transportation of oil and gas from wellhead or field to refining and then to customer markets is a critical component of the global oil industry value chain. Transportation, even in the earliest years of the industry, was often the link in the value chain that was the source of competitive pressure and sometimes significant profit. Ultimately, it has become a highly capital-intensive linkage in the global oil and gas marketplace, which is often driven as much by geopolitics as it is by traditional business concerns. Oil may be transported for thousands of miles by truck, tanker, barge, train, pipeline, or a combination of these means before it reaches its final destination.

The challenge of course, is that oil and gas are found in some of the most hard-to-reach places on earth, and moving the oil and gas from field to refinery is in many cases nearly an insurmountable challenge. This chapter provides an overview of the oil and gas transportation segment of the global value chain—a segment that is increasingly critical given the lengths and depths the industry goes to find its reserves.

Fundamentals of Transportation

The transportation of oil and gas is a highly specialized operation that requires meticulous coordination across the entire supply chain. Principle issues of transportation for oil and gas include the following:

- **Distance.** The distance that crude needs to be moved often dictates the mode of movement. Shorter distances may be dominated by trucking, medium distances by barge or rail, and the longest distances by tanker or pipeline.

- **Oil versus gas.** Oil has always been transportable as a heavy liquid. Gas, however, has always been distinguished by its lack of portability. Crude oil in the field may be moved a variety of ways. Natural gas is typically gathered and moved via pipelines. Pipelines, however, are inflexible and costly. If gas is to find its way to global markets, it must be liquefied and then moved by most traditional transportation means available to oil—but refrigerated.

- **Ownership and geopolitics.** Probably no other link in the global oil value chain is more diversified in its ownership structure, and often more politically debated. Thousands of different owners, organizations, governments, and interests are involved in the complex web of transport means employed globally.

- **Environmental safety and security.** Truck rollovers, railroad derailings, pipeline leakages and accidents, and supertanker accidents have all led to an intense focus on safety, security, and environmental protection to a degree not seen in many other industries.

- **Impact on prices paid producers.** As noted in the previous chapter, transportation costs have an enormous impact on the price paid to the producer at the wellhead. If market prices are relatively competitive for similar crudes delivered to any specific refining facility, the price paid the producer is backed out of the refining purchase price for that specific grade of crude, including transportation cost.

$$Price_{Producer} = Price_{Crude\ refinery} - Transportation\ cost$$

The fundamental structure of oil transportation is illustrated in figure 11–1. Transportation is segmented into upstream and downstream. In the *upstream segment*, oil is moved from production area to refinery typically by crude oil tanker or crude oil pipeline. There are, of course, exceptions, but these two major modes cover the vast majority of all crude oil transportation today. After refining, the *downstream segment*, the crude oil derivative products and petrochemical products are then moved via one of the traditional four transport modes used—truck, rail, tanker, and pipeline.

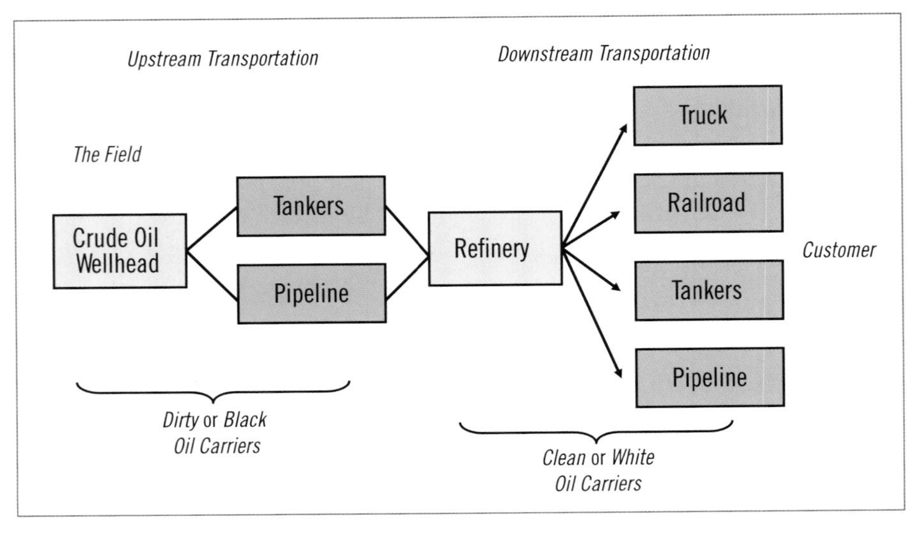

Figure 11–1. Transportation of crude oil

Although it appears relatively simple in figure 11–1, each segment has its own challenges. As the following discussion describes, the upstream transportation segment is one that could be characterized as challenging—it must overcome both land and water physical challenges, including distance, terrain, depth, climate, and geopolitical issues. The downstream process is what we would characterize as complex, in which the increasingly specialized petrochemical commodity products are now closer to market, but must be moved into a network that reaches down to the individual customer. As a result, it is a process and industry with multiple layers of product and service providers globally. In the following sections, we will describe the business, economic, and historical issues and practices in transporting oil and gas.

The barrel

It was so simple: use wooden barrels to carry crude oil from the field to the refinery. It would just take a lot of them. The first commercial oil developments in Pennsylvania and Baku (some imported from Pennsylvania) used wooden barrels and then moved the barrels via wagon and railroad to refineries.

The discovery of oil in Pennsylvania led to not only a boom in oil, but a boom in the demand for the casks or barrels in which to store it for transportation. In the very earliest stages of the industry, any barrel, whether it was for whiskey or wine (generally 40 gallons in size) was used to carry oil. This also meant that one never really knew the exact volumes, as barrels were of all sizes and capacities. It is generally agreed that the standardization of the barrel came from Standard Oil itself, which moved oil in the Standard Oil blue barrel, 42 gallons in capacity, successfully branding the product (figure 11–2). It is also believed that this same blue barrel led to the common abbreviation used for a barrel today—bbl.[1] The blue barrel was 42 gallons in size as opposed to the common whiskey barrel to account or allow for leakages or evaporation during shipment.

Although the industry quickly replaced the barrel with tanks and containers of larger, more economic size, the barrel has remained the unit of measurement of the oil industry. By 1866, the standard was established that is still used globally today—a standard barrel of oil being 42 US gallons (34.972 Imperial gallons or 158.987 liters).

Figure 11–2. The Standard Oil blue barrel
Source: U.S. Department of Energy.

The strategic chokepoint

> *The chief refining competitor of Oil Creek in 1872 was Cleveland, Ohio. Since 1869, that city had done annually more refining than any other place in the country. Strung along the banks of Walworth and Kingsbury Runs, the creeks to which the city frequently banishes her heavy and evil-smelling burdens, there had been since the early sixties from twenty to thirty oil refineries. Why they were there, more than 200 miles from the spot where the oil was taken from the earth, a glance at a map of the railroads of the time will show: By rail and water Cleveland commanded the entire Western market. It had two trunk lines running to New York, both eager for oil traffic, and by Lake Erie and the canal it had for a large part of the year a splendid cheap waterway. Thus, at the opening of the oil business, Cleveland was destined by geographical position to be a refining center.*
>
> — Ida Tarbell, 1903[2]

Strangely, the transportation segment of the oil and gas industry has always led a shadow life. Transportation, both its availability and cost, has proven critical to competitiveness since the very beginning, but it has always been treated as something of a noncore activity, a service to be acquired or provided but not a core source of value. Nothing could be further from the truth.

Both Pennsylvania and Baku in 1860 experienced very similar crude oil booms, with production very quickly outstripping transportation capabilities. Crude oil produced from the hundreds of well sites in the immediate region was hauled in barrels on horse-drawn carts on bad roads to refineries (many of which were nothing more than stills in the early years). Barrels were in short supply, spills and losses high, labor costs always on the rise, and labor unrest a growing problem.

Moving freight like oil by wagon, water, and rail was an easy and obvious development for the industry. The oil industry growth rate became a driver for transportation development as massive increases in crude oil volumes required scale and cost efficiencies. Railroads, like wagons, could easily move volumes of barrels over increasing distances rapidly and at low cost. The railroads quickly realized that shipping wood barrels on "flats" or in converted boxcars was not cost-effective, and they rapidly developed steel cylindrical tank cars capable of transporting bulk liquids like oil. Starting in late 1865, the Densmore two-tank oil tank car (two vertical oil tanks on a flat car) was invented and was copied by others immediately. Starting in 1867, the horizontal iron tank car came into use, and the vertical tank car shortly became obsolete.

By the 1930s and 1940s in the United States, custom-built tanker trucks were the primary means of transporting crude oil from production sites in the US to refineries. After refining, derivative petrochemical products were then shipped via both truck and rail to final customer destinations. In the years that followed, tankers of all kinds for transporting oil and petrochemicals continued to evolve, becoming increasingly complex in order to provide growing safety, security, and environmental protection. Tanker trucks remain a component of the oil supply chain today, primarily in downstream petrochemical and refined products.

Upstream transportation

Moving crude oil ("black oil") and natural gas from the field to the refinery today is about pipelines across land and supertankers across water. The primary complexity for the transportation of most of the world's crude oil and gas, however, is location. Many production sites on land or water are relatively inaccessible for the construction of economic transportation of the crude product.

As shown in figure 11–3, with increasing distances there is no real competitive choice other than the modern supertanker. Although every case is different, the sheer scale economies, flexibility, and speed to market that the world's crude oil tanker fleet provides is insurmountable. Crude oil pipelines have a critically important role to play in the global oil industry, specifically in those land-locked areas of the globe in which new (or old) oil and gas reserves are now being developed. The railroad tank car, the lesser third leg of the three-legged transportation stool, still plays an integral role in many different oil transportation arrangements. Although significantly more expensive, railroads still offer the ability to move oil over existing infrastructure on a selective or opportunistic basis. In some cases, there are no other choices.

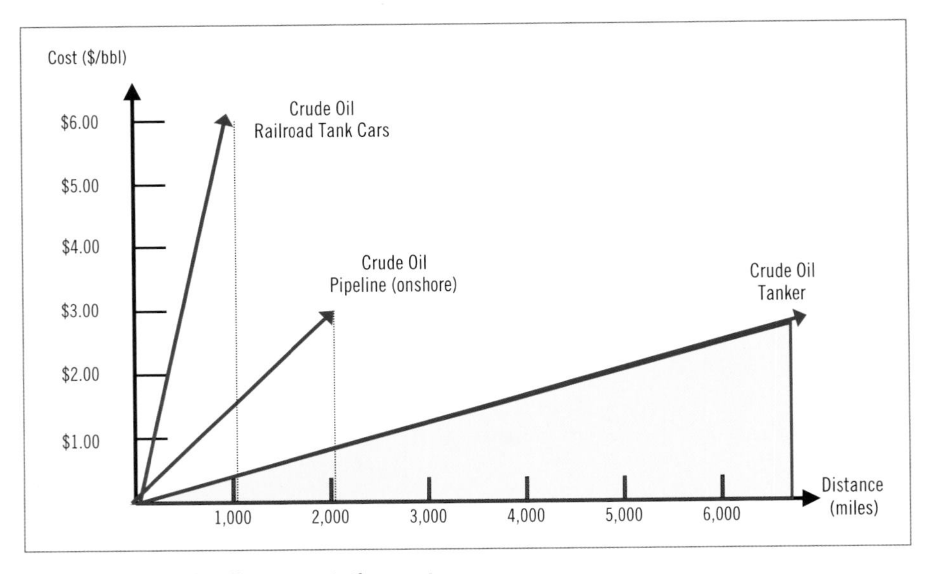

Figure 11–3. Crude oil transportation costs

Pipelines

Pipelines are critical from the first millimeter of oil and gas production. As many land and subsea pipeline engineers frequently say "Nothing happens until we get our lines in place."[3] Beginning at the wellhead, on land or sea, the oil and gas produced from the well must be first contained and moved via pipeline.

Gathering systems

A production field with multiple wells must first combine or gather the crude oil before initial processing and shipping. This initial network of pipes, the *flowline network*, moves the oil to a central processing plant and/or shipping point. The entire *gathering system* will include the pumps, separators, treatment facilities, tanks, regulators, compressors, dehydrators, valves, and all other associated equipment. The expertise and technology at work within this first critical step has been an area of increasing company competence and competitiveness.

Within gathering systems there are two categories, radial systems and trunkline systems. A *radial system* integrates multiple flow lines to a central header where the oil fluids are collected.[4] The *trunkline system* is an advanced radial system which integrates multiple remote headers into a single collection and processing facility, used in the largest fields. The various lines running from individual pipeline webs are then integrated at a field-level gathering point. From here they move to larger volume crude oil pipelines or tankers.

Some of the most complex and critical forms of gathering systems today are those for deepwater collection. Although offshore oil production is always pictured with the jack-up rigs or floating platforms of enormous size, these are typically only the drilling rigs. Once production has commenced, it is increasingly a subsea system of multiple wellheads at extraordinary depth requiring gathering systems for the oil and gas produced, and then consolidated and processed for further transport to refining facilities either by pipeline or tanker.

Pipeline construction

Oil pipelines are constructed today from steel or plastic tubing, with inner diameters ranging from as small as 4 inches (10 cm) for refined/downstream products to 48 inches (120 cm) for crude oil. When possible, most pipelines are buried 3 to 6 feet below ground. Buried lines provide improved safety and security, although they add considerably to construction cost. These higher costs are usually easily recovered in fewer repair and maintenance costs over the pipeline life.

Pipeline construction costs vary dramatically depending on diameter, length, terrain (both onshore and offshore), and environmental conditions, including weather. Pipeline companies use calculations based on cost per pipe diameter per distance to estimate pipeline project costs. For example, the calculation for a gathering system may be $1,000 per millimeter diameter per kilometer. A 50-kilometer system consisting of 50-millimeter pipe may be roughed out as:

$$50 \text{ millimeters} \times 50 \text{ kilometers} \times \$1{,}000 \text{ per kilometer} = \$2.5 \text{ million}$$

If the proposed gathering system (note the smaller diameter pipe at 50 millimeters) appears economical using this formula, a more detailed cost estimate will be performed. Pipelines, because they often traverse long distances and a variety of political units like states, provinces, or even countries, typically require a number of years to organize and build. Once governmental approvals and legal rights-of-way for the pipeline path have been obtained, pipeline construction follows the path described in figure 11–4.

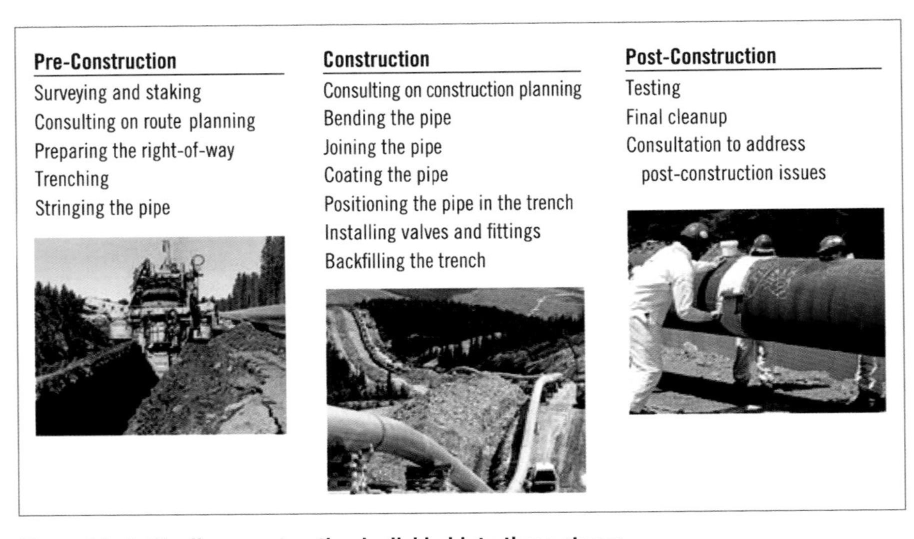

Figure 11–4. Pipeline construction is divided into three stages

Reprinted with permission of the Canadian Energy Pipeline Association (CEPA).

Pipelines are highly capital intensive, and once completed operate at very low cost per barrel per mile transported. Although all pipelines are different, recent pipeline investments in North America have broken down as follows: Pipeline construction—45%; line pipe and fittings—19%; pump stations and equipment—16%, land and ROW—5%; other—15%. Pipeline construction costs, as the largest single component, are 38% labor and 36% materials, ROW and damages make up about 7%, and miscellaneous expenses 20%. Obviously construction, pipe, and pumps and compressors dominate investment.

Pipeline operation

The movement of crude oil through a pipeline is maintained by pump stations located along the length of the line. The oil moves through the pipe at a variety of speeds depending on the type of crude oil (specifically viscosity), the diameter of the pipe, the length of the pipeline, elevation change in the pipeline segment, etc., but typically ranges between 1 and 6 meters per second.

Crude oil pipelines require maintenance both inside and out. Because many crude oils contain a variety of wax components that may adhere and buildup on the internal surfaces of a pipeline, a variety of automated cleaning devices called *pigs* (or *scraping devices* or *go-devils*) are run through the pipelines on a regular basis for inspection and cleaning. These inspection robots are also capable of detecting small leakages or other possible pipeline failures before they happen.

Natural gas pipelines today are constructed nearly exclusively of carbon steel, and may vary anywhere from 50 millimeters to 1,500 millimeters in diameter. The gas is moved via pump stations like crude oil, but instead of pumps there are compressors. Because natural gas often has high levels of both sulfur and water when initially produced, it is critically important for the long-term preservation of pipelines that the sulfur and water be removed from the gas before entering transit pipelines. The gas itself is often colorless and odorless unless an additive like mercapatan (having an odor resembling rotten eggs) has been added to allow instant detection in the event of a leak.

As pipelines have become the most efficient means of transporting oil over long distances, they have also become safer and more technologically advanced. Pipelines now facilitate the transport of crude oil over vast distances, traversing rugged mountains, valleys, and rivers and utilizing precisely engineered pump systems to regulate the movement of products through them. The intersection of a number of pipelines creates a *hub* or *marine terminal*. Because most hubs provide a number of alternative oil or gas supplies to a single point, they often serve a very important role—price setting in the industry. For example the Cushing Hub in Oklahoma in the United States has been the benchmark quoted price for the West Texas Intermediate (WTI) oil price for decades. It is also where the New York Mercantile Exchange's (NYMEX) WTI crude oil futures contracts have physical delivery.[5]

As one would expect, the cost of constructing pipelines in the oil and gas industry depends on many things, but one thing is certain: they are expensive. But cost has not proven to be the biggest obstacle to pipeline construction over the past century; the biggest obstacle has been aligning the interests—business and political—of field developers and host country governments.

The chicken or the egg?

One of the more esoteric questions regarding oil projects is which should be developed first, the oil or the pipeline? As one author suggested, it is something of a "chicken or egg?" dilemma.[6] One is relatively valueless without the other. Key issues include the following:

- **Capital cost and risk.** The capital cost of major field development is significantly larger than pipeline construction costs, depending on the field. Therefore, the pressure to earn returns on field development would urge pipeline readiness sooner rather than later.

- **Funding sources.** Many pipelines are constructed using debt financing, often using project financing consortiums. Debt service requires regular payments but may be stretched out over long periods of time. Field development, however, is typically funded with corporate funds and poses a significant source of risk for the investor. Again, once the field is operational, owners wish to get the product to market as quickly as possible. Having the pipeline ready and available is important.

- **Field development and production.** A counterargument to pipeline first is that many fields get delayed, repeatedly, and even when developed, early production can be substantially below what is expected later in the project life.

- **Pipeline option value.** Theoretically, once the pipeline is constructed, the oil field developer has a number of production strategy alternatives, the so-called option values associated with differing levels and geographic distributions of drilling, development, and production.

Ultimately, the value is in the oil or gas. The production and value of any specific field may be stunted, delayed, or depressed as a result of inadequate transportation routes to market. But it is value that remains, waiting. A pipeline is an extremely expensive fixed asset investment, immovable and incredibly inflexible. Controlling transportation is indeed an element of power in international oil and gas development negotiations, but he who holds the field is ultimately in the position of greater influence.

In the end, which comes first may reside in the hands more of government than business. Pipelines are usually constructed by consortiums of organizations, government and development banks included. They may provide support, access, and rights of eminent domain at a different time scale than what some individual companies, or small groups of companies, require to develop an oil field. Or as one colleague once noted, "Pipelines are like college funds. You know its going to take a lot of years, so you better get started now."

US Pipeline System

In the United States, there are more than 2.43 million miles of pipelines carrying natural gas and hazardous liquids (primarily crude oil, refined products, and chemicals). The pipeline system consists of three major elements:

- Pipelines that collect product from oil and gas wells on land or offshore and from ships. These gathering systems move oil and gas to storage or processing.

- Transmission pipelines that transport liquids or gas over long distances. Transmission lines deliver gas to power plants, industrial customers, and municipalities for further distribution to homes. Petroleum transmission lines deliver crude oil to refineries or refined products to markets, such as airports or storage depots where fuel oil and motor fuels are loaded on trucks for local delivery.

- Distribution lines as part of natural gas systems. These lines consist of main lines that move gas to industrial customers and smaller service lines that connect to businesses and neighborhoods.

The main business of US pipeline companies is the transportation of product owned by producers and distributors. Kinder Morgan, one of the largest pipeline companies in North America with 37,000 miles of pipelines, says that it "operates like a giant toll road, receiving a fee for our services and providing connectivity for our customers."[7] In the United States pipeline companies are what are known as *common carriers* and cannot refuse to deliver products that meet the conditions of service of publicly posted tariffs. The fees or tariff rates for oil and gas pipelines that transport product across state borders are regulated by the Federal Energy Regulatory Commission (FERC). The rate-setting process is extremely complex and involves various rate categories, including initial rates, indexed rates, grandfathered rates, settlement rates, market-based rates, and cost-of-service rates. The rates do not fluctuate with the price of crude or natural gas.

Liquid pipeline sector

Pipelines are used to transport almost 70% of the crude oil and refined products that are transported in the United States.[8] Pipelines are a very efficient and safe way to move liquid products. For example, it costs only a few cents to move a gallon of gasoline from the Gulf Coast to Chicago, a distance of about 1,600 kilometers.

Pipeline companies ship petroleum products of the same quality in sequence through a pipeline, with each product or "batch" distinct from the one before or after. Each pipeline has specific requirements for minimum batch size based on factors such as pipe size and flow rate. With a product like gasoline, pipeline batches are impacted by both product demand and by regulations such as regional and summer and winter vapor pressure requirements. Fuels oils must be segregated based on sulfur content and dyed for specific markets. Jet fuel requires segregated batches to meet different military and domestic aviation specs.

An obvious operational question is how pipeline companies deal with the mix of different liquids that flow through their pipelines. A large petroleum pipeline might have 30 to 50 different products regularly moving through the pipeline over a specific cycle of days (a *cycle* is the period of time beginning with the pumping of a product until all the product grades are pumped and the initial product is pumped again, beginning the new cycle). Figure 11–5 illustrates the batch process of multiple products in a single pipeline.

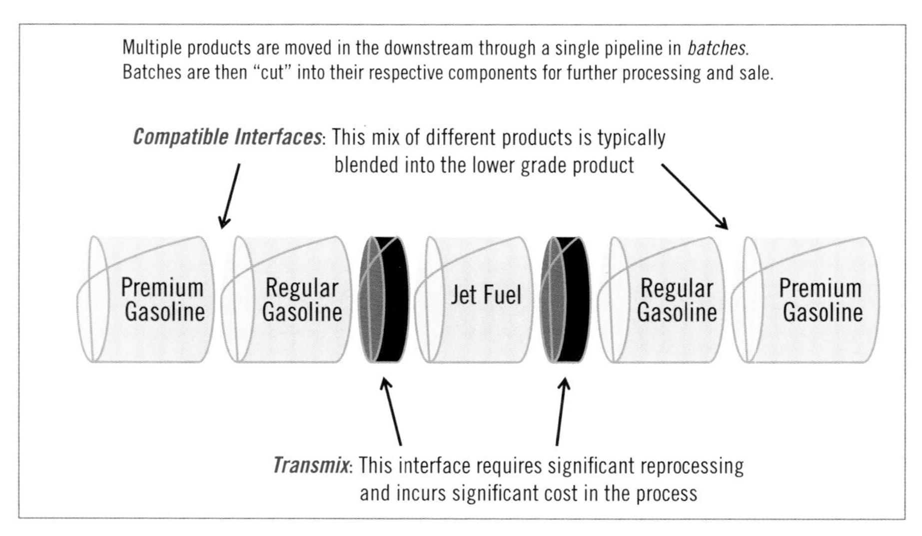

Figure 11–5. Downstream pipeline operations

Different products such as regular gasoline, premium gasoline, or jet fuel, are moved through the same pipeline in batches. Once the different batches arrive at their respective destinations they are pulled out of the line—*batch cutting*—into other lines or tankage through a complex set of valves. Some mixing of product occurs at the interface of adjacent product grade batches. This volume can sometimes be mixed into one or both of the adjacent batches and still meet product specifications.

However, all fluid interfaces are not the same. Interfaces between common products such as regular and premium gasolines are relatively simple in chemical structure and are typically cut into the product of lesser quality or value, the *downgrade*.[9] Interfaces between regular and premium gasoline would therefore be cut into the regular gasoline, downgrading or regrading the interface and part of the premium gasoline batch into the regular gasoline batch.

When products are of significantly different grade and value, for example that between jet fuel and regular gasoline, the product interface, called *transmix*, must be diverted to tanks and either reprocessed or moved via truck to a reprocessing center or returned to a refinery. The transmix then has to be reprocessed via a costly procedure into marketable qualities of the two products. Interfaces between gasolines, diesels, and jet fuels are all commonly redirected to transmix tanks to avoid contamination. Components of transmix include trucking, reprocessing, refining (where crude oil must often be replaced with transmix), labor, and administrative cost (measurement and ticketing costs).

With more regulations, the pipeline companies have to deal with increased complexity and a greater number of fuel grades. However, the number of different products does not translate into higher revenue because pipeline companies are paid based on the volume of product flowing through their pipeline. In addition, because the basic technology and processes used for interface processing have been largely unchanged for decades, there are significant opportunities for improving business results by reducing regrade and transmix losses. Many companies today are working to reduce *overwash*, the tendency to cut too much of the premium batch interface into the lower grade batch "just to be safe." Given the sizeable price and value differentials between downstream products, reducing wastage of premium value products improves profitability. An obvious extension of this business savings is the reduction of transmix as well. The development of a variety of devices and technologies for the implementation of precision batch cutting (PBC) shows significant promise in improving the financial results of downstream pipeline transportation.[10] *PBC* is a general category of technologies that improve decision making and execution of the batch cut process.

As with any capital asset, maximizing a pipeline's capacity utilization is a constant objective. Pumps and motors can be added to increase pressure and flow rates. Chemical compounds, called *drag reducing agents*, can be injected to increase overall flow at the same pressure. Sequencing of products is a key operational issue along with ensuring that product integrity is maintained. Failure to maintain federally mandated product specifications can lead to enforcement actions like that described in the *Industry Insight: Undersea Pipelines and Corrosion*.

Industry Insight: Undersea Pipelines and Corrosion

Pipelines play a critical role in oil and gas production and transportation both onshore and offshore. For example, most offshore oil that has been produced must reach collection and processing facilities somehow, and undersea pipelines are the primary method. Although the fixed and floating platforms are the visual symbol of offshore oil production, the platform is first the drilling rig, and then if oil is found, the production platform. Oil produced must be either loaded onto a floating storage and offloading vessel for shipment by transport or transferred to an undersea pipeline to be moved to onshore collection and processing facilities. Large offshore oil field developments like those in the North Sea and Gulf of Mexico are a veritable spiderweb of undersea pipeline systems.

These same pipeline systems are not perpetual. They do require regular maintenance, and eventually replacement, as a result of corrosion. Although crude oil itself is in some ways an anticorrosive compound, water and sulfur that are also present in the crude will eventually cause corrosive deterioration of the pipe. The lower the water vapor content, and the higher the flow pressure, the less corrosive the impacts. Regardless, eventually, corrosion will occur.

Chevron's Cook Inlet operations in Alaska serve as a case in point. Inspections of a 130-foot *riser* (a pipeline connecting the production platform to a subsea pipeline) in 2010 indicated that more than 60% of the pipeline's wall thickness had been lost to corrosion. US regulations require that pipelines losing 50% of their thickness must be replaced or shut down. Chevron, after an appeal for continued operation as is, was informed that the riser had to be replaced. Chevron immediately shut down the Anna Platform, halting its 900,000 b/d output. Once again, continual investment is required in order to sustain production capability.

Gas pipeline sector

Based on miles of pipelines, the natural gas system is by far the largest part of the US pipeline system because it connects to cities and neighborhoods. The geographic imbalance between producers and consumers (producers mainly in the South and West; most consumers in the Northeast, Midwest, and Pacific West) means that natural gas must be transported long distances across the country.

The regulation of the gas system has been tumultuous from the first days of the industry. Excessive regulation is blamed for gas shortages in the 1970s. Until 1985, pipeline companies purchased natural gas from producers, transported it to its customers (mostly local distribution companies), and sold the gas for a regulated price. In 1985, the first of several regulatory changes occurred that resulted in the unbundling of gas transportation and sales. FERC Order No. 636 in 1992 was the culmination of deregulating the interstate natural gas industry. The order gave all natural gas sellers equal footing in moving natural gas from the wellhead to the end user. It allows the unbundling of transportation, storage, and marketing, and the customer can choose the most efficient method of obtaining gas.[11]

Deregulation had a major impact on the US natural gas market and the business of pipelines. A competitive wholesale gas market resulted, and gas marketing emerged as a new segment of the industry. Many new companies entered the wholesale gas market, and there was a substantial decrease in real gas prices. Deregulation also resulted in a shift of market transactions to the hubs where major pipelines interconnect. Hubs allow market participants to acquire gas from several sources and ship it to different markets. In the US, there are 24 operational hubs/market centers (9 in Canada), with Henry Hub being the largest. The Henry Hub is owned and operated by Sabine Pipe Line, LLC, which is a wholly owned subsidiary of Chevron Pipeline Company. The Sabine pipe Line starts in eastern Texas near Port Arthur, runs through south Louisiana, and ends in Vermillion Parish, Louisiana, at the Henry Hub near the town of Erath. The Henry Hub is physically situated at Sabine's Henry Gas Processing Plant.

The Henry Hub interconnects nine interstate and four intrastate pipelines, including: Acadian, Columbia Gulf, Dow, Equitable (Jefferson Island), Koch Gateway, LRC, Natural Gas Pipe Line, Sea Robin, Southern Natural, Texas Gas, Transco, Trunkline, and Sabine's mainline. These pipelines provide access to markets in the Midwest, Northeast, Southeast, and Gulf Coast regions of the United States. The Henry Hub can handle about 3.0% of average daily US gas consumption.

Figure 11–6 provides a basic overview of the structure of both gas pipeline sales and physical distribution. Although the gas pipeline companies actually move the gas to end user, they do not typically own the gas that they are moving. Gas producers and spot market participants—marketers and distribution companies—handle most of the sales of gas, with the pipeline companies focusing on delivery.

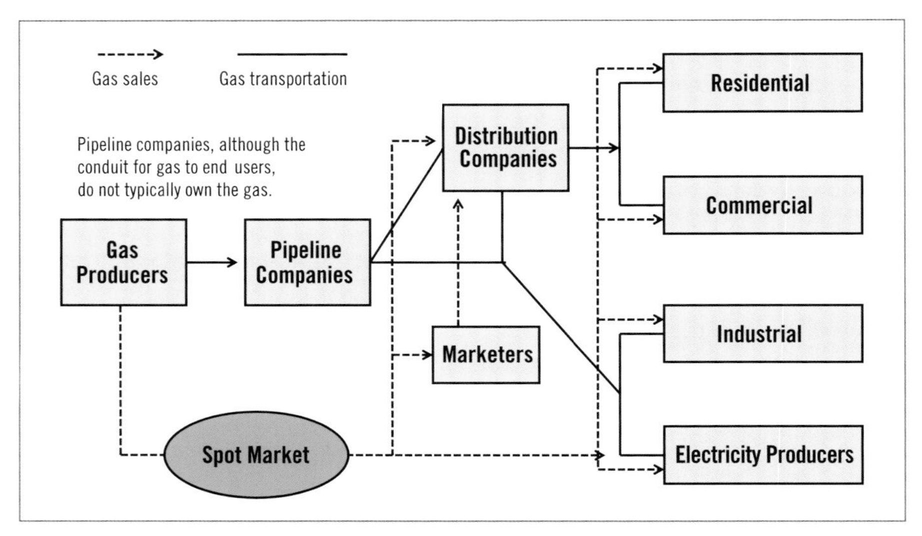

Figure 11–6. Structure of the US gas market

Source: Andrej Juris, "Development of Competitive Natural Gas Markets in the United States," Public Policy for the Private Sector, World Bank Group, Note 141, April 1998.

An Overview of Selected Pipelines

Pipelines are the skeletal infrastructure of the global oil and gas industry. Oil and gas pipelines traverse thousands of miles across land and below water, through jungles and over mountain ranges in some of the most difficult and distant places on earth. Table 11–1 shows a sample of pipelines that crisscross the globe. Each of these pipelines was controversial in a variety of ways, most commonly because of concerns over environmental impact.

Table 11–1. Selected major oil and gas pipelines

Characteristics	Trans-Alaska Pipeline System	Druzhba Pipeline	Baku-Tbilisi-Ceyhan Pipeline	Chad-Cameroon Pipeline	Bolivia to Brazil Gas Pipeline	Langeled Underwater Pipeline
Oil or gas carrier	Oil	Oil	Oil	Oil	Gas	Gas
Geography	Alaska, USA	Russia to Eastern Europe	Azerbaijan, Georgia, Turkey	Chad to Cameroon Atlantic coast	Bolivia to Purto Alegre, Brazil	Norway to the United Kingdom
Pipeline length (km)	1,300	5,327	1,768	1,070	3,150	1,166
(miles)	(808)	(3,310)	(1,099)	(665)	(1,957)	(725)
Below ground	48%		100%	100%		
Above ground	52%		0%	0%		
Capacity (mmb/d)	2.136	1.2–1.4	1	0.25	11.0 bcm/yr	25.5 bcm/yr
Pipe diameter (inches)	48	1st: 11–33	42 (narrowing to 34)	42	32 (24 to 16 last leg)	44 (largest in North Sea)
Pumping stations	12	173	8	3	16	
Completion date	1977	1964	2005	2003	2000	2006
Construction period	1974–1977	1st: 1960–1964 2nd: 1965–1966	2003–2005	2000–2003	1998–2000	2004–2006
Construction cost	$8.0 billion	$12.7 million (400 million rubles)	$3.9 billion	$2.2 billion	$2.1 billion	£1.7 billion

Trans-Alaskan pipeline

The Trans-Alaska Pipeline System (TAPS), completed in May 1977, is an 800-mile-long, 48-inch-diameter pipeline designed to move oil from the North Slope of Alaska to the southern Alaska port of Valdez. Oil was first discovered at Prudhoe Bay on the North Slope in 1968. Then, following the oil crisis of 1973, the pipeline was constructed over the 1973 to 1977 period. TAPS crosses three mountain ranges and traverses more than 800 rivers and streams. It maintains roughly 1,180 psi pressure over its 800-mile link by using 12 separate pumping stations. When completed, at $8 billion, it was the largest privately funded construction project in human history. By end-of-year 2008, nearly 20,000 tankers had been loaded at Valdez, Alaska over 30 years, with oil transported by the pipeline.

Langeled underwater pipeline

Originally named the "Britpipe," the Langeled pipeline runs 725 miles (1,166 km) across the North Sea from the Nyhamna terminal in Norway to the Easington Gas Terminal in the United Kingdom. Pipeline construction began in 2004, and first-gas was received in Easington in October 2006.[12] It is the world's longest underwater pipeline. A gas pipeline, it has a capacity of 25 billion cubic meters (bcm) of gas, 70 million cubic meters of gas per day, capable of supplying roughly 20% of the United Kingdom's total gas needs. The pipeline was completed in two segments, a northern and southern. The southern section, 44 inches in diameter and 540 km long, runs between the Sleipner platform in the North Sea and Easington, and was completed in 2005. The north section, 42 inches in diameter and 626 km in length, runs from the Nyhamna terminal to the Sleipner platform. It was completed in 2006.

Chad-Cameroon pipeline

The Chad-Cameroon pipeline is a 1,070 km pipeline constructed to move crude oil from the Kome oil production area in southern Chad in central Africa to a floating storage and offloading (FSO) vehicle off the coast of Kribi, Cameroon. Construction was started in 2000, with first oil reaching the FSO in late 2003. The Chad production area is owned and operated by a consortium of companies led by ExxonMobil (US), with Petronas (Malaysia) and ChevronTexaco (US), as well as the Chad and Cameroon governments. The pipeline's construction cost of $2.2 billion was largely funded by the consortium companies, with some debt provided by multilateral and bilateral credit financing provided by Western governments. Debt-based financing came from the International Finance Corporation (private sector arm of the World Bank) in the amount of US$100 million, while the French export credit agency, COFACE, and the US-Exim Bank each provided US$200 million. IFC-coordinated private lenders added US$100 million.

Camisea pipeline

The Camisea, Peru pipeline begins at the Camisea Gas Field in the Amazon Rainforest, traverses through the steep Andes mountains, and terminates within the Paracas National Reservation near San Martin, the port of Pisco. A second pipeline runs from near Pisco north along the coast to Lima for distribution to residents and industries in the capital city. In April 2007, Suez Energy and Kuntur Gas Transport unveiled rival proposals for pipelines to carry Camisea gas to southern Peru. The Camisea gas project had been planned for many years but was finally executed under the government of President Alejandro Toledo, starting operations in August 2004.

Druzhba pipeline

The Druzhba (the name means "friendship") pipeline, also called the *Friendship pipeline* and the *Comecon pipeline*, is the world's longest oil pipeline. The pipeline, completed in 1964, begins in Southeastern Russia and travels more than 4,000 km (2,485 miles) to a variety of destinations in the Ukraine, Belarus, Poland, Hungary, Slovakia, the Czech Republic, and Germany. The Druzhba line splits in Belarus into a northern route (the larger of the two lines) that crosses Poland ending in Germany, and a southern route that crosses the Ukraine branching further across eastern Europe.

First oil arrived in Czechoslovakia in 1962, with subsequent destinations operational in 1963 and 1964. It has a daily capacity of between 1.2 and 1.4 million b/d and has 20 pumping stations over its length. It is today the largest single transportation conduit for Russian and Kazakh oil across Europe. Germany receives roughly 500,000 b/d via the northern Druzhba line, nearly 20% of its daily consumption needs, with Poland taking roughly 360,000 b/d, nearly 70% of its daily needs. Refineries belonging to Total, Shell, and BP are among the biggest buyers of crude from Druzhba. The southern pipeline branch has a capacity of 400,000 b/d, which has historically not always been fully utilized. The Czech Republic receives nearly 50% of its daily needs (100,000 b/d), with the Slovak Republic (76,000 b/d) and Hungary (135,000 b/d) receiving essentially all of their total daily oil consumption needs via the Druzhba line. The Russian part of the pipeline is operated by the Russian transportation conglomerate Transneft through its subsidiary OAO Druzhba. Operations in other countries are, in turn, conducted by individual country subsidiaries.

Case Study: The Baku-Tbilisi-Ceyhan (BTC) Pipeline

The Baku-Tbilisi-Ceyhan (BTC) pipeline serves as a modern case study of the many costs and complexities with the construction and operation of a crude oil pipeline. The BTC pipeline links the world's oldest oil city, Baku, Azerbaijan, with the Turkish Mediterranean port city of Ceyhan. It is 1,100 miles in length (1,768 kilometers), can carry up to 1 million b/d, and follows a twisting path from Azerbaijan to Georgia to Turkey. The indirect path, as illustrated in figure 11–7, reflects the ongoing hostilities between Azerbaijan and Armenia. The pipeline was constructed to follow a circuitous path around Armenia. Although buried at least 1 meter deep for its entire length, it still requires more than 100 surface stations and associated pumping facilities and plants along its path.[13]

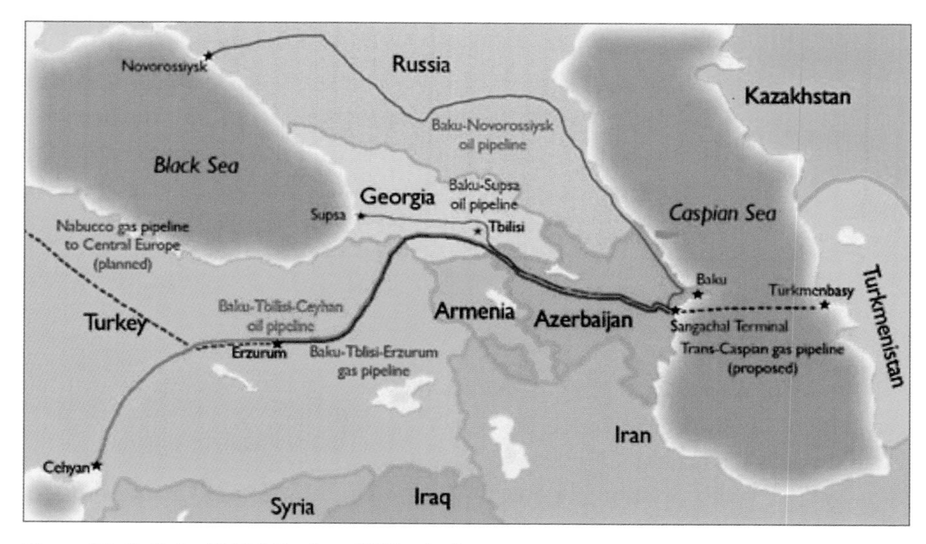

Figure 11–7. Baku-Tbilisi-Ceyhan (BTC) pipeline

Source: South Caucasus Pipeline's map, Copyright © 2010 AY Deezy. Graphics by AY Deezy, 2010. Reprinted with permission.

Completed in May 2005, the BTC pipeline was constructed by a consortium of companies led by BP and financed using a very high level of debt via project financing. The debt financing for the pipeline included the International Finance Corporation (IFC) of the World Bank, as well as loans from the European Bank for Reconstruction and Development (EBRD), and a variety of other development finance institutions.

The BTC pipeline, a project whose origins go back more than two decades, received much global attention because of the potential social and environmental impacts of the project. Ironically, it was not the first major oil pipeline in the region, as illustrated in figure 11–7. Both the Baku-Novorossiysk and Baku-Supsa pipelines were already operating at the time of BTC's construction. But at the heart of the political issues surrounding the BTC pipeline was the fact that it would be the first to move crude oil from the landlocked Caspian to a non-Russian-controlled port on the Mediterranean, bypassing the Bosporus and the Turkish Straits. The Turkish Straits themselves were the subject of a growing debate over safety and environment due to the ever-growing tanker traffic already passing through the narrow and over-used straits.

BP's own description of the financial structure leading up to final construction is helpful in understanding the complexity of pipeline construction and financing in many of the world's emerging markets today:

The reasons for considering public funds or public loans is to enable the participation of local partners who might lack the financial resources possessed by BP. The consortium is a joint venture, the joint venturers being obliged to fund the project in proportion to their respective equity shares. Partners such as the State Oil Company of the Azerbaijan Republic (SOCAR) do not have sufficient funds available to make their investment without a loan. Consequently, the project is taking the expensive step of borrowing 70% of the money for the investment. BTC Co. intends to raise $1.7 billion of external debt, mostly from commercial banks, but also from multilateral agencies such as the World Bank's International Finance Corp. and the European Bank for Reconstruction and Development. Political and commercial risk insurance is being sought from a number of export credit agencies around the world. If BP were running the BTC project alone, it would largely use its own funds and rely upon relatively minor participation by financial institutions, solely for the purpose of reducing political risks and to ensure an extra dimension of expert, external monitoring. Now, far from enjoying a "public subsidy," the project will receive loans from financial institutions that charge interest rates reflecting the risks of the countries. The loans and interest are repaid in full by the project.[14]

The BTC pipeline was always classified as a "speculative project" because it did not secure oil revenues of sufficient size to assure financial viability. The only oil that was actually committed to the BTC prior to construction startup was Phase I of BP's Azeri-Chirag-Guneshi (ACG) project in Azerbaijan. With a pipeline carrying capacity of 1 million barrels per day, there was little reason to believe that the ACG project, particularly in its early years, would provide sufficient crude for positive pipeline economics. The BTC pipeline did have one very strong motivating political force behind it: the prospect that some of the world's largest oil and gas reserves outside of Russia and the Middle East could now find their way to the western markets. In addition to avoiding the traditional stranglehold on oil and gas reserves of the Caspian Sea, it would move the oil to the Mediterranean, clear of the commercially congested and highly environmentally sensitive Bosporus and Turkish Straits (table 11–2).

The BTC also served to prove a point on one of the longer-standing debates over crude oil and natural gas pipelines in general—whether the pipeline itself was to be more of a common-carrier service or a source of profitability for the investors and operators who had taken the risks associated with the pipeline's development. Clearly, for most of the crude oil developers around the world, pipelines have always been considered literally a "means to an end," but for the actual pipeline investors and operators, still a business. As we will discuss in a subsequent section, the business activities and profitability associated with pipelines for oil and gas have been recognized as more stand-alone businesses on the downstream side of the industry.

Table 11–2. The Baku-Tbilisi-Ceyhan pipeline

Characteristic	Total	Azerbaijan	Georgia	Turkey
Length (km)	1,768	443	249	1,076
Profile	Summit 2,800 m	Flat, rising to 300 m at border	Rising sharply to 2,500 m, falling to 1,000 m, then rising	Mostly 2,000 m, several 2,500 m summits before falling to coast
Diameter (inches)	42, 46, 34	42	46	42, 34
Pump stations	8 + 1 pressure	2	2	4 + 1 reduction
Operating staff	850	250	250	350
Hydraulic power (MW)	130	40	40	50

Source: "The Baku-Tbilisi-Ceyhan Pipeline and BP: A Financial Analysis," Claros Consulting, May 2003.

Although the IOCs had argued from the very beginning that the BTC would be overly expensive and more political than business in scope, as more and more of the agreements surrounding its organization and construction were reached, the actual transportation charges appeared increasingly attractive. BTC transport charges were estimated to range between $2.80 and $3.30/bbl.[15] Assuming an average $3.00/bbl, this means that the total transport cost is $0.0017 per barrel per kilometer.

Along the way, the crude passes over three different countries, two of which—Georgia and Turkey—require payment of a transit tariff. (Azerbaijan does not charge a transit tariff as it is a member of the consortium that is developing the ACG field, the primary feedstock for the early years of the BTC's business. If the BTC were to begin carrying crude oil for a non-Azerbaijani party, for example, oil from the Tengiz field in Kazakhstan, Azerbaijan would levy a transit tariff.)

Table 11–3 shows how the transit tariff is scheduled to change over the operating life of the pipeline, rising from $0.32/bbl in the early years of operations to a high of $0.54/bbl in the last years of the envisioned 40 year pipeline life span. These are considered relatively small by transit tariff standards, although many market analysts fear that over time, Georgia may wish to renegotiate the rate.

Table 11–3. BTC transit tariffs (US$/bbl)

Country	2005–09	2010–20	2021 on
Azerbaijan	$0.00	$0.00	$0.00
Georgia	0.12	0.14	0.17
Turkey	0.20	0.30	0.37
Total	$0.32	$0.44	$0.54

The "multilayer monitoring plan" described in figure 11–8 provides additional detail on the degree of management attention that BTC attracted. The monitoring plan, designed to increase cooperation and communication between the pipeline operators, the three country hosts, and the other users and stakeholders, was designed to provide a continuous oversight structure both during construction and in the following years of operation. Given the size and length and country-span of the pipeline, BTC was subject to extensive international environmental concern from the very beginning. Although the BTC has never had any significant environmental events, it did suffer a significant explosion in May 2006, and to date, has never been utilized up to its capacity of 1 million barrels per day.

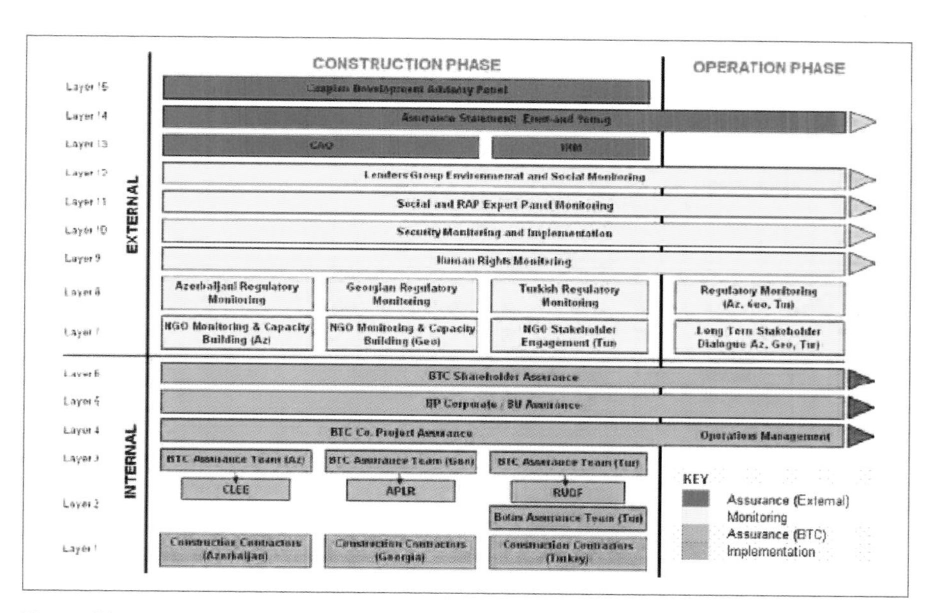

Figure 11–8. Monitoring, assurance, and oversight of BTC

Source: "BTC Project Environmental and Social Annual Report, 2005 (Construction Phase)," BTC Pipeline Project, figure 8.1, p. 24.

By most measures the BTC pipeline construction and operation have been relatively successful. (And yes, we are aware that the BTC was the subject of the James Bond movie *The World is Not Enough* (1999). The construction of an oil pipeline through the Caucuses to the Turkish Mediterranean is at the core of the plot. The pipeline backdrop, although logically including Azerbaijan in principle, also somehow includes Kazakhstan. Regardless, actual filming was primarily in the United Kingdom, Spain, and the Bahamas—representing the waters of Istanbul!)

Oil Tankers

Most of the world's crude oil today is moved via crude oil tankers. A *crude oil tanker* is a ship constructed specifically for moving crude oil, usually long distances from production areas to refineries. *Product tankers*, typically much smaller in size, carry petrochemical products of refineries closer to final consumer markets—the downstream market.

The history of oil tankers is one of evolving size. As illustrated by table 11–4, size classes today have reached more than 500,000 deadweight tons (DWT), the *ultra large crude carriers* (ULCC), which may be more than 1,300 feet long (400 meters) and require nearly 7 miles to stop when traveling at an average speed of 20 knots.

Table 11–4. Oil tanker classifications and size

Tanker Class	Size (deadweight)	Price
Product Tanker	10,000–60,000	$43m
Panamax	60,000–80,000	
Aframax	80,000–120,000	$58m
Suezmax	120,000–200,000	
Very Large Crude Carrier (VLCC)	200,000–320,000	$120m
Ultra Large Crude Carrier (ULCC)	320,000–550,000	

Note: Tanker classes shown are those of the flexible market scale introduced in the mid-1960s and widely used and recognized today.

Tanker sizes vary by use—inland waterways and coastal movements often require the use of smaller sizes due to draft, waterway width, and navigational restrictions, as well as the availability of appropriate loading and handling facilities. The other critical component of size is which tankers today may pass through the Suez and Panama canals, the *Panamax* and *Suezmax* sizes listed in table 11–4. The Panama Canal, because it uses locks to move ships from one ocean to the other, when oceans are of a different sea level, cannot accommodate larger than an 80,000 DWT tanker. The Suez Canal, although not using locks, does limit the size of tankers to 200,000 DWT due to their draft.

A note on tanker sizes and carrying capacity: ships—tankers—are classified by deadweight tonnage (DWT). This is the total weight of the ship when fully loaded, including fuel, cargo, passengers and crew, and provisions—including fresh water and ballast water. A fully loaded tanker is indicated by the water level rising to what is called the *Plimsoll mark*, the horizontal line indication on the external hull. (This allows immediate inspection as to whether a ship is under or over its design carrying capacity.) Although imprecise, a rough estimate of the crude oil carrying capacity is to estimate 7.5 barrels per DWT. This means that a VLCC tanker of 300,000 DWT is capable of carrying 2.25 million barrels of oil. Greater precision would require knowing a variety of other factors, including the characteristics of the crude and the temperatures associated with the vessel, loading, and voyage.

A brief history of oil tankers

The first oil-carrying ships were built in the 1860s.[16] They quickly moved from wind-driven (sailing) ships to steam-driven ships. The first steam-driven tanker, the *Vaderland*, was built by the Palmers Shipbuilding & Iron Company of Jarrow, England in 1873. At the same time, the early oil producers in Pennsylvania in the United States were using barges on inland waterways to move crude oil. Barges, at that time, were nothing more than floating platforms, having no power of their own and pulled by a variety of other engines or ships.

The oil tanker industry escalated rapidly with the construction of the *Zoroaster* by the Nobel Brothers, Ludvig and Robert, of Sweden, in 1878. Once again we return to our traditional oil core city of Baku, in which the Nobel Brothers were then operating. Ludvig and Robert were the principal owner-operators of their own oil company, Branobel, founded and based in Baku, but were frustrated in their ability to move their crude oil to the marketplace beyond the Caspian Sea. (A third brother, Alfred, pursued other interests such as the invention of dynamite and the granting of funds for the establishment of the Nobel Prizes.) Robert Nobel's engineering challenges in 1878 included many of the same safety and security issues tankers face today. Nobel struggled to design a crude oil–carrying ship that could prevent crude fumes from igniting near the ship's power source, allow the cargo to shrink and expand with changes in temperature, yet could be of such size and tank rigidity to assure safety and security.

The *Zoroaster* was a 242 ton kerosene carrier with two tanks onboard linked by pipes. It was followed by two identical ships, the *Buddha* and the *Nordenskjöld*. They were, for all intents and purposes, steam-powered barges with two large tanks atop the hull. The *Nordenskjöld* became infamous as one of the first major tanker disasters. During loading in Baku in 1881, the fixed kerosene loading pipe was torn away from the ship, and the kerosene spilled and exploded, killing half the ship's crew. Nobel immediately introduced a flexible loading pipe that could withstand the constant movement of a ship while at dock.

Evolution of tanker design

The grounding of the tanker *Glückauf* in 1893 is one of the memorable events in tanker history. Designed by Colonel Henry F. Swan, a designer for the Nobel brothers' firm, for operation in the Caspian Sea, the ship is considered the design origin of the modern tanker. The *Glückauf* featured cargo valves operable from the deck, cargo main piping, a vapor line, cofferdams for added safety, and the ability to load seawater ballast when empty of cargo. Wilhelm Anton Riedemann, an agent for the Standard Oil Company, later purchased the *Glückauf* and several of her sister ships. The *Glückauf* ran aground on Fire Island, New York, in heavy fog.

In 1880, Robert Nobel moved a critical step closer to the oil tankers of today by constructing an oil tanker that used its own hull as the containment vessel for the kerosene or crude oil. These early ships, the first of which was named the *Moses*, were the first single-hulled tankers. The *Moses* design was advanced soon after when Colonel Swan designed a single-hull tanker which divided the ship's hull into a series of subtanks. The subtanks were divided laterally (port and starboard) and longitudinally, spanning the entire width and length of the ship. This subdivision eliminated one of the biggest early problems with the single-hull design—when oil or kerosene would slosh from side to side in ocean transit, destabilizing the ship and contributing to a number of major ships capsizing. The subdivisions reduced the sloshing, the *free surface effect*.

Soon after the turn of the 20th century, a number of new tankers were introduced with internal combustion engines, moving from steam power to diesel power and diesel-electric power. The greater engine power in turn allowed the introduction of larger and larger sizes in tanker capacity.

Oil tankers shift the competitive balance

Standard Oil (US) dominated a number of the world's growing markets for crude oil in the 1880s and 1890s. With Standard's near-monopoly ownership of major ocean-going crude oil tankers, and the distance and difficulty of moving Russian or European oil to the growing markets of Asia (particularly Japan), all major competitors were effectively locked out. That all changed, however, when two key individuals in history, Marcus Samuel and ship owner/broker Fred Lane, lobbied tirelessly for approval for oil tankers to pass through the Suez Canal. Fears of fire and explosion had led many, including the Suez Canal Company, to not allow tanker passage.

Marcus Samuel, a British importer and entrepreneur, saw the great opportunities in the global oil markets of the 1890s, but continued to be frustrated by his inability to get Russian oil (Baku and the surrounding region) competitively to many of the rapidly growing markets in the Far East. After repeated rejections by the Suez Canal Company, he took a different tack: "Tell me the specific requirements of an oil tanker in size and safety from the Canal Company's perspective, and I will have an oil tanker designed to meet those requirements." The William Gray shipbuilding company of England then built the *Murex*, the *Conch*, and the *Clam* (note that all the names are of shells), which were finally allowed passage through the Suez Canal—the *Murex* passing through on August 24, 1892. These three ships were owned by a newly created organization, the Tank Syndicate, later the core business unit of the company named Shell. This new routing for crude oil allowed Russian oil from Baku to compete in the growing Asian market.

The First and Second World Wars had significant impacts on the development of crude oil tankers. One such development was the introduction of *underway replenishment*—refueling of other ships at sea—which had a significant beneficial impact on the British and American navies' combat readiness in the First World War. Both wars saw the massive build-up of shipbuilding and tanker building, increasing the infrastructure, knowledge base, technological developments, and skills needed for bigger and better crude oil tankers.

The modern supertanker

Most crude oil today moves via VLCCs (very large crude carriers) or ULCCs (Ultra Large Crude Carriers). It is a service involving large volumes and long distances. VLCCs and ULCCs move massive quantities of crude oil over extensive trade routes circling the globe, typically loading at offshore platforms and discharging at designated lightering zones constructed miles off the coast of destination ports. The economic benefits have proven enormous. The average cost of transporting crude oil today varies between $0.80/bbl and $1.20/bbl, but may vary widely from these values depending on special circumstances. (To add perspective to these transportation costs, railroad transport costs for a barrel of oil averaged roughly $1.50/bbl in the 1870s.) These are costs second only to pipelines, and tankers are much more flexible in terms of their utilization and application. Take special note of these costs; there is no truly competitive cost alternative to supertanker transit costs.

Like any land route, the major sea routes used today are well traveled and tightly scheduled. The economic benefits of utilizing the supertanker fleet is so strong that it also proves economic to use the VLCCs and ULCCs for major routes, then off-load their crude to smaller tankers, a practice termed *lightering*, when specific receiving terminals are inaccessible by the larger tankers. These smaller tankers are then capable of much intercoastal trade and movement. One of the continuing problems with tanker use is that most tanker movements are one-way transits, the ship having to move to its next port or return to its origin port for its next cargo.

Chartering a vessel

Roughly 75% of the world's crude oil tanker fleet is independently owned and chartered. These independent owners use major brokers in Singapore, New York, London, and a few other major shipping centers, to charter ships, a process called *fixing a charter*.

A broker tries to match crude that is available for shipping at a specific port or terminal with an appropriate ship available at the same time and place.[17] It is a pretty dicey problem in many instances to find the appropriate matches, with many ports having no crude available for lifting—*tight avails*—while others are awash in crude with no transportation—*plenty of avails*. If there is crude available and in need of transport, the broker contacts other brokers or ship owners directly to check on tonnage available. Ships that are *laid up* are awaiting charter. The charterer or broker will then discuss with the potential charter vessel operational tolerance (volume within accepted deviations), the product type for transport, the size of the vessel, the route, and the ports for loading and discharging.

For many brokers and vessels, this is not a random process. Many brokers will have a very narrow window of when they expect crude to be available at the port, the *lifting window*, and therefore can schedule vessels for chartering out into the near future. Once a vessel reaches a port for a scheduled or assigned layover, it will contact the port authorities and file a notice of readiness (NOR). The actual loading of the vessel will then be scheduled within the previously noted lay days.

There are three basic types of tanker charter: spot charters, contract of affreightment, and period charters. *Spot charters* are exactly what they sound like, single voyage charters to move a single cargo load for a one-time journey. Spot charters designate a specific vessel and route, and charter rates must cover all expenses at port and during the voyage, including bunkering (fueling), port charges, crew costs, repairs, insurance, and canal tolls, if applicable. Spot charter charges are quoted and paid in US dollars per metric ton (tonne). Similar to spot charters are *contract of affreightment* (COA) charters, in which a specific crude quantity is named and time and places for loading and delivery are specified, but the specific vessel is not specified. Like spot charters, COA agreements are priced in dollars per tonne. *Period charters* are roughly equivalent to a longer term contract and may specify a series of charters of a specific vessel for the same or multiple routes, an extended period of time sometimes approaching 8 to 10 years, and are priced on a day-rate basis rather than on a metric ton basis.

Shipping contracts

There are a number of different types of contracts commonly in use in shipping today: free on board (FOB); cargo, insurance and freight (CIF); and delivered ex-ship (DES) being the most common.

Free on board (FOB). A free on board contract obligates the buyer to be responsible for shipping the LNG from the liquefaction facility to receiving terminal. Ownership of the crude passes from the producer to the buyer at the export loading point facilities. All shipping and insurance costs are born by the buyer.

Cargo, insurance, and freight (CIF). A CIF sale agreement makes shipping costs the responsibility of the seller. An important feature of CIF sales is that ownership of the crude is transferred in the midst of the voyage, typically in international waters (and not at the point of delivery). This transfer of ownership is sometimes of value to buyers who for one reason or another do not like taking possession of the crude in the home waters of the exporter's loading facilities.

Delivered ex-ship (DES). Under a DES sale the seller is responsible for all shipping and delivery costs, the transfer of crude ownership not occurring until transferred to the buyer at the receiving terminal.

Tanker charter rates

Most of the world's tankers today are priced using a rather unique system known as *Worldscale*. Developed largely from different United Kingdom and United States government rate systems designed for pricing government vessels following World War II, the Worldscale system has evolved into a globally accepted pricing standard. The system has been successful because it provides a standardized pricing mechanism for a highly dispersed global business in which actual charter rates need to be standardized in convention but flexible enough to accommodate a market that changes daily.

Worldscale, a nonprofit organization, publishes a 500-page rate book annually. Worldscale NYC publishes the vessel rates for the Americas, with Worldscale London publishing appropriate rates for the rest of the world. Worldscale first establishes a *flat rate*, a baseline rate on a US dollar per metric ton basis, for more than 320,000 individual routes globally. These flat rates, which are called WS100 or Worldscale 100%, are revised annually to reflect changes in fuel costs (bunkering), port tariffs, and exchange rates. The flat rates are established both on a region-to-region basis and a port-to-port basis, allowing for flexibility in many contracts in which loading and unloading may occur at multiple ports within an origin or destination region.

Worldscale's standard flat rate charter rates, WS100, are based on the following:

- Standard vessel size of 75,000 tonnes
- Average service speed of 14.5 knots
- Average bunker (fuel) consumption of 55 tonnes per day
- Port time of four days per voyage
- Fixed hired rate, a daily charge, of $12,000
- Port costs converted to US dollars at average exchange rate
- Panama canal transit time of 24 hours
- Suez Canal transit time of 30 hours
- Bunker price (tanker fuel) as calculated from the mean of October 1–September 30 of the previous year

The rates are then scaled across differing tanker sizes and routes to create the complex and detailed published rates. One of the biggest changes from year to year is average cost of bunker fuel. For example, bunker fuel costs, which are themselves a product of crude oil and therefore crude oil prices, rose dramatically in the October 1, 2007 to September 30, 2008 period. When WS announced their adjustments for rates for the 2009 year, rates rose 35% as a result of bunker fuel rising from $328.75/tonne to $554/tonne for the period.

Table 11–5 provides a sample of selected Worldscale rates for 2008. (Worldscale establishes rates for a year, such as 2008, based on the observed costs from October 1 through September 30 of the previous year. This means that the 320,000 individual flat rates for 2008 are established in October and November of 2007.) The specificity of tanker rates is clearly established from the beginning, as the rates are distinguished by tanker class (VLCC, Suezmax, etc.), the specific size of the individual tanker (280,000 DWT, 260,000 DWT, 80,000 DWT, etc.), and the exact route by both region and port cities of loading and discharge. The routes are often reduced to the codes shown, TD1 and TD3 being two of the most widely watched and quoted routes as they apply to some of the largest cargos and longest routes and most valued crude oil trades in the world.

Table 11–5. Selected Worldscale rates, 2008

VLCC				WS100		
Route	Cargo (mmt)	Load / Discharge Region	Load / Discharge Port	2008	2007	Change
TD1	280	Arabian Gulf / US Gulf Coast	Ras Tanura / LOOP	$30.86	$30.27	1.9%
TD2	260	Arabian Gulf / Singapore	Ras Tanura / Singapore	$10.33	$10.13	2.0%
TD3	260	Arabian Gulf / Japan	Ras Tanura / Chiba	$18.05	$17.72	1.9%
TD4	260	West Africa / US Gulf Coast	Offshore Bonny / LOOP	$15.50	$15.20	2.0%
TD15	260	West Africa / China	Serpentina & Off Bonny / Ningbo	$26.56	$26.08	1.8%
						Average 1.9%
Suezmax				WS100		
Route	Cargo (mmt)	Load / Discharge Region	Load / Discharge Port	2008	2007	Change
TD5	130	West Africa / US Atlantic Coast	Offshore Bonny / Philadelphia	$14.19	$13.93	1.9%
TD6	135	BSEA - Mediterranean	Novoryssiysk / Augusta	$7.13	$6.95	2.6%
						Average 2.2%

Table 11–5. (*Continued*)

Aframax				WS100		
Route	Cargo (mmt)	Load / Discharge Region	Load / Discharge Port	2008	2007	Change
TD7	80	Inter UKC	Sullom Voe / Willhelmshaven	\$5.40	\$5.09	6.1%
TD8	80	Arabian Gulf / Singapore	Mina Al Ahmadi / Singapore	\$10.83	\$10.61	2.1%
TD9	70	Caribbean / US Gulf Coast	Jose Terminal / Corpus Christi	\$7.31	\$7.18	1.8%
TD11	80	Cross Mediterranean	Banias / Lavera	\$6.96	\$6.72	3.6%
TD14	80	Southeast Asia / EC Australia	Seria / Sydney	\$12.65	\$12.24	3.3%
TC1	75	Arabian Gulf / Japan	Ras Tanura / Yokohama CPP	\$17.80	\$17.47	1.9%
						Average 3.1%

Panamax				WS100		
Route	Cargo (mmt)	Load / Discharge Region	Load / Discharge Port	2008	2007	Change
TD10	50	Caribbean / US Atlantic Coast	Aruba / New York	\$6.32	\$6.17	2.4%
TD12	55	ARA / US Gulf Coast	Antwerp / Houston	\$15.25	\$14.81	3.0%
TC5	55	Arabian Gulf / Japan	Ras Tanura / Yokohama CPP	\$17.80	\$17.47	1.9%
						Average 2.4%

Source: McQuilling Services, LLC, December 2007.

The key to understanding Worldscale quotes and points is that they represent a base rate at which the charterer and ship owner *begin* their negotiations. WS100 flat rates are established on an annual basis, and therefore will not reflect the day-to-day changes in operating costs and market conditions that need to be included in any negotiated charter.

For example, consider the first route shown in table 11–5, TD1 for a VLCC tanker. TD1 is one of the largest volume route, from the loading in the Persian Gulf (African Gulf or AG) port of Ras Tanura (Saudi Arabia), to the discharge port of LOOP, the Louisiana Offshore Oil Port, the deepwater offshore mooring and discharging facility that is used by supertankers on the US Gulf Coast. The flat rate quoted for 2008, \$30.86 per tonne, is then used to establish the base rate for the voyage charter. If the specific charter on some date in 2008 was agreed upon as "WS100 or Worldscale 100%," the applicable charge would be:

$$\$30.86 \times \text{WS}1.00 = \$30.86 \text{ per metric ton}$$

for a 280,000 DWT VLCC supertanker. Most likely, however, the specific charter rate on that date in 2008 will reflect changing market conditions, including changing bunkering costs and general avails, so that the parties may agree to a Worldscale rate of, for example, WS140. (In recent years the actual charter rates per year have varied anywhere from WS40 to WS300.) For this charter, the charge would then be:

$$\$30.86 \times \text{WS}1.40 = \$43.20 \text{ per metric ton}$$

For the 280,000 DWT vessel noted, this would translate into a total charge of a little more than $12 million for this shipment of crude oil:

$$280{,}000 \text{ DWT} \times \$43.20 = \$12{,}096{,}000$$

If we assume an average of 7.5 barrels of crude oil per metric ton, then the actual transportation cost of this load of crude oil is $5.76/bbl.

$$\frac{\$12{,}096{,}000}{280{,}000 \text{ DWT} \times \$7.5 \dfrac{\text{bbls}}{\text{DWT}}} = \frac{\$12{,}096{,}000}{2{,}100{,}00 \text{ bbls}} = \$5.76/\text{bbl}$$

Not bad for a voyage of more than 9,800 nautical miles taking 29 days and 5 hours' sailing time![18]

Worldscale flat rates obviously change from year to year. Figure 11–9 shows several of the more widely quoted Worldscale rates for the 2000 to 2010 period. Note that the 2010 flat rates, established in October 2009, showed dramatic decreases from the 2009 flat rates (which had been established in October 2008, only months following the peak of crude oil prices at above $140/bbl.)

Like crude oil prices themselves, Worldscale flat rates clearly rose throughout the period, jumping significantly in 2008 and 2009. One of the biggest single cost drivers to Worldscale over the period was bunkering charges themselves, as fuel prices rose from $86.50/mt in 2000 to $554.05/mt in 2009.[19] The Worldscale system's real flexibility, is in its ability to withstand the severe changes that occur within the year, the percentage variances to the WS100 scale. The market has traditionally been highly seasonal, with the first and fourth quarters chartering at premiums to the flat rate, and the second and third quarters averaging significant discounts from WS100.

One of the most widely used measures of bunkering costs is IFO 380, an intermediate fuel oil, a blend of gasoil and heavy fuel oil (maximum viscosity of 380 centistokes). But as figure 11–10 makes clear, the ability of a flat rate, even for IFO 380, is to serve as a benchmark or marker for starting negotiations, and it is the primary role Worldscale plays, not in the setting of the specific rates for a specific charter in a rapidly changing market.

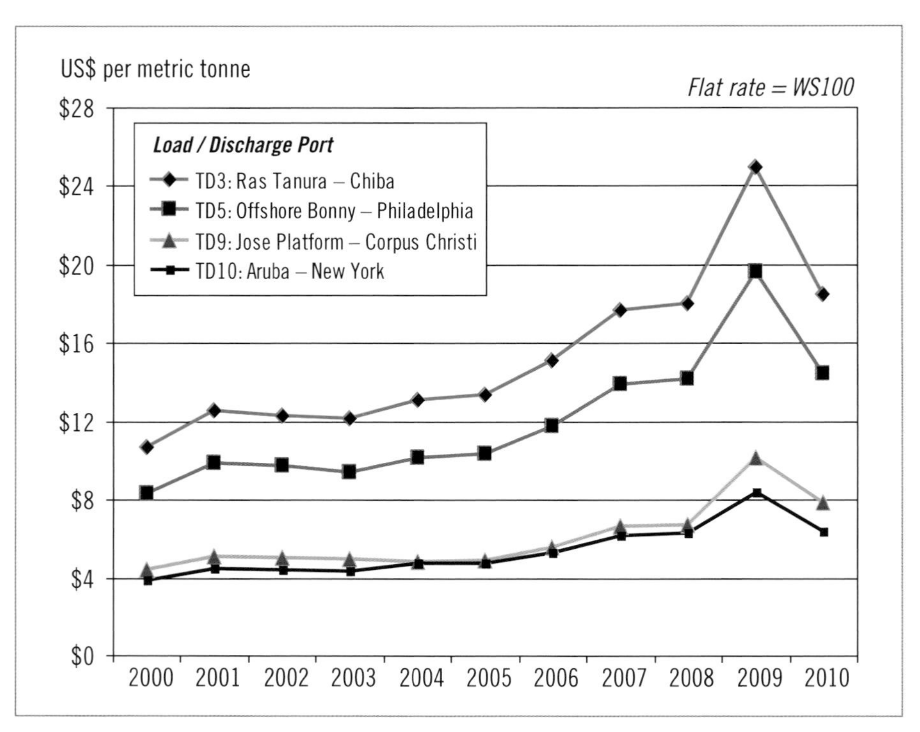

Figure 11–9. Worldscale flat rates for selected routes

Source: McQuilling Services LLC, "Tankers, No. 26, Flat Rate Forecast for 2010," October 15, 2009.

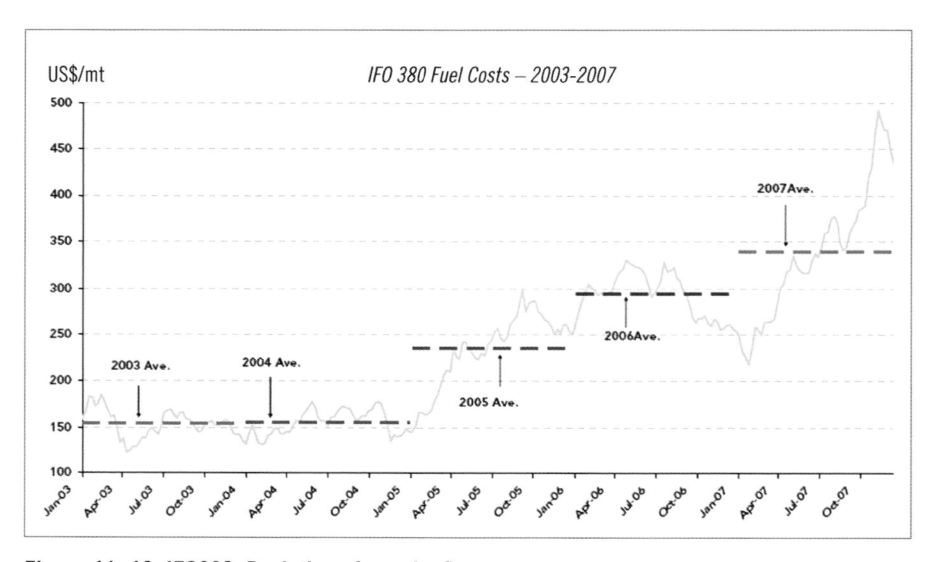

Figure 11–10. IFO380: Deviations from the flat rate

Source: "Flatter Flat Rates," Poten & Partners, December 21, 2007. Copyright © 2007 Poten & Partners, Inc. Reprinted with permission.

Tanker operators

The world's crude oil supertanker fleet is predominantly owned, roughly 75%, by independent owners (nonoil companies). It is, they will anxiously relate, a cyclical business of feast and famine. The risks, as conveyed by Frontline, the world's largest tanker operator, are numerous:

> *The operation of an ocean-going vessel carries inherent risks. These risks include the possibility of a marine disaster, piracy, environmental accidents, cargo and property losses or damage; and business interruptions caused by mechanical failure, human error, war, terrorism, piracy, political action in various countries, labor strikes, or adverse weather conditions.*[20]

Companies like Frontline (table 11–6 provides a list of major oil tanker owners globally) face a business environment in which the borders between regulatory and political regimes are indeed vague and liquid. Depending on the company and its particular fleet, the biggest challenges to current investments and returns combine single-hull components of the portfolio, rising fuel and labor costs, and the proportion of the total business portfolio subject to spot rate chartering.

Table 11–6. Largest crude tanker operators

Teekay Corporation	Maersk Tankers
Frontline	BP Shipping
MOL Tankship Management	Sovcomflot
Overseas Shipholding Group	Novorossiysk Shipping Company
Euronav	National Shipping Co of Saudi Arabia
Tanker Pacific Management	Shipping Corporation of India
Kristen Navigation	Thenamaris
Nippon Yusen Kaisha	TORM
MISC Berhad	Chevron Shipping
Tsakos Group	COSCO Group
Vela International Marine	Kuwait Oil Tanker Co.
NITC	Titan Ocean
Hyundai Merchant Marine	China Shipping Development Tanker
BW Shipping	SK Shipping
Dynacom Tankers Management	Minerva Marine

Environmental risk

Yeah, this is the Valdez back. We should be on your radar there. We've fetched up hard aground north of Goose Island off Bligh Reef. Evidently we're leaking some oil and we're going to be here for a while.

Captain Joseph J. Hazelwood, Initial radio message from ExxonValdez to the US Coast Guard, 12:27 a.m., March 24, 1989[21]

With these first words a new era in crude oil transportation was born. The grounding of the *Exxon Valdez* ushered in a new age of heightened awareness regarding the potential environmental degradation possible from the transportation of crude oil. But as illustrated by figure 11–11, the actual tonnage of oil spilled on an annual basis has trended downward for the past 40 years.

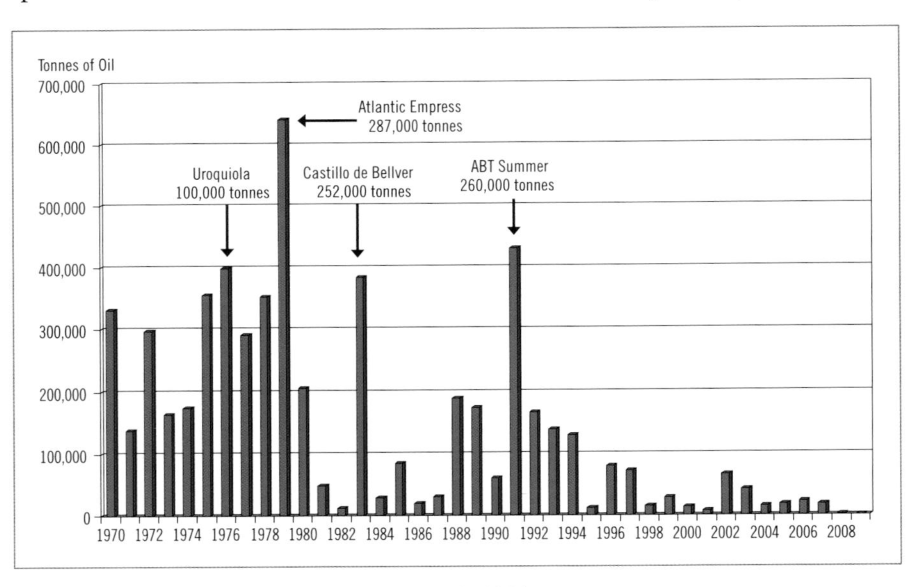

Figure 11–11. Quantity of tanker oil spilt, 1970–2009

Source: International Tanker Owners Pollution Federation LLC. Oil spill measurement includes all oil spilled, burned, and remaining in ships; all oil released to the environment.

Since 1970, a total of 5.66 million tons of oil have been spilled by tankers globally.[22] Figure 11–11 also drives home a critical element of the environmental risk associated with crude oil tanker traffic: it is a risk based on single events. The four largest spill years since 1970—1976, 1979, 1983, and 1991—were all largely a result of singular tanker spills. In fact, the top 20 largest oil spills since 1970, as listed in figure 11–11, make up 2.52 million tons of the total 5.66 million tons, 45%. Ironically, the oil spill that is the most infamous, that of *Exxon Valdez*, is not even within the top 20 spills (see table 11–7), but actually is ranked number 35 over the past 40 years of crude oil tanker traffic and history. And it was the grounding of the *Torrey Canyon* in 1967 that led to much of the legislation and conventions in action today to reduce ocean pollution from oil.

Table 11–7. Largest oil tanker spills, 1967–2009

Rank	Shipname	Year	Location	Spill Size (bbls)	Hull Type
1	*Atlantic Empress*	1979	Off Tobago, West Indies	287,000	Single
2	*ABT Summer*	1991	700 nautical miles off Angola	260,000	Single
3	*Castillo de Bellver*	1983	Off Saldanha Bay, South Africa	252,000	Single
4	*Amoco Cadiz*	1978	Off Brittany, France	223,000	Single
5	*Haven*	1991	Genoa, Italy	144,000	Single
6	*Odyssey*	1988	700 nautical miles off Nova Scotia, Canada	132,000	Single
7	*Torrey Canyon*	1967	Scilly Isles, UK	119,000	Single
8	*Sea Star*	1972	Gulf of Oman	115,000	Single
9	*Irenes Serenade*	1980	Navarino Bay, Greece	100,000	Single
10	*Urquiola*	1976	La Coruna, Spain	100,000	Single
11	*Hawaiian Patriot*	1977	300 nautical miles off Honolulu	95,000	Single
12	*Independenta*	1979	Bosphorus, Turkey	95,000	Single
13	*Jakob Maersk*	1975	Oporto, Portugal	88,000	Single
14	*Braer*	1993	Shetland Islands, UK	85,000	Single
15	*Khark 5*	1989	120 miles off Atlantic coast of Morocco	80,000	Single
16	*Aegean Sea*	1992	La Coruna, Spain	74,000	Double
17	*Sea Empress*	1996	Milford Haven, UK	72,000	Single
18	*Nova*	1985	Off Kharg Island, Gulf of Iran	70,000	Single
19	*Katina P*	1992	Off Maputo, Mozambique	66,700	Single
20	*Prestige*	2002	Off Galicia, Spain	63,000	Single
				2,520,700	

Source: International Tanker Owners Pollution Federation, Ltd., and author research.

Oil tanker spills result from many causes, often a combination of events, circumstances, or sequence of events. The International Tanker Owners Pollution Federation, which groups spills into operational or accidental in causation, has concluded that:[23]

> *Most spills from tankers result from routine operations such as loading, discharging and bunkering which normally occur in ports or at oil terminals; the majority of these operational spills are small, with some 90%*

involving quantities of less than 7 tonnes; accidental causes such as collisions and groundings generally give rise to much larger spills, with at least 84% of incidents involving quantities in excess of 700 tonnes.

What is also observed from figure 11–7 is a common characteristic—a single hull (19 of the 20). The double-hull tanker, in which there is a second watertight hull that is constructed further into the ship (typically 1 to 2 feet), preventing most spills in the event of an outer hull rupture. The construction and use of double hull tankers has been the focus of much of the global effort to reduce oil tanker spillage, spills like that of the *Torrey Canyon* described in the Industry Insight that follows.

Industry Insight: The Wreck of the *Torrey Canyon*

In 1967, the tanker *Torrey Canyon* ran aground while entering the English Channel and spilled her entire cargo of 120,000 tons of crude oil into the sea. This resulted in the biggest oil pollution incident ever recorded up to that time. The incident raised questions about measures then in place to prevent oil pollution from ships and also exposed deficiencies in the existing system for providing compensation following accidents at sea.

First, IMO [International Maritime Organization] called an Extraordinary session of its Council, which drew up a plan of action on technical and legal aspects of the *Torrey Canyon* incident. Then, the IMO assembly decided in 1969 to convene an international conference in 1973 to prepare a suitable international agreement for placing restraints on the contamination of the sea, land, and air by ships.

In the meantime, in 1971, IMO adopted further amendments to OILPOL 1954 to afford additional protection to the Great Barrier Reef of Australia and also to limit the size of tanks on oil tankers, thereby minimizing the amount of oil that could escape in the event of a collision or stranding.

Source: International Maritime Organization, www.imo.org/conventions.

One of the primary international efforts over the past 30 years to reduce oil spills is the International Convention for the Prevention of Pollution from Ships (commonly referred to as *Marpol* for "marine pollution"). With initial conventions in 1973 and 1978, Marpol was initiated to reduce the pollution of the seas—including dumping and spillage of oil. Marpol today counts 136 signatory countries, representing 98% of the world's shipping tonnage. One of Marpol's strongest initiatives has been the drive to eliminate single-hull tankers.[24]

In the United States, the Oil Pollution Act of 1990 following the wreck of the *Exxon Valdez*, required the phasing out of all oil shipment in single-hull vessels in US waters from 1995 to January 1, 2015 (the largest and oldest ships would be phased out first). After January 1, 2015, only double-hull vessels may be used. The European Parliament and European Council/Commission on Legislative Documents have also taken steps phasing in the double-hull or an equivalent design for single-hull oil tankers.

Although continued study of all shipping operations globally has obviously increased shipping safety and spill reduction over time, no single measure such as double hulls can prevent accidents and oil spills on its own. One recent study summarized it quite effectively: Well maintained, diligently operated, high quality tankers whatever the construction, are the answer.[25]

Downstream Transportation

Whereas upstream crude oil transportation is dominated by pipelines and super-tankers, the transportation of downstream products—the refined derivatives and petrochemical products of crude oil—uses nearly every form of transportation known to man. Tanker ships of smaller tonnage and shallower draft may be used in many markets, for example intercoastal movements in the United States, for moving large volumes to other transportation nodes or modes. That said, moving petroleum products from refinery to market is generally dominated by pipeline, railroad, and tanker truck in most major markets today.

Railroad transport today

The railroad tanker car has evolved today to be a structure of increasing sophistication, complexity, and cost. Modern railroad tank cars include many notable features.

Railroad tankers today can be insulated, lined, pressurized, single, or multiple load in capability. Tank cars are specialized pieces of equipment, with the interior of the car often lined with a material such as glass to isolate the car's structure from the contents. This may prove critical given the corrosive or acidic qualities of many loads. Tank cars, unlike the traditional freight or boxcars, are typically one-way cars owing to their specialized structure. This, however, makes them *dead weight half-time*, a cost feature that is unappealing to most of the major railroads of the world. These cars must then be financed—invested in—by separate companies specializing in the transportation of crude oil products. This is usually indicated by the signage on rolling stock; a railcar designation that ends in **X** means that the owner is not a common carrier.

Industry Insight: Railroads and Rebates

Railroad transportation of crude oil became the centerpiece of public debate early in the industry's growth as the cost of oil transportation and the rate-setting practices of railroads came under public scrutiny.

As the single largest client of the railroads, Standard Oil was in an advantaged bargaining position with them. As a result, it often received extremely favorable rates. For each destination, a public rate was published, which was almost never used. The members of the Standard Oil scheme got a substantial rebate, like 30%–50% off the public price.

Companies operating outside of the scheme, Standard's competitors, not only had to pay full price, like $2.56 a barrel when their competitor paid only $1.56 but the railroad paid back to Rockefeller himself the difference with the public price, here $1! This deal strangled the competitors who had to compensate for this inflated costs elsewhere. (See figure 11–12.)

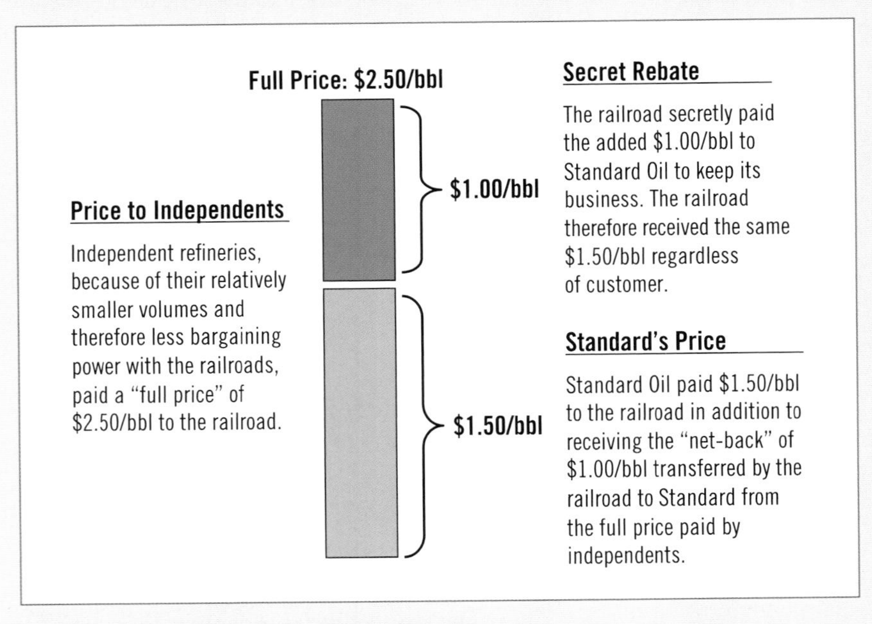

Figure 11–12. The secret railroad rebate system
Source: Ida Tarbell, *The History of the Standard Oil Company*, 1904.

In the railroad tanker industry, tank cars are grouped by the nature of their interior linings, not specifically their cargo. They may be lined with rubber or coated with specialized coatings for protection or the preservation of product purity. Tank cars are regularly inspected for corrosion and damage. This increased regulation and monitoring is somewhat of a result of many train accidents and incidents over the years in which dangerous chemical or product spills posed significant danger to communities along rail lines. One example of the regulatory requirements for environmental and safety concerns is the US requirement that all tank cars have type-F couplers, which prevent the cars from disconnecting during an accident. Disconnections pose the largest risks as they may then puncture neighboring cars, releasing potentially harmful product.

Summary Points

- The transportation of oil and gas, in both the upstream and downstream segments of the global industry, is often critical to both the conduct of business and its possible competitiveness in delivered fuel costs.

- Both pipelines and tanker ships can trace their heritage to the very earliest years of the oil and gas industry, specifically the 1860s in both the United States and the countries surrounding the Caspian Sea.

- The key to minimizing crude oil transportation cost is to get the crude from the well to the ocean tanker in the shortest distance possible. Super tankers today are by far the lowest-cost and most flexible means of moving crude oil around the world, from field to refinery.

- Crude oil pipelines enjoy significant operating cost advantages over other nontanker-based transportation, such as rail and truck.

- Pipelines, because of their physical presence of crossing vast terrain and political jurisdictions, are often the subject of intense political and social debate. They are also quite costly, requiring massive capital outlays that then must be amortized over time through consistent use and high utilization.

- Downstream transportation of oil and gas derivative products utilizes nearly every form of transportation, and often combines elements of sea, rail, and truck to reach the final consumer.

Notes

1. Morgan Downey, "A Brief History of Oil," *Oil 101*, New York: Wooden Table Press, p. 4.
2. Ida M. Tarbell, *The History of the Standard Oil Company*, chapter 2, New York: McClure, Phillips, 1904.
3. Bera, www.bera.org.
4. Canadian Energy Pipeline Association (CEPA), www.cepa.com/pipelines101.
5. Morgan Downey, "Transporting Oil," *Oil 101*, New York: Wooden Table Press, p. 257.
6. Mark Mansley, "Building Tomorrow's Crisis?" Claros Consulting, May 2003.
7. http://www.kindermorgan.com/.
8. http://www.api.org/aboutoilgas/sectors/pipeline/flexibility.cfm.
9. Timothy R. Harbert, "ULS and Transmix/Regrade Reduction, The Last Profit Center for Pipelines," PetroChem Automation and Integration Specialist, undated, p. 2.
10. Ibid, p. 6.
11. http://www.naturalgas.org/regulation/history.asp.
12. Andrew Cargill, "Special Pigs Provided for the Langeled Pipeline," *Pipeline and Gas Journal*, August 2007.
13. "The Baku-Tbilisi-Ceyhan (BTC) Pipeline Project," *Lessons from Experience*, International Finance Corporation, World Bank Group, September 2006, No. 2, pp. 1–2.
14. "The Baku-Tbilisi-Ceyhan Pipeline—A Briefing for BP Staff," BP, February 4, 2003.
15. "The Baku-Tbilisi-Ceyhan Pipeline and BP: A Financial Analysis," Claros Consulting, 2003.
16. This section draws upon a variety of sources, including the Rigzone.
17. Morgan Downey, "Transporting Oil," *Oil 101*, New York: Wooden Table Press, p. 248.
18. Calculation from searates.com.
19. McQuilling Services LLC, "Tankers, No. 26, Flat Rate Forecast for 2010," October 15, 2009.
20. Frontline, Ltd., Form 20-F, Filing with the US Securities and Exchange Commission, December 31, 2008, p. 10.
21. Stan Jones and Patti Epler, "Hazelwood Risked Sinking Ship at Reef Tapes Reveal Captain Tried to Free Valdez," *Anchorage Daily News*, April 25, 1989. http://www.adn.com/evos/stories/EV72.html.
22. Compiled by authors from the statistical data compiled by the International Tanker Owners Pollution Federation.

23. International Tanker Owners Pollution Federation, www.itopf.com/information-services/data-and-statistics/statistics.
24. http://www.imo.org/Safety/mainframe.asp?topic_id=155.
25. "Double Hull Tankers: Are They the Answer?" OCIMF, 2003, p. 10.

Chapter 12

REFINING

I've been pessimistic on refining for 30 years, and I've run the damn places.

—Lee Raymond,
former CEO, ExxonMobil[1]

A petroleum refiner, like most manufacturers, is caught between two markets: the raw materials he needs to purchase and the finished products he offers for sale. The prices of crude oil and its principal refined products, heating oil and unleaded gasoline, are often independently subject to variables of supply, demand, production economics, environmental regulations, and other factors. As such, refiners and non-integrated marketers can be at enormous risk when the prices of crude oil rise while the prices of the finished products remain static, or even decline. . . . Because refiners are on both sides of the market at once, their exposure to market risk can be greater than that incurred by companies who simply sell crude oil at the wellhead, or sell products to the wholesale and retail markets.

—New York Mercantile Exchange, *Crack Spread Handbook*, 2000, p. 4[2]

The refining process is critical to the petroleum industry value chain because crude oil has no value until it is transformed into products such as gasoline, diesel, heating oil, propane, asphalt, and petrochemical inputs. Thus, to a refiner, the value of crude is nothing other than the value of its derivative products. This chapter examines the competitive aspects of refining and discusses some of the key trends impacting the industry.

The first refinery opened in 1861 to produce kerosene for heating and lighting. For the next 30 years, kerosene was the primary refined product. With the invention of the electric light and the automobile, the refining sector shifted its emphasis to motor fuels—mainly gasoline. The introduction of thermal cracking in 1913 significantly increased the amount of gasoline that could be produced. Catalytic cracking and polymerization processes developed in the mid- to late 1930s further improved gasoline yields. Since the 1950s, refining technology has continued to become more efficient. In the United States, many small and inefficient refineries (the so-called tea kettle refineries) were closed because they were too costly to operate, could not keep up with new environmental requirements, or they suffered a decline of local crude oil supplies.

Despite the fact that no new greenfield refineries have been built in the United States for several decades, refinery output continues to grow via expansion of existing refineries. The number of operating US refineries dropped from 195 in 1987 to 149 in 2009 but during that period, US production capacity increased from just under 15 million b/d to over 17 million b/d. On a global basis, refining capacity continues to grow and evolve, with China and India making significant investments in greenfield sites. The growth in demand and capacity must be considered in light of the ongoing cyclical nature of the industry. Whereas a few years ago refining was said to be in its Golden Age, the future promises to bring new challenges and competitive shifts that require new skills and capabilities to succeed.

Global Refining

Figure 12–1 shows that global refining capacity has been steadily growing over the past decade, largely driven by demand from emerging markets. What is initially striking about figure 12–1 is that global capacity peaked in 1980 and then fell. Global capacity did not reach 1980 levels again until 1998—a span of 18 years. Since the mid-1990s, refining capacity has grown, albeit rather slowly given the multitude of other global factors related to energy use, liquids demand, and general oil consumption.

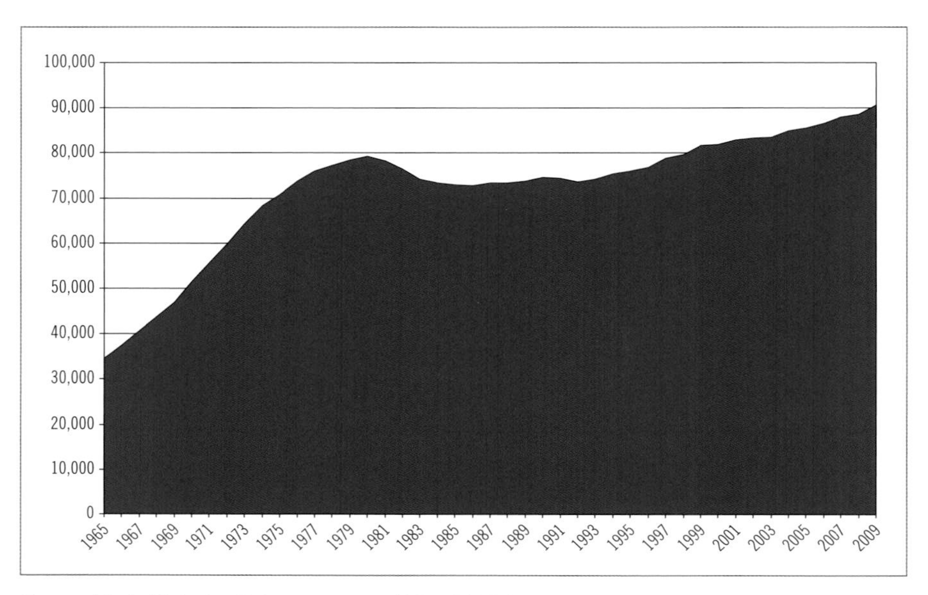

Figure 12–1. Global refining capacity, 1965–2009 (thousand b/d)
Source: *BP Statistical Review of World Energy*, June 2010.

Figure 12–2 shows that by far the fastest growth has been in the Asia Pacific region. Two new refineries have recently been completed in China: the 200,000 b/d Qingdao refinery (owned 75% by Sinopec and 25% by Saudi Aramco) and the 240,000 b/d Huizhou refinery (100% owned by CNOOC). CNOOC also announced a second phase for the Huizhou refinery of 200,000 b/d with a planned startup in 2011 and a cost of $6.6 billion. The $5 billion, 240,000 b/d expansion of the Quanzhou refinery in Fujian Province began operations in 2009 (owned 50% by Sinopec, 25% by ExxonMobil, and 25% by Saudi Aramco). In India, Reliance Industries Ltd. recently started operations at its new Jamnagar refinery in Gujarat Province (see *Industry Insight: Reliance Industries' Jamnagar Refinery*).

In contrast, Europe and North America have had limited growth in recent years. In the US, gasoline demand over the decade following the mid-1990s increased faster than refinery output, straining fuels markets and leading to major disruptions when refinery outages occurred. In response, several major refining expansions are underway. Motiva, the joint venture between Royal Dutch Shell and Aramco, is in the midst of a $7 billion refining expansion in Port Arthur,

Texas that will create the largest refinery in the United States. The 325,000 b/d expansion at Port Arthur is equivalent to building the first new refinery in the United States in more than 30 years. Marathon is adding a 180,000 b/d expansion to its 245,000 b/d Garyville, Louisiana refinery.

Adding refining capacity through debottlenecking also allows refiners to increase capacity. For example, through debottlenecking since 1995, ExxonMobil has effectively added the equivalent capacity of a new industry-average-size refinery to its portfolio every three years at only a fraction of the greenfield cost.

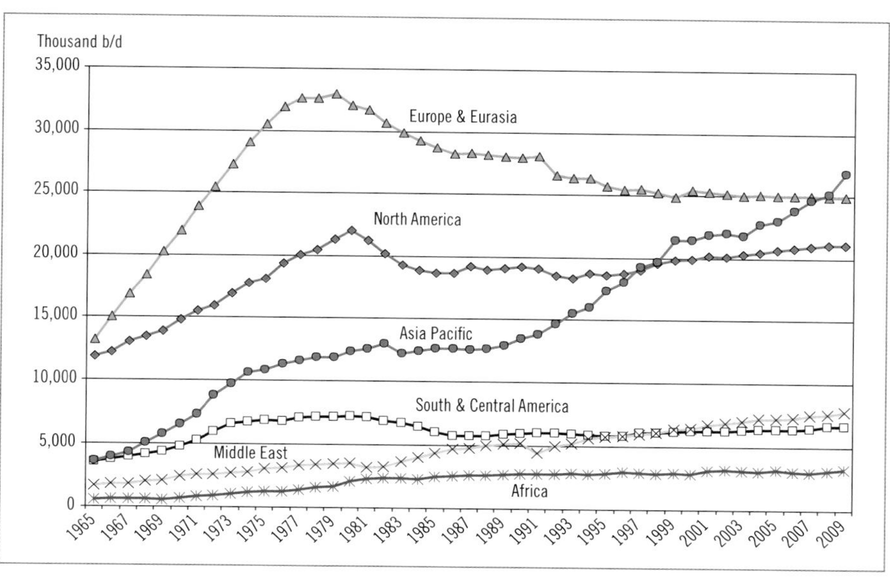

Figure 12–2. Refining capacity by region, 1965–2009 (thousand b/d)
Source: *BP Statistical Review of World Energy*, June 2010.

Industry Insight: Reliance Industries' Jamnagar Refinery

Although no new US and European refineries have been built for many years, refineries are being built in emerging market countries. Consider the case of Reliance Industries. In late 2008, India's Reliance Industries Ltd. started operations at its new 580,000 b/d Jamnagar refinery in Gujarat Province. The new refinery, along with Reliance's neighboring refinery that was completed in 2000, make up the world's largest refining complex, with a capacity of 1.24 million b/d. The new refinery is owned by Reliance's Reliance Petroleum Ltd. unit. To raise capital for the project, about 20% of the equity was sold in an IPO (seven times over subscribed) and 5% was sold to Chevron.

The capital cost of the project was estimated at about $10,000 per barrel, about a third lower than that estimated for other announced refinery projects. According to Reliance, "RPL's low capital cost is a result of the benefits of 'intelligent repeat' of design and engineering aspects of RPL's existing refinery, proactive procurement strategy and faster implementation of the refinery project."* During the construction phase of about 36 months, there were as many as 150,000 construction workers on site.

The refinery is one of the most complex in the world and was designed to process some of the toughest crudes in the world. The Nelson Complexity Index for the new Jamnagar refinery is over 14.0. However, finding a market for the refined products was expected to be a challenge because the refinery was not authorized to market in India. In order to develop export markets, Reliance set up trading desks in Singapore and London and planned to open one in Houston.

According to Mukesh Ambani, Reliance chairman, "We will leverage our competitive advantages of scale, complexity and capability to process a wide range of crude oils and flexibility to produce high-quality transportation fuels."**

---

* http://www.reliancepetroleum.com/html/aboutus.html ** G. Singh, "India's Reliance Refinery Could Weigh on Oil Margins," *Wall Street Journal*, December 24, 2008.

Table 12–1 shows the world's largest refining companies. Not surprisingly, the top of the list is dominated by the major oil and gas firms (both IOCs and NOCs). Although there are some small independent refiners that have managed to compete in refining, most of the world's refining is concentrated in a small set of global players. These global players have refining capacity in various parts of the world and use their global networks to optimize refining operations. In addition, nine of the top ten largest refiners are integrated firms. Only Valero in the top ten is not involved in upstream activities.

Table 12–1. The world's largest refiners

Rank Jan 1, 2010	Rank Jan 1, 2009	Company	Country	IOC or NOC	Crude Capacity (barrels per day)
1	1	ExxonMobil Corp.	US	IOC	5,797,000
2	2	Royal Dutch Shell PLC	UK	IOC	4,630,239
3	3	Sinopec	China	NOC	3,971,000
4	4	BP PLC	UK	IOC	3,328,390
5	6	ConocoPhillips	US	IOC	2,778,200
6	8	Valero Energy Corp	US	IOC	2,704,660
7	6	Petroleos de Venezuela SA	Venezuela	NOC	2,678,000
8	9	China National Petroleum Corp.	China	NOC	2,615,000
9	7	Total SA	France	IOC	2,594,600
10	12	Chevron Corp.	US	IOC	2,492,000
11	10	Saudi Aramco	Saudi Arabia	NOC	2,433,000
12	11	Petroleo Brasileiro SA	Brazil	NOC	1,997,000
13	13	Petroleos Mexicanos	Mexico	NOC	1,703,000
14	14	National Iranian Oil Co.	Iran	NOC	1,451,000
15	17	Nippon Oil Co. Ltd.	Japan	IOC	1,317,000
16	15	Rosneft	Russia	IOC	1,293,000
17	—	Reliance Petroleum	India	IOC	1,240,000
18	16	OAO Lukoil	Russia	IOC	1,217,000
19	20	Marathon Oil Corp.	US	IOC	1,196,000
20	18	Repsol YPF SA	Spain	IOC	1,105,000
21	19	Kuwait National Petroleum Co.	Kuwait	NOC	1,085,000
22	20	Pertamina	Indonesia	NOC	993,000
23	22	Agip Petroli SPA (Eni)	Italy	IOC	904,000
24	23	Sunoco Inc.	US	IOC	825,000
25	25	SK Corp.	Korea	IOC	817,000

Source: *Oil & Gas Journal* and authors.

Table 12–2 shows the world's largest refineries. Of the top ten, only two are in the United States. What is interesting in this table is that the largest IOCs tend not to have the largest refineries, which suggests there is perceived value in having a larger network of smaller plants. Also, most of the biggest refineries are outside the largest Western markets.

Table 12–2. World's largest refineries

Name of Refinery	Location	Barrels/day
Paraguana Refining Complex (CRP)	Amuay and Cardón Venezuela	940,000
SK Energy Co., Ltd.	South Korea	817,000
LG-Caltex	South Korea	680,000
Reliance Industries I*	Jamnagar, India	660,000
ExxonMobil	Singapore	605,000
Reliance Industries II	Jamnagar, India	580,000
ExxonMobil	Baytown, TX, USA	572,000
Saudi Aramco	Ras Tanura, Saudi Arabia	550,000
S-Oil	South Korea	520,000
ExxonMobil	Baton Rouge, LA, USA	503,000
Hovensa LLC	Virgin Islands	500,000
BP Texas City	Texas City, TX, USA	460,000
Shell Eastern	Singapore	449,000
Kuwait NPC	Kuwait	442,700
Citgo Lake Charles	Lake Charles, LA, USA	440,000
Shell Pernis Refinery	Netherlands	406,000
Sinopec	Zhenhai, China	403,000
Saudi Aramco	Rabigh, Saudia Arabia	400,000
Saudi Aramco-Mobil	Yanbu, Saudia Arabia	400,000

* The two Reliance refineries located in Jamnagar are adjacent to each other, making this the largest single refining complex in the world. Source: *Oil & Gas Journal.*

Independent and integrated refiners

As mentioned, the leading refiners in table 12–1 are primarily large integrated oil and gas companies. Given the predominance of IOCs and diversified companies on the list, does this mean that independent refiners have a competitive disadvantage in refining because they are not part of a larger diversified or integrated firm? This is a complex question and involves the fundamental corporate strategy question of value creation and diversification: Is a firm better off as an independent, focused, and stand-alone competitor, or is it better off as part of a larger diversified corporation? For an example of one firm that has managed to successfully compete for many years in only the downstream and as a family-owned firm, see *Industry Insight: Irving Oil—Can a Family-Owned Oil Company Succeed in a World of Giants?*

Industry Insight: Irving Oil—Can A Family-Owned Oil Company Succeed in a World of Giants?

In 1922, Kenneth (K.C.) Irving landed a job selling cars in Bouctouche, New Brunswick, Canada.* A few years later K.C. Irving began building an eastern Canadian business empire that would eventually encompass gas stations, forestry products, potato farms, hardware stores, shipyards, newspapers, food processing, building supplies, transportation, and engineering and construction companies. Along the way, K.C. became one of the richest Canadians, and in doing so, ensured that the family businesses remained in the family, as they have to this day. The various Irving companies have never gone public and there are now third and fourth generation Irvings in control.

As a staunch believer in the power of vertical integration, K.C. wanted to control the supply of gasoline for his 1,900 gas stations and reduce his reliance on buying crude from the Seven Sisters. In 1960, Irving Oil, in partnership with Socal (now Chevron), opened a refinery in Saint John, New Brunswick. The plant was built to allow expansions, which were done in 1971, 1974, and 2000. The 1974 expansion during the first oil crisis was a huge gamble. When the expansion was completed in 1977, Irving Oil had a 250,000 b/d refinery, the largest in Canada. In 1989, Irving Oil bought out Socal's share. In 2009, the refinery had a capacity of 300,000 b/d. The Irving oil refinery has had a number of firsts in Canada:

- First Canadian refiner to make high-octane gasoline without the use of lead additives
- First to offer low-sulfur gasoline, five years ahead of regulation
- First Canadian refinery to offer low-sulfur diesel ahead of regulations
- First Canadian oil company to own and operate double hulled tankers

Not surprisingly, for a family as wealthy and powerful as the Irvings, the growth of Irving Oil and the other companies has not been without a fair amount of controversy. The company has been accused of being many things, including monopolists, bullies, and polluters. Nevertheless, in an industry dominated by huge multinational firms, the family-owned Irving Oil has shown that it is possible to compete against the best and thrive. Perhaps by virtue of keeping the business away from the short-termism of financial markets the Irvings can make longer term and better strategic decisions.

---

* John DeMont, *Citizen Irving: K. C. Irving and His Legacy,* 1992, Toronto: McClelland & Stewart.

It is often said that IOCs, unlike independent refiners, have a natural hedge against adverse price movements of the refining spread components because they control their entire supply chain. In contrast, an independent refiner exposed to the risk of increasing crude oil costs and falling refined product prices runs the risk that refining margins will be less than anticipated.

While it is true that IOCs are involved in multiple activities, to efficiently manage their complex network of assets they buy and sell crude with other refiners. A closer look at ExxonMobil, the world's largest refiner, shows that this natural hedge argument only partially holds true. ExxonMobil refines more than twice as much crude as it produces, which means even if the company refined all of its own crude (which would be highly unusual), the firm would still have to buy crude for its refineries. It is therefore most likely that the primary advantage the IOCs have relative to stand-alone refiners is that the integration between refining and upstream activities allows the IOCs to minimize short-term cyclical effects in either branch of the business, rather than a theoretical argument over their ability to hedge crude prices.

Some analysts argue that unlike an IOC, an independent refiner without any upstream ties can operate with greater flexibility. Some independent refiners go a step further: with no marketing ties to other parts of the firm, the refiner can operate as a pure merchant refiner. A *merchant refiner* is a stand-alone and not part of an integrated distribution system. For example, Hess and PDVSA jointly operate a merchant refinery in St Croix, US Virgin Islands. With a crude oil processing capacity of 500,000 b/d, it is one of the largest in the world.

A merchant refiner can buy crude from any supplier and sell refined product to any customer. The absence of upstream or marketing ties creates flexibility and the potential for fast response to market changes. The absence of the need to optimize a network means the refinery is essentially a stand-alone business. An IOC must optimize a complex network of upstream, refining, and marketing assets. Done efficiently, this optimization can lead to superior performance relative to independent refiners. Done poorly, the IOC could have a slow-moving, cumbersome network of underutilized assets. When it comes to trading its own crude, an IOC always has the option to "bring its own crude home" if it feels it is not getting a square deal in the marketplace. This gives IOC crude traders a strong negotiating position.

The Refining Process

Table 12–3 shows the major products produced by a refinery, including naphtha, gasoline, kerosene, diesel, petrochemical feedstock, lubricants, and other products. A refiner will always be adjusting the yield of products in response to market demands. For example, as discussed later, refiners in the US have been shifting to more diesel production.

Table 12–3. Major refined products

Product	Description
Gasoline	The most important refinery product in the United States; blend of hydrocarbons with boiling ranges from ambient temperatures to about 400°F.
Distillate Fuels	Includes diesel, the most important refinery product in Europe, and heating oils; boiling ranges of about 400°–700°F.
Kerosene	A middle-distillate petroleum product; fuel for jet engines and tractors; starting material for making other products; used around the world in cooking and space heating. Commercial jet fuel has a boiling range of about 375°–525°F, and military jet fuel 130°–550°F.
Liquified Petroleum Gas (LPG)	LPG, mainly propane and butane, is used as fuel and as a petrochemical input.
Naphtha	Intermediate that will be further processed to make gasoline.
Residual Fuels	Used by ships, power plants, commercial buildings and industrial facilities for heating and processing, often in combination with distillate fuels.
Coke and Asphalt	Coke is almost pure carbon with a variety of uses from electrodes to charcoal briquettes. Asphalt, used for roads and roofing materials, must be inert to most chemicals and weather conditions.
Solvents	A variety of products whose boiling points and hydrocarbon composition are closely controlled and are produced for use as solvents; includes benzene, toluene, and xylene.
Petrochemicals	Products such as ethylene, propylene, butylene, and isobutylene primarily intended for use as petrochemical feedstock in the production of plastics, synthetic fibers, synthetic rubbers, and other products.
Lubricating Oil Base Stocks	Additives such as demulsifiers, antioxidants, and viscosity improvers are blended into the base stocks to provide the characteristics required for motor oils, industrial greases, lubricants, and cutting oils.

The modern refining processes that separate these different components are extremely complex. The processes are continuously evolving as new technology is developed and refiners seek to lower their operating costs, reduce emissions, and increase their profitable yields from a barrel of crude. A schematic view of a refinery is shown in figure 12–3. Some of the more important processes are discussed in the sections that follow.

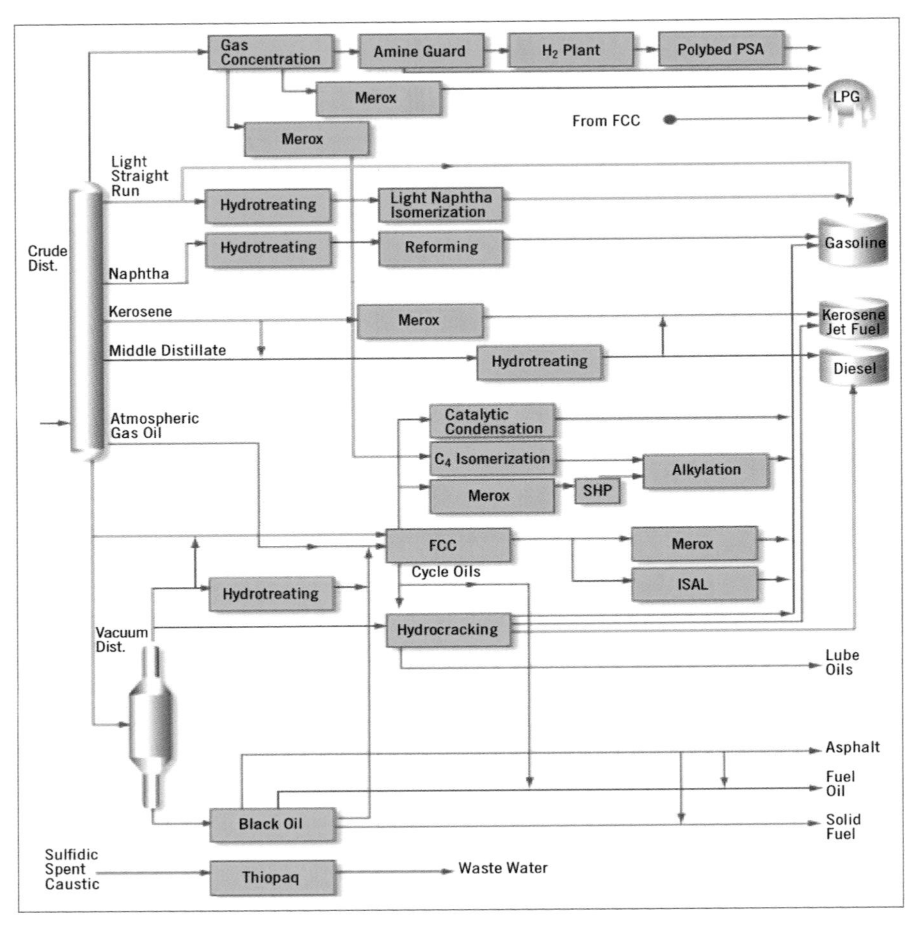

Figure 12–3. The refining process moves from crude oil on the left to derivative products on the far right, each through its own complex chemical process.

Source: "Petroleum Refining Flow Diagram," © 2004–2009 UOP LLC. All rights reserved.

Distillation

Crude oil often contains water, inorganic salts, suspended solids, and water-soluble trace metals. As a first step in the refining process, to reduce corrosion, plugging, and fouling of equipment and to prevent poisoning the catalysts in processing units, these contaminants must be removed by desalting (dehydration). The next step is distillation. The desalted oil passes through a furnace where it is heated to temperatures ranging from 650°F to 700°F (340°C to 370°C). The majority of the crude vaporizes and enters a cylindrical atmospheric distillation column, getting cooler as it goes up. When a hydrocarbon cools below its boiling point, it reverts to liquid form. Inside the distillation column is a series of trays stacked on top of each other. Each tray contains numerous holes that allow the vapor to pass through. A tray's location in the distillation column determines which fraction or cut it will collect. Fractions, arranged from most volatile to least volatile, fall into the following categories: gases, light distillates, middle distillates, gas oils, and residuals.

The trays collect the liquid hydrocarbons, which have now been sorted into several distinct streams. Trays toward the upper end of the column, farther away from the boiler and exposed to cooler temperatures, collect lighter hydrocarbons (i.e., hydrocarbons with fewer carbon atoms) such as naphtha and gasoline. Heavier hydrocarbons such as lubricating oil and heavy gas oil collect on trays on the lower, warmer end of the column. Ultimately processed into products such as asphalt, waxes, and coke, residuals or bottoms are the heaviest hydrocarbons and remain in the liquid state at the lowest part of the distillation tower. Several other processes will also be used for further distillation. Vacuum distillation towers are typically used to separate catalytic cracking feedstock from surplus residuum. Smaller distillation towers called *columns* are designed to separate specific and unique products.

The lighter products from the atmospheric distillation process must be further processed because the simple distillation of crude oil produces amounts and types of products that are not consistent with those required by the marketplace. The catalytic cracking process is used to change the product mix by altering the molecular structure of the hydrocarbons. Cracking breaks or cracks the heavier, higher boiling-point petroleum fractions into more valuable products such as gasoline and diesel. The two basic types of cracking are thermal cracking, using heat and pressure, and catalytic cracking.

Hydrotreating, cracking, and reforming

Hydrotreating is a hydrogenation process used to remove contaminants such as nitrogen, sulfur, oxygen, and metals from liquid petroleum fractions (the amount of contaminants removed is impacted by fuel standard regulations). These contaminants, if not removed from the petroleum fractions as they travel through the refinery processing units, can have detrimental effects on the equipment, the catalysts, and the quality of the finished product. Hydrotreating is usually done prior to processes such as catalytic reforming so that the catalyst is not contaminated by untreated feedstock. Hydrotreating is sometimes used prior to catalytic cracking to reduce sulfur and improve product yields and to upgrade middle-distillate petroleum fractions into finished kerosene, diesel fuel, and heating fuel oils. In addition, hydrotreating converts olefins and aromatics to saturated compounds.

Visbreaking, a mild form of thermal cracking, is used to reduce the pour point of waxy residues and reduce the viscosity of residues used for blending with lighter fuel oils. *Steam cracking* is a petrochemical process sometimes used in refineries to produce olefinic raw materials (e.g., ethylene) from various feedstocks for petrochemicals manufacture. *Coking* is a severe method of thermal cracking used to upgrade heavy residuals into lighter products or distillates. The residue from the process is a form of carbon called "coke."

Fluid catalytic cracking (FCC) breaks complex hydrocarbons into simpler molecules in order to increase the quality and quantity of lighter, more desirable products such as gasoline and kerosene and decrease the amount of residuals. The FCC process breaks large hydrocarbon molecules into smaller molecules by contacting them with powdered catalyst at a high temperature and moderate pressure, which first vaporizes the hydrocarbons and then breaks them.

Use of a *catalyst* (a material that assists a chemical reaction but does not take part in it) in the cracking reaction increases the yield of improved-quality products under less severe operating conditions than in thermal cracking. FCC is the most widely used secondary conversion process in the refining industry.

Hydrocracking is a process in which heavier feedstocks are cracked in the presence of hydrogen to produce more desirable products. The process employs high pressure, high temperature, a catalyst, and hydrogen. Hydrocracking takes feedstocks similar to those used in FCC but yields more diesel instead of gasoline. The hydrogen and catalyst cost make this a reasonably costly process.

Catalytic reforming is an important process used to convert low-octane naphthas into high-octane gasoline blending components called *reformates*. Reforming does not break up molecules; rather, it improves the quality of gasoline by changing chemical characteristics.

Other processes

Isomerization is important for the conversion of n-butane into isobutane, to provide additional feedstock for alkylation units, and for the conversion of normal pentanes and hexanes into higher branched isomers for gasoline blending. *Polymerization* is the process of converting light olefin gases including ethylene, propylene, and butylene into hydrocarbons of higher molecular weight and higher octane that can be used as gasoline blending stocks. Polymerization may be accomplished thermally or in the presence of a catalyst at lower temperatures. *Alkylation* combines low-molecular-weight olefins (primarily a mixture of propylene and butylene) with isobutene in the presence of a catalyst, either sulfuric acid or hydrofluoric acid. The product is called *alkylate* and is a premium blending stock because it has exceptional antiknock properties and is clean burning. *Merox* is an acronym for mercaptan oxidation.

Oil refinery processes that remove mercaptans or hydrogen sulfide are commonly referred to as *sweetening processes* because their products no longer have the sour, foul odors of mercaptans and hydrogen sulfide. Sweetening can be accomplished at an intermediate stage in the refining process or just before sending the finished product to storage.

Safety and environmental issues

Given the nature of a refinery's raw materials and the products produced, refining has many safety and environmental challenges. The refining process releases various chemicals into the air and uses large amounts of water. Refineries can be noisy and may produce a foul odor. Flaring by refineries is an ongoing issue in many communities. There is always the risk of industrial accidents, spills, and explosions. As OSHA's technical manual for refineries states, "The safe and orderly processing of crude oil into flammable gases and liquids at high temperatures and pressures using vessels, equipment, and piping subjected to stress and corrosion requires considerable knowledge, control, and expertise."[3] The various safety and environment issues surrounding refineries is one reason why there have been no greenfield refineries built in the United States for several decades.

For the most part, refineries today operate safely and without incident. Unfortunately, accidents do occur—see *Industry Insight: BP and Texas City Explosion*. In the United States and other industrialized nations' strict environmental regulations cover both the refining process and the products produced. For example, according to the Department of Energy's Office of Energy Efficiency and Renewable Energy, by the late 1990s, US refineries were recycling more than 60% of the hazardous and nonhazardous wastes they created. On the product side, the phase-in of new ultra low sulfur diesel (ULSD) in the United States will be complete by 2010. Used in combination with cleaner-burning diesel engines and vehicles, ULSD yields significant reduction in sulfur emissions. Likewise, removal of lead from gasoline during the refining process began in the 1970s in response to passage of the 1970 Clean Air Act by the US Congress.

The first unleaded gasoline, and the first catalytic converter, appeared in 1975. In 1970, 220,869 short tons of lead were emitted to the air in the US, most of it by automobiles. That number was reduced by 66% in 1980, 97% in 1990, and 98% in 2000.[4] In 2009, leaded gasoline was still sold in some emerging market countries, including China.

Industry Insight: BP and Texas City Explosion

BP's Texas City Refinery in Texas City, Texas is the second-largest oil refinery in the state and the third largest in the United States. Texas City is a major industrial center and in 1947 suffered one of the worst industrial accidents in the United States when a ship full of fertilizer exploded, killing nearly 600 people. Texas City Refinery has a capacity of 437,000 b/d. BP acquired the Texas City Refinery when it bought Amoco in 1998.

Another major explosion occurred in Texas City in an isomerization unit at the site on March 23, 2005, killing 15 workers and injuring more than 170 others. The isomerization (ISOM) unit is used to convert raffinate, a low octane blending feed, into higher octane components for unleaded regular gasoline. The unit has four sections including a splitter that takes raffinate and fractionates it into light and heavy components. The splitter consists of a surge drum, fired heater reboiler, and a fractionating column 164 feet tall.

According to BP's initial internal report, operators overfilled and overheated a processing tower at the unit that housed hydrocarbon liquid and vapor. The liquid and vapor mix was overpressurized, flooded into an adjacent stack, and escaped into the atmosphere around the unit. The resulting vapor cloud was ignited, causing the explosion. The source of ignition is not known. The report said supervisors did not verify that correct procedures were followed and were absent from the unit at key times. "The failure of [the] managers to provide appropriate leadership and the failure of hourly workers to follow written procedures are among the root causes of this incident."*

The subsequent investigation by the US Chemical Safety Board alleged that high-level decisions to defer overhauls, cut staff, and rein in costs at the Texas City, Texas plant helped cause the accident. BP's internal investigation identified several causes, including:**

- Over the years, the working environment had eroded to one characterized by resistance to change, and lacking in trust, motivation, and a sense of purpose. Coupled with unclear expectations around supervisory and management behaviors, this meant that rules were not consistently followed, rigor was lacking, and individuals felt disempowered from suggesting or initiating improvements.

- Process safety, operations performance, and systematic risk reduction priorities had not been set and consistently reinforced by management.

- Many changes in a complex organization had led to the lack of clear accountabilities and poor communication, which together resulted in confusion in the workforce over roles and responsibilities.

As a result of the incident, BP set aside $700 million to compensate victims of the explosion. BP also agreed to pay several fines totaling about $70 million and agreed to a number of corrective actions, including the hiring and placement of process safety and organizational experts at the refinery.

---

* C. Cummins & T. Herrick, "BP Takes Blame for Lethal Blast, Citing Mistakes by Its Employees," *Wall Street Journal*, May 18, 2005, p. A3, ** http://www.bp.com/liveassets/bp_internet/us/bp_us_english/STAGING/local_assets/downloads/t/final_report.pdf.

All the major refiners view safety as a core value that is critical for competitive success. Aside from the fundamental issue of what is socially acceptable, an unsafe refinery will open a company up to regulatory and legal problems and will substantially raise the cost of doing business. In tough times refiners may defer maintenance, which inevitably leads to problems. A major challenge for refiners is convincing the public and their local communities that refineries can be operated safely and that refinery shutdowns are not tied to a desire to suppress supply in order to raise prices. As the executive VP of the refining industry's main trade group, the National Petrochemical and Refiners Association, commented, "Refiners want to keep running in today's economic environment. But when they shut down they are accused of gouging the system. When they don't, they are criticized for overrunning their facilities."[5]

One of the challenging environmental issues for the future involves the processing of heavier crudes, which are becoming increasingly important to the industry. When processing heavy crude, a refiner must deal with additional environmental considerations. Because heavier crudes contain higher volumes of nickel and vanadium, the facility must be equipped with catalysts that can handle the higher content of these potentially harmful metals. Also, the increased volume of water used to process heavy oil means more water that needs to be treated. Modifying the facility's feedstocks requires the refiner to add scrubbers or, quite likely in the future, purchase carbon offsets to account for increased airborne emissions.

Refineries and exit barriers

Refineries are costly to build and often difficult to close. For example, in 2004, Shell announced that it would shut its 70,000 b/d Bakersfield, California refinery because of a decline in production in the Kern River oil field, the source of the refinery's crude. The refinery, which began production in 1932 as a 1,500 b/d facility, was eventually sold after a public outcry about the loss of capacity in California's supply-constrained market (a subsequent 2008 bankruptcy and shutdown by the buyer, Flying J, generated new controversy involving Shell, other suppliers, and access to crude supplies).

In 2009, Valero closed its 182,000 b/d Delaware City refinery when it was unable to find a buyer. In early 2010, Valero announced that it would sell the refinery to a closely held partnership controlled by refining investor Thomas O'Malley. O'Malley has been buying and selling refineries in the United States and Europe for several decades. In the 1990s, he built Tosco into one of the largest US refiners and fuels marketers (see *Industry Insight: The Tosco Story*). The sale of Delaware City to O'Malley was the second time he had acquired the refinery. Premcor, a company run by O'Malley, purchased the refinery from Shell in 2004 for about $455 million. Premcor was acquired by Valero in 2005, and the Delaware City refinery was valued at about $1.8 billion. In his second purchase of Delaware City, O'Malley paid $220 million, less than half of what he paid in 2004 and a small fraction of what Valero paid in 2005. And, the state of Delaware provided a number of financial concessions for O'Malley's company.[6]

Industry Insight: The Tosco Story

The Tosco story is evidence that it is possible to create significant shareholder value in the refining sector. In 1990, Tosco (formerly The Oil Shale Corporation) was a financially struggling, independent oil exploration and production firm with a modest refining presence on the US West Coast. Its sole revenue-producing asset was its Avon refinery, which was located in the San Francisco Bay area and served the wholesale petroleum market in Northern California. By 2000, Tosco had revenue approaching $25 billion (US) and 37,000 employees. Tosco was the third-largest refiner (behind ExxonMobil and BP) and the second-largest C-store chain in the US, easily ranking ahead of such firms as Phillips and Sunoco in retail gasoline sales. Tosco was added to the Standard and Poor's 500 Index in September 1999.

Led by CEO and former oil trader Thomas O' Malley, Tosco utilized a strategy of timely acquisitions. Tosco was able to acquire refining assets at prices below replacement cost largely because many of the major firms were disposing of what they considered to be surplus assets. During much of the 1990s, the return-on-capital figures for refining were well below the cost of capital of the major oil firms and the federal government was mandating the disposition of some refining assets as a condition for approval of some of the many mergers and acquisitions taking place.

Utilizing its strategy of growth by acquisition and taking the risk of building capacity during a down cycle in the industry, Tosco was able to grow and prosper in an industry that has historically been dominated by firms many times its size. In 2001, Tosco was acquired by Phillips Petroleum for $7.49 billion in stock. When the deal was announced, Tosco shares were valued at $46.40, a price that represented about a 34% premium. In 1991, the closing share price for Tosco was $7.46, and in 1992 it was $6.13.

The environmental challenges associated with refining also create large exit barriers for refiners. Because of environmental remediation costs, it is usually financially advantageous to sell a refinery rather than close it. When refineries are sold, the transaction price is usually substantially less than the estimated replacement cost. For example, when Clark Refining (now part of Valero) acquired Chevron's 200,000 b/d Port Arthur refinery in 1995, Morgan Stanley estimated that the $75 million purchase price was only 9% of replacement cost.

The Refinery Product Mix

The value of crude oil is a function of the value of the products that are refined. Refiners evaluate crude oils by what they can earn from the products the different crude oils produce. Any crude can be turned into any set of derivative products (gasoline, asphalt, jet fuel, naphtha, etc.), depending on the specific refinery and the state of its technological flexibility. That said, no two refineries will achieve the same margin for a given crude oil or the same mix of refined products. The mix of products is driven by several factors

Refinery complexity

Refinery complexity refers to an oil refinery's ability to process crude feedstocks into value-added products. Table 12–4 shows different types of processes within a refinery. The simplest refining process is distillation. As more expensive units like cokers and hydrotreaters are added, the complexity of the refinery increases. Refineries that are configured to have a high conversion and desulfurization capacity can achieve better yields of higher value-added petroleum products by processing heavier and lower priced crude. Heavier crudes account for about 25% of global supply, and some analysts predict that by 2030, the figure could be 50% or more. More complex refineries can also produce higher margin products designed to meet specific local specifications, such as gasoline complying with California's strict environmental requirements. According to some analysts, the addition of new complex refining capacity over the past few years could create a situation where there is too much of the wrong products on the market.

Table 12–4 Types of refinery and Nelson Complexity

- The *topping refinery* separates the crude into its constituent petroleum products by distillation, known as *atmospheric distillation.* The topping refinery produces naphtha but no gasoline.
- The *hydroskimming refinery* is a refinery equipped with atmospheric distillation, naphtha reforming, and necessary treating processes. The hydroskimming refinery is more complex than a topping refinery and it produces gasoline.
- The *cracking refinery* is, in addition to the above, equipped with vacuum distillation and catalytic cracking. The cracking refinery adds one more level of complexity to the hydroskimming refinery by reducing fuel oil by conversion to light distillates and middle distillates.
- The *coking refinery* can process the vacuum residue into high-value products using the delayed coking process. The coking refinery adds further complexity to the cracking refinery by high conversion of fuel oil into distillates and petroleum coke.

Catalytic cracking, coking, and other conversion units are referred to as *M-secondary processing units.* The Nelson Complexity Index captures the proportion of the secondary conversion unit capacities relative to the primary distillation or topping capacity. The Nelson Complexity Index typically varies from about 2 for hydroskimming refineries, to about 5 for the cracking refineries, and over 9 for the coking refineries.

Source: Reliance Industries Limited.

Refinery complexity is commonly measured by the *Nelson Complexity Index*, which is calculated by the *Oil & Gas Journal* annually. The Nelson Complexity Index assigns a complexity factor to each major refining unit based on its complexity and cost in comparison to crude distillation, which is assigned a complexity factor of 1.0. The complexity of each refining unit is then calculated by multiplying its complexity factor by its throughput ratio as a percentage of crude distillation capacity. Adding the complexity values assigned to each unit, including crude distillation, determines a refinery's complexity. A refinery with a complexity of 10.0 on the Nelson Complexity Index is considered 10 times more complex than crude distillation for the same amount of throughput. The Nelson Complexity Index provides insight into refinery complexity, replacement costs, and the relative value addition ability and allows different refineries to be ranked.

The higher the index number, the greater the cost of the refinery and the higher the value of its refined products (see table 12–4).

The most sophisticated refineries can use almost any crude and are capable of producing an attractive and profitable product mix. These refineries might use as many as 15 different crudes in a month and more than 30 over the course of a year. Older refineries that have not been upgraded have limited ability to refine many of the more difficult crudes; the costs and reliability of doing so are prohibitive. As a point of comparison, about 80% of US refineries can handle sour crude oil of the kind that comes from Iran, Saudi Arabia, and other large producers.[7] In 2006 in China, only about 15% of Chinese refineries could process sour crude. As a result, China's refining sector was rapidly expanding and augmenting its capacity to refine a wider variety of crude oils. Given China's crude oil consumption, if China's ability to refine lower quality crudes increases significantly, the price discounts for lower quality crudes could diminish.

Location

The location of an oil refinery has an important impact on refining margins, since it determines how well it can access feedstocks and distribute its products. Location also dictates whether feedstocks and products can be transported by sea or via pipelines, rail, or road. If products cannot be moved to market, there is no value in refining them. Although a landlocked refinery has fewer crude options than one with deepwater access, some landlocked refineries are very profitable because they have regional monopolies.

Product demand

Market demand has a huge impact on refinery profitability, and demand is constantly changing for refined products. Miles driven and average fuel consumption are critical factors impacting demand for motor fuels. Demographics play a key role: older people tend to drive less than soccer moms. The economic downturn in the US and Western Europe beginning in 2007 resulted in a reduction in total miles driven in those markets, driving down motor fuel demand. The 2007–2008 fuel price shock motivated US consumers to trade larger vehicles for those with more fuel efficiency, also decreasing demand. In China, miles driven have skyrocketed as car sales increase at a double-digit pace. Regulatory policies that mandate greater fuel efficiency or raise fuel taxes can lead to a reduction in demand or a shift to more fuel efficient vehicles.

Motor fuel preferences can change. Figure 12–4 shows how the consumption of diesel is increasing much faster in Europe relative to the United States. The US Energy Information Administration (EIA) has predicted that US demand for diesel fuel will grow about four times faster than that of gasoline through 2015. Looking further out, toward 2030, diesel demand is expected to increase about 14 times faster than that of gasoline. The Paris-based International Energy Agency has said that middle distillates (jet fuel, kerosene, diesel, and other gasoil products) have become—and will remain—the main growth drivers of world oil demand. Of course, if electric cars successfully penetrate the car market in the next decade, all of these projections could be completely wrong.

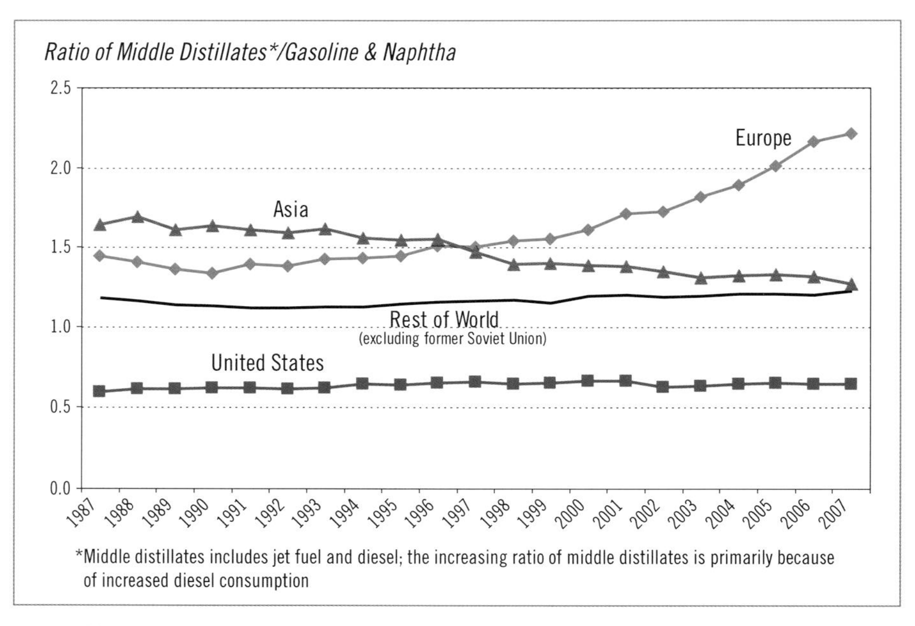

Figure 12–4. World consumption mix shift toward diesel
Source: *BP Statistical Review of World Energy*, June 2008.

There is also seasonality in the demand. In the US, refiners tend to increase gasoline production for the summer and then reset to produce more heating oil for winter. In late winter, refiners will typically do much of their maintenance to get ready for the summer driving season.

Refining Margins and Profitability

The factors identified above will influence both the mix of products that can be refined and refining margins. The laws of supply and demand have a major impact on refinery margins. On the supply side, as refined product prices rise, refiners will react as basic economics suggests: they will increase production to capture the high prices before they dissipate under competitive pressure that can be regional, national, or global. For example, in recent years in the United States, diesel margins have been much higher than gasoline margins, leading refiners to shift production away from gasoline. On the demand side, as discussed above, consumption is influenced by an array of factors largely beyond the control of refiners, such as miles driven, preferences for diesel versus gasoline, consumer choices for vehicles, weather, the volume of air travel and sea shipments, and macroeconomic cycles. This section discusses some further issues that influence refinery margins and profitability.

To understand refining profitability several factors need to be considered: 1) Refining is a capital-intensive process, which means that circumstances impacting the availability or allocation of capital to the industry can have a major impact on industry profitability; 2) crude oil is the major variable cost in refining, and this cost is set by world markets; 3) labor, energy, and other operating costs are a small percentage of total costs but receive a lot of attention because they are controllable by refiners; 4) regulatory costs, such as environmental mandates, play a major role in overall refining costs; and 5) most of the products produced by a refinery are commodities or commodity-like.[8]

The determination of refining profitability starts with the *gross margin*, which is the difference between the wholesale composite price (i.e., the weighted average of the various products produced) and the composite refiner acquisition cost of crude oil. In the income statement, this shows up as revenue minus crude oil cost. The *net margin* is the gross margin minus operating costs such as labor, energy, maintenance, and petroleum marketing. Both margins are usually expressed on a per-barrel basis. As discussed in the following sections, refining margins are volatile and unpredictable. For example, if crude prices go up, refiners may be able to pass this increase on to their customers. At other times, crude prices could rise because of an event that reduces demand for refined products, meaning the refiner has higher prices and reduced demand for its products. An example would be the wars in Iraq, which increased crude prices and reduced the demand for air travel.

The cost of crude

Although the cost of crude oil is the major input cost for refiners, the relationship between crude cost and the final price for a petroleum product produced from that crude, such as gasoline or diesel, is not as direct as one would think. Prices are not just a function of cost (i.e., costs pushed through to the end user) but are also strongly influenced by demand, as discussed above. The demand and supplies of other petroleum products, such as heating oil, strongly influence prices for motor fuel. In the United States, cold winter weather can increase demand for heating oil, which can shift production away from gasoline, thus reducing gasoline inventory and raising gasoline prices.

The result is that high crude prices do not necessarily mean lower refining margins. Figure 12–5 shows US refinery margins and WTI prices over the past 30 years. With the exception of the unusual correlation between high prices and high margins during 2003 to 2006, no real trend is discernible. Clearly, other factors besides the price of crude impact the profitability of refining.

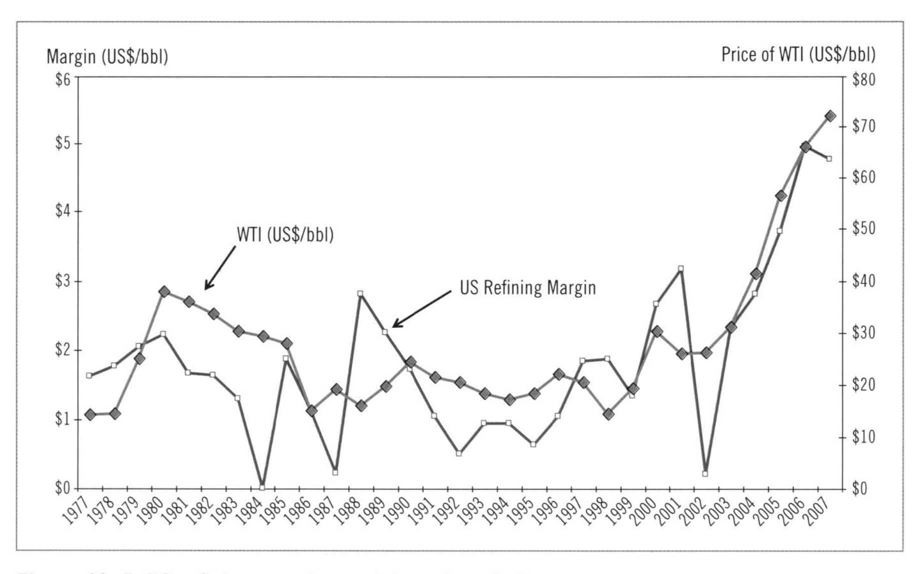

Figure 12–5. US refining margins and the price of oil

Source: Energy Information Administration.

Refining value and crack spreads

The refining value of crude oil, sometimes referred to as *technical value*, is the value a refiner expects to realize for the refined products less the operating costs for processing the crude. The refining value less the cost of the crude equals the refining margin.[9] To facilitate greater transparency in financial markets with regard to refining margins, the concept of crack spread was created.

Refiners' profits are tied directly to the spread, or difference, between the price of crude oil and the prices of refined products. *Crack spread* is a term used by the refining industry for the difference between the price of crude oil and refined petroleum products. The crack spread represents the theoretical refining margin and is quoted in dollars per barrel. The most common US crack spread is known as the 3:2:1 because it uses a ratio of three crude to two gasoline plus one heating oil. It is calculated by starting with the combined value of 2 times gasoline and heating oil. This value is then compared to the price of crude. Since crude oil is quoted in dollars per barrel and the products are quoted in cents per gallon, heating oil and gasoline prices must be converted to dollars per barrel by multiplying the cents-per-gallon price by 42 (there are 42 gallons in a barrel). If the combined value of the products is higher than the price of the crude, the gross cracking margin is positive. Conversely, if the combined value of the products is less than that of crude, then the gross cracking margin is negative. This sum is then divided by the number of barrels of crude to reduce the spread value to a per-barrel figure.[10]

The 3:2:1 gross cracking margin is calculated as follows. Assume the wholesale price of gasoline is $1.00 per gallon, the wholesale price for heating oil is $0.90 per gallon, and crude is $38.00 per barrel.

$1.00 per gallon × 42 = $42.00 per barrel of gasoline × 2 bbl	$84.00
$0.90 per gallon × 42 = $37.80 per barrel of heating oil	+37.80
The sum of the products is	$121.80
Three barrels of crude costs ($38.00 × 3)	– 114.00
Therefore, the gross cracking margin ($121.80 – $114.00)	$7.80
The 3:2:1 crack spread is $1.80/3 bbl (margin per barrel)	$0.60

If crude oil prices rise above $40.60, the crack spread becomes negative.

Trading crack spreads. Adverse price movements in crude and finished products can present a significant economic risk. Given a target optimal product mix, an independent oil refiner can attempt to hedge itself against adverse price movements by buying oil futures and selling futures for its primary refined products according to the proportions of its optimal mix. Various financial intermediaries have created products to facilitate trading crack spreads. In 1994, NYMEX created the crack spread contract to helps refiners lock in a crude oil price and heating oil and unleaded gasoline prices simultaneously in order to establish a fixed refining margin. For example, if a refiner expects crude prices to hold steady, or rise somewhat, while product prices fall (a declining crack spread), the refiner would "sell the crack;" that is, the refiner would buy crude futures and sell gasoline and heating oil futures.

NYMEX provides some specific examples of how refiners may use futures markets. When refiners are forced to shut down for repairs or seasonal turnaround, they often have to enter the crude oil and product markets to honor existing purchase and supply contracts. If they are unable to produce enough products to meet supply obligations, the refiner must buy products at spot prices for resale to customers. Furthermore, if the refiner has inadequate storage space for incoming crude supplies, the refiner must sell the excess on the spot market. If the refiner's supply and sales commitments are substantial and if it has to sell on the spot market, it is possible that prices might move against the refiner. For protection from increasing product prices and decreasing crude oil prices, the refinery can use a short hedge against crude and a long hedge against products. The outcome is the same as purchasing the crack spread, i.e., sell crude futures and buy gasoline and heating oil futures.

In another example, a refiner experiences mechanical failures at its plant in February, shutting down production. The company has ongoing obligations to purchase crude and now lacks the capacity to process it. The company has a month of sales commitments for refined products, but only a 20-day inventory. The company also has several cargoes of crude oil in transit that must be sold on the spot market. The company decides to sell the crude oil on the futures market to reduce the risk of falling crude prices that result from increased crude supply on the market. At the same time, the refining company needs to purchase heating oil and gasoline in order to meet sales obligations. A late February cold snap could drive up heating oil prices and an early inventory drawdown could boost gasoline prices. The company also wants to protect against its open market purchases of products causing a strengthening of prices. If the refiner does nothing it could find itself quite exposed. The refinery's use of the futures market gives it a variety of options for reducing its risk.

Figure 12–6 shows crack spread data over a number of years in the US and Europe. The 3-2-1 margin is not a good indicator for either European or Asian refineries, where distillate output is larger than gasoline and the contribution from residual fuel must be considered. The 6-2-3-1 NWE spread is used in Europe and refers to 6 barrels of crude, 2 of gasoline, 3 of distillates (diesel), and 1 of residual fuel and is a means of approximating the European region to compare with the United States.

Figure 12–6 shows that from 2000 to 2004 the spreads stayed in roughly the same range. Starting in 2005, the spreads became much higher, and this period corresponds with the unusual profitability in the refining sector that occurred in the mid- to late part of the decade shown in figure 12–5.

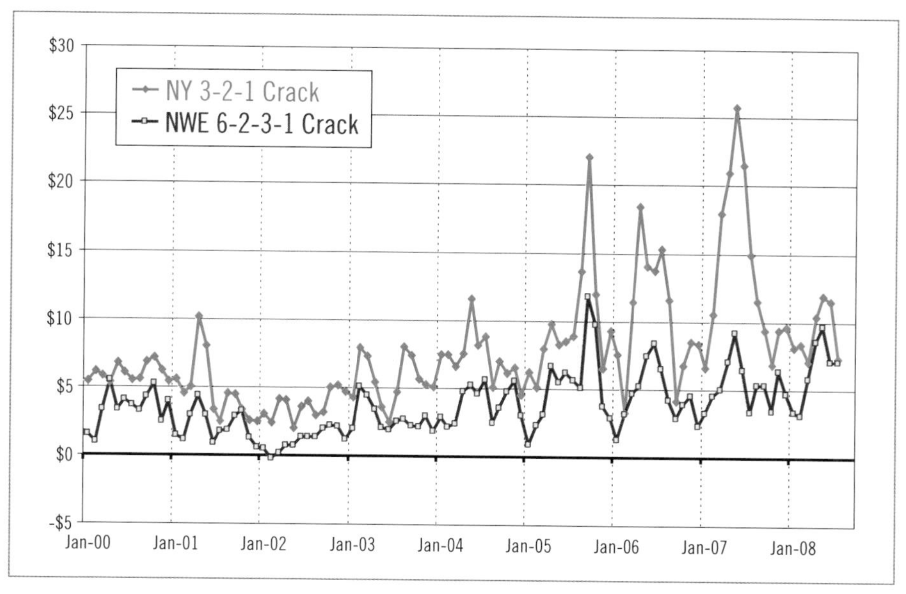

Figure 12–6. US and European crack spreads (US$/barrel)
Source: Energy Information Administration.

Other refinery cost drivers

Besides crude oil and other petroleum-based products that are added to the refining process, refineries have a variety of other inputs, such as electricity, labor, chemicals, and water. Managing these costs, and especially labor and electricity, is critical in order to run an efficient and profitable refinery.

Technological and managerial capabilities. Refining is manufacturing and like all manufacturing industries, some firms will have greater capabilities in terms of what can be manufactured for a given set of raw material inputs. In some cases, different refining capabilities will be the result of technological differences, such as proprietary catalysts. In other cases, the differences will be the result of superior managerial skills that can generate better employee productivity and capacity utilization. As a result, even though the slate of refined products from one refiner to another may be almost the same given a similar input crude, refining profitability will vary widely, just as it does in other manufacturing industries such as cars, computers, or electronics.

Technological innovation has played an important role in refining over the past century and a half. Table 12–5 shows some of the most important innovations. While most of the key innovations occurred many years ago, refiners continue to develop new technology that leads to greater complexity (as discussed earlier) and higher yields per barrel of crude. Recently, refiners have been developing new technologies that can be used to process residual fuel oil, which can be as high as 35% in a barrel of heavy crude and 50% in bitumen crude from Venezuela. A process called *slurry hydrocracking* was being developed by companies such as Eni and UOP, a division of Honeywell.[11] This process uses catalysts to help break down the heaviest oil. Firms that can implement these technologies have the opportunity to work with oil producers that produce heavy oil and that need access to refining capacity to exploit their reserves.

Table 12–5. Selected examples of refining innovations

Year	Process Name	Purpose	By Products
1862	Atmospheric distillation	Produce kerosene	Naphtha, tar
1870	Vacuum distillation	Lubricants (original) Cracking feedstocks (1930s)	Asphalt, residual coker feedstocks
1913	Thermal cracking	Increase gasoline	Residual, bunker fuel
1916	Sweetening	Reduce sulfur and odor	Sulfur
1932	Hydrogenation	Remove sulfur	Sulfur
1937	Catalytic cracking	Higher octane gasoline	Petrochemical feedstocks
1940	Isomerization	Produce alkylation feedstock	Naphtha
1960	Hydrocracking	Improve quality and reduce sulfur	Alkylation feedstocks
1975	Residual hydrocracking	Increase gasoline yield from residual	Heavy residuals

Environmental mandates. Environmental requirements for gasoline are another factor in the refinery margin mix and have at times pushed product prices significantly higher than the movements in crude prices would suggest. In the United States, the probability of state and/or federal officials mandating additional investments in existing refineries for environmental or regulatory reasons is high. These investments often end up being "no return" projects in the sense that they result in either the same or less capacity and can cost upward of $50 million to $100 million for a large refinery.

Capacity utilization. Like any manufacturing business, effective utilization of plant capacity is critical to refining profitability. The US Energy Information Administration (EIA) tracks refinery capacity utilization and generates detailed data. In the years 1947 to 2007, US capacity utilization on an annual basis ranged from a high of 97.5% in 1951 to a low of 68.6% in 1981. (Some caution with the data is warranted because full capacity utilization becomes more difficult as conversion complexity is added to a refinery. Today's refineries are much more complex than those of 50 to 60 years ago.) In more recent years, capacity utilization has averaged about 90%. Of course, this average data says nothing about the variance between companies and between refineries. Western European capacity utilization is similar to that in the US. Asia, excluding China, is slightly lower than the US and Europe. Refineries in China and the former Soviet Union lag significantly behind their counterparts in Asia, Europe, and the US in capacity utilization.

A refinery's capacity utilization theoretically can range from 0% to 100%. In reality, 100% utilization for a refinery or for any plant is extremely difficult to achieve. The optimum rate of capacity utilization in the US is considered to be 90%–95%, with a 95% utilization rate considered to be full capacity. Rates below 90% suggest many units are down for maintenance or that refining margins are so depressed that capacity was taken offline. At low levels of utilization, refineries lose some economies of scale and are likely to have higher costs. At utilization rates over 95%, costs may also increase because of process bottlenecks.[12]

Many factors can reduce utilization rates, including scheduled maintenance, fires, mechanical breakdowns, environmental mandates that require shutdowns and upgrades, and storms. In the US, the industry utilization rate in August 2005—just before Hurricane Katrina hit—stood near a record high of 97%. The catastrophic damage caused by Hurricanes Katrina and Rita in 2005 may have caused more than 20 times the reduction in effective Gulf Coast refining capacity in recent months than was experienced during the previous two years. For the three months ending in February 2006, 733,000 barrels per day of refining capacity were idled in the Gulf Coast refinery area, a huge increase over the amount of capacity that was temporarily idled in the Gulf Coast area for the same period in 2005 (33,000 barrels per day) and 2004 (32,000 barrels per day).

Scheduled maintenance that requires total or partial refinery shutdowns is necessary to ensure refinery efficiency and safety. See *Industry Insight: Refinery Turnarounds* for a description of the turnaround process.

Industry Insight: Refinery Turnarounds

A *turnaround* is a planned, periodic shutdown (total or partial) of a refinery process unit or plant to perform maintenance, overhaul and repair operations, and to inspect, test, and replace process materials and equipment. Most turnarounds incorporate the following:

- Turnarounds allow for necessary maintenance and upkeep of operating units and are needed to maintain safe and efficient operations.
- Turnarounds are scheduled at least one to two years in advance and do not necessarily focus on the same operating units.
- In assessing whether to delay a turnaround, a refiner has to include the opportunity cost of a possible unplanned shutdown resulting from the decision to delay the turnaround.
- Depending on the process unit and the amount of maintenance needed, the length of the turnaround can vary from one week to four weeks or more.
- A major turnaround usually will involve the crude unit or the catalytic cracking unit and will result in a more significant decrease in the utilization rate than a minor turnaround that may involve units such as the alkylation unit, isomerization unit, or sulfur plant.
- Not every unit is impacted during every turnaround. For example, the industry average is about four years between turnarounds for catalytic cracking units.
- A turnaround is not necessary to enable a refinery to shift from gasoline mode to distillate mode. However, if a turnaround is planned, it is possible to take steps, such as catalyst upgrades, that could improve the distillate yields.

- In general, the less often units are started up and taken down, the better (safer) it is, since refinery incidents are more likely to occur during these occasions.

Source: API, www.api.org.

Refinery scheduling. Maximizing capacity utilization is highly dependent on planning and scheduling. Refinery scheduling is one of the most complex operational tasks, involving a multitude of steps, in all of manufacturing. The goal is optimization of crude oil inputs and refined product outputs.

The first step is scheduling the delivery of crude. Crude delivery scheduling involves the coordination of pipelines, tanker shipments, marine terminals, and crude tanks with the firm's crude buyers. Because crude differs from cargo to cargo, initial processing requires specific unit scheduling and blend scheduling. The various refined products will have different economic models, technical specifications, and different channels to market. Batches of specific products have to be blended, stored, shipped, and sold in a complex market environment where demand is changing rapidly because of factors such as seasonality, competitor actions, consumer behaviors, and government regulations. Environmental rules require a multitude of exact specifications that must be adhered to, and these rules vary region by region. For example, US gasoline, a complex mixture of various hydrocarbons, is produced in different grades (regular, midgrade, and premium) and many different specs, depending on the specific region and season. The goal of a refiner is to make the highest value products with the right specs and incur the lowest cost. To make the scheduling work effectively, refiners spend huge amounts of money on software and human talent.

Refinery complexity and efficiency. As discussed previously, more complex refineries can produce better yields of higher value-added petroleum. In the last decade, the most complex refineries have generated the highest margins, primarily because these refineries can run the lowest cost crude.[13] Older, less complex refineries have been left out of the so-called Golden Age. Interestingly, as refineries become more efficient, demand for crude will drop. New refineries such as the Reliance Jamnagar refinery may be able to use as much as 20% less crude than older refineries to produce the same amount of gasoline and diesel. According to Deutsche Bank:

> *The implied crude conversion efficiency of the new refineries coming on stream is some 20% more gasoline/distillate light products per barrel of crude. Assuming that the 2 mb/d of new refining capacity displaces inefficient refining capacity in the world's 75 mb/d refining capacity, then 2.6% of the world's refining capacity will be 20% more efficient, representing a 0.5% lower demand for crude oil at the margin, or some 400 kb/d less oil demand for the same product output, simplistically analysed.*[14]

As demand shifts from gasoline to diesel, refiners have had to respond to the shifting market. Some analysts have suggested that investing in refinery upgrades to produce more diesel could actually hurt a refiner. The reason? Because diesel engines produce better fuel economy than gasoline engines, the demand for diesel relative to gasoline will be lower, which means refineries could end up with too much capacity. In reality, this issue is much more complex than just looking at the volume of diesel versus gasoline. Diesel fuel is denser than gasoline, so there are issues associated with "volume swell" as refiners switch to lower gasoline production.

In Europe, largely because of tax policy, refiners have found themselves with surplus gasoline, much of which has been exported to the United States. The result is that US refiners have been able to shift their yields to greater emphasis on diesel. As new refineries come onstream in Asia, there will likely be new competition for the US import gasoline market, which will put pressure on older, less efficient European refiners with surplus gasoline.

Given the price differentials for lower quality crudes, refiners in the past had a strong economic incentive to increase imports of sour and high-acid crudes. The discounts on lower quality crude oils can be substantial. For example, high-acid Doba blend from Chad has sold at a discount of as much as $17/bbl relative to Dated Brent, a key international crude pricing benchmark. Lowering crude feedstocks price by incorporating such oils can have major bottom line benefits. A few years ago Valero reported, "Our complex refineries were able to take advantage of the wide sour-crude discounts in the fourth quarter, when the Maya discount to WTI averaged about $15/b and the Mars discount averaged nearly $9/b." As more refineries are upgraded to handle challenged crude, the differentials are shrinking, which is what basic supply and demand economics would predict.

Refiners use complex economic models to decide which crude to purchase. Crude oils differ significantly in their characteristics and potential yield of specific derivative products. The value of crude to a refiner depends on refinery complexity, market demands, and various other factors. For a specific refinery the most valuable or profitable choice is not necessarily the lowest cost crude oil nor the lightest or sweetest crude. Consider the example in table 12–6. A refinery is debating whether to purchase Crude A or Crude B. Crude A is more expensive, selling at $50/bbl, whereas Crude B is only $48/bbl. Crude A's refining costs are slightly lower than Crude B's, $4.00/bbl to $4.25/bbl. Refinery management performs an analysis of the derivative products each crude will yield at current market prices.

As table 12–6 illustrates, although Crude B is cheaper, it yields less of the higher-value petroleum derivatives such as motor gasoline, jet fuel, and diesel fuel. Because these higher-end derivatives are valued at higher dollar per barrel market rates, the resulting gross revenue of Crude B is lower, $51.75/bbl compared to Crude A's $55.25/bbl. After including the cost of crude and the operating costs, the actual profit per barrel for crude A—although more expensive to purchase—is positive, while Crude B does not yield a refinery profit under current market conditions.

Table 12–6. Sample exercise in the economics of refining crude oil

	Crude A		
Component	Crude Yield (%)	Value (US$/bbl)	Refined Value (US$/bbl)
Fuel Gas	10%	$45	$4.50
Motor Gas	30%	$65	$19.50
Unfinished Motor Gas	5%	$55	$2.75
Jet Fuel	5%	$65	$3.25
Diesel Fuel	25%	$65	$16.25
Unfinished Diesel Fuel	5%	$55	$2.75
Asphalt	5%	$40	$2.00
Fuel Oil	5%	$35	$1.75
Other	10%	$25	$2.50
Gross revenue	100%		$55.25
Cost of crude			($50.00)
Refining costs			($4.00)
Profit per barrel of crude			$1.25

Table 12–6. (*Continued*)

	Crude B		
Component	Crude Yield (%)	Value (US$/bbl)	Refined Value (US$/bbl)
Fuel Gas	10%	$45	$4.50
Motor Gas	15%	$65	$9.75
Unfinished Motor Gas	20%	$55	$11.00
Jet Fuel	0%	$65	$0.00
Diesel Fuel	20%	$65	$13.00
Unfinished Diesel Fuel	10%	$55	$5.50
Asphalt	5%	$40	$2.00
Fuel Oil	10%	$35	$3.50
Other	10%	$25	$2.50
Gross revenue	100%		$51.75
Cost of crude			($48.00)
Refining costs			($4.25)
Profit per barrel of crude			($0.50)

Inventory management. Over the past several decades, global manufacturing of cars, electronics, and other products has been revolutionized by a shift to JIT (just-in-time) inventory management and the implementation of lean thinking in the way that manufacturing processes are coordinated and implemented. The roots for these ideas come primarily from leading Japanese companies, and in particular, from Toyota. Unlike traditional supply arrangements, manufacturers relying on JIT purchase the bulk of their inputs from a few suppliers who have intimate knowledge of the manufacturer's operations. In many cases, the supplier actually has direct access to, and manages, his customer's inventory. Inventory levels are kept to the bare minimum and deliveries are made as late as possible, hence the term *JIT*. Relationships between manufacturers and suppliers, especially in Japan, are more like alliances than traditional arm's-length contracting.

JIT processes, coupled with lean manufacturing, allow companies to reduce inventories and pursue additional goals such as improved customer service, shorter lead times, and zero-defect quality standards. Although low inventories are a characteristic of companies employing JIT inventory management, real lean management involves a philosophical shift away from managers to the shop floor. In a nutshell, the concept of lean means that all employees are problem solvers and are seeking better ways to do things.

In recent years, many refiners have reduced crude inventory to reduce working capital and tank maintenance costs. However, true JIT inventory management of crude oil has not occurred. Although conceptually possible for a refiner, major crude producers in the Middle East, Europe, and Latin America have not initiated formal JIT relationships with their nondomestic refining customers. The major producing nations are unable or unwilling to meet the immediate demand needs of refiners attempting to create a continuous flow of feedstocks and finished products. One challenge to a refining JIT model is the fact that most US, Asian, and European refiners depend either on waterborne crude supplies or pipeline shipments. The time lag associated with these shipping methods limits the ability of refiners to establish true JIT relationships. In addition, the commodity nature of the crude oil business and the extreme cost focus of the industry mean that refiners depend on a variety of suppliers to ensure access to a wide spectrum of crude oils, rather than risk profitability by depending on a select few. Finally, because certain crude oils complement each other when blended, refineries can use a variety of sources for their crude oil. For example, Angola's acidic Kuito blends well with sour Saudi Arab Light because the Saudi dilutes the Kuito's acidity, while the sweet Kuito reduces the product stream's sulfur content and maximizes valuable middle distillate yields.[15]

Refinery profitability: Putting it all together

As the previous discussion has emphasized, the profitability of the refining sector is the result of a complex set of factors. It is possible to make money when demand is high and crude prices are high and it is also possible to lose money under the same scenario.

Analysts called the four-to-five-year period ending in 2007 the *Golden Age of Refining*. Prior to this period the industry consistently failed to earn its cost of capital.[16] By 2010, all the major refiners were struggling financially, and firms such as Shell and Valero were considering either refinery closures or asset sales. The future of refining definitely looks challenging but as decades of experience have shown, the future is impossible to predict in the oil and gas industry. Although rates of return in all sectors of the oil and gas industry are highly cyclical, data from the EIA shows that US refining and marketing rates of return lag behind non-US returns, which in turn lag behind upstream returns. (*Note:* the data from the EIA lumps refining and marketing together so there is less precision in the data than we would like.) However, there are periods such as the late 1980s and the Golden Age when refining and marketing had the highest returns. Relative to other industrial sectors, over the 28-year period from 1977 to 2005, US durable goods industries earned a 9% average ROI, compared with 7% in refining. US durable goods industries also had lower variability in rates of return.[17]

So, who will earn the highest profits in the future? With respect to specific refineries (as opposed to refining companies) and according to Deutsche Bank, using data from Wood McKenzie, the 25 most profitable non-government-owned refineries in the world are all in the United States. These refineries are more complex, more crude advantaged, and essentially have been forged by years of bad US refining margins into a system that can make money in the worst of times.[18] However, new refineries coming onstream in Asia and the Middle East will be state-of-the art and will put significant pressure on older and smaller refineries in the US and Europe.

Refiners will also have to deal with the quality gap shown in figure 12–7—the growing demand for higher quality products by consumers and regulators coupled with the declining quality of crude oil. Meeting market expectations and dealing with lower quality crude will require investments in technology, processes, and people. Passing on the cost of these investments to customers will be a challenge.

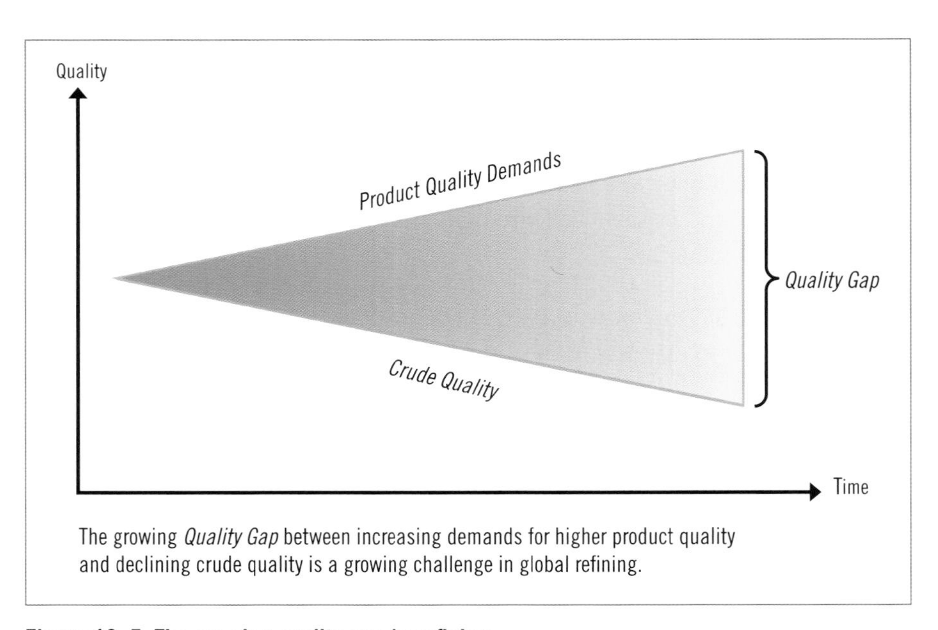

Figure 12–7. The growing quality gap in refining

As to which firms will earn the highest profits, that is impossible to predict. The largest refiners in the world are the large publicly traded IOCs, and these firms have a reputation for tight cost control and disciplined management. However, Middle Eastern and Chinese firms are getting stronger and are building new capacity and have growing domestic market demand. Clearly, competition in the refining sector will continue to evolve over the coming decades.

Summary Points

- Although steady demand growth for refined products is likely over the coming decades, short-term refining margins will continue to reflect the forces of supply and demand.
- The world's crude oil supply is shifting towards greater emphasis on sour, heavy crude, which means older refineries must invest in costly upgrades or risk higher premiums sweeter, lighter crude.
- Diesel demand will increase in Europe and the United States, and gasoline demand will be flat or declining.
- The US regulatory environment effectively precludes new greenfield refinery capacity and in the near term, US clean fuel regulations limit the ability of foreign refiners to meet US demand.
- Until a few years ago, two decades of underinvestment in refining capacity in the United States and Europe had created bottlenecks in refining. New refineries coming onstream in Asia and the Middle East will add significant new complex capacity over the coming decades and reduce bottleneck threats in Western markets.
- Product prices are no longer a compound of raw material price plus refining and transportation costs. Instead, the price of crude is increasingly becoming a function of the price of products that can be made from it (especially at the heavy end of the barrel).
- In most years refining profitability lags behind the profitability from the upstream sector.
- Although new and more complex refineries are coming onstream, there are many highly inefficient refineries that stay in business because of government ownership in countries such as China and Russia.

- Complex refineries with high conversion capacities will earn the highest margins.

- Future environmental mandates that require major investments by refiners are a high probability.

- Although the global petroleum industry is heavily integrated across the value chain, there are advantages to being part of an IOC and there are advantages to being an independent refiner.

Notes

1. Gregory Zuckerman and Ann Davis, "Who Is Hurt by Oil's Fall? Driller, Ethanol Makers Lead Pack of Stocks that Could Be Hit Hardest," *Wall Street Journal*, 2007, January 19, pp. C1, C3.
2. New York Mercantile Exchange, *Crack Spread Handbook*, 2000, p. 4.
3. *OSHA Technical Manual.*
4. http://www.api.org/ehs/performance/refining/morerefining.cfm.
5. Jad Mouawad, "Record Failures at Oil Refineries Raise Gas Prices," *New York Times*, July 22, 2007, www.nytimes.com.
6. Reuters, "PBF to Buy Shut Valero Refinery in Delaware," April 8, 2010.
7. Gabe Collins, "China's Refining Expansions to Reshape Global Oil Trade," *Oil & Gas Journal*, February 18, 2008, pp. 22–29.
8. Robert Pirog, "Petroleum Refining: Economic Performance and Challenges for the Future," CRS Report for Congress, May 9, 2005.
9. Colin Birch, "Achieving Maximum Crude Oil Value Depends on Accurate Evaluation," *Oil & Gas Journal*; January 14, 2002, Vol. 100, No. 2, pp. 54–59.
10. *Crack Spread Handbook*, New York, Mercantile Exchange, 2000.
11. Ana Campoy, "Bottoms UP: Refiners Say New Technology Lets Them Get More Clean Fuel Out of Every Barrel of Heavy Crude," *Wall Street Journal*, February 9, 2009, p. R8.
12. Carol Dahl, "What Goes Down Must Come Up: A Review of the Factors behind Increasing Gasoline Prices, 1999–2006," Division of Economics and Business, Colorado School of Mines Golden, CO, April 2007.
13. Booz, Allen, Hamilton, *Refining Trends: The Golden Age or the Eye of the Storm*, Part II, 2004, www.boozallen.com.
14. Deutsche Bank, "Global Refining Efficiency: The Mega-Bear Case for '09 Oil," November 18, 2008, p. 5.
15. Gabe Collins, 2008.

16. Booz, Allen, Hamilton, *Refining Trends: The Golden Age or the Eye of the Storm*, Part III, 2006, www.boozallen.com.

17. Carol Dahl, 2007.

18. Deutsche Bank, November 18, 2008, p. 14.

Chapter 13

SALES AND MARKETING OF PETROLEUM PRODUCTS

Kilometers are shorter than miles. Save gas, take your next trip in kilometers.

—Comedian George Carlin

I figured out Karl Rove's political strategy—make gas so expensive, no Democrats can afford to go to the polls.

—John Kerry, 2004 Democratic Party presidential nominee

This chapter examines the marketing and sales of refined products. In the developed nations, motor fuels (gasoline and diesel) are by far the most visible and politicized elements of the global oil and gas industry. Local gas stations and their highly visible price displays are unavoidable by motorists, making motor fuels unique among consumer products in that customers are always aware when prices are rising and falling.

The daily movements of fuel prices are debated and discussed by politicians, pundits, and everyone who drives a car. Other refined products are much less visible and are sold primarily into B2B or industrial markets. Although motor fuels eventually end up as consumer products, refiners may sell motor fuel to industrial and wholesale markets or they may sell to company-owned or company-franchised retail sites. Other products, such as asphalt or marine fuel, are rarely purchased by individual consumers. Lubricants serve both business customers and end-consumer customers. This chapter discusses the various products, their channels to market, and the competitive dynamics of the various products. Because there is a vast array of refined products, specifications, end uses, and markets, we will discuss only the more well-known products, such as motor fuels, aviation fuel, and lubricants

Motor Fuel

The marketing of motor fuels changes significantly from one market or geographic region to another. In the US market, there are many different channels and a very open market. In countries like Mexico and Malaysia, NOC-owned retail outlets dominate the landscape. In France, hypermarkets have the majority share, and in Italy and Japan, the oil majors have the largest market shares.

As evidence of the differences across markets, table 13–1 provides data on retail sites in Western Europe. The table shows that throughput per site varies widely, with Luxembourg the highest and Greece and Italy the lowest. The proportion of sites with convenience stores also varies by country, from 100% in Luxembourg, to 84.8% in the UK, to 23.9% in Italy. These country differences suggest some significant opportunities for consolidation and regional strategy implementation.

Table 13–1. Retail site market data for Western Europe, 2007

Country	Number of Retail Sites	Fuel Volume (million liters)	Throughput Per Site (million liters)	Proportion of Sites with Shops
Austria	2,800	10,139	3.62	65.5%
Belgium	3,200	5,848	1.83	75.3%
Denmark	2,000	3,791	1.90	46.8%
Finland	2,072	5,119	2.47	45.6%
France	13,030	42,567	3.27	35.9%
Germany	14,902	54,115	3.63	98.4%
Greece	7,955	8,118	1.02	70.7%
Ireland	1,289	3,237	2.51	73.5%
Italy	22,789	37,559	1.65	23.9%
Luxembourg	240	2,653	11.06	100.0%
Netherlands	4,203	10,648	2.53	60.1%
Norway	1,877	4,128	2.20	79.4%
Portugal	2,510	4,842	1.93	59.8%
Spain	8,974	41,712	4.65	78.1%
Sweden	3,559	9,890	2.78	41.8%
Switzerland	3,470	10,631	3.06	37.0%
UK	9,271	38,075	4.11	84.8%

Source: "Oil and Gas Insights: Opportunities in Western European Retail Marketing," Ernst & Young, September 2009, p. 8. Original data from Datamonitor.

Channels to market

Refining companies use a variety of methods to sell their motor fuels.[1] On a global basis there are probably more than a hundred variations in how motor fuels are brought to the market around the world, from roadside stalls in West Africa selling gasoline in bottles and jars to the giant motorway stations in Europe and the United States. Refiners may own and operate their own retail outlets. They may sell their product to a dealer who operates a branded franchise. Product may be sold to a jobber that operates its own retail outlets. Refiners may sell to large retailers, such as Costco in the United States or Carrefour in Europe. Finally, refiners may occasionally sell their fuel into spot markets. Thus, refiners have a variety of ways to sell their product and often find themselves competing in a retail channel with the same companies to which they supply fuel on a wholesale basis.

Refiner-owned-and-operated retail outlets

At one time, retail outlets owned and operated by large IOCs were common in Europe, the US, and other developed markets. The integrated oil companies used their company-owned outlets to promote their brands and to ensure that they controlled the vertical market from refining to retail.[2] The brands were supported by national advertising and slogans such as Shell's "You can be sure of Shell" and Exxon's "Put a tiger in your tank" became iconic. From the customer's perspective, the quality of the brand was a function of the gasoline and diesel quality and the retail experience at the gas station, such as cleanliness, hours of operation, location, and service attendant experience. Gasoline sold by the IOCs was often differentiated through proprietary gasoline additives, such as Chevron's additive Techron. According to Chevron's website, "Techron® Concentrate Plus delivers a number of unsurpassed benefits and one unique benefit—the ability to eliminate faulty fuel gauge readings."[3]

Although owning retail stations provides the opportunity for quality control by an IOC, this type of vertical integration is rapidly disappearing. Most of the oil majors decided after 2000 to divest much of their company-owned retail businesses. ExxonMobil announced in 2008 that it would sell 2,200 company-owned gas stations in the United States. Shell, BP, and ConocoPhillips also announced their intent to divest company-owned stores. The stated rationale for divesting the stations was usually the same: the market is very challenging and margins are falling.

The decision to sell retail stations reflects the challenge facing all diversified firms: how does the firm make the various businesses in the portfolio better off by being within the firm and not stand-alone? For many years, the IOCs had a number of advantages over stand-alone operators and potential entrants into fuels retailing. The IOCs:

1. Had accumulated managerial knowledge in retailing fuel
2. Had access to fuel supply
3. Owned a fuel brand that was valued in the market
4. Had locational advantage by virtue of having been in the business for a century or more

Over the past decades each of these advantages has deteriorated because of various changes in the competitive environment. The first advantage, accumulated knowledge, has eroded with time. In the US focused convenience store/fuels marketers such as QuikTrip, WaWa, Sheetz, and Couche-Tard (owner of the Circle K brand) have become efficient and profitable competitors. These focused competitors do not have the allocated overheads of the IOCs and, some will argue, operate with less emphasis (i.e., lower cost) on safety and control than the IOCs. (For an overview on QuikTrip see *Industry Insight: QuikTrip*.) The focused competitors also see merchandising as their core business, whereas the IOCs struggle to make merchandising a core competence. Moreover, once a firm makes merchandising a core business, its competitors become Wal-Mart, Carrefour, 7-Eleven, and Circle K, not other oil companies. It is a reality in the current fuels marketing environment that convenience store sales of products like beer and snack foods yield higher margins than fuel sales. Companies like QuikTrip have elevated the quality of the convenience store offerings with new initiatives such as company-owned bakeries. It is unlikely that an IOC would see a bakery as a core business.

Industry Insight: QuikTrip

In the United States, there are a variety of well-run and profitable fuels marketing companies sometimes referred to as *super convenience stores*. One of these is privately owned QuikTrip. Tulsa-based QuikTrip is a successful fuels marketing firm located primarily in the South, Midwest and Southwest. With over 500 stores, all QuikTrip gas station locations are company owned and operated. Most of the stores are larger than the average gas station, with as many as 24 pumps. The company has a standardized approach to service and store layout, not unlike that used by McDonald's and other successful fast food chains. The convenience stores are well lighted and the company offers a gasoline guarantee. Bakery products sold by QuikTrip are produced in QT Kitchens, a division of QuikTrip. Over 70% of the products sold in QuikTrip stores are shipped through the QT warehouse system.

One interesting aspect of QuikTrip is its approach to employees and their development. Most Americans would probably describe working in a gas station as a "dead-end job" with limited growth opportunities. On QuikTrip's website, the company says with regard to employees, "We give them ongoing training, pay the best wages for their jobs in the convenience store industry, and promote to new jobs from within. With rapid expansion of convenience stores, there are great jobs and employment opportunities for full- and part-time jobs. For six years, QuikTrip convenience stores have been named by Fortune Magazine as one of the 100 Best Companies to Work For."*

* http://www.quiktrip.com/aboutqt/aboutqt.asp

The second former advantage, access to fuel supply, is no longer a valid reason to be integrated given that, as we discussed in chapter 12, the distribution of motor fuels is now a global business. As well, antitrust laws make it almost impossible for a refiner to restrict supply to its own retailers. The fuel brand, the third historic advantage, has also lost much of its value. See *Industry Insight: Can a Differentiation Advantage Be Created in Fuel Marketing?* for a more detailed discussion.

Industry Insight: Can a Differentiation Advantage Be Created in Fuel Marketing?

When gasoline and diesel leave a refinery, they are commodities. However, by the strict definition of a commodity, gasoline and diesel sold through a service station are not commodities. A real commodity is a product like copper or silver that cannot be differentiated (in chapter 5, we introduced the concepts of competitive advantage and differentiation). Every producer's copper is exactly the same as that of the other producers. Copper is traded on commodity exchanges and its price fluctuates significantly based on supply and demand. Selling retail motor fuels involves a range of dimensions that potentially could be the basis for differentiation and brand-building: fuel additives, store location, store cleanliness, the quality and speed of the pumps, the efficiency of the payment transaction, the attitude and behavior of the store employees, and convenience store attributes. For any of these dimensions to lead to differentiation (i.e., not a commodity), they must be valued by customers. For the retailer seeking a differentiation advantage, the attributes offered to the customer must be different from those of rivals. Thus, if the retailer can offer unique attributes and the buyer places a high value on the attributes, the retailer will be able to differentiate its product and command a premium price.

Lowering customer costs can be a powerful driver of differentiation. For example, Chevron promotes the cleaning capability of its gasoline, which could extend the life of the engine or reduce maintenance costs. Shell, on the other hand, focuses on performance with its additive V-Power, which Shell says "is the most advanced fuel Shell has ever developed—a high-octane, nitrogen-enriched premium gasoline that actively cleans while you drive to help improve engine performance."* Shell maintains that customers buying "discount" gasoline risk hurting their vehicle's performance. BP says that its Ultimate brand of fuel is cleaner than other brands.

The sustainability of differentiation depends on 1) continued perceived value to customers and 2) the lack of imitation by competitors.** Sustainability of differentiation can be examined in more detail by considering the following four questions:

1. Does the customer have many choices?

2. Is the customer sensitive to price?

3. Does the customer have full access to information about product price, product availability, competitors, and product specifications?

4. If a customer switches from one type of motor fuel to another, does the customer incur significant switching costs?

For motor fuels, the answer to the first three questions would be yes. The answer to the fourth would be no, despite the best efforts by Chevron, Shell, and others to convince users that their product is unique. Today, most people in the fuels marketing business will admit that despite years of advertising and brand promotions, gasoline is essentially a commodity product at the retail level. Customers care most about the price, and because of regulatory standards for gasoline and diesel, there is actually little difference between the various products. As a result, we can say that motor fuels exhibit commodity-like characteristics.

A statement from QuikTrip's website confirms the commodity-like characteristics that lead to low cost as the source of advantage in fuels marketing: "All gasoline marketers in the United States get their gasoline from various terminals in their cities. This gasoline has been mixed in pipelines with gasoline from a number of refineries. Although we may purchase our gasoline from a specific refiner, the gasoline we actually pull out of the pipeline is really generic. Even retail outlets of major oil companies end up drawing their gasoline out of the same pipeline as everyone else."***

---

* http://www.shell.us/home/content/usa/products_services/on_the_road/fuels/shell_vpower /about_vpower/about_vpower.html ** M. E. Porter, *Competitive Advantage: Creating and Sustaining Superior Performance*, New York: Free Press, 1985. *** http://www.quiktrip.com/aboutqt/faqs.asp.

Finally, while it is true that ExxonMobil, BP, Shell and others are divesting company-owned stations, some IOCs have chosen to maintain and even expand company-owned stores. In 2005, Total bought a large number of African stores from ExxonMobil. Lukoil recently purchased European stores from ConocoPhillips. There are also downstream companies like Valero and Sunoco that own and operate their own stores. Valero, in particular, has been aggressively expanding its retail presence and expanding its retail network.

Franchisers and distributors

Although the IOCs are exiting the company-owned side of the business, they seem committed to maintaining a strong brand presence in motor fuels. The predominant model used in most markets for branded retail fuels is a franchising arrangement based on the refiner's brand. With a franchise, the dealer either owns the site or, more commonly, leases the site from the refiner. The franchised dealer agrees to purchase its fuel supply from the refiner and the refiner manages the delivery of fuel to the site.[4] The franchise agreement, which could run for as long as 20 years, establishes a set of operating conditions, and as a partnership, both the brand owner and the franchisee have a stake in the business outcome.

For the franchisee there are several potential benefits to a partnership with a large refiner/brand owner: branded gasoline, access to an oil company credit card, access to loans and subsidies, marketing assistance, rebates based on incremental volume, training and support in running a gas station, technical support and station startup design, and security of supply.[5]

For a refiner, a franchised model means that the brand can be maintained and the operating expenses and responsibilities are carried by the franchisee. The refiner establishes national standards for the forecourt, customer relationships, and cleanliness, with penalties if the standards are not maintained. The franchisee is usually a family-owned company and is expected to operate as an entrepreneur and small businessman in a local community. In times past, the dealer would build a local clientele of regular customers. Given the earlier discussion on branding, it is questionable if a community-centric service station model has much value in most markets, since few gas stations actually perform oil changes or automotive repair, most customers are highly price sensitive, and fuel prices have generally been rising for several decades. That said, there are franchise dealers (and independent dealers—see below) who have managed to build strong customer relationships, primarily through their repair services.

Jobber-operated outlets

A *jobber* or distributor is an independent operator that buys refined products from a refinery, or possibly from a larger jobber. The jobber may own retail outlets or possibly has a franchise deal with other independent operators. A jobber will often own a terminal and a fleet of trucks to make deliveries to its stations. If the jobber is buying from a refinery, it is likely that the jobber has an agreement to display the refinery's brand and sell that refinery's fuel products at a particular set of stations. As the oil majors divest their company-owned stores, they will likely look for jobbers to buy their stations on a regional basis. Jobbers are typically privately held companies and may also operate multiple brands. For example, Giant Oil, a jobber based in Florida, operates sites with BP, Chevron, and Marathon brands. As evidence of the close link between the convenience store side and the branded fuel business, Giant Oil also has Subway and Dunkin' Donuts in many of its stores.

Independent dealers

Independent dealers are owner-operators who are not tied into a particular brand or larger corporate entity. Independents account for about 35% of the stations in the United States. Independent dealers may sign supply contracts with a particular brand but can shift their allegiance when the contract expires. The following quote from the book *Oil on the Brain* captures the position of the small independent dealer that has no relationship with any of the major brands:

> *Independent stations see themselves as underdogs. They buy gasoline from wholesalers called jobbers, who buy wholesale gas on the spot market and truck it to them. The gasoline they sell is chemically identical to that of the branded station—it comes from the same refineries, travels through the same pipelines, and sits around in the same tanks. But while the brands advertise that their gas contains special detergent formulas, the independents all use a generic formula and discount the wholesale gas by 2 to 3 cents a gallon. Without national advertising or a strong image, independents often try to keep their prices lower than the branded stations, which means they have to skillfully navigate both the wholesale market and the retail market.*[6]

Super convenience stores

The super convenience store category includes companies like QuikTrip and Sheetz, discussed earlier. These companies are able to leverage economies of scale in the retail convenience store area and have built brand names around some of the convenience store offerings. Selling fuel is critical for these retailers because fuel stops help to get customers into the stores.

Supermarkets/hypermarkets

For many years location was considered a barrier to entry in retail fuels. Customers want convenience, and the IOCs and large independent chains have decades of experience in finding and acquiring locations for their stations. In recent years, new entrants to fuels retailing in the form of supermarkets and hypermarkets have eroded location advantage and provided further evidence that fuels brand have limited value. The entry of these alternative fuels retailers began in earnest in France. In the mid-1960s the French government imposed limits on the amount of refined petroleum products that could be imported by the oil companies. The supermarkets realized that they were able to get around these regulations and set up retail sites near Benelux and German refineries. By selling petrol a few cents cheaper than the branded retail outlets, the supermarkets were able to build market share quickly and demonstrate that location and brand were less important than most industry experts believed. In fact, customers were willing to drive several miles out of their way to save a few cents per liter.

Over the next several decades the petrol business of French supermarkets and hypermarkets like Carrefour and E. Leclerc pushed many independent dealers out of business. In 2009, hypermarkets and supermarkets accounted for about 60% of the French petrol market. The supermarket/hypermarket fuels retailing model spread across Europe and then to the United States. In the United States, Costco and other large supermarket/hypermarkets have become important fuels retailers. A similar phenomenon can be seen in other countries such as Canada, Australia, and the UK.

Australia is a particularly interesting market. Two large supermarkets, Coles Myer and Woolworths, have almost 50% of the fuels market. The branded companies' competitive response to retailer entry into fuels marketing has been to partner with the supermarkets. Woolworths has a joint venture with Caltex Australia, a company 50% owned by Chevron, and Shell has partnership with Coles Myer. The lack of a partnership with a major retailer may have been what prompted ExxonMobil to exit the retail fuels business in Australia in the last few years. ExxonMobil sold its Australia business to Caltex.

The response of the oil majors to the entry of hypermarkets has been the subject of several studies. A UK-based study analyzed Esso's decision to wait 10 years before responding to UK hypermarket entry. The study suggests that Esso had two choices when hypermarkets first appeared: 1) accept losses in market share in parts of the country affected by the presence of hypermarkets and maintain prices in other areas, or 2) retaliate by matching or undercutting hypermarkets prices, which likely would have resulted in the spread of lower prices to all areas across the country, including those unaffected by the presence of hypermarkets. Esso's delayed response appears to have been based on a decision to maintain profits by keeping prices higher than those of the hypermarkets. However, the study argues that the pricing strategy may have been rational in the short term but gave the hypermarkets an opportunity to gain a foothold in the market, with disastrous long-term consequences for the oil majors.[7] By 2008, hypermarkets in the UK accounted for more than 40% of retail gasoline sales and more than a third of retail diesel sales.

Where does the gasoline at a particular station come from?

According to the US Energy Information Administration:

> *The name on the service station sign does not tell the whole story. The fact that you purchase gasoline from a given company does not necessarily mean that the gasoline was actually produced by that particular company's refineries. While gasoline is sold at about 167,000 retail outlets across the nation, about one-third of these stations are "unbranded" dealers that may sell gasoline of any brand. The remainder of the outlets are "branded" stations, but may not necessarily be selling gasoline produced at that company's refineries. This is because gasoline from different refineries is often combined for shipment by pipeline, and companies owning service stations in the same area may be purchasing gasoline at the same bulk terminal. In that case, the only difference between the gasoline at station X versus the gasoline at station Y may be the small amount of additives that those companies add to the gasoline before it gets to the pump. Even if we knew at which company's refinery the gasoline was produced, the source of the crude oil used at that refinery may vary on a day-to-day basis. Most refiners use a mix of crude oils from various domestic and foreign sources. The mix of crude oils can change based on the relative cost and availability of crude oil from different sources.*[8]

The truck-based distribution system used to transport fuel from bulk storage terminals to gas stations is a source of much of the flexibility in the retail system. The tanks in these trucks, which can hold up to 10,000 gallons, usually have several compartments, enabling them to transport different grades of gasoline or petroleum products. The truck tank is where the special additive packages of gasoline retailers get blended into the gasoline to differentiate one blend from another. Thus, a truck could pick up a load at a terminal, blend the additives into the different compartments, and then make deliveries to several different branded retailers.

Fuel standards and specifications

Although many consumers believe that gasoline and other motor fuels products are age-old products that have been around for a century or more with little change to the basic specifications, products refined and marketed today are far different from what they were just a few years ago. Gasoline and diesel have undergone continuing changes to burn cleaner in response to state and federal environmental regulations since the 1970s. For example, as discussed in chapter 12, governmental environmental mandates, which vary by state, impact the specifications of refined products. The recent shift to ultra-low-sulfur diesel (ULSD) required significant capital investment by European and US refiners. In the European Union, the Euro V standard is 10 ppm of sulfur in diesel fuel. In the US and Canada the new standard is 15 ppm. The cost to US refiners to produce and distribute diesel with the new standard was estimated by the API to be about $8 billion in capital investment.[9] To the extent possible, refiners try to pass this cost on to their customers, although that is not always possible.

As a result of the various environmental regulations, new blends of gasoline have been developed, and in the United States there are about 17 different kinds of gasoline sold across the country. In order for fuel standards to be effective, all industry stakeholders must agree on the characteristics of gasoline necessary for satisfactory performance and reliable operation. This consensus is reached in the United States under the auspices of the American Society for Testing Materials (ASTM) International. The API describes ASTM's role in standard setting:

> *The ASTM specification for gasoline, D 4814—Standard Specification for Automotive Spark-Ignition Engine Fuel, is widely recognized. ASTM International Committee D-2 on Petroleum Products and Lubricants is responsible for gasoline specifications and test methods. Specifications changes are required to be made in a balanced process, incorporating the positions of the "Producers," such as the members of API, "Users," such as the automobile manufacturers and "General Interest" parties, such as consumers or government representatives. Their positions and viewpoints are brought to the D-2 forum by representatives who also are members of ASTM, including: members of the American Petroleum Institute, individual refiners, petroleum distribution, pipeline, and marketing companies, vehicle and engine manufacturers, automotive equipment suppliers, and state and federal government officials.*[10]

Standards in the refining industry are a mixed blessing from a competitive perspective. On the one hand, they level the playing field for all refiners and create a level of industry transparency. However, once end customers learn that all refiners must produce to the same standard, this opens the door for hypermarkets and other nonbranded fuel retailers to sell the same products as the branded companies. Because discount pricing for motor fuels usually does not mean lower quality, customers have an incentive to look for the lowest priced product, as discussed above.

Should an IOC (oil major) compete in fuels marketing?

With the IOCs reducing their ownership and operation of gas stations, a logical question is whether IOCs should compete in fuels marketing in any form. For a century or more large oil companies have competed across the value chain from exploration to retail. Retail was traditionally seen as necessary because it provided an outlet for refining and allowed IOCs to control all activities in the value chain. In today's global world, the refinery outlet argument loses much of its validity from a long-term business perspective. Arguments against having a retail fuels marketing business are:

- As the motors fuels business has globalized and become closer to a commodity business, access to supply is no longer an issue for retailers.
- The retail business is as much, or more, about convenience stores and sales through the backcourt. As a result, a successful fuels marketer must also be a fast food, beer, and snack food merchandiser.
- A retail business creates enormous visibility because of the politicization of fuel prices (discussed in the section, "The politicization of motor fuels prices"). Without a retail business, IOCs possibly would have a different (perhaps more positive) corporate reputation.
- Competitive advantage based on differentiation is not easy in retail, and flexible independent retailers have the lowest costs.

That said, it could be argued that having an outlet for refined products helps ensure ongoing stability in the short to medium term. In the longer term and given the evolving nature of the retail fuels marketing business, participation by the IOCs seems to be far from a necessary value chain activity. With a value proposition increasingly driven by low prices, convenient location, and a well-stocked convenience store, the IOC's ability to enhance this proposition looks questionable. Thus, the rationale for maintaining a retail business may have to be justified on the basis of one or more of the following: connection to a broader corporate reputation, barriers to exit tied to issues such as environmental remediation or dealer contracts, or expected new opportunities in retail that remain to be exploited.

The retail price of motor fuels

Like crude oil, motor fuels prices are volatile and unpredictable. Figure 13–1 shows US gasoline and diesel prices over a four-year period. Prices for gasoline ranged from less than $2.00 per gallon to more than $4.00. Prices for diesel followed the same trends as gasoline, but the highest prices for diesel were quite a bit higher than the highest gasoline prices. The 2008 difference between the highest gasoline and diesel prices occurred when refiners were bringing low-sulfur diesel to the market and there were shortage in some areas.

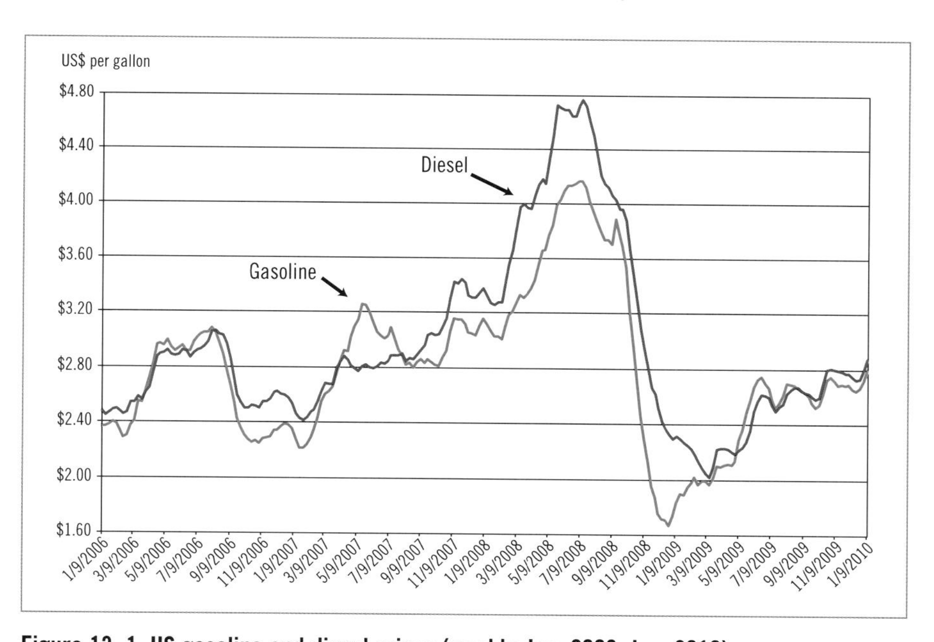

Figure 13–1. US gasoline and diesel prices (weekly, Jan. 2006–Jan. 2010)
Source: Energy Information Administration, accessed January 15, 2010. Data is weekly all-grade average.

In the United States, the main components of the retail price of gasoline are crude oil, federal and state taxes, refinery costs, and margins, distribution, marketing, and retail dealer costs and profits.[11] Figure 13–2 shows the different percentages for the components.

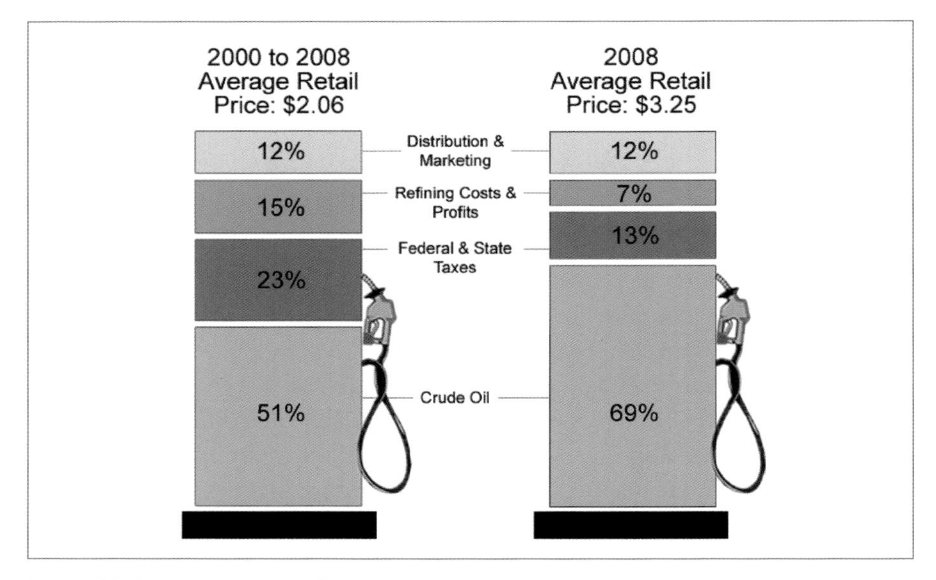

Figure 13–2. What do we pay for in regular grade gasoline?
Source: Energy Information Administration.

The percentage of price in distribution, marketing, and retail dealer costs and profits is lower now than in the previous decade, which is consistent with the decision by the IOCs to divest company-owned stations. Profit margins have been falling for several decades as a result of competition from hypermarkets, increased credit card transaction fees, and the limited ability to capture brand value.

On a global basis, taxes on motor fuels play a significant part in most government budgets. Because motor fuels are, for most people and businesses, a largely nondiscretionary outlay, fuel taxes create a predictable and steady flow of government revenue. There are often a range of taxes, such as federal and state levies in the US. For example, in South Africa there are the following taxes: an equalization fund levy, a fuel tax, a custom and excise levy, a road accident fund, and a state levy.

In examining gasoline retail prices around the world, the most obvious difference is the percentage of tax in the price. Figure 13–3 shows the percentage of taxes in retail prices for a selected set of countries. Although these numbers vary from period to period as the retail price is impacted by other costs, such as the price of crude, it provides some insight into the impact of taxation.

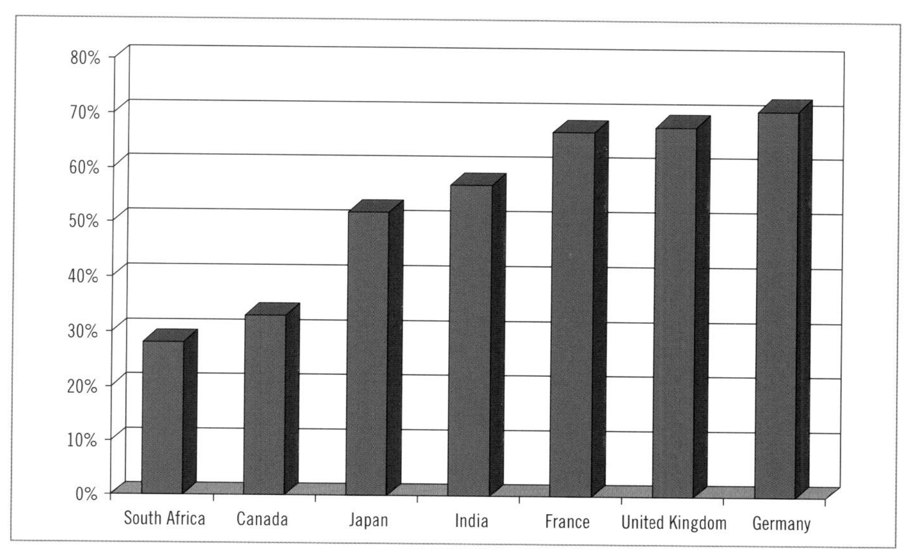

Figure 13–3. Percentage of retail gasoline price from tax
Source: International Energy Agency.

In many countries, including China, Egypt, India, Indonesia, Malaysia, Nigeria, and Venezuela, fuel prices are government subsidized. The countries with the highest subsidies and lowest retail prices are oil producers: Venezuela (retail price for a gallon of gasoline is 6¢), Iran (32¢), Saudi Arabia (45¢), and Kuwait (85¢).[12] In recent years, subsidies have created enormous financial difficulties for governments. During the huge increase in crude prices in 2008, Malaysia was projecting that it could spend about half of its federal government revenue on fuel price subsidies. This prompted the Malaysian government to cut subsidies, albeit not without a public outcry. Unfortunately for many emerging market countries, a decision to subsidize fuel prices is one that has become politically difficult to end and extremely expensive to maintain (the proverbial "between a rock and a hard place").

Zone pricing

Zone pricing is a controversial pricing practice used in the US by refiners selling their gasoline to dealers and franchisees. Zone pricing involves selling the same brands and grades of fuel to retail sellers at different prices depending on the "price zone" in which the retail seller is located. The refiners charge dealers a wholesale price, known as the dealer tank wagon price, based on a map of price zones derived from economic modeling. Zone pricing means that prices will be lower in some areas than others.

According to one consumer group, zone pricing is anticompetitive and should be outlawed because it:[13]

1. Limits the dealer's ability to lower prices and can be used to drive dealers out of business

2. Creates high-priced ghettos

3. Creates the illusion of competition when it actually stifles competition

Oil companies argue that zone pricing reflects the different competitive conditions in different geographic markets. Zone pricing allows refiners to adapt to local market conditions by, for example, lowering prices to dealers who face stiff competition from high-volume sellers such as Costco.[14]

What are the most important factors in the retail price of motor fuels?

A detailed study by the US Federal Trade Commission identified various factors that influence the retail price of gasoline in the US:[15]

1. The world price of crude oil is the most important factor in the price of gasoline. Over the last 20 years, changes in crude oil prices have explained 85% of the changes in the price of gasoline in the United States.

2. Since 1973, production decisions by OPEC have been a very significant factor in the prices that refiners pay for crude oil.

3. Gasoline supply, demand, and competition produced relatively low and stable annual average real US gasoline prices from 1984 until 2004, despite substantial increases in US gasoline consumption.

4. For most of the past 20 years [before 2005], real annual average retail gasoline prices in the US, including taxes, have been lower than at any time since 1919.

5. Increased environmental requirements since 1992 have raised the retail price of a gallon of gasoline by a few cents in some areas.

6. Regional differences in access to gasoline supplies and environmental requirements for gasoline affect average regional prices and the variability of regional prices.

7. Boutique fuels and differential access to gasoline supplies also can contribute to the variability of gasoline prices—that is, the fluctuation of gasoline prices—in particular circumstances.

8. State and local factors, as well as the extent of vertical integration among firms, can affect retail gasoline prices.

9. Other things being equal, retail gasoline prices are likely to be lower when consumers can choose, and switch purchases, among a greater number of gas stations.

10. Over the past three decades, the format of retail gas stations has changed to include convenience stores to increase sales volumes per station. Examples suggest that the largest-volume stations, so-called hypermarkets, lower local retail gasoline prices.

Price elasticity of motor fuels

Historically, both miles driven by consumers and the efficiency of the automobile fleet have changed in response to consumer priorities and external environmental changes. As the previous section shows, the external environmental changes include factors such as fuel prices, government regulation, technological innovation, and environmental consciousness.[16] According the United States Federal Trade Commission (FTC):

> *The price of a commodity, such as gasoline, reflects producers' costs and consumers' willingness to pay. Gasoline prices rise if it costs more to produce and supply gasoline, or if people wish to buy more gasoline at the current price—that is, when demand is greater than supply. Gasoline prices fall if it costs less to produce and supply gasoline, or if people wish to buy less gasoline at the current price—that is, when supply is greater than demand.*[17]

The relationship between fuel prices and demand raises questions about the price elasticity of demand. The *price elasticity of demand* for a product refers to the responsiveness in the quantity demanded for a product as a result of changes in the price of the same product. Research and anecdotal evidence shows that consumers do not respond to short-term fluctuations in price. Empirical studies show that prices must increase significantly to cause even a relatively small decrease in the quantity of gasoline demanded by consumers. The studies suggest that in the short run, a gasoline price increase of 10% will result in a reduction in demand of just 2%.[18] For a single case study illustrating consumer reaction to rising prices, see *Industry Insight: Gasoline Prices, Demand, and the Phoenix Gasoline Market.*

Industry Insight: Gasoline Prices, Demand, and the Phoenix Gasoline Market*

The US Federal Trade Commission (FTC) monitors the price of gasoline and diesel more closely than any other product. The weekly average retail gasoline and diesel prices are monitored in 360 cities nationwide by the Gasoline Price Monitoring Project. In 2003, the FTC observed some unusual price movement in the Phoenix market. At the beginning of August 2003, the average price of regular gasoline in Phoenix was $1.52 per gallon. By the third week of August the price peaked at $2.11 per gallon. Over the next few weeks the price dropped $0.31 per gallon and was $1.80 at the end of September.

Arizona has no refineries of its own. Gasoline is sourced primarily through two pipelines: one starting in Los Angeles and the other starting in El Paso, Texas. On July 30, 2003, the Tucson-to-Phoenix section of the El Paso pipeline ruptured. Temporary repairs initially failed, closing the Tucson-to-Phoenix section of the line from August 8 until August 23, when partial service resumed. The outage reduced the volume of gasoline delivered to Phoenix via pipeline by 30%. However, even though prices rose significantly, consumers did not reduce their consumption of gasoline by 30%. Phoenix gas stations replaced some of the lost supply by purchasing gasoline from West Coast refineries for which they had to pay a higher price than the normal West Coast customers. This additional supply was delivered by truck, resulting in higher costs.

Based on the Phoenix case, the FTC concluded that "for the most part, consumers do not substantially reduce their demand for gasoline in response to either short- or long-run price increases. The relative inflexibility of consumer demand for gasoline makes consumers more vulnerable to substantial gasoline price increases. . . . Consumers often lack adequate short-run substitutes for gasoline to power their cars.Thus, prices may have to rise substantially to reduce consumer demand in order to restore the balance between the quantity supplied and the quantity demanded.

With respect to producer responses to the pipeline outage, the FTC found that "producer supply responses work with consumer demand responses to result in a new equilibrium price. Together, consumer and producer responses to changes in market conditions will produce the new market equilibrium price. . . . If prices are not allowed to increase in reaction to a supply reduction, producers have no incentive to provide additional supplies to alleviate the supply reduction. Shortages are therefore likely to be prolonged."

---

*Federal Trade Commission, *Gasoline Price Changes: The Dynamic of Supply, Demand, and Competition*, 2005.

During the significant increase in crude prices in 2008, there was evidence of changes in consumer behavior. The demand for smaller cars increased significantly and public transportation ridership rose. The degree to which these events represent long-term behavioral change is not yet clear.

The politicization of motor fuels prices

In many countries, the price of motor fuels is a hot button political issue, and the large oil companies have become fodder for many politicians. The visibility of gasoline prices, the essential quality of fuel purchases, and the size and scale of IOCs combine to put fuel prices in the public eye. Very few people in developed countries have no opinion about the oil industry, and the opinions are usually shaped by what is happening with retail prices. Regardless of how often analysts, the industry, and the oil companies try to explain market dynamics, inevitably the large firms are accused of price gouging and profiteering. Social activists like Ralph Nader in the US often use the oil industry as their foil, as can be seen in Nader's statement that "the use of solar energy has not been opened up because the oil industry does not own the sun."

Economist Thomas Sowell had some interesting observations on the term *price gouging*. According to Sowell:[19]

> *What all this boils down to is that prices higher than what observers are used to are called "gouging." In other words, prices under normal conditions are supposed to prevail under abnormal conditions. This completely misunderstands the role of prices. When hurricanes knocked out both oil drilling sites and refineries around the Gulf of Mexico, there was suddenly less supply of oil. That meant higher prices and higher profits. What do higher prices do? Force people to restrain their own purchases more so than usual. What do higher profits do? Cause more money to be invested in producing whatever is earning higher profits, and this in turn expands output. Isn't a larger supply of oil and a reduced consumption of it what we want?*

Regardless of the reality of how fuel prices are set in the market, oil companies will continue to be accused of manipulating prices and markets. Over the last century, there have been many investigations in the United States and other countries into the retail pricing practices of oil companies. As an example of one such investigation, after Hurricane Katrina in 2005, the Federal Trade Commission (FTC) investigated whether gasoline prices nationwide were "artificially manipulated by reducing refinery capacity or by any other form of market manipulation or price gouging practices" by refiners, large wholesalers, and retailers in the aftermath of Hurricane Katrina. The FTC found:[20]

- No evidence to suggest that refiners manipulated prices through any means, including running their refineries below full productive capacity to restrict supply, altering their refinery output to produce less gasoline, or diverting gasoline from markets in the United States to less lucrative foreign markets. The evidence indicated that these firms produced as much gasoline as they economically could.

- No evidence to suggest that refinery expansion decisions over the past 20 years resulted from either unilateral or coordinated attempts to manipulate prices. Rather, the pace of capacity growth resulted from competitive market forces.

- No evidence to suggest that oil companies reduced inventory to increase or manipulate prices or exacerbate the effects of price spikes generally, or due to hurricane-related supply disruptions in particular. Inventory levels have declined, but the decline represents a decades-long trend to lower costs that is consistent with other manufacturing industries. In setting inventory levels, companies try to plan for unexpected supply disruptions by examining supply needs from past disruptions.

- No situations that might allow one firm—or a small collusive group—to manipulate gasoline futures prices by using storage assets to restrict gasoline movements into New York Harbor, the key delivery point for gasoline futures contracts.

The FTC also noted that federal gasoline price gouging legislation, in addition to being difficult to enforce, could cause more problems for consumers than it solves, and that competitive market forces should be allowed to determine the price of that drivers pay at the pump.

Another FTC report stated, "In no other industry does the FTC maintain a price monitoring project such as its project to monitor retail gasoline and diesel prices. . . . The vast majority of the FTC's investigations have revealed market factors to be the primary drivers of both price increases and price spikes."[21]

Aviation Fuel

In the early days of the commercial aviation industry in the United States, kerosene was chosen as the primary fuel because of its widespread availability and because gasoline was required for the war effort. Today, five basic grades of kerosene-type jet fuel with actively maintained specifications are used worldwide in civilian aircraft:[22]

- Jet A: used at civil airports throughout the US and in parts of Canada.
- Jet A-1: used worldwide outside of North America, the former Soviet Union, and mainland China. This grade was developed by major international oil companies to alleviate problems experienced by international air carriers when purchasing fuel in other countries.
- Jet TS-1: used within the former Soviet Union and in some Eastern European countries.
- Jet TH: used at all civil airports in Romania.
- Jet Fuel No 3 (formerly known as Grade RP-3): similar to Jet A-1 and used at all civil airports in mainland China.

Unlike motor fuels customers, who are primarily interested in price and station location, jet fuel customers are concerned with an array of fuel properties, including energy content and combustion quality, storage stability, thermal stability, lubricity, fluidity, volatility, noncorrosivity, fuel system icing inhibitors, antioxidants, and cleanliness. Also, whereas the branded motor fuels compete on the basis of proprietary additives, only additives specifically approved may be added to jet fuel, the result being that Jet A in the United States usually contains no additives.[23]

Airlines purchase most of their fuel through term contracts based on a projected volume for a given period. In the US, prices were typically adjusted once a week; in Europe once a month. As a US example, airline X might agree with supplier Y to supply its requirements for a one-year term estimated at 5 million gallons per year on a Platts Gulf Coast index (based on the week prior to delivery) plus or minus a fixed differential (usually stated in cents per gallon). In contrast to some other markets in the oil and gas industry, the spot market is relatively unimportant. After term contracts and hedging, spot-market purchases are a small part of the industry's jet fuel consumption.

Once a batch of refined aviation fuel has been tested to ensure it meets applicable specification requirements, it must be transported to an airport and pumped into an aircraft. Although the fuel could be shipped directly to an airport fuel storage facility, usually the distribution chain includes one or more intermediate storage terminals. Transportation may be via pipeline, ship, barge, railroad tank car, and tanker truck. Quality checks are performed on the fuel at each point in the distribution system to guard against contamination.[24]

In recent years, as oil and gas supply chains have globalized and companies have retreated from some of their less profitable aviation fuel distribution options, some airlines have struggled to ensure reliable supply. For example, Air Canada is building its own fuel storage depots, pipelines, and docks and leasing railcars, trucks and barges. The company is entering into direct purchases with suppliers in countries like Saudi Arabia, Nigeria, and Venezuela. According to a spokesperson at Air Canada, "We want to own it, control it. The only thing we are not doing is buying the crude and processing it [into jet fuel]."[25]

Airline pricing and fuel costs

> *Southwest Airlines Co. Chairman and Chief Executive Gary Kelly, when asked recently to name the greatest risk for airlines in 2010, said: "That's easy. It's energy prices."*[26]

With the significant increase in oil prices in the late 2000s, fuel became the largest single cost for most airlines. According to the chief economist of the Air Transport Association in testimony before a US Congress subcommittee, "No other industry is as conscious of energy consumption as passenger and cargo airlines. In the best of times, conservation and efficiency are a way of life. In the worst of times, they are a matter of survival."[27] In recent years, airline fuel costs have surpassed labor costs as the largest component in airline cost structures. Small movements in fuel prices can be very costly to the airline industry. The Air Transport Association estimates that a $1 increase in the jet fuel cost per gallon would cost the US airline industry $18.8 billion.[28]

Although fuel prices have increased, the airlines have struggled to pass the cost increases on to their customers. Given that airlines share many of the same commodity-like characteristics as motor fuels, customers usually seek the lowest price and may choose not fly if prices are too high. A further problem for airlines is the way the pricing models work. Airlines must price their tickets several months in advance, but their fuel contracts could reset the price on a weekly basis. As a result, airlines could find themselves with negative margins if fuel prices increase significantly between the time the ticket prices are set and the date of the flight.

To manage fuel price uncertainty many airlines engage in hedging. Hedging requires a relatively healthy financial condition (i.e., investment-grade credit), a willing counterparty, and often a large up-front transaction cost. Without investment-grade credit, it's very difficult to secure a good hedging contract. Airlines in bankruptcy, which has included almost all of the large US companies at one time or another, cannot hedge because they will not have investment-grade credit.

Unlike other fuels, such as gasoline, heating oil, and natural gas, jet fuel is not traded on any of the major exchanges. To hedge their fuel costs, airlines must use financial instruments based on traded commodities, such as crude oil, gasoline, or heating oil. Given the volatility of crude oil markets, companies involved in hedging can expect to occasionally experience what is known as ineffectiveness in their hedging programs. Accounting for hedges is a complicated technical area; *ineffectiveness* is defined by NYMEX as "the derivative instrument's ability to generate offsetting changes in the fair value or cash flows of the hedged item… Any difference between the change in fair value of the derivative and the change in fair value of the hedged item will result in hedge ineffectiveness. All hedge ineffectiveness is recognized currently in earnings."[29]

Lubricants

The lubricants sector produces products for just about every industrial segment imaginable. Lubricants are used in countless ways in the home, the office, and for recreational uses. Petroleum-based lubricants are essential for windmills. Everything from bicycles to roller blades to artificial joints requires lubricants to function properly. In the home, lubricants can be found in washing machines, air conditioners, refrigerators, windows, door hinges, garage doors, fans, and all small electrical appliances. In short, without lubricants to reduce friction and wear between moving parts, many products we take for granted would cease to work. Anyone who has tried to operate a car without sufficient motor oil or ride a bicycle with a rusty chain can testify to this fact.

Although there is a wide range of synthetic lubricants made from nonpetroleum materials, most lubricants are made primarily from petroleum products. A supplier of industrial lubricants must consider customer needs and the drivers of the customer's purchasing decision. These drivers include: product specifications, original equipment manufacturer (OEM) approval, product performance and longevity, operational tolerance (i.e., under what environmental conditions the product will perform or not perform), economic factors, environmental friendliness, composition, and quality.

In contrast, the B2C side of the lubricant business is much less about specific product attributes because consumers are unlikely to measure product performance in the same way as a B2B customer. Lubricants are also one of the few areas in the oil and gas industry where it is possible to create a brand-based differentiation advantage. (See *Industry Insight: Synthetic Lubricants as Differentiated Consumer Brands.*)

Industry Insight: Synthetic Lubricants as Differentiated Consumer Brands

Although the vast majority of products sold in the oil and gas industry are commodities, there are a few products that have achieved unique brand positions. Various petrochemical products sold in the B2B market have achieved strong market positions based on limited supply and technology-driven barriers to entry for competitors. On the consumer side, most products, such as motors fuels, heating oil, and propane, are close to commodities (although with a product like heating oil, service attributes like delivery reliability can sometimes be used as a basis for differentiation). The lubricants business probably offers the greatest opportunities for creating differentiated brands. In the market for synthetic motor oil for cars and trucks, the two leading products are Mobil 1 from ExxonMobil and Castrol EDGE from BP. Both products command premium prices relative to conventional oils. Mobil 1, introduced in 1974, is usually priced a bit higher than the Castrol products.

The brand promise for synthetic motor oils is along the lines of, "The product features advanced technology that provides performance beyond conventional motor oils." The marketing message for Castrol EDGE says that Castrol EDGE delivers:

- An advanced proprietary formula that provides performance and protection in extreme conditions such as towing, hauling, high and low temperatures, rapid acceleration, and stop-and-go traffic. Guaranteed extended drain interval protection for up to 15,000 miles, or one year, whichever comes first.

- Performance and protection that outperforms conventional, high-mileage and synthetic blend motor oils.

Since it is almost impossible for the average motorist to technically verify the producer's claim of superior performance, the fact that the products are purchased at premium prices is testament to the success and value of the brand. The brands are supported in a variety of ways. Both Mobil 1 and Castrol EDGE are major sponsors for car racing such as Formula One and NASCAR. Castrol EDGE has even done advertising during the Super Bowl, where prices for advertising space can cost millions of dollars.

Fuel Oils for Heating and Power Generation

There are a wide variety of refined products that have uses for heating or power. Fuel oil is any refined petroleum product used for generating heat or for energy to propel a ship, train, or other mode of transport, or to run a machine. Fuel oil is part of the distillate family that includes diesel. Because heavy fuel oil requires heating before use, it cannot be used in road vehicles, boats, or small ships. The heating equipment takes up a large amount of space and would make the vehicle size impractical. Heating the oil is also a complex procedure and would not work in small, fast-moving vehicles.

Heating oil as a source of residential heat is common in the northeastern US and eastern Canada where historically there was limited access to natural gas, as well as high electricity prices. With some gas discoveries off the coast of eastern Canada, the use of gas for heating has increased in the northeast. Heating oil is used in parts of Europe, with Germany the largest consumer. Heating oil has historically been a low-cost fuel for heating, although it does require specialized equipment for burning. With the increased growth in natural gas supply it is likely that heating oil will slowly decline in importance.

On a regional basis, heating oil is brought into oil storage terminals by refiners and other suppliers. The oil is then redistributed by barge or truck to other consuming areas. Once heating oil is in the consuming area, it is redistributed by truck to smaller storage tanks closer to a retail dealer's customers, or directly to residential customers. According to the Energy Information Administration, heating oil prices break down as follows:[30]

1. Cost of crude oil: 62%
2. Refinery costs and margins: 16%
3. Distribution, marketing, dealer costs, and profits: 22%

The fuel oil used to power ships is called *bunker fuel*. Bunker fuel may also be known as *residual fuel*, *heavy oil*, or *#6 fuel oil*. The hydrocarbon chains in bunker fuel are very long and the fuel is highly viscous as a result. Bunker fuel produces thick black smoke when it burns, which make it unsuitable for most uses on land. As indicated above, the thick fuel is difficult for most engines to burn since it must be heated before it will combust, so it is primarily used in ship engines. Large ships have sufficient space to heat bunker fuel before feeding it into their engines. Bunker fuel is both cheap and essential for the maritime industry. It could be argued that bunker fuel is the lifeblood of both international seagoing trade and the world's navies. Without access to cheap fuel, the world of international trade, and especially the use of China as the world's center for low-cost manufacturing, would surely be very different. Some futurists have speculated that once peak oil becomes reality, the cost of shipping will start to increase dramatically, forcing manufacturers to relocate much closer to their customers.

Because of the concerns over air and water pollution, the maritime industry has been looking at alternatives to bunker fuel, using terms such as *sustainable shipping*. For example, a few years ago the large Norwegian shipping company Maersk announced that its ships calling on California ports would switch to low-sulfur marine distillate fuel instead of bunker fuel within 24 miles of the coastline. Realistically, however, given the low cost of bunker fuel and the difficulties of regulating international waters, bunker fuel will likely be used for many years.

Residual fuel's use in electricity generation has decreased globally in recent years, particularly as LNG has become a viable energy source and environmental mandates have been imposed on emissions. In 1970, petroleum (which includes diesel and various distillate fuel oils, jet fuel, kerosene, petroleum coke, and waste oil) produced about 12% of electricity in the United States. In recent years, the percentage has dropped below 2%.

Asphalt and Propane

Asphalt and propane are two products familiar to most people: asphalt because we drive on it and propane because we use it to cook our steaks and hamburgers. Both products are representative of the ubiquitous nature of petroleum products and their impact on the lives of billions of people on a daily basis.

Asphalt

Asphalt, a North American term, or *bitumen* (outside North America) is residue from the distillation process primarily used in road construction. Asphalt has the consistency of tar and is used as the glue or binder for the aggregate (i.e., crushed rock and sand) particles. The road surfacing material is usually called asphalt, which creates some confusion because the term bitumen is also used. Over 90% of the world's total road network has an asphalt surface, and asphalt is one of the most recycled materials by volume. In addition to roads, asphalt is used for driveways, industrial floors, trails, tennis courts, roofing products and, with cement added, a product called *asphalt concrete*. Most of the major refiners are also major asphalt producers. In the US firms such as BP, Shell, Marathon, Valero, and ExxonMobil are major producers. There are also many small companies involved in the industry.[31] In Europe, there are over 4,000 asphalt production sites, and over 10,000 companies are involved in production and/or laying of asphalt.[32]

Even in a product sector like asphalt, technology and innovation play an important role. The major asphalt producers make more than 100 different formulations tailored to national specifications and customer needs. New technology is helping to create harder and more durable roads. For example, using polymer-modified asphalt, thinner and tougher layers of pavement can be used, which helps maintain roads in areas of wide temperature fluctuations. New mixing techniques are reducing the energy used by road builders in the blending process.

A challenge facing the industry is that as more refineries add conversion units for processing residuals, some refineries have announced that they will stop asphalt production in favor of more highly valued products. Innovations in concrete roads are also a competitive threat to the asphalt business.

Propane

Propane or liquefied petroleum gas (LPG) is a product that comes from refining and from natural gas processing (see chapter 8). In the refining process, about 3% of a typical barrel of crude oil is refined into propane, although it could be as much as 40% of a barrel. Because propane is portable, it can serve many different product markets, such as heating homes, heating water, cooking, drying clothes, fueling gas fireplaces, and as an alternative fuel for vehicles. Almost half of the global sales of propane are to individual consumers.

The price of propane is influenced by many factors, including the prices of competing fuels in each market, the distance propane has to travel to reach a customer, and the volumes used by a customer.[33] Because propane is a by-product of refining and gas production, the available supply cannot easily be adjusted when prices and/or demand for propane fluctuate. Since consumer demand for propane is seasonal and production is not, inventory imbalances are common. Figure 13–4 shows that US propane prices follow crude prices but the correlation is not 1:1.

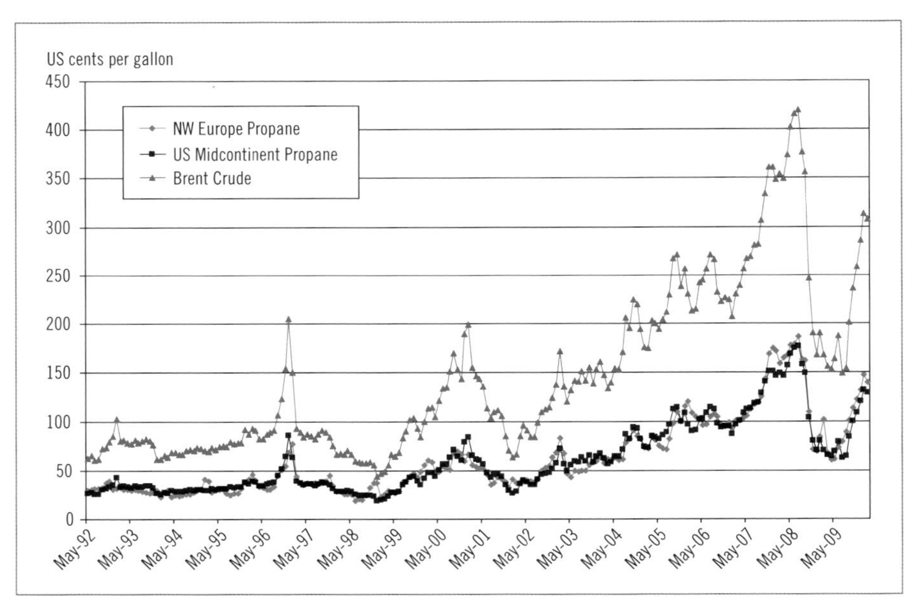

Figure 13–4. Propane prices follow crude oil (monthly, 1992–2009)
Source: Energy Information Administration.

In the US, the primary propane transportation mode is 70,000 miles of interstate pipelines. The pipeline system is most developed along the corridors between production areas and petrochemical consumers along the Gulf Coast and the agricultural-industrial consumers in the Midwest. The Northeast and South Atlantic states each are served by a single pipeline. The upper Midwest also is served by two lines from Canada. The petrochemical sector, which is primarily located near major propane supply sources, uses large volumes of propane delivered by pipeline. Individual consumers must be supplied by truck deliveries.

Summary Points

- Refined products are sold at the downstream end of the oil value chain. Without refined products and customers willing to buy them, there would be no oil industry.

- The oil business is heavily dependent on the transportation sector as a market. More than 75% of transportation-related energy demand comes from cars, light trucks, and commercials vehicles that run mainly on gasoline and diesel.

- The retail price of fuels depends on various factors beyond the crude price, such as the amount of tax and the degree to which the product is subsidized by the government.

- Although most motor fuels are sold under corporate brands, consumers tend to view motor fuels as commodity products. Competitive advantage based on brands has virtually disappeared from the retailing sector but cost-based advantage derived from scale and efficiency will play an important role regardless of the type of energy.

- Today's downstream competitors in the marketing and distribution segments are always evolving (witness the entry of hypermarkets into retailing), but the core elements of the value chain remain quite similar to that of many decades ago.

- The retail motor fuels industry has seen major changes in the large developed countries, with IOCs increasingly exiting company-owned operations because of market penetration by hypermarkets and other new entrants. Further consolidation and change is likely in many countries in the future.

- In addition to motor fuels, there is a diverse set of products produced as derivatives from refining, such as aviation fuel, heating fuel, propane, and asphalt. Each of these products has its own industry structure, bases for competitive advantage, customer value propositions, and intense competitive dynamics.

Notes

1. A. N. Kleit, "The Economics of Gasoline Retailing: Petroleum Distribution and Retailing Issues in the US," The Pennsylvania State University, December 2003.
2. Even with the IOCs selling their company-owned stations, they still dominated the market. Including sales from company-owned stations and franchised dealers, in the late 2000s the top four positions in gasoline sales in the US were held by ExxonMobil, ConocoPhillips, BP, and Shell. These four companies accounted for about 50% of total US gasoline sales.
3. http://www.chevron.com/products/ourfuels/prodserv/additives/tcp.aspx.
4. Kleit, 2003.
5. www.consumerist.com.
6. Lisa Margonelli, *Oil on the Brain*, New York: Nan. A. Talese, 2007, p. 15.
7. Marcel Cohen, "Pricing Peculiarities of the UK Petrol Market," *Journal of Product & Brand Management*, Vol. 8, No. 2, 1999, pp. 153–162.
8. http://www.eia.doe.gov/bookshelf/brochures/gasoline/index.html.
9. http://www.api.org/aboutoilgas/diesel/upload/May-2009_diesel_prices_differ_from_gasoline_prices.pdf.
10. http://www.api.org/aboutoilgas/gasoline/gasoline-octane.cfm.
11. "A Primer on Gasoline Prices," Energy Information Administration.
12. S. Johnson, "Motorist's Dream: Gas at 6 cents a Gallon," MoneyTalksNews, August 22, 2010, finance.yahoo.com.
13. http://www.ucan.org/gasoline_autos/gas_prices/redlining_why_zone_pricing_of_gasoline_must_be_outlawed.
14. E. Douglass & G. Cohn, "Zones of Contention in Gasoline Pricing," *Los Angeles Times*, June 19, 2005, www.latimes.com.
15. Federal Trade Commission, *Gasoline Price Changes: The Dynamic of Supply Demand and Competition*, 2005.
16. Booz Allen Hamilton, *US Refining Trends: The Golden Age or the Eye of the Storm*, 2007.
17. Federal Trade Commission, 2005.
18. Hilke A. Kayser, "Gasoline Demand and Car Choice: Estimating Gasoline Demand Using Household Information, Energy Economics, Vol. 22, No. 3, 2000, pp. 331–348.
19. http://www.capmag.com/article.asp?ID=4473.
20. http://www.ftc.gov/opa/2006/05/katrinagasprices.shtm.
21. Federal Trade Commission, *Gasoline Price Changes: The Dynamic of Supply Demand and Competition*, 2005.
22. *Air BP Handbook of Products*.

23. *Aviation Fuels Technical Review*, Chevron Global Aviation.
24. Chevron Global Aviation.
25. Susan Carey, "Air Canada Learns Fuel Self-Reliance," *Wall Street Journal*, December 1, 2008, pp. B1–B9.
26. Ann Keeton, "Southwest Airlines Hedges Its Bets," *The Wall Street Journal*, January 22, 2010.
27. J. P. Heimlich, "Commercial Jet Fuel Supply: Impact on US Airlines," Testimony before the Aviation Subcommittee of the Committee on Transportation and Infrastructure of the House of Representatives, February 15, 2006.
28. http://www.airlines.org/economics/energy/fuel+QA.htm.
29. http://www.nymex.com/ath_main.aspx?pg=4.
30. Energy Information Administration, "Residential Heating Oil Prices," December 2008.
31. http://www.eapa.org/default_news.htm.
32. To see how an asphalt plant works, go to: http://www.beyondroads.com/index.cfm?fuseaction=page&filename=tourPlant.html.
33. For more information on propane prices, see "Propane Prices: What Consumers Should Know," US Energy Information Administration, January 2008.

Chapter 14
PETROCHEMICALS

Cheap, disposable plastic objects are so omnipresent in our lives it is easy to be blind to what plastic can become in good hands. "It is a perfect material," says designer Piero Lissoni, "It is a marvel with all the incredible configurations it makes possible. It offers extremely high quality at a reasonable price." Plastic can take nearly any shape, consistency, and colour. It can be rigid, spongy or pliable. It can be opaque or transparent, shiny or matte, colourful or colorless.

—Emily Backus, "Clearly Inspired," *Financial Times*, April 3/4, 2010, p. 7

Mr. McGuire: *I want to say one word to you. Just one word.*
Ben: *Yes, sir.*
Mr. McGuire: *Are you listening?*
Ben: *Yes, I am.*
Mr. McGuire: *Plastics.*
Ben: *Just how do you mean that, sir?*
Mr. McGuire: *There's a great future in plastics. Think about it. Will you think about it?*
Ben: *Yes I will.*
Mr. McGuire: *Shh! Enough said. That's a deal.*

—From *The Graduate*, Embassy Pictures/United Artists, 1967

Petrochemicals are chemical products produced from petroleum and natural gas feedstocks. The products made from petrochemicals are amazingly diverse and touch our daily lives in many ways. Petrochemicals are the building blocks for plastics and rubbers and are used to make soaps, detergents, solvents such as paint thinner, paints, drugs, fertilizer, pesticides, explosives, synthetic fibers, flooring, insulating materials, and product packaging. Petrochemicals and their derivative products are found in many common products such as aspirin, cars, clothing, compact discs, video tapes, electronic equipment, furniture, and sports equipment. The use and consumption of petrochemicals is closely tied to industrialization and economic development. The fastest growth in consumption is in countries that are rapidly developing, such as China and India.

This chapter provides an overview of the petrochemical industry and its main competitive characteristics. The industry is based on the reactivity of the carbon molecule and its ability to create a diverse range of products that have very different properties. The products span a range from undifferentiated commodities to high-margin specialty chemicals. The oil and gas industry plays a fundamental role in providing hydrocarbon feedstocks for the petrochemical industry. Several of the IOCs (ExxonMobil, Shell, and Total) are among the world's largest chemical companies (see table 14–1) and compete with other chemical companies like BASF, Dow Chemicals, and SABIC.

Table 14–1. World's largest chemical companies

Company	Country	Sales (million US$)	Operating Income (million US$)
BASF	United States	87,832	9,111
ExxonMobil	United States	58,062	4,325
Dow Chemical	United States	57,514	1,321
LyondellBasell*	The Netherlands	50,706	–5,928
Royal Dutch Shell	The Netherlands	49,085	–55
Ineos	United Kingdom	40,986	–356
SABIC	Saudi Arabia	40,198	9,753
Sinopec	China	35,366	NA
DuPont	United States	30,529	2,391
Mitsubishi Chemical	Japan	29,905	84
Total	France	28,406	1,231

Source: ICIS Top 100 Chemical Companies. * In bankruptcy in 2010

Chemical Industry Overview

The global chemical industry is enormous and consists of more than 100,000 companies around the world. In the United States, chemical companies produce more than 70,000 diverse compounds.[1] The industry is also quite fragmented, with many small niche players. The top five global companies listed in table 14–1 account for about 10% of total output.

A first cut at understanding the chemical industry requires a division between the inorganic and organic sectors. In the inorganic sector, chemicals are produced from noncarbon elements such as phosphates and nitrogen. In the organic sector (the focus in this chapter), hydrocarbon raw materials are used to produce about 10 base products, which are then used to produce many different products. About 95% of organic products are produced from oil-and-gas-derived feedstocks, with a small and declining percentage produced from coal and an increasing percentage from biomass.[2] The base materials are further processed to produce intermediate products, which are the raw materials for many final products.

Within the industry there is enormous variety in the product market scope of chemical companies. Some companies produce only basic or primary chemicals; others produce basic chemicals and then convert them into basic polymers such as polypropylene; some purchase all their basic chemicals and produce intermediates; and some focus on highly specialized products or industry solutions. The huge company BASF produces just about everything before the final products, from oil and gas to basic chemicals to intermediate products to plastics and a vast array of products for many different industries. Hence, the BASF slogan "We don't make a lot of the products you buy. We make a lot of the products you buy better."

Petrochemical Production

Feedstocks for most petrochemical plants are provided by refineries and include petroleum gases, naphtha, kerosene, and light gas oil. Natural gas processing plants also produce feedstocks in the form of methane, ethane, and liquid petroleum gases (LPGs). Powerful operating synergies can be gained when petrochemical plants are colocated next to the refineries that produce feedstocks. In recent years, closer integration between refining and petrochemicals has become an important source of operating efficiency (discussed later in the chapter).

Organic chemicals are usually manufactured using a process that involves heating primary agents to encourage a variety of chemical reactions. These chemical reactions are initiated through the addition of high-temperature steam or by raising the temperature of process streams. Figure 14–1 shows some of the key petrochemicals and includes: 1) base chemical building blocks that comprise olefins and aromatics; 2) intermediates that are often derived from base chemicals; and 3) polymers that are combinations of 1) and 2). Figure 14–2 shows the base chemicals and derived products from both intermediate chemicals and polymers.

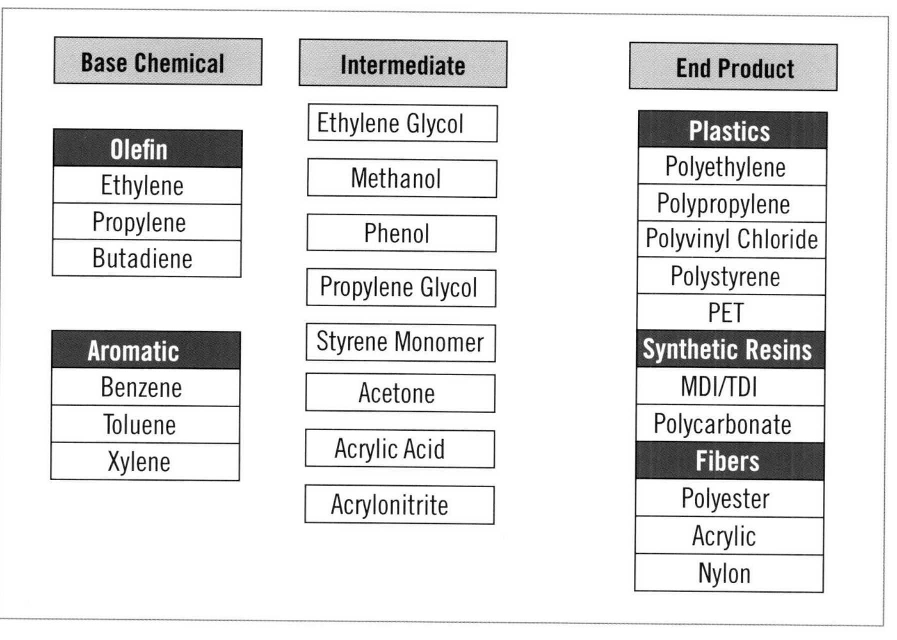

Figure 14–1. Key petrochemicals

Source: JP Morgan, "Chemicals Primer: An Investor's Guide to the World of Chemicals," June 23, 2008, JPMorgan Chase & Co. All rights reserved.

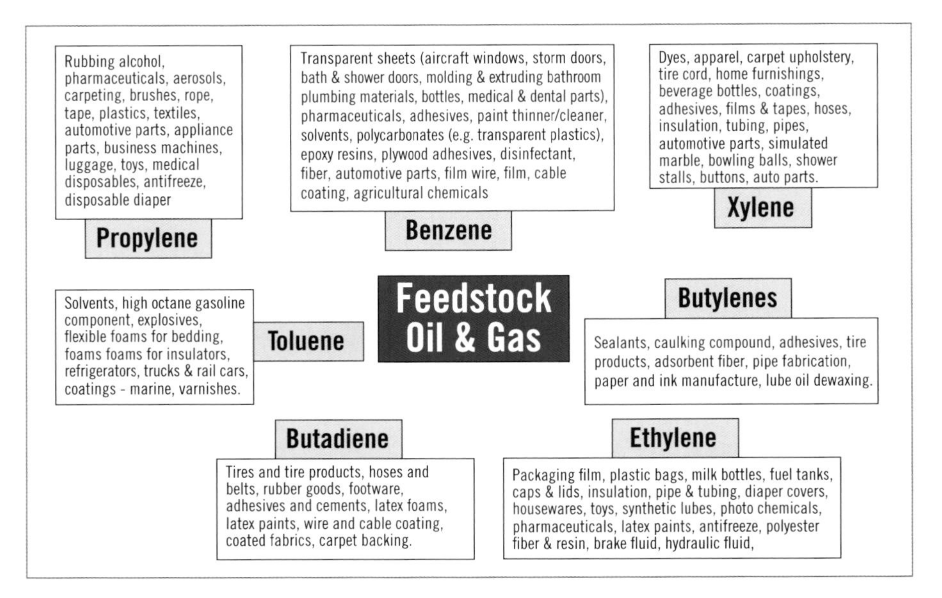

Figure 14–2. Base chemicals and the product derivatives

Source:

Stage 1: Base chemicals

In the first stage, the first column on the left in figure 14–1, basic olefins and aromatics are produced from hydrocarbon feedstocks. Olefins and aromatics are the building blocks of the petrochemical industry. Ethylene, propylene, and butadiene are the most significant olefins and are the foundation for about 75% of all chemicals produced. Ethylene and propylene are important raw materials for a number of polymers and intermediate chemicals. Butadiene is a raw material used in the production of synthetic rubber and is produced as a by-product of ethylene. Figure 14–3 shows the three main olefins and their most important derivative products.

The primary feedstock used for olefin production depends on location. In Europe and Japan naphtha, a low-octane form of gasoline made by fractional distillation of crude oil is predominant. In the US and Middle East, light gas is the main feedstock. US cracking complexes use mainly ethane and propane, available as by-products of oil and gas production. Both naphtha and gas feedstock are subject to wide price swings, which means chemical companies must deal with volatile input costs.

Thermal and steam cracking of light gases or naphtha produces olefins (ethylene, propylene, and butadiene) with the proportions dependent on feedstock, temperature, and pressure. The cracking reactions are very energy intensive, creating clear advantage where low-cost energy is available. Cogeneration, the production of steam and electricity, is often integrated as an energy option, especially with an adjoining cracker and refinery present.

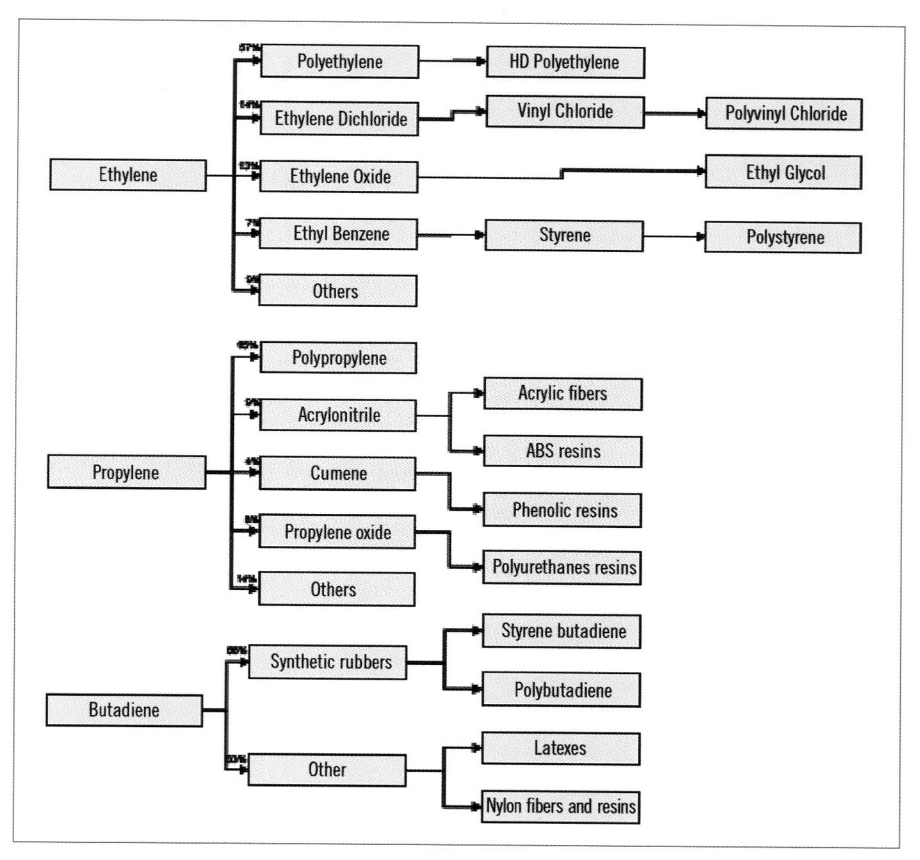

Figure 14–3. Olefins and the derivatives

Source: "Olefins and Their Derivatives," from *Chemicals Primer*, JP Morgan June 23, 2008, Fig. 71, p. 56.

Benzene, toluene, and xylene are the three most important aromatics (the name *aromatic* comes from the strong odors of most chemicals in this family). Almost all benzene is produced from oil-based inputs, making the price of benzene as volatile as oil. Toluene prices also depend on the price of oil and refining margins. Because toluene is also used as a gasoline additive, price is heavily impacted by gasoline prices. Xylene is mainly produced from petroleum and is used primarily as a solvent and an additive in gasoline.

Stage 2: Base to intermediate

In the second production stage, the base chemicals are converted into various intermediate products. This industry sector produces a wide variety of chemical products, including ethylene glycol, styrene, and acetone. As the name suggests, intermediate chemicals are inputs for other products. For example, ethylene and oxygen are combined to create ethylene oxide, which is then processed to produce ethylene glycol. Ethylene glycol is used primarily in the manufacture of polyester (81%) and antifreeze (12%).

Most of the intermediate chemicals are produced in olefin plants resulting in an overlap between the key players in olefins and intermediates. Methanol is one of the few organic intermediate chemicals not produced in an olefin plant. As an example of the remarkable versatility of organic chemicals, crude methanol can be converted to low-sulfur, low-benzene gasoline that can be sold directly or blended with conventional refinery gasoline. Methanol can be produced from natural gas and also from coal and biomass. ExxonMobil operates a plant in New Zealand that converts methanol to gasoline.

Stage 3: Consumable products

Both base chemicals and intermediates are further processed or converted into products used directly by industry or consumers. The petrochemical sector associated with the oil and gas industry is primarily concerned with stages one and two because that is closest to the downstream refining sector.[3] The process of converting basic and intermediate petrochemicals into plastics, fibers, and resins is termed *polymerization*. Polymers are created through the linking of monomers (such as ethylene), into long chains, typically using heat, pressure, and a catalyst.

Four plastics polymers produced on a major scale are polyethylene, polypropylene, polyvinylchloride, and polystyrene. Within each class of polymer, such as polyethylene, there are different types. For example, there is low-density polyethylene, linear low-density polyethylene, and high-density polyethylene. Most polyethylene is used for packaging in consumer and institutional products (mainly plastic bags). The primary competitors for major products like polyethylene tend to be large, vertically integrated chemical companies. The six largest global firms in polyethylene are Dow Chemical, ExxonMobil, Lyondell-Basell, SABIC, Sinopec, and Ineos (formed via acquisition, including much of BP's former chemical division).

Polypropylene is the second largest volume plastic resin and has a multitude of uses, such as product packaging, car and truck parts, containers, and fibers for carpet, clothing, and textiles. The top 10 polypropylene producers account for more than 50% of global capacity. The rankings of the top polypropylene producers have changed quite a bit in recent years as emerging market firms such as Reliance Industries (India) and PetroChina move up.

Polyvinyl chloride (PVC) is the most versatile thermoplastic and is used for pipes, building siding, flooring, gutters, windows, and food packaging. PVC is tough and resistant to water and most chemicals. PVC is also gaining popularity as a substitute product for wood and metals used in construction. Ties to the construction industry make the PVC market highly cyclical. The PVC industry has been under close environmental scrutiny because the manufacturing process produces small quantities of toxic dioxins.[4]

Polystyrene is a rigid plastic used for packaging and where hard plastic is required, such as disposable cutlery, CD cases, appliances, and electronic housings. Extruded polystyrene can be used as foam insulation. Styrofoam, produced by Dow Chemical, is a well-known foam insulation brand (in fact, Styrofoam is so well known that many people use the term *Styrofoam* as a generic name for polystyrene foam). In the fiber area, polyester is by far the most important product and is discussed in more detail in *Industry Insight: Polyester*. Acrylic and nylon are also important fibers, with many industrial and consumer uses.

Commodity and Specialty Chemicals

The terms *commodity* and *specialty chemicals* are widely used in the chemical industry. Strictly speaking, a chemical is not a commodity in the same sense as crude oil is a commodity. Chemicals are not traded on futures markets and producers have the ability to offer value-added services, enhanced delivery, branding, and other product attributes as a means of capturing customer demand. In contrast, an oil producer sells its products into a global market with prices established by traded benchmarks. Crude oil cannot be branded or enhanced by the seller to increase its value.

Industry Insight: Polyester

Polyester was invented in the 1940s and commercialized by DuPont. The product was introduced to consumers in 1951 and advertised as a "miracle fabric." Fabrics woven from polyester are used in clothing, bedding, and home furnishings. Industrial polyester yarns have many uses, such as tire chords, ropes of all types, conveyer belts, and heavy duty canvas-like materials. Polyesters are also used to make bottles, films, insulation, and as a finish for wood products. Along with polyethylene and polypropylene, polyester is one of the largest volume polymers and most well-known plastic materials. Polyester's success as a chemical product is the result of several factors: widely available raw materials, a simple chemical process used in manufacturing, low toxicity level of the product, enormous variety of intermediate and final product applications, and recyclability.

There are many possible chemical variations of generic polyester fiber. Two that are currently produced commercially are polyethylene terephthalate (PET) and poly-1,4, cyclohexylene dimethylene (PCDT). The production of polyester shows how chemical products move along a value chain from hydrocarbon raw materials, to base chemicals, to intermediate chemicals and polymers, and then to products that become inputs to a huge range of industries. The three main chemical intermediates (sometimes referred to as fiber intermediates) for polyester are PTA (purified terephthalic acid), DMT (dimethyl terephthalate), and MEG (mono ethylene glycol). P-xylene (paraxylene), an aromatic hydrocarbon, is an important base material used to produce PTA and DMT. Liquefied petroleum gas (LPG) is the base material used to produce MEG. Large PTA producers include BP, Reliance, Sinopec, SK-Chemicals, Mitsui, and Eastman Chemical. Invista, a subsidiary of privately owned Koch Industries, is the world's largest DMT supplier. The largest MEG producer is MEGlobal, a joint venture between Dow Chemical and Petrochemical Industries Company (PIC) of Kuwait.

There are many polyester producers, including some of the largest and most well known chemical companies, such as BASF, DuPont, Eastman, Invista, Mitsui, Mitsubishi, Reliance, SABIC, and Sinopec. Within the polyester segment, brand names are often used by chemical companies for their polyester products, such as DuPont's Rynite PET polyester resin and Mylar, Melinex, and TeijinTetoron polyester films. BASF produces Luquafleece, a polyester nonwoven material, and Ecoflex, a biodegradable polyester. The use of brand names is an attempt to create differentiation for a product that most people assume is generic and commodity-like. Not surprisingly, given how ubiquitous polyester is in our daily lives, China is the largest producer and consumer of polyester.

The objective in commodity chemicals is to carry out chemical transformations on a massive scale, producing relatively simple molecules at the lowest possible cost. Commodity chemicals include both base chemicals (olefins and aromatics) and plastics such as polyethylene and polypropylene plastics. By definition commodity products cannot be differentiated. Buyers will be indifferent between one product and another and will prefer the lowest price supplier since the products are all the same (assuming no differences in shipping cost and delivery options).

The specialty market is at the other end of the spectrum from commodity chemicals. Specialty chemicals are sold on the basis of their performance in customer applications, not chemical composition. Patented products or technologies can enhance the value of specialty chemicals. Customers of specialty chemicals may have limited bargaining power because there are few choices on the market.[5] Producers can develop niches in which they excel in meeting specific customer requirements. Specialty chemicals are used in the manufacture of a wide variety of products, including fine chemicals, additives, advanced polymers, adhesives, sealants and specialty paints, pigments, and coatings. The specialty chemical market can be broken down into segments based on industrial usage, such as plastics additives, textile chemicals, and specialty polymers.

Finally, it is worth noting that relative to the refining sector, chemical products are produced to very tight specifications. As primary inputs to other products, chemical products must perform in chemical processes consistently within a narrow set of properties. In contrast, a product like gasoline is a blend of many hydrocarbons and additives and must meet regulatory specifications for the particular market in which it is sold. Chemical products specifications are ultimately driven by customers and their product input needs.

Industry Structure

The petrochemical industry structure mirrors the production process described above. The industry value chain begins with oil and gas feedstocks and then progresses from basic chemicals to final products for business and consumer use.

Strategic diversity

The consulting firm ATKearney classifies chemical firms into three different business models:[6]

Asset players. The first sector is asset-driven players primarily focused on the production of commodity chemicals. These firms operate in the upstream sector of the chemicals value chain. Scale economies, the cost of oil and gas feedstock, and cost of energy are key drivers of cost-based competitive advantage.

Integrated players. The second sector is integrated players. These firms produce base chemicals and also venture downstream into the intermediate and polymer areas. Some of the integrated players also have specialty divisions.

Specialty players. The third sector is made up of the specialty chemical firms. These firms buy intermediate chemicals and process them into products with specific functionalities. Some specialty firms are focused on a niche market and others are more broad-based.

Figure 14–4 shows the three business models. The IOCs dominate the left side of the chart. The production of base chemicals plays to IOC strengths in capital-intensive commodity businesses. BASF, the largest chemical company, competes across the entire value chain but does not have a refining business (integration between refining and chemicals can lead to cost-based feedstock synergies when refineries and chemical plants are colocated in the same area). To manage its access to gas feedstocks, BASF operates a natural gas trading and distribution business and has a joint venture with Gazprom. In the intermediate product area, BASF produces 600 products, which is the world's most comprehensive range of intermediates. BASF is also one of the world's leading suppliers of engineering plastics, polyamides, and polyamide intermediates, foams, and specialty plastics.

Moving to the right in figure 14–4, companies like SABIC and Mitsubishi Chemicals are strong in base chemicals but do not have access to internally produced feedstocks. On the right side of the graphic are specialty chemical companies. These firms range in size and product line breadth. Some focus on only one area, such as paint and coating additives, whereas others provide products for a range of primarily industrial customers. Because there are thousands of different chemicals derived from oil and gas feedstocks, no two chemical companies have the same strategies. This makes it difficult to draw comparisons across competitors.

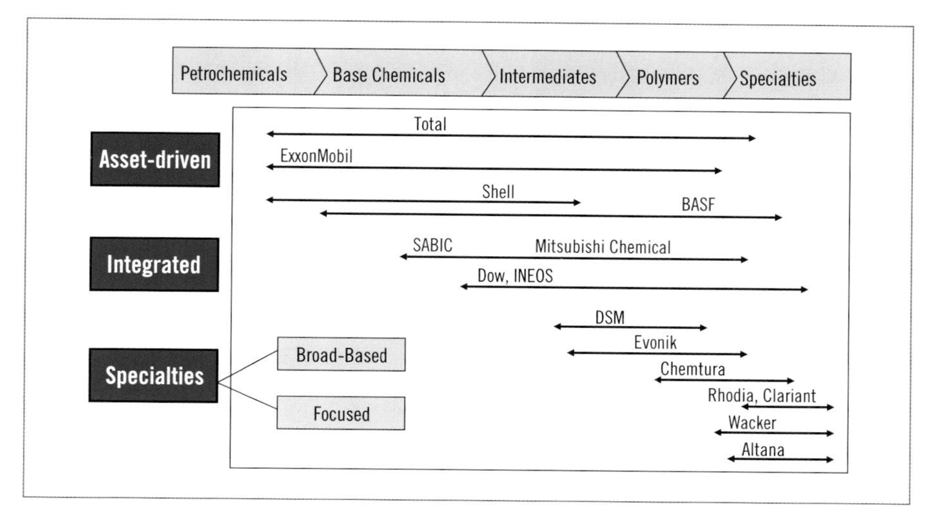

Figure 14–4. Chemical company business models

Source: ATKearney Management Centre Europe, "Chemical Industry: Developing Your Managers to Deal With the Changes in the Chemical Industry," *The Executive Issue* 36, Quarter 1, 2010. Copyright © 2010 Management Centre Europe. All rights reserved.

Consider the chemical strategies of three IOCs. (Figure 14–5 illustrates the percentage of corporate earnings contributed by the chemical businesses of the largest IOCs.)

Shell. Shell is primarily in the commodity chemical business and tries to leverage synergies between refining and chemicals. Shell describes the strategy of its chemical business as "delivering bulk petrochemicals to large industrial customers, through standardized global processes and at the lowest possible total cost."

Total. Total's chemical business covers base chemicals (olefins and aromatics) and their derivatives (polyethylene, polypropylene, styrenics). Total also produces a range of polymer and specialty chemicals including rubber processing, resins, adhesives, and electroplating (Atotech). Several of Total's specialty chemical lines include consumer brand names such as Bostik (glues and adhesives) and Hutchinson (bicycle tires).

ExxonMobil. Based on revenue and operating profit, ExxonMobil's primary focus is basic chemicals ,and the firm is a top global producer of ethylene, propylene, benzene, and paraxylene. ExxonMobil also produces polyethylene and polypropylene and a number of less cyclical specialty products such as synthetic rubbers, solvents, and adhesives. The specialty products have in recent years contributed as much as one third of ExxonMobil's chemicals operating profit. ExxonMobil also has a polypropylene films business, a product primarily used for flexible packaging applications such as snack food packaging.

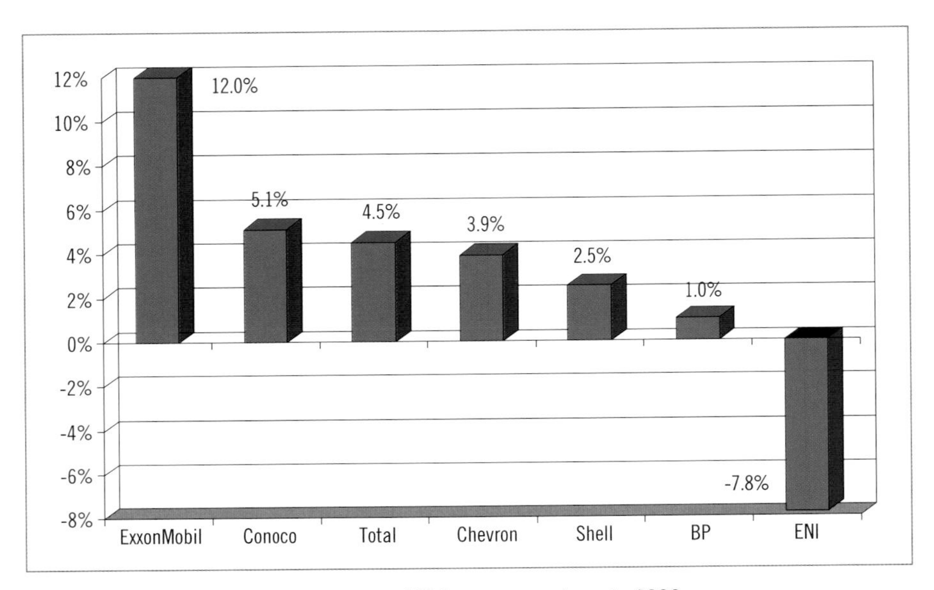

Figure 14–5. Chemicals as a percent of IOC group earnings in 2009
Source: Company reports.

Case Example: BASF and Dow Chemical—A Strategic Shift Away from Commodities

Two of the world's largest chemical companies, BASF and Dow Chemical, are trying to reduce their dependency on commodity chemicals. Commodity chemicals are much more cyclical than specialty chemicals. As evidence, sales in some BASF divisions fell by almost half in 2009.[7] In addition, base chemicals require massive expenditures of energy for their production. Chemical companies in the Middle East have a significant low-cost energy advantage, putting pressure on companies based in Europe and North America that do not have access to low-cost energy and/or advantaged feedstocks from company-owned refineries.

Both BASF and Dow have been acquiring specialty chemical companies. In 2009, BASF acquired Swiss specialty chemicals company Ciba, a producer of pigments, plastics additives, and dyes. In 2010, BASF announced that it was buying Cognis, also a specialty chemicals company. Cognis' products are used in various industries, including cosmetics and household cleaning products. In 2008, Dow acquired specialty chemical manufacturer Rohm and Haas. At the time of the acquisition, Dow Chemical CEO Andrew Liveris commented:

> *The acquisition of Rohm and Haas is a defining step in our transformational strategy to shape the "Dow of Tomorrow"—a high value, diversified chemicals and materials company, creating the largest specialty chemicals company in the United States with a leading global position in performance products and advanced materials.*

Dow's developing new strategy was further explained in its *2009 Annual Report*:

> *Dow is reinventing itself into an agile scientific innovator with the capability to solve major global issues. . . . Our strategy is to preferentially invest in a portfolio of technology-integrated, market-driven businesses that create value for our stockholders and growth for our customers. . . . The Dow of Today operates three integrated business models with more high-growth, high-margin, technology-rich businesses than ever before. The 2009 acquisition of Rohm and Haas shifted our business portfolio, which is now two-thirds comprised of Advanced Materials, AgroSciences and Performance businesses, where science and research and development (R&D) capabilities fuel our growth.*

Figure 14–6 examines the chemical industry segments and product areas where Dow holds an industry leading position. Moving from top to bottom on the figure shows the segments shifting from specialty to commodity. The bottom three segments (basic plastics, basic chemicals, and hydrocarbons and energy) are the areas that Dow is shifting away from. These are also the segments where Dow is most likely to find itself competing against a vertically integrated oil company with refinery colocation advantages or a Middle Eastern chemical company with access to cheap feedstock. In 2009, the basic segments accounted for 37% of total sales for Dow, and Dow had strong positions in a number of commodity product areas.

Dow Advanced Materials

- **Electronic and Specialty Materials:** CMP Pads and Metallization Chemistry, Reverse Osmosis Membranes, Ion Exchange Resins, Specialty Cellulosics and Biocides
- **Coatings and Infrastructure:** Architectural Binders, Architectural Additives, Acrylic and Styrene, Acrylic Emulsions,Epoxy Resins, Additives and Solvents Infrastructure, Extruded Polystrene Foam Insulation
- **Health and Agricultural Sciences:** Green Chemistry/Insecticides, Silage Corn and Omega-9 Naturally Stable Oils
- **Performance Systems:** Automotive Glass Bonding, Specialty High-Performance Sealants, Wire and Cable Compounds, Polyurethane Systems
- **Performance Products:** Propylene Oxide, Polyether Polyols, Epoxy and Intermediates, Ethyleneamines, Ethanolamines, E-series Glycol Ethers, P-series Glycol Ethers, Acetone Derivatives, Alcohols

Basic Segments

- **Basic Plastics:** Polyethylene (the most commonly used plastic in the world), polypropylene, polycarbonate, and polystyrene
- **Basic Chemicals:** Chlorine, chlorinated organics, ethylene dichloride, ethylene oxide, vinyl chloride monomer
- **Hydrocarbons and Energy:** Ethylene, propylene, benzene, various olefins and aromatics

Figure 14–6. Industry segments and product areas where Dow holds industry leadership
Source: Dow Chemical, *2009 Annual Report*.

Industry Profitability

The petrochemical industry exhibits significant cyclicality. As such, the industry fluctuates between periods of supply tightness and slack, with product prices and margins varying accordingly.[8]

Much of the petrochemical industry can be described as capital-intensive, deeply cyclical, and heavily exposed to mature, low-growth, commodity businesses. Margins and profitability for commodity chemicals depend on scale, capacity utilization, operating cost discipline, and access to low-cost feedstocks. Commodity chemical producers usually have little or no pricing power. Since commodity/base chemicals provide the building blocks for many industries, when economic downturns occur, this sector is always hit hard. As a result, the earnings associated with commodity chemicals tend to be so volatile that only the largest or most diversified players can maintain sustainable businesses.[9]

The factors influencing profitability in commodity chemicals are similar to those in refining. Production costs are heavily dependent on the cost of feedstocks and energy, which are not controllable by chemical companies (although, as discussed later, the integration of refining and chemical production provides more flexibility in managing feedstock costs). To improve the profitability of commodity chemicals, firms must drive down the cost curve and widen the wedge between their costs and the market price. Figure 14–7 shows the major cost components and strategic drivers that impact costs. The cost leader will be the firm that can move down the furthest on the cost curve. In the following sections, we consider some of the drivers and major costs elements, beginning with a discussion of ethylene.

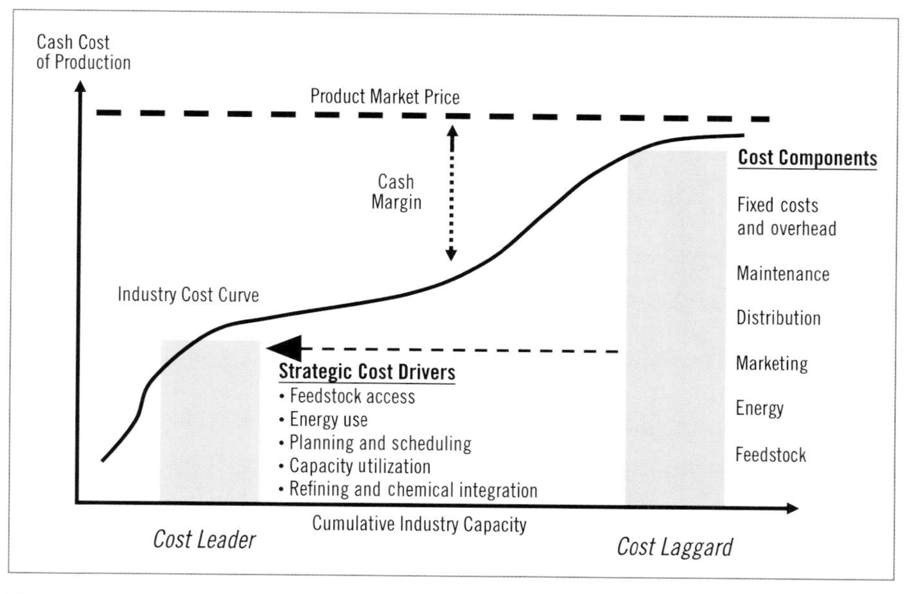

Figure 14–7. Commodity cost competitiveness

Source: Adapted from KBC. © 2009 KBC Advanced Technologies plc.

Ethylene production and costs

To understand commodity chemicals it is useful to consider the case of ethylene, a basic building block for chemicals manufacturing. Ethylene, a colorless gas, is the lightest and most used hydrocarbon chemical. Although ethylene has no end use, it is a basic chemical input for a variety of industrial products. Ethylene is the raw material used in the manufacture of polymers such as polyethylene, polyester, polyvinyl chloride, and polystyrene, as well as fibers and other organic chemicals. Because of its importance, ethylene is often used as a surrogate for the performance of the petrochemical industry at large.[10] Figure 14–8 examines the largest ethylene producers. Three of the largest producers are IOCs (ExxonMobil, Shell, and Total) and two are NOCs (Sinopec and NPC-Iran).

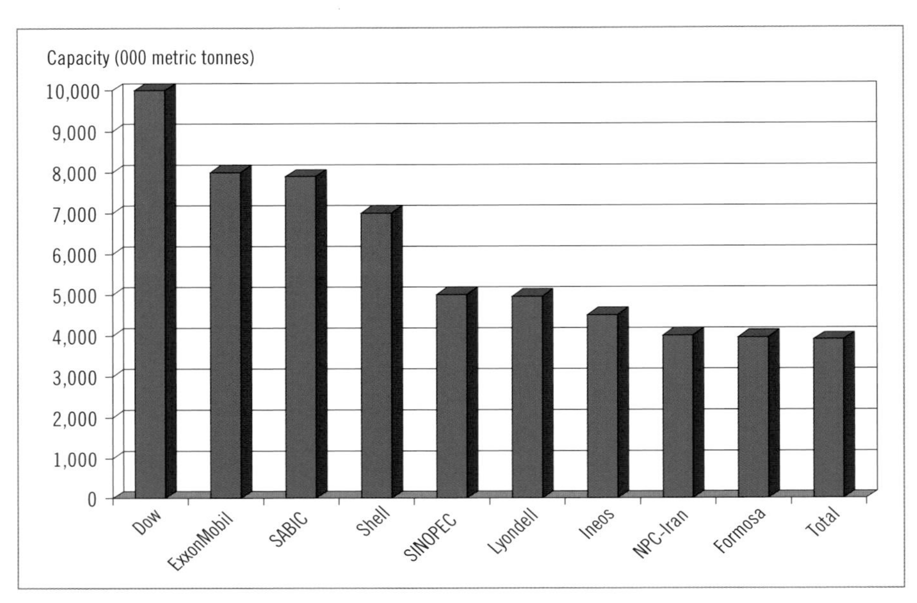

Figure 14–8. Top ethylene producers, 2007

Source: "Oil & Gas for Beginners," Deutsche Bank, January 7, 2008, Fig. 226, p. 170. Copyright © 2008 Deutsche Bank AG.

Ethylene is produced by steam cracking a wide range of hydrocarbon feedstocks. Ethylene can be obtained from cracking oil-derived feedstock like naphtha, gasoil, and condensates. Plants using oil-derived feedstocks are predominant in Europe and Asia, accounting for two-thirds of the world's ethylene capacity.[11] Ethane and propane feedstocks for ethylene that is produced from natural gas are found primarily in the US, Canada, and the Middle East. Until 2000, North American olefins producers dominated the market for ethylene and its derivatives. Since that time, Middle East and Asian production has increased significantly.

Using the ratio of energy produced per volume of energy sold (i.e., a barrel of oil versus a million British thermal units) the price of oil should be six times the price of gas. Given oil's greater versatility, the long run difference is about eight times. In late 2009, the difference was 14 times. In an industry producing commodity products, the highest cost producers set the marginal price of the product (until the point at which they are so unprofitable they must exit the industry).

Figure 14–9 shows the global ethylene cost curve. Based on gas prices in early 2009, US chemical plants using gas as feedstock could produce ethylene for between $550 and $600 a ton.[12] Ethylene prices were about $1,100 a ton because of the high price of oil relative to gas, which means European and Asian oil-based ethylene producers were the price setters, and the US and Asian producers were the most profitable. Over the next decades, many analysts predict that European producers of base chemicals will experience limited growth because of cost pressures from Middle East plants.

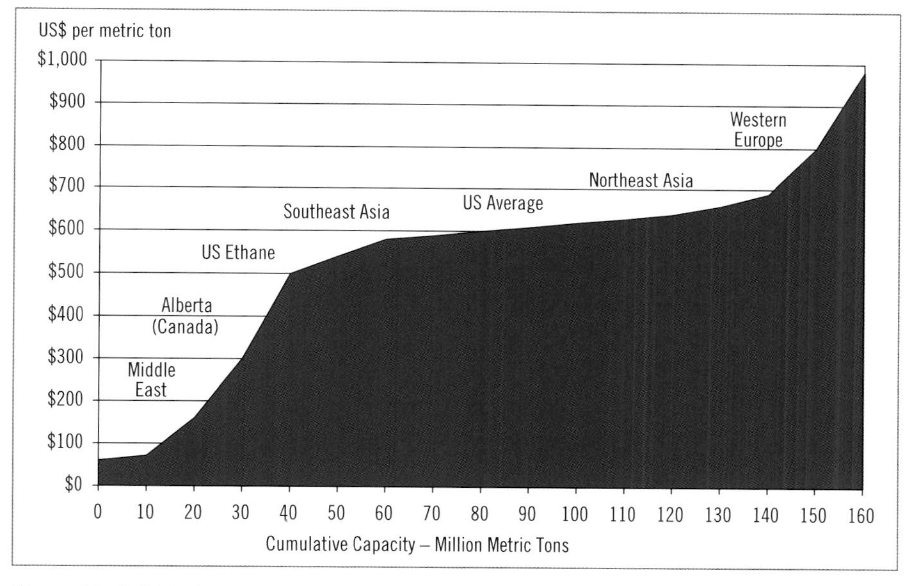

Figure 14–9. Global ethylene cost curve
Source: Morgan Stanley, "The Dow Chemical Company," January 10, 2010, p. 48. © 2010 Morgan Stanley.

Because an output from one sector of the chemical industry is often an input to a sector down the vertical chain, low-cost ethylene in the Middle East will have a major impact on competition for other products. Consider the production of polyethylene. The first step in the production of polyethylene is the production of ethylene, the primary input for polyethylene. Since ethylene from gas feedstocks is much cheaper to produce than oil-based ethylene, polyethylene firms with access to low-cost ethylene will have an advantage.

As illustrated in figure 14–9, the lowest cost ethylene is in the Middle East, and more specifically, probably Saudi Arabia. Saudi Arabian ethylene production is estimated to be approximately 50% cheaper than US gas-based producers and 70% cheaper than European oil (naphtha)-based producers.[13] Figure 14–10 shows ethylene capacity by region. The figure shows North American and European capacity shrinking and Asian and Middle East capacity growing. New capacity is being built in the polyethylene sector in Saudi Arabia to take advantage of the commodity chemical cost differences. However, because major markets for polyethylene are outside the Middle East, after additional costs involving logistics, duties, handling, and transferring from ship to railcar have been added, the feedstock advantage is diluted.

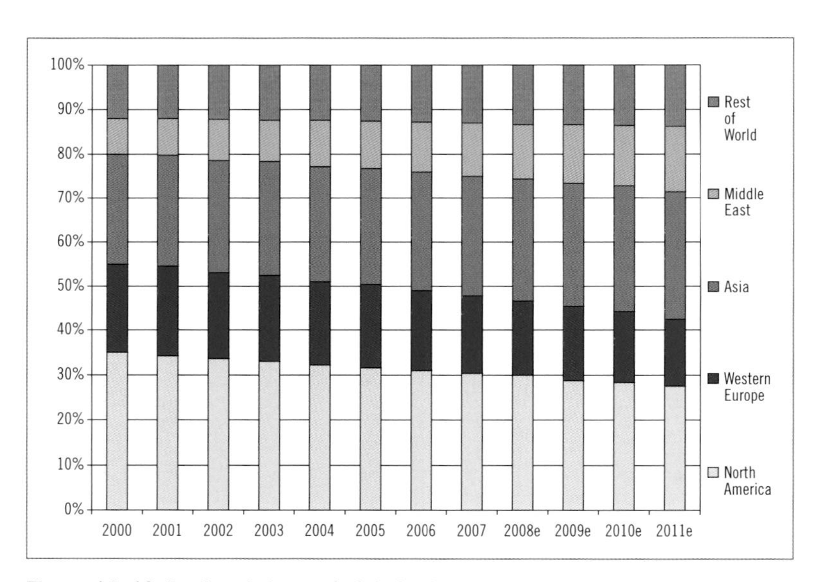

Figure 14–10. Regional share of global ethylene capacity

Source: "Oil & Gas for Beginners," Deutsche Bank, January 7, 2008, Fig. 227, p. 171. Copyright © 2008 Deutsche Bank AG.

Capacity utilization

As in refining, plant capacity utilization (i.e., operating rate) is an important factor driving performance, especially for commodity products like ethylene. Figure 14–11 shows global ethylene operating rates over a multiyear period. Operating rates ranged from as high as 94% to as low as 80%. The figure shows that as consumption flattens or drops, as it did in 2001 to 2002 and 2007 to 2008, operating rates also drop. Because of the high fixed costs in chemical plants, a drop in operating rates will correspond with a drop in profitability.

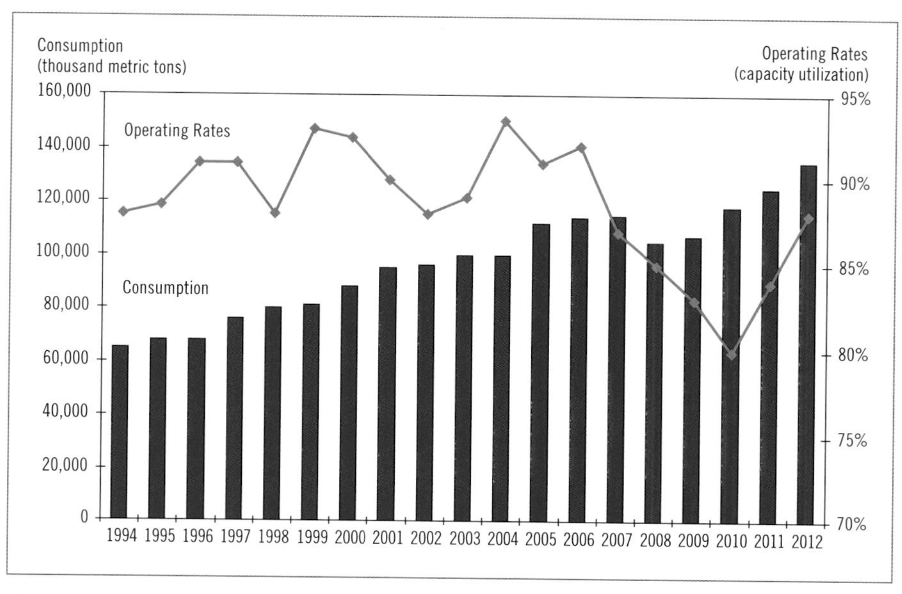

Figure 14–11. Global ethylene operating rates
Source: Morgan Stanley, "The Dow Chemical Company," January 10, 2010, p. 47.

Despite some volatility, the figure also shows steadily increasing consumption. To meet this growing consumption, producers will need to add capacity. Inevitably, some producers will miss the timing on new capacity and bring new plants onstream during down cycles, pushing down operating rates even further. Given the slow demand growth expected in most developed markets, most analysts expect chemical companies in the US and Europe to be very cautious about adding new capacity (see the discussion later in the chapter on capital investment in the chemical industry).

Specialty chemicals performance

On the specialty side of the business, the key factors driving profitability are innovation and customer relationships. Feedstocks and energy costs are important but advantage is really about differentiation, which means creating unique value for customers that is difficult to imitate. As discussed, differentiation may be the result of proprietary technologies such as unique catalysts or chemical processes, and it can also be the result of brand names, marketing, customer service, and delivery. Regardless of the method used to create a differentiation advantage, the key is to ensure that customers do not use price as the primary factor in their buying decision. If price is the determining factor, the product is not a specialty chemical.

Refining and chemicals integration

In recent years, petrochemical firms, and especially several of the IOCs, have been emphasizing cost synergies that result from integration between refining and chemicals. The main synergies come from cost-effective availability of feedstocks and the ability to direct product streams from refineries to petrochemical manufacture. Synergies can also be captured by sharing utilities and overhead costs. For an example of integration see *Industry Insight: Shell Chemicals and Integration.*

Industry Insight: Shell Chemicals and Integration

According to Shell Group, integration is core to Shell Chemical's strategy. A Shell executive explained the strategy:

> *Oil-chemical integration is a key tenet of Shell's downstream strategy, which we believe gives us a competitive edge. Most integration value comes from directing hydrocarbons to the highest-value application, irrespective of traditional refining—chemical boundaries. Secondary or by-product streams from refining units can have their highest value as feedstock for chemical units. Likewise, by-products from chemical units can be most cost-effective as refinery feeds or fuel blending components.*
>
> *Investing in hydrocarbon integration has also improved our operational flexibility, and helped maximise returns as feedstock economics shift. We're developing and deploying new optimisation tools that day-to-day enable us to maximise and coordinate hydrocarbon value across our integrated refining-chemical locations. This feedstock flexibility also brings customers the benefits of product supply security.*
>
> *Extensive process design work at the Shell Global Solutions laboratories in Amsterdam and Houston has paved the way for innovation in heavy feed cracking, which is crucial to facilitating enhanced production. In Singapore, our Shell Eastern Petrochemicals Complex will use a higher percentage of heavier feeds compared to other liquid crackers. And nearer to home in The Netherlands, our Moerdijk cracker is being reconfigured to enable it to crack hydrowax from the Pernis refinery.*

Source: Aslam Moola, "Creating Petrochemical Growth Platforms: Leveraging Upstream and Existing Assets with World Class Technology," Petchem Arabia 2009, October 2009, Abu Dhabi.

As noted earlier, Dow Chemical has struggled with its commodity business, with the cost of feedstocks being one of the firm's primary challenges. In 2007, Saudi Aramco and Dow Chemical Company signed a memorandum of understanding (MOU) on construction, ownership, and operation of a world-scale chemicals and plastics production complex in Saudi Arabia named the Ras Tanura Integrated Project (RTIP). The plan was for RTIP to be one of the largest petrochemical facilities in the world with integration between the Ras Tanura Refinery (the largest refinery in the Middle East), Ju'aymah Gas Plant, and various chemical plants. The complex was to include:

- A 1.2 mt/y ethane/naphtha steam cracker, fed with 70 mcf/day of ethane supplied from the Ju'aymah gas plant and 40,000 b/d of naphtha from the 550,000 b/d refinery

- A fluid catalytic converter (FCC) consisting of an 80,000 b/d vacuum gas oil hydrotreater, a 70,000 b/d high olefins FCC, and associated sulfur and amine plants

- An aromatics complex fed with 70,000 b/d of reformate and pyrolysis gasoline (pygas) to produce benzene, toluene, and paraxylene to feed downstream derivative units

- A chlor alkali facility producing about 640,000 t/y of chlorine to feed downstream process units

- Downstream process units comprising 28 different world-scale chemical units, including a specialty mix of purefied terephthalic acid (PTA), polyethylene teraphthalate (PET) resins, toluene di-isocyanate (TDI) or methyl diphenyl di-isocyanate (MDI), styrene butadiene rubber (SBR) or acrylonitrile butadiene styrene (ABS) resins, and acrylonitrile (ACN)

- Feedstock pipelines, an ethane pipeline from the Ju'aymah system, a new import/export jetty, and a materials handling facility

The advantages cited for integration at RTIP were numerous: reliability of feedstocks supply with less transport cost; supply chain optimization resulting in faster delivery of products and optimum distribution; significant reduction in shared utilities systems; more flexibility in reprocessing, storing, and transporting off-specification products; more outlets for high-value by-products; energy savings in well-integrated hydrocarbon processes; feedstock flexibility to capitalize on available low-cost crude oils and intermediates; significant savings in storage requirements; centralized support services, engineering, maintenance, laboratory, EH&S, security, etc.; and independence of feedstocks and supply security.[14]

Several challenges of integration were also acknowledged for RTIP. These included increased complexity in running the complex; limited operational flexibility; conflicting planning and operational objectives for the various products and outputs; the mix of disparate cultures; difficult technical issues; and the potential for a diffused business focus.

In 2009, Dow and Aramco decided to restructure their proposed deal, citing increased costs associated with the original deal. Integrated petrochemical facilities are also being planned for other Middle East countries and also in various other oil- and gas-producing countries such as Russia and Kazakhstan.

Marketing

In the crude oil business, commodity markets set prices, and deliveries involve a mix of spot and long-term supply agreements with a relatively small set of refinery customers. Crude oil moves from production to refinery customers in huge quantities via pipelines and ships. There is no real concept of a "small crude oil customer." Product innovation and brand names do not exist in commodity oil markets. In the natural gas business and especially the LNG sector, marketing primarily involves negotiating long-term contracts with large customers. Although there is not a true commodity pricing model for natural gas, most pricing is tied to oil prices.

Moving downstream, refineries produce a range of products that must be marketed and delivered to a wide range of customers. In that respect, there are some similarities with the chemical industry. However, the pricing and distribution of motor fuel, the main refinery product, is cost and commodity oriented. With its emphasis on scale and cost control, the commodity side of the chemical business has some similarities with refining marketing and distribution. In contrast, specialty chemicals looks more like a mainline manufacturing business than the scale/cost model used in refining. With specialty products, and to a lesser degree with commodity chemicals, sales leads must be developed, customers must be supported, and areas like service delivery and customer responsiveness are critical.

The significant differences in chemical consumption patterns from country to country are a further difference between refining and chemicals. For example, in Germany and Japan, the automotive and manufacturing sectors account for a large share of consumption. In Brazil, industries such as agriculture and pulp and paper are very important. The multidecade shift from manufacturing to service industries in the US has contributed to relatively slow growth in chemical consumption. These country differences in consumption require chemical companies to tailor the distribution and marketing approaches by geographic region and, ultimately, by specific customer.

Distribution

Depending on the customer and the type of chemical, chemicals will be shipped and distributed in various ways: pipelines, railway tank cars, container ships, trucks, and even air freight for high-value products. The distribution method will depend on the type of chemical and size of customer. Chemicals such as ethylene and propylene are usually distributed by pipeline, whereas the small customer with a single injection molding machine will buy materials by the pallet or maybe by the bag (as an aside, there are more than 25,000 polymers that can be used in injection molding and many more are developed each year).

Customers for chemicals vary from huge global firms to small mom-and-pop operations. According to Boston Consulting Group (BCG), between 20% and 40% of chemicals are consumed by companies buying amounts of $100,000 or less. Small customers are a challenge for large chemical firms geared towards high-volume customers and large bulk shipments. Within the industry the 80:20 rule is often cited: 80% of business is done with 20% of the customers. To deal with small- and medium-sized customers, many chemical companies use third-party distributors to provide field staff and technical support.[15] The distributors add value in several ways: managing complexity for producers and customers, physically handling the chemical products, and providing financing and support. A distributor performs a valuable middleman role by buying product from multiple producers, taking physical delivery and warehousing the product, mixing and blending to customer needs, and shipping and delivery to customers. There are more than 10,000 distributors worldwide, with the five largest (Brenntag, Univar, Ashland, Ravago, and Helm) controlling about 20% of the market.[16]

BCG surveyed customers on the most important criteria in their purchasing decisions. For large companies, the four most important criteria were price, product quality, speed of delivery, and flexibility of delivery. For small customers, the most important criteria were flexibility of delivery, speed of delivery, product quality, and price. Following from these criteria, BCG suggested that there were significant opportunities for innovative thinking in chemical distribution, particularly with regard to moving beyond transactional relationships with customers and distributors to more partner-based networks with more emphasis on cooperation and mutual value creation. We can speculate that in the commodity side of the chemical business dominated by huge players (including IOCs), sales and marketing skills have lagged behind other areas such as manufacturing and operations.

Capital investment in the chemical industry

One of the challenges in the chemical industry is the timing of new manufacturing capacity. If a chemical company is to grow, it must make capital investments for new capacity, via either new plants or additions to existing plants (capacity can also be added through acquisitions, but since no new capacity is added to the industry, we will focus on the capital investment option.) The dilemma faced by chemical companies concerns the choice between commitment and flexibility.

Commitment and flexibility lie on opposite ends of a firm's investment spectrum. Flexibility means keeping options open by postponing strategic investments and waiting for demand or technical uncertainty to subside. The other strategic choice is commitment—making the capital investment with the goal of securing future market growth and discouraging rivals from investing. Whereas the flexibility option could mean lost opportunities, the commitment option may mean adding capacity at the wrong time in the economic cycle.

The time-to-build is a key factor in any capital-intensive industry. Time-to-build is the product of multiple factors, including technological complexity, access to suppliers, and governmental licenses and permitting. In the chemical industry, time-to-build is two to three years for a plant of average size and complexity. The longer the time-to-build and the more time it takes to accumulate permits, resources, and technologies prior to entry into a market, the more slowly firms will enter that market. In industries where it takes a long time to accumulate resources, managers will be unable to quickly adjust their strategies to new market and competitive information. For example, in an industry like financial services, new products can be created and marketed very quickly. In capital-intensive industries like refining and chemicals, managers must weigh the costs and benefits of early investment. The opportunity cost of waiting is the forgone profit from the investment, which depends on product margins during the delay. The danger of early investment is that firms time the cycle wrong, as happened in the refining and chemicals sectors for firms that committed to investment during the 2005 to 2007 period and then brought new capacity on-stream in the US and Western Europe during the 2007 to 2009 recession.

Conventional wisdom suggests that when there is high uncertainty associated with future market conditions, firms tend to be conservative and not build. A research study in this area contradicts the view of uncertainty as a strong disincentive for investment.[17] The study found that an increase in uncertainty may encourage rather than dissuade investment. The study argues that the longer it takes to plan and bring new investments onstream, the more time firms have to learn about conditions in a market (demand, prices, competition, technology, regulation, etc.) before starting operations in that market. Long resource accumulation lags are often associated with staged investment projects. If new unfavorable market information comes to light during the execution of the stages, firms can abandon construction before moving to the next stage. Because the investment can be abandoned at multiple stages, its downside is less severe. The overall argument is that an increase in uncertainty can augment the upside potential of a project more than its downside, boosting its expected profits and leading to a greater and earlier likelihood of investment.

The Future of Petrochemicals

From the perspective of end users, the future of petrochemicals looks very bright. The number of products derived from petrochemicals grows exponentially every year. Even if the demand for fossil fuels declines, the demand growth for petrochemical products will never stop. It is possible that petrochemical feedstocks may come from hydrocarbons that are not fossil fuels. Regardless of the source of feedstocks, plastics and polyesters and all the other amazing products produced from petrochemicals are integral parts of modern life and will always be in demand. At some point in time, we will probably look back and bemoan the fact that we burned fossil fuels instead of making them into durable and long-lasting petrochemical products.

In the near term, the industry will evolve in some fairly significant ways. New petrochemical capacity in Asia and the Middle East is driving one of the most important competitive shifts. The new capacity in Asia is the result of growing markets, especially in China and India. In the Middle East, new capacity is driven by two related factors. One, the Middle East region has abundant supplies of competitively priced gas and liquids, meaning reduced costs for petrochemical facilities as well as lower costs of electric power generation. It is typical in chemical companies for the cost of hydrocarbon feedstocks and energy to account for one-third or more of total operating expenses. In commodity businesses (such as ethylene) low energy costs help support low-cost competitive advantage. For example, compound annual growth in ethylene capacity in the Middle East was almost 20% for 2007 to 2010, compared to almost no growth in the United States and Western Europe.[18] In 1995, the Middle East was producing about 5% of global ethylene output, and it is estimated that by 2015, the share will increase to 20%. Two, Middle East countries want their NOCs and private companies to participate in more of the downstream and chemical value chain areas. Their access to energy and feedstocks makes it possible to be quickly competitive in commodity chemicals. It is likely that Middle East petrochemical companies will also try to build positions in specialty chemicals and even push further down the value chain to end products.

Another important consideration is that the world has become a single global market for many chemical products. Innovations in shipping have allowed chemical producers to target markets far from their manufacturing plants. The result is that commodity manufacturers must be highly competitive to survive. The industry may also see new entrants emerge from feedstock-rich countries such as Russia. Finally, volatile prices for petroleum and natural gas are creating an incentive for US and European companies to evaluate alternative feedstocks. These feedstocks could include unconventional processing technologies such as coal gasification and coal liquefaction; unconventional reserves such as stranded natural gas and heavy oil from tar sands or oil shale; and novel resources such as biomass or algae. As an example, see *Industry Insight: Rubber from Bacteria.*

Industry Insight: Rubber from Bacteria

In 2008, Genencor, the biotech division of Danisco, announced a partnership with the tire company Goodyear. The goal of the partnership was to create BioIsoprene®, a renewable, cost-competitive alternative for petroleum-derived isoprene. Isoprene is a chemical that is found naturally as latex, a material obtained from rubber trees. Isoprene can also be produced from petroleum feedstocks. The polymerization of isoprene using catalysts yields a synthetic rubber, polyisoprene, which closely resembles natural rubber. Isoprene is primarily used as a feedstock to manufacture other industrial chemicals. Most high-purity isoprene is used to manufacture polymers, such as polyisoprene, styrenic thermoplastic elastomer block copolymers, and butyl rubber. Automobile tires are made from a combination of natural and synthetic rubber. About 60% of worldwide synthetic isoprene production is used by the tire industry.

Genencor uses a genetically modified form of *E. coli* bacteria to produce BioIsoprene. Using gene splicing techniques, Genencor is able to use the bacteria to produce isoprene from sugars found in plant material such as corn cobs and switchgrass. The modified organism ferments a gas that can be collected and purified as isoprene. In December 2009, the first demonstration tires made with BioIsoprene technology were produced. Genencor plans to have a pilot plant ready in about two years followed by commercial production.

Source: "Greener Tyers: A Radial Brew," *The Economist*, April 22, 2010, p. 79.

Summary Points

- The petrochemical industry touches the lives of almost everyone on earth. Innovation in coming years will result in many new products for industrial and consumer purposes.

- The chemical industry is capital intensive, highly cyclical, and for the commodity sector, limited in its ability to pass costs on to customers. In that respect, the commodity sector is similar to refining.

- The vast number of organic chemicals means that chemical companies can specialize in only a few areas, making it difficult to compare the strategy of one chemical firm with another.

- The basis for advantage in chemicals depends on whether the product is a commodity or specialty chemical. In commodity petrochemicals, integration between refining and chemicals will be increasingly important as a driver for cost advantage. In specialty chemicals, the main drivers of sustainable competitive advantage are innovation and customer relationships.

- To the extent that oil and gas companies invest in new petrochemical capacity, the focus will be on facilities close to the major demand centers in China or in Middle Eastern countries with an advantaged supply of feedstocks and access to low-cost energy.

Notes

1. Maarten Neelis, Ernst Worrell, and Eric Masanet, "Energy Efficiency Improvement and Cost Saving Opportunities for the Petrochemical Industry," Ernest Orlando Lawrence Berkeley National Laboratory, June 2008.
2. Neelis, Worrell, and Masanet, 2008.
3. Deutsche Bank, January 2008.
4. JP Morgan, "Chemicals Primer: An Investor's Guide to the World of Chemicals," June 23, 2008.
5. Data Monitor, Global Specialty Chemicals, March 2010.
6. ATKearney Management Centre Europe, "Chemical Industry: Developing Your Managers to Deal with the Changes in the Chemical Industry," *The Executive Issue,* 36, Quarter 1, 2010.
7. BASF Seeks a Stable Formula," *The Economist*, June 23, 2010.
8. Deutsche Bank, "Oil & Gas for Beginners," January, 2008.
9. JP Morgan, "Chemicals Primer," 2008.
10. JP Morgan, "Chemical Primer," 2008.
11. L. Denning, "Natural Gas and Exxon's Chemical Romance," *Wall Street Journal*, December 28, 2009, p. C10.
12. Denning, *Wall Street Journal*, 2009.
13. Morgan Stanley, "The Dow Chemical Company," January 20, 2010.
14. Hussain A. Al-Qahtani, "Refining & Petrochemicals Integration: Drivers and Challenges," 27th JCCP International Symposium, Japan, January 29, 2009.
15. Bernd Elser, Udo Jung, and Yves Willers, "Opportunities in Chemical Distribution: Optimizing Marketing and Sales Channels, Managing Complexity, and Redefining the Role of Distributors," Boston Consulting Group, January 2010.
16. Boston Consulting Group, January 2010.
17. Goncalo Pacheco-De-Almeida, James E. Henderson, Karel O. Cool, "Resolving the Commitment Versus Flexibility Trade-Off: The Role of Resource Accumulation Lags," *Academy of Management Journal*, 2008, Vo. 51, pp. 517–536.
18. Deutsche Bank, January 2008.

Chapter 15

THE FUTURE OF THE GLOBAL OIL AND GAS INDUSTRY

The world's energy resources are adequate to meet the projected demand increase through to 2030 and well beyond. . . . Fossil fuels remain the dominant sources of energy worldwide, accounting for 77% of the demand increase in 2007–2030. . . . The world's remaining resources of natural gas are easily large enough to cover any conceivable rate of demand increase through to 2030 and well beyond.

—International Energy Agency, World Energy Outlook 2009

2005 could be the peak year of OECD oil demand. The long-term demand outlook has dimmed under ongoing demographic and socioeconomic changes.

—IHS CERA, September 29, 2009

At the end of chapter 1 we offered a few "safe" predictions as to where the oil and gas industry was headed: the global demand for oil and gas will continue to rise over the next few decades; NOCs will continue to expand beyond their home markets; finding new sources of oil and gas will get harder and require innovative new technologies; investment in nonhydrocarbon energy sources will continue; the industry will remain one of the most vital for the global economy; and, despite the high prices of recent years, the industry will continue to go though up-and-down cycles.

In this chapter, we take a deeper look at the future and discuss some of the key trends that will impact the oil and gas industry. We present our thoughts in three segments—the products, the markets, and the players and their strategies. As we discuss the trends, we identify questions that must be considered by the firms that compete in the industry. In doing so, we continue with a core theme that runs throughout the book—the oil and gas industry is a dynamic and evolving industry. Although the basic commodity products are little changed from the earliest day of the industry, the core activities of the industry value chain change constantly. It is a safe bet that a decade from now, the industry will be vastly different from today.

The Products

We begin where all debates begin with regard to oil, the peak oil debate. We then confront the growing challenge of finding more oil, the emergence of natural gas, and the pursuit of alternative energy sources.

Peak oil demand

In contrast to the peak oil argument that the world is going to run out of oil, the opposite looks likely to occur: the global demand for oil will begin to drop in the coming decades. The demand for oil in the OECD developed economies will probably peak in the coming decade. The OECD peak in demand will be a function of several factors, including: mature economies, an aging population that drives less (especially in Japan and European countries like Italy and Spain); greater fuel economy in cars and trucks (hybrids, advanced diesel engines, a shift to smaller vehicles); the introduction of electric cars; greater commitment to efficiency and conservation; and the continued shift away from heating oil to gas and electric.

Assuming shifts to new energy sources occur, the overall demand for fossil fuels will also peak in the coming decades, although specific projections are difficult. The demand for fossil fuels will fall because of innovations in how energy is produced, transported, and used. Whether it is algae, hydrogen, or some other undiscovered technology, innovative ideas will eventually supplant the need for oil and gas (if it is algae, the existing downstream infrastructure will remain viable). Given the energy density of crude oil and its ease of transportation and storage, if a viable alternative liquid form of energy is invented, it is likely that the IOCs of today will continue to dominate the private sector side of the energy business. At some point, the rationale for NOCs will disappear (when oil and gas development is no longer economically viable) and the private sector energy companies will substantially expand their role.

Finally, while oil demand will peak, it likely will not happen on a global basis for several decades at the earliest. The demand for gas is another story and will likely continue to grow for at least three or four decades and remain sustainable for a very long time.

Crude oil: More distant, greater depths, lower yield, higher costs

Over the coming decades as demand for oil reaches its peak, E&P firms will need to find new sources of oil. Some of that oil will come from new technologies that allow for enhanced recovery from existing oil fields. As was discussed earlier in the book, enhanced recovery plays a major role in ensuring that older fields like Kern River in California continue to be productive. A number of large oil-producing nations, including Iran, Mexico, and Russia, need the capital and technological expertise of the IOCs and large oil field service firms to upgrade their poorly managed oil fields.

Although new technology associated with enhanced recovery will provide some increased oil supply, significant new sources of oil will need to be discovered. The new oil wells will be at greater depths beneath the earth's surface and they will have lower yields and higher costs than the oil fields of the past. Some of the new oil will come from onshore wells in countries that are geographically, culturally, and politically distant from consuming nations. Assuming security issues are manageable, Iraq will become a much more important oil producer. In Africa, historically unstable countries like the Democratic Republic of the Congo will see exploration activity.

Most of the new discoveries will be in offshore areas such as Africa, Brazil, Canada, and Norway. The Arctic region could be very productive as could the East and West coasts of the United States (although politically, most of the United

States offshore regions are off limits and will likely stay that way for a while). To exploit these challenging fields, new technology will be required and most of it will come from the supermajors working in conjunction with their major contractors. The potential risks of offshore drilling have become magnified as a result of the *Deepwater Horizon* oil spill (discussed later). However, as oil demand increases, the risks will become more acceptable. The costs will also be acceptable until the point when peak oil demand occurs, at which point technological innovation will slow down and costs will plateau.

A shift to gas

In recent years some of the largest oil and gas companies, including ExxonMobil and Shell, have substantially increased the gas portion of their product portfolio. Most of the increase has involved LNG and megaprojects in countries such as Qatar, Papua New Guinea, and Australia. The increased emphasis on gas has occurred for several reasons. One, as access to oil reserves becomes more difficult, E&P companies have shifted resources to gas. Two, mega LNG projects require huge amounts of capital, technology, and project management expertise—all of which the supermajors can provide. Three, gas is a cleaner burning energy and there is an ongoing shift from coal to natural gas as a source of energy for electricity.

In recent years, the financial performance of gas projects has suffered because of low gas prices. As the world comes out of recession and energy demand grows, gas prices should rise and the megaprojects should pay off. If prices stay flat, the supermajors could find themselves with depressed returns on capital employed for many years. There is also the possibility that more GTL projects will be developed. If GTL costs become competitive relative to oil, there could be a surge of interest in new developments.

Shale gas

Shale gas is transforming the natural gas industry in North America, and a global impact is likely. Commercial quantities of shale gas will probably be found in China, Australia, the Middle East and North Africa regions, Latin America, and Western Europe. Shale gas will impact both the operations of oil and gas firms and gas prices. On the operations side, shale gas wells require different skills relative to onshore and offshore conventional gas projects. Since shale oil wells cost as little as a few million dollars each, gas producers have much more flexibility in matching production with demand. They can drill wells when demand increases and pull back during down cycles. At the other end of the spectrum, megaprojects involving multi-billion dollar LNG trains cannot easily ramp production up or down.

With respect to prices, the new shale gas production in North America will likely contribute to an excess supply of gas. A significant increase in shale gas supply will result in gas prices remaining depressed for some time and will put pressure on suppliers to Europe and Asia to adjust their prices downward. That said, there is still uncertainty as to the minimum price that makes shale gas unprofitable. As a corollary, there is also uncertainty as to the marginal cost of shale gas production and how fast it will decline with learning and the introduction of new technology.

Substitute products and renewable/alternative energy

Every industry has the threat that substitute products will erode the value created by an existing set of products. The oil and gas industry is faced with the inevitable threat that new energy sources will shift demand way from core oil and gas products. Projections from the International Energy Agency (IEA) under its 450 Scenario (greenhouse gas emissions stabilizing at 450 parts per million (ppm) of CO_2-equivalent) see fossil fuels peaking in use by 2020. Under this scenario, by 2030 zero-carbon fuels will make up a third of the world's primary sources of energy demand. A significant increase in the use of renewables and biofuels will be required to meet the emissions target. In addition, the IEA projects that the energy sector will require incremental investment of $10 trillion between 2010 and 2030, with much of the spending needed to increase energy efficiency. In the developing world, huge investments will be needed for clean power, energy-efficiency measures in industry and buildings, and next-generation motor vehicles.

What the IEA does not address is the source of the investment capital. The economic downturn and the lingering fiscal deficits in the OECD world make it unlikely that governments will be able to make the investments necessary to achieve the ambitious targets found in the 450 Scenario. In addition, government mandates to shift demand away from fossil fuels have resulted in programs that are widely viewed as inefficient, such as the ethanol program in the United States and solar programs in Germany. On the subject of government mandates, the chairman of Gazprom said the following:

We in Russia are well familiar with such a mind set: it is akin to the Soviet-style planned economy. And the result is also very well known: under mandatory distribution, natural incentives to improve competitiveness evaporate. This is a disservice not only to consumers but also to the new alternative energy.[1]

Realistically, significant developments in renewable energy will not happen because of governments. Governments may help in establishing basic conditions but innovative technologies have always come from the hard work and pioneering genius of entrepreneurs. One new energy company started by a couple of entrepreneurs is Sapphire Energy, a company that recruited its president from BP (see *Industry Insight: BP's Head of Global Refining Joins an Algae Company*).

Industry Insight: BP's Head of Global Refining Joins an Algae Company

In 2008, Cynthia Warner, a BP refining executive, met with a team from Sapphire Technology. Sapphire's goal is to produce what the company calls Green Crude: renewable fuels and petrochemical products produced on a large scale from algae, sunlight, and carbon dioxide (CO_2). Sapphire has a pilot facility in New Mexico and intends to build a larger demonstration plant that can produce 100 b/d. If successful, the next step is a commercial plant producing 10,000 bpd by 2018, a large enough quantity for an oil refinery to consider purchasing.

The Sapphire team visiting BP included the founder, Jason Pyle, and one of the company's venture capital investors. They were interested in Warner becoming a board member for the young company. Warner declined and later in 2008, she was promoted group vice president for global refining at BP. In 2009, Warner decided to quit BP and join Sapphire as the company's president. According to Warner, "Most of the alternative choices I could see were short-term fixes with a lot of resource trade-offs . . . it was essential to find a solution that drops in to the existing energy infrastructure, from pipelines to refineries to tanker trucks . . . I had that state of readiness that enabled me to recognize that it [algae] was the solution. It's not going to compromise food production, it's not going to compromise potable water, it doesn't require land that is in high demand for alternative uses, and it's very low carbon."

Source: A. Kamenetz, from "Big Oil to Big Algae: How a High-Ranking Veteran of BP Was Won Over by the Potential of Pond Scum," *Fast Company*, July/August 2010, pp. 84–89, 116.

The Markets

Following the previous section's discussion of industry products, we now consider trends changing the world's energy markets.

Growing demand for energy in emerging markets

According to the IEA, 2009 was the first year that China used more energy than the United States. Although most of China's energy consumption is from coal, the huge increase in the use of automobiles (China is now the largest car market by unit volume) means that oil imports will rise substantially in the coming years. India is also experiencing significant increases in energy demand. If China and India continue to experience 5% to 10% annual economic growth, their demand for energy will grow at similar or higher rates. Most other emerging market economies are also growing and they will need more energy. Although energy users will become more efficient, as personal incomes grow there will be growing demand for automobiles, air travel, and products that consume electricity. The demand for energy in emerging markets will be the main driver of growth for energy over the coming decades. However, even with the demand for energy growing globally, the demand for oil as a source of energy is still expected to peak, as we indicated earlier in the chapter.

China's energy security

Without energy, it is impossible to create a modern economy, so energy security is an essential goal for all countries. The United States manages its security through a variety of means such as domestic oil and gas sources, industry regulation, the maintenance of a military presence in key areas such as the Persian Gulf, diplomacy, and the Strategic Petroleum Reserve. Absent in the US approach to energy security are national oil companies. The US relies on private companies and markets to meet its energy demand. China's approach to energy security is very different from that of the United States and other large developed countries. As discussed earlier, China has enormous energy needs. Although China is currently consuming less than half as much oil as the United States, the IEA projects that after 2025, China will become the world's biggest importer of oil and gas, while India will surpass Japan soon after 2020 to take third place.[2]

China's government appears not to trust markets for its energy needs. The large Chinese NOCs are the arm of the government's energy policy, and they operate very differently than IOCs. In the future, the IOCs and the Chinese NOCs will find themselves in competition for energy reserves and also for government influence in the resource-rich countries. The Chinese NOCs have already made a number of acquisitions of oil and gas companies. Although CNOOC's 2005 attempt to take over Unocal failed, Chinese companies have made oil and gas acquisitions in Central Asia, Africa, South America, Canada, and the Gulf of Mexico. OPEC has reacted to the increasing demand from China by increasing their investment in storage and refining assets in Asia. Saudi Arabia, the world's biggest crude exporter, now ships more oil to China than to the United States.[3] China recently completed a 1,100-mile gas pipeline to connect to the vast gas reserves of Central Asia.

Price, supply, and demand volatility

Although volatile crude oil prices have been the norm for the past few decades, many analysts would argue that volatility has increased because of increased speculator activity and greater variation in supply/demand factors. In our view, increased volatility is not just the result of speculator activity. Volatility is occurring because of a range of factors. First, there is shifting consumer behaviors associated with energy consumption. Events of the past few years have demonstrated that consumer behavior can change rapidly. When the rapid increase in crude prices occurred in 2008, consumers in the United States bought small cars and hybrids at record levels. During the ongoing economic downturn many consumers have opted for less driving or public transport. From these recent events, it would appear that consumers are much more concerned about energy prices than in the past.

Second, as we discussed previously, rapid growth in energy demand in large emerging markets has contributed to rising prices. Unlike that of developed economies, energy demand in emerging markets is very volatile. For example, the ups and downs of the Chinese car market reflect the Chinese government's interventions in areas such as car taxes, fuel prices, and sales incentives. At various times the Chinese government's steps in to stimulate car demand and at other times tries to dampen overheated markets. The result is that rather than a steady and predictable market demand for cars, there is huge variation from month to month and year to year, which corresponds with variation in motor fuel demand.

Third, unpredictable government behaviors tied to resource nationalism in large producing nations such as Venezuela and Iran are likely to continue. Often, these nationalistic behaviors lead to reduced product supply or higher prices, or both. Venezuela's reduced production over the past decade is a case in point: PDVSA in 2010 is much less productive than when the company was run autonomously by skilled managers not beholden to the government. As another example, the Tengiz project in Kazakhstan operated by Chevron with partners ExxonMobil and Lukoil is "periodically squeezed by the Kazakh government for additional taxes and fines to prop up the national budget—something that has become more common during the recession. This month [July 2010] . . . the Kazakh authorities announced a new export tax of $2.73 per barrel, which will cost Chevron and its partners $1.6 million a day. The government also said it was investigating illegal drilling, which could bring huge fines."[4]

Fourth, uneven development of non-fossil-fuel forms of energy creates uncertainty in energy markets. A few years ago in the United States, ethanol was touted as the fuel that would "reduce America's dependence on Middle Eastern oil." Despite government subsidies, ethanol (derisively called moonshine by ExxonMobil CEO Rex Tillerson) is now viewed as an unsuccessful first generation biofuel that inefficiently uses valuable farmland, water, and other resources. As the *Wall Street Journal* wrote, "Given these realities, the only mystery is how an industry that produces a fuel that no one would willingly buy has managed to be subsidized over four decades at costs that are higher than anyone ever imagined."[5] Without subsidies, ethanol production would likely disappear. Further attempts to stimulate demand for alternative energies will impact energy markets in unpredictable ways from supply and price perspectives.

Global market and pricing for natural gas

Unlike the oil market, gas is not bought and sold in a global market. Gas markets have operated primarily as three large regional markets: the Americas; Europe, Africa, and the CIS; and the Middle East and East Asia. Each region has its own distinctive supply/demand and pricing dynamics.[6] As the shift to gas continues and gas markets become more interconnected, regionalization is breaking down. Although a true global gas market similar to the oil market may never happen and certainly does not exist in 2010, a more interconnected global gas market has several implications:

- Interconnected markets are more transparent and liquid, which should mean a convergence in gas pricing models across the regions.

- Spot markets for gas will become more important. For spot markets to exist there must be supply without customers or contracts. A few years ago major investment in regasification terminals in the United States and Canada looked probable. Now, with the increased US focus on shale gas, these investments have been deferred or canceled. The few regasification terminals that have been completed are operating below capacity. With less

LNG needed in the Americas, there is greater supply available for Europe. In the event that a larger spot market for LNG in the Atlantic Basin opens up, there could be pricing pressure on pipeline gas in Europe. Russian suppliers, in particular, could be hit by reduced demand or pressure to lower prices.

- A global market for gas could occur in conjunction with an OPEC for gas—Organization for Gas Exporting Countries (OGEC). In our view, an OGEC is low probability in the near future, but more coordination between the major gas producers looks likely.

- In most industries, greater global connectivity increases the incentives for new entrants and the choices for customers. The overall impact is more innovation, heightened competition, and industry growth.

Will Russia become accessible to international oil companies?

Russia's gas reserves are the largest in the world, and its oil reserves are among the largest. The Russian economy is highly dependent on oil and gas exports. However, with the exception of TNK-BP and the Sakhalin projects, there is almost no international company participation in "equity oil" E&P activities in Russia. The Russian state has shown a willingness to actively intervene in the industry and taxes on exports and production are very high. The Russian state has also shown an aversion to foreign investment in oil and gas E&P. A number of the oil field services companies, such as Schlumberger and Halliburton, are operating in Russia, but they are contractors and do not have capital at risk.

Russian companies, despite their global aspirations, are not nearly as technically proficient as their Western counterparts. Nor do they have the project management skills necessary for the largest mega projects. As a rational actor, the Russian government should be seeking foreign investment because 1) the industry has many old oil fields that need to be revitalized with modern technology that Russian companies lack; 2) the state is hugely dependent on the industry and its cash flows; 3) Russian companies lack the technology and capital to maximize the potential tied to new discoveries and developments; and 4) partnering with foreign oil companies and allowing foreign companies to operate upstream projects would create learning opportunities for Russian firms and individuals. Unfortunately, rational behavior from a non-Russian perspective is often absent inside Russia and especially inside the Kremlin. It should also be noted that several other countries are as restrictive as Russia or more so when it comes to foreign investment in oil and gas, including Mexico and Bolivia.

Although opening up the Russian oil and gas industry to foreign investment would provide major opportunities for Western E&P firms, the political winds in Russia will have to shift before the opportunities become reality. The xenophobic Russian government of Vladimir Putin and his cronies appears committed to keeping out foreign investment, a strategy that will inevitably harm Russia in the long term.

The Players and Their Strategies

The changing fuels and markets will have at their center the actors—the players and their strategies—forcing change, action, reaction, and a diverse set of new opportunities.

New entrants and evolving competitive environment

The oil and gas industry continues to grow more diverse in terms of the number and variety of competitors. In the publicly traded sector, one analyst noted that a decade ago, three supermajors—BP, ExxonMobil, and Shell—accounted for about half of the oil and gas industry's stock market capitalization. Today, they account for less than a third, having lost ground to faster-growing exploration-and-production, services, and infrastructure stocks. As a result, some investors are creating their own integrated portfolios rather than owning the stocks of the integrated supermajors.[7]

A fairly recent development is the emergence of private-equity-backed, pure-play exploration firms with no oil and gas production. One such company is Cobalt International Energy Inc. (Cobalt). Cobalt was formed by a group of industry veterans, including the former COO of Unocal. Cobalt was formed in 2005 and went public in May 2010. The company startup was funded primarily by private equity firms, including Goldman Sachs and Carlyle Group. In 2010 Goldman Sachs owned 21% of Cobalt. According to Cobalt's website, "Cobalt's strategic objective is very clear: to create distinctive value for our investors by exploring for oil in the largest hydrocarbon-rich plays in the deep offshore waters of the US Gulf of Mexico and West Africa, with an emphasis on sub-salt and pre-salt exploration, development, and production."[8] That is extremely specific.

Kosmos Energy is another private equity backed company. Formed in 2003 and backed with about $300 million from private equity firms such as Warburg Pincus and Blackstone Capital, Kosmos discovered the large Jubilee Field in Ghana in 2007. Kosmos's exploration strategy divides the basins of the world into four categories: frontier, emerging, growing, and mature. Kosmos focuses on emerging areas, where drilling has established the presence of oil and gas but where economic discoveries have never been made. As an article reported, Kosmos goes after "plays, basins and petroleum systems that it believes are under explored, under-evaluated or generally misunderstood; its targets are typically unpopular with the rest of the industry."[9] With the success in Ghana, Kosmos was able to raise an additional $500 million in capital from Blackstone Capital and Warburg Pincus.

Finally, as discussed in chapter 2, many NOCs continue to expand their scope of activities into upstream and downstream areas. Going forward, we expect the industry to evolve in ways that, when they happen, will seem surprising to the incumbent players. As evidence to support this statement, go back 10 to 15 years and consider what has happened over this period. Some examples of surprising events include: Petrobras has becomes one of the most valuable oil and gas firms; some of the largest discoveries of recent years were by small companies such as Kosmos and Tullow Oil; Reliance has built the world's largest refinery complex; and many of the largest gas station operators have decided to exit the company-owned sector because of new entrants such as hypermarkets.

Downstream and chemicals: Slow-growth and high-growth regions

The IOCs will see a divergence in their global downstream and chemical strategies. For historical reasons, the IOCs have major downstream and chemical investments in North America, Japan, and Western Europe. The organizations that manage these downstream assets will have to become accustomed to a slow-growth environment not likely to change much over the next several decades. In contrast, Middle Eastern and Asian regions will become the high-growth markets because of growing energy demand and economic development.

In most industries, when growth slows, assets are divested and capital is shifted to different regions or industrial sectors. In the oil and gas industry, high exit barriers make it difficult to close refineries and chemical plants. As a result, some assets in the slow growth areas will continue to operate at suboptimal utilization rates. IOC strategies in their global regions will require different skills and different leadership abilities. In the high-growth areas, the challenges will involve managing growth, building capable management teams, recruiting high-quality young people, and ensuring sufficient capital is available to fund expansion. In the low-growth areas, the managerial challenges will include job retention, ageing workforces, improving plant productivity, and selective divestitures.

Refining and fuels marketing remain core to the industry

There is a tendency in the oil and gas industry to view the downstream sectors of refining and fuels marketing as necessary evils. They are necessary because crude oil must be refined into saleable products, and fuels marketing provides the main customer channel for the most important refined product. However, in recent years neither sector has achievable profitability levels close to those earned in the upstream, raising questions about the majors' commitment to the downstream. If higher profits can be earned in the upstream, why tie up so much capital in refineries, terminals, and distribution channels?

In our earlier discussions of oil company corporate strategy, we made it clear that decisions about core businesses in which to operate are complex. Every corporate strategy has unique conditions that must be managed. For some firms, exiting downstream sectors and focusing on the upstream may be the best decision. For others, an integrated strategy is more appropriate. Regardless of the decision made by a specific firm, refining and fuels marketing are core industry segments. At the time of writing, the refinery sector was suffering through excess capacity and slow growth in the major developed markets, whereas expansion was occurring in the Middle East and Asia. Fuels marketing was seeing a global shift away from branded products to supermarket/hypermarket channels. In both sectors, historical experience suggests that cycles will eventually shift back to more respectable profitability. Thus, using a period like 2007–2010 to make long-term strategic decisions about integration would appear to be shortsighted.

Availability of capital

As a capital-intensive industry throughout the value chain, availability of capital is a critical ingredient for industry growth. As the upstream industry looks farther afield with ever more complex projects, the industry's appetite for capital will increase substantially in the coming years. The industry will also have to invest more heavily in risk management and technology associated with safety and the environment. In the downstream and chemicals, capital will be required to upgrade old assets and to take advantage of feedstocks located in countries like Angola and Qatar.

As we discussed in chapter 7, a variety of capital sources are available to oil and gas firms. Going forward, continued capital availability is essential. Although public policy initiatives around the world will most likely channel more capital towards alternative energy sources, private capital will continue to flow to the oil and gas industry. That flow, however, differs dramatically depending on where the individual commercial firm falls in the pyramid—the IOCs enjoying ready access to cheap and affordable capital, while the smallest exploration firms need to beg and borrow. This will not change any time soon.

The overall industry continues to be viewed by shareholders as a stable industry with long-term potential, but not expected to generate the types of returns seen in many high-tech industries today. Of course that could change if radical technological innovation renders fossils fuels uncompetitive with alternative sources of energy. In our view, that scenario is unlikely for decades to come.

Deepwater Horizon oil spill

The *Deepwater Horizon* oil spill will have wide-ranging implications for offshore drilling and the global oil and gas industry. Although investigations into the spill continue, initial reports from the National Commission on the BP *Deepwater Horizon* oil Spill and Offshore Drilling provide important insights. In examining efforts to contain the blowout, the National Commission concluded that the industry was unprepared to contain a deepwater blowout and the federal government was unprepared to provide meaningful supervision. Although BP's efforts to design, build, and deploy new containment technologies were described as impressive, those efforts were necessitated by the failure to anticipate a blowout. Likewise, the government's new oversight capabilities were the result of a lack of advanced preparation for overseeing a major blowout.[10]

Some of the possible longer term implications of the spill include: a permanent moratorium on offshore drilling on the East and West coasts of the United States and along the Florida coast; increased regulatory oversight on offshore drilling in United States waters; closer scrutiny of offshore drilling in all countries with deepwater reserves; rising costs for offshore drilling because of regulatory changes and new mandatory and voluntary safety processes; the exit of small E&P companies from deepwater because of increased costs and risks; and a rethinking inside companies as to how to manage the risks associated with deepwater drilling.

Deepwater discoveries are essential for the global reserves base. Since 2000, global deepwater production capacity has more than tripled, and total deepwater production exceeds that of every other country except Saudi Arabia, Russia, and the United States. In addition, deepwater discoveries are, on average, significantly larger than onshore discoveries.[11] Hopefully, the *Deepwater Horizon* spill will become a major learning opportunity for the industry.

Increased focus on safety and the environment

Following the discussion above, oil and gas companies should expect greater societal and regulatory pressure to increase the resources they devote to safety and environmental issues. The *Deepwater Horizon* spill will have immediate consequences. For example, one of the US legislative proposals is for drilling rigs in the Gulf of Mexico to use an enhanced and more expensive type of blowout preventer. Upgrading drilling rigs to accommodate the new blowout preventers will be costly, especially for the smaller rigs. Whether or not these new safety requirements become global or not is not clear. In all likelihood, regulators in all countries with offshore drilling will examine their safety practices and requirements.

More challenging for the industry is the expectation that firms will voluntarily increase their focus on safety. Adding a new blowout preventer to a rig is a costly but straightforward safety change. Moving from equipment and facilities changes to management systems and sustainable culture changes around safety is much more challenging. Resistance to change is inevitable and in the short term, firms will struggle with the cost/benefit ratio. As BP learned over the past few years, trying to create a new safety culture is extremely difficult. In order for the change to be successful, firm management must believe that safer organizations are also better-performing organizations. Attention to safety forces work processes to be carefully planned and executed and in the long term, the safest organizations will be the highest performers. Those firms viewed as lagging in safety performance will suffer in terms of reputation, which has implications for raising capital, obtaining licenses to operate, hiring talented employees, and becoming a sought-after partner.

On the environment side, the oil and gas industry has always been under heavy scrutiny. The *Deepwater Horizon* spill will add to that scrutiny. As with safety, firms that are viewed as poor environmental stewards will damage their corporate reputations. Given that the industry as a whole is viewed as a poor environmental steward, firms that can enhance their environmental performance will strengthen their competitive position. Green/alternative energy pressures are intensifying in the developed economies, and oil and gas firms need to make choices that allow them to be viewed as "part of the solution and not part of the problem."

Ongoing talent shortage

A number of studies have recently confirmed what everyone in the oil and gas industry already knew: a growing shortage of technical talent. Booz Allen recently noted that "more than 50% of workers for oil companies and for contractors will retire in the next five to ten years (the average age of operator technicians is over 45, while the average for contractor technicians is over 50)."[12] Difficulty in attracting new talent has long been a problem in the oil and gas industry in the United States, Canada, and other developed countries. The problem is not just in the engineering and managerial ranks. Understaffing and inexperience is a serious problem with offshore rig workers and may have contributed to safety issues on the *Deepwater Horizon.*

There are various factors driving the industry's failure to attract, develop, and retain new talent:

- **Industry perception.** Young college graduates see the oil and gas industry as an old, mature, sunset industry. The industry is often characterized as being seniority based, meaning it may be many years before new recruits are allotted opportunities for advancement and development. Employment in the industry is actually forecast to decline in coming years.

- **Engineering shortage.** The shortage of students interested in studying science and engineering fields in the United States has been repeatedly documented in recent years. While the US is graduating less than 60,000 engineers annually, Europe is producing nearly 140,000, India 150,000, and China nearly 200,000 annually.

- **Project basis.** The "single project" mentality leads to job insecurity. Companies are seen to hire and fire employees depending on the specific project's life cycle (particularly true for engineering, procurement, and Construction firms that maintain staff at the whim of oil and gas companies).

- **People development.** Many oil and gas companies have failed to create formal career paths and people development structures. These firms have often used on-the-job training as the primary development process. This is not necessarily a problem with the oil majors but is a definite problem with the independents, oil field service firms, and other contractors.

- **Lifestyle choices.** As the E&P sector shifts to distant locales for upstream development, the impact on families based in the developed countries has become harder. Although firms have tried to compensate their professionals for this lifestyle disruption, it is proving either insufficient or not desirable. Many argue that societal expectations for the work/life balance have changed, but the industry's expectations for its people have not.

In contrast to the challenges faced by companies in the developed countries, in countries like Nigeria, Angola, Saudi Arabia, and Qatar, the oil industry is viewed as a highly desirable industry in which to work. In these countries international oil companies and their contractors face a different challenge: they must compete for talent with NOCs that may have higher salaries, a less-demanding work environment, and greater job security. Increasingly, some NOCs can even promise the opportunity for an international career in a growing and dynamic company.

The future of the large oil and gas firms

As noted earlier in this chapter, the stock market capitalization of the three supermajors (BP, ExxonMobil, and Shell) is declining as a percentage of the total industry market capitalization. At various points in the book, we have discussed the increasing strength of the NOCs and the challenges of resource access for firms outside the major oil producing regions. Does this mean the era of "Big Oil" is over and that the dominant position of the supermajors and other large IOCs is waning? There is no question that resource access will remain a challenge for the supermajors. On another competitive front, independents such as Occidental Petroleum and Apache Corporation have shown that they can acquire unwanted oil fields from the supermajors and make them profitable. Independents have had superior financial performance in recent years relative to the supermajors. Oil field services companies like Schlumberger and Halliburton have upgraded their skills and can offer attractive value propositions to NOCs.

That said, the large IOCs (the supermajors plus Chevron, ConocoPhillips, Total, and a few other large firms) remain financially robust and very profitable (although BP's financial position is uncertain because of the *Deepwater Horizon* oil spill). The large IOCs are the only firms capable of developing and operating the most complex megaprojects such as those in Sakhalin, Papua New Guinea, Western Australia, and Kazakhstan. The IOCs have industry leading technology, deep reservoirs of talented employees, and accumulated experiences in the most difficult and challenging regions of the world. Many of the large oil-producing countries will need IOC competencies and technologies to rejuvenate their mature fields. Should alternative forms of liquid energy become commercially viable, the downstream businesses of the IOCs will be well placed to manage refining and distribution activities.

To maintain viable industry positions, the IOCs are no different than other firms—they must create strategies that are unique and difficult to imitate and they must produce products and services that customers are willing to pay for. In a largely commodity business like oil and gas, as long as oil and gas are in demand, the companies producing oil and gas will be able to sell their products. The real question is whether the IOCs can create unique strategic positions relative to their independent E&P competitors and especially the NOCs. To gain access to equity oil, the IOCs must have skills and knowledge the NOCs do not have or cannot purchase elsewhere. History suggests that the skills/knowledge gap between the IOCs and most, if not all, resource-rich NOCs will remain significant for some time. However, if NOCs from China and India are willing to accept lower returns in their quest for national energy security, the IOCs will have a tough road ahead.

Growing power of national oil companies

The national oil companies (NOCs) have been growing in power for decades. This trend will continue as long as there is low-cost oil in countries like Iran, Iraq, Mexico, Nigeria, Angola, Russia, and Venezuela. There is no going back to the old model of the NOC as the custodian of its nation's energy resources and the IOC as the operator and manager of the E&P business. The NOCs fully expect to manage their own resources, and private sector firms should expect continued resource nationalism to impact their upstream operations. Some of the upstream impact may be mitigated by downstream projects in nationalistic countries, although to date, the IOCs have been reluctant to make refining and chemical capital investments in countries like Nigeria and Venezuela.

Many NOCs also want to be global players with fully integrated operations. A few, such as Petrobras and Statoil with their mix of public and private ownership, have managed to build competitive international businesses. For other NOCs with global ambitions, definitive conclusions about the success of their international strategies is not yet possible. What is clear is that many NOCs want to grow and see international expansion as an integral element in their long-term strategies.

Finally, when global oil demand slows, as mentioned in the discussion about peak oil demand, the role of the NOCs will change significantly. Can NOCs evolve from oil and gas companies managing a national resource to international energy companies focused on producing products for a diverse customer base? Our view is that with a few exceptions, most NOCs will always be oil and gas companies, and they will continue to exert a powerful influence over the industry. When peak oil demand occurs, the NOCs will still control the lowest marginal cost oil, so their power in the industry will be maintained. However, their power over the broader energy industry will eventually decline.

Notes

1. Alexey Miller, "Natural Gas: Energy of the XXI Century," The European Business Congress, Cannes, France, June 6, 2010.
2. http://www.worldenergyoutlook.org/docs/weo2009/fact_sheets_WEO_2009.pdf.
3. S. Swartz and S. Oster, "China Tops US in Energy Use: Asian Giant Emerges as No. 1 Consumer of Power, Reshaping Oil Markets, Diplomacy," *Wall Street Journal*, July 18, 2010.
4. A. E. Kramer, "A Big Oil Field in Central Asia Isn't Earning What Chevron Planned On," *New York Times*, July 22, 2010.
5. "Survival of the Fattest: What a Deal: Ethanol Reduces Carbon for Only $754 a Ton," *Wall Street Journal*, July 26, 2010.
6. Booz & Company, "A Journey from Regional Gas Markets to a Global Gas Market," 2008.
7. Liam Denning, "Rethinking Oil Investing May Be on Tap," *Wall Street Journal*, May 17, 2010.
8. http://www.cobaltintl.com/about-us/overview-2.
9. P. Williams, "Kosmos and Jubilee Field," *Oil & Gas Investor*, September 2008, pp. 105–108.
10. "Stopping the Spill: The Five-Month Effort to Kill the Macondo Well," National Commission on the BP *Deepwater Horizon* Oil Spill and Offshore Drilling, Staff Working Paper No. 6, November 2010.
11. IHS CERA, http://press.ihs.com/article_display.cfm?article_id=4267.
12. EYGM limited, *The 2009 Ernst & Young Business Risk Report—Oil and Gas*, 2009.

INDEX

A

B

C

D

E

F

G

H

I

J

K

L

M

N

O

P

Q

R

S

T

U

V

Z

ABOUT THE AUTHORS

ANDREW C. INKPEN

Andrew Inkpen is the J. Kenneth and Jeanette Seward Chair in Global Strategy at Thunderbird School of Global Management in Glendale, Arizona. Dr. Inkpen holds a B. Comm. degree from St. Mary's University, an MBA degree from the University of Western Ontario, and a Ph.D. in Business Policy and International Business from the University of Western Ontario. Prior to entering academe, Dr. Inkpen worked in public accounting and qualified as a Chartered Accountant in Canada. He is co-director of the Thunderbird Center for Global Energy Studies.

Before joining the faculty at Thunderbird, Dr. Inkpen was on the faculty of Temple University. He has also taught at the University of Western Ontario, National University of Singapore, and Nanyang Technological University. Dr. Inkpen has taught a variety of courses, including Competitive Strategy, Global Strategy, and Corporate Strategy. Dr. Inkpen is actively involved in a variety of Executive Development Programs at Thunderbird. He is the Academic Director for the ExxonMobil General Leadership Program. This program is delivered 27 times a year in three global locations. He has also taught on many different oil and gas programs for clients such as Baker Hughes, Smith International, ExxonMobil Gas and Power Marketing, TNK-BP, and Integra. He has also worked other companies, including Ericsson, Pfizer, CEMEX, General Motors, LG Electronics, Teleflex, Integra, Goodyear, DENSO, Cisco Systems, Solar Turbines, McDonald's, TRW, Alticor, Vitro, Warner-Lambert, Pharmacia & Upjohn, and Caremark.

Dr. Inkpen's research and teaching interests focus on the management of multinational firms, with a particular focus on strategic alliances, mergers and acquisitions, organizational learning, and global strategy. Articles by Dr. Inkpen have been published in various academic and practitioner publications including *Academy of Management Review*, *Academy of Management Executive*, *California Management Review*, *Strategic Management Journal*, *Journal of International Business Studies*, *Journal of Management Studies*, *Long Range Planning*, *Organizational Dynamics*, *Organization Science*, *Decision Sciences*, *Journal of Applied Behavioral Science*, and *European Management Journal*. He is the author of a book examining automotive supplier alliances titled *The Management of International Joint Ventures: An Organizational Learning Approach* (Routledge) and a book on global strategy called *Global Strategy: Value Creation and Advantage in the International Arena* (Oxford). He is actively involved in teaching case development and has written more than 35 cases. He is a coauthor of the fifth edition of *International Management: Text and Cases* (Irwin/McGraw-Hill). He is on the editorial boards of seven journals, including the *Strategic Management Journal* (the leading academic journal in strategic management), *Journal of International Business Studies* (the leading journal in international business), *Organization Science, Journal of International Management, Management International Review*, and *Asia Pacific Journal of Management.*

MICHAEL H. MOFFETT

Michael H. Moffett is the Continental Grain Professor of Finance at the Thunderbird School of Global Management in Glendale, Arizona. Formerly Associate Professor of Finance at Oregon State University, he has held teaching or research appointments at the University of Michigan, Ann Arbor, the Brookings Institution, Washington, D.C., the University of Hawaii at Manoa, the Aarhus School of Business (Denmark), the Helsinki School of Economics and Business Administration (Finland), the International Centre for Public Enterprises (Yugoslavia), and the University of Colorado, Boulder. He is co-director of the Thunderbird Center for Global Energy Studies.

Professor Moffett's primary areas of teaching and research expertise are in multinational financial management, focusing on the financial demands of the global oil and gas industry and its continual development. Professor Moffett received a B.A. (Economics) from the University of Texas at Austin (1977), an M.S. (Resource Economics) from Colorado State University (1979), an M.A. (Economics) from the University of Colorado, Boulder (1983), and Ph.D. (Economics) from the University of Colorado, Boulder (1985).

He has authored, coauthored, or contributed to a multitude of journal articles, books, and other publications. Professor Moffett is coauthor of several books in multinational business and finance, as well as the author of more than 50 case studies in international business, strategy, and financial management.

Professor Moffett is the coauthor of several books in multinational business and finance, *Multinational Business Finance, 12th Edition*, with Arthur Stonehill and David Eiteman (Addison Wesley 2009), and *Fundamentals of Multinational Finance, 3rd Edition*, (Addison Wesley 2009). *He also coauthor of International Business, 7th Edition* (Southwestern Publishing 2005), with Michael Czinkota and Ilkka Ronkainen, as well as *Global Business, 4th Edition* (2002), and *Fundamentals of International Business* (2004). He is the author and coauthor of more than 50 case studies in international business, strategy, and financial management.

He has acted as a consultant and educator with a multitude of global businesses including Adams, Allied Signal, American Express, AT&T, BP, Brasil Telecom of Brazil, Briggs & Stratton, Cemex de Mexico, Delphi, Delta Airlines, Discount Tire, Dow Chemical, East Asiatic Corporation of Denmark, EDS, Englehard, ExxonMobil, Fluor Corporation, Gate Gourmet, General Motors, Gudme Raaschou of Denmark, Honeywell, IBM, Kellogg's, Kimberley-Clarke, Legrand of France, Lincoln Electric, Mattel, Mobil Oil Corporation, ONGC of India, Parker Hannifin, Pfizer, Phelps Dodge Corporation, Solar Turbines, State Farm, Statoil of Norway, SK of Korea, Teleflex, Texaco, Vitro de Mexico, Warner Lambert, and Woodward Governor.